T0220951

Advances in Embedded and Fan-Out Wafer-Level Packaging Technologies

Advances in Embedded and Fan-Out Wafer-Level Packaging Technologies

Edited by

Beth Keser, Ph.D.

and

Steffen Kröhnert

This edition first published 2019

Registered Office
John Wiley & Sons, Inc., 111 River Street, Hoboken, NJ 07030, USA

Editorial Office
111 River Street, Hoboken, NJ 07030, USA

For details of our global editorial offices, customer services, and more information about Wiley products visit us at www.wiley.com.

Wiley also publishes its books in a variety of electronic formats and by print-on-demand. Some content that appears in standard print versions of this book may not be available in other formats.

Library of Congress Cataloging-in-Publication Data

Names: Keser, Beth, 1971– editor. | Kröhnert, Steffen, 1970– editor.
Title: Advances in embedded and fan-out wafer level packaging technologies / [edited by] Beth Keser, Ph.D., Steffen Kröhnert.
Description: 1st edition. | Hoboken, NJ, USA : Wiley, [2018] | Includes bibliographical references and index. |
Identifiers: LCCN 2018034374 (print) | LCCN 2018037473 (ebook) | ISBN 9781119313977 (Adobe PDF) | ISBN 9781119313984 (ePub) | ISBN 9781119314134 (hardcover)
Subjects: LCSH: Chip scale packaging. | Integrated circuits–Wafer scale integration.
Classification: LCC TK7870.17 (ebook) | LCC TK7870.17 .A38 2018 (print) | DDC 621.39/5–dc23
LC record available at https://lccn.loc.gov/2018034374

Cover Design: Wiley
Cover Images: Foreground: Provided by Volker Mai, Background: © KTSDESIGN/SCIENCE PHOTO LIBRARY/Getty Images

Set in 10/12pt Warnock by SPi Global, Pondicherry, India

10 9 8 7 6 5 4 3 2 1

Contents

Preface *xvii*
List of Contributors *xxiii*
Acknowledgments *xxvii*

1 **History of Embedded and Fan-Out Packaging Technology** *1*
Michael Töpper, Andreas Ostmann, Tanja Braun, and Klaus-Dieter Lang
1.1 Introduction *1*
1.2 First Embedding Technologies Based on MCM-D Concepts *5*
1.2.1 GE High Density Interconnect *5*
1.2.2 Fraunhofer IZM/TU Berlin *8*
1.2.3 Thin Chip Integration *15*
1.2.4 Irvine Sensors *18*
1.2.5 3D Plus *20*
1.2.6 Casio and CMK *21*
1.2.7 Panel-Level Molding *21*
1.3 First Embedding Technologies Based on Organic Laminates and Flex *24*
1.3.1 GE Chip-on-Flex *24*
1.4 Helsinki University of Technology and Imbera Electronics Embedded Chips *25*
1.5 Fraunhofer IZM/TU Berlin Chip-in-Polymer (CiP) *27*
1.6 HiCoFlex, Chip-in-Flex, and UTCP *31*
1.7 Conclusion *33*
References *35*

2 **FO-WLP Market and Technology Trends** *39*
E. Jan Vardaman
2.1 Introduction *39*
2.2 FO-WLP: A Disruptive Technology *39*

2.3 Embedded Die Packaging *40*
2.4 FO-WLP Advantages *42*
2.5 FO-WLP Versions *42*
2.6 Challenges for FO-WLP *48*
2.7 Drivers for FO-WLP *49*
2.7.1 Markets and Applications for FO-WLP *51*
2.8 Strong Demand for FO-WLP *53*
References *53*

3 Embedded Wafer-Level Ball Grid Array (eWLB) Packaging Technology Platform *55*
Thorsten Meyer and Steffen Kröhnert
3.1 Technology Description *55*
3.2 Basic Package Construction *59*
3.2.1 Cost Performance *60*
3.2.2 Electrical Performance *60*
3.2.3 Interconnect Density *60*
3.2.4 Thermal Performance *61*
3.2.5 Board-Level Reliability *61*
3.2.6 System Integration *61*
3.3 Manufacturing Process Flow and BOM *63*
3.4 System Integration Capability *66*
3.5 Manufacturing Format and Scalability *68*
3.6 Package Performance *69*
3.7 Robustness and Reliability Data *71*
3.8 Electrical Test Considerations *73*
3.9 Applications and Markets *74*
References *76*

4 Ultrathin 3D FO-WLP eWLB-PoP (Embedded Wafer-Level Ball Grid Array-Package-on-Package) Technology *77*
S.W. Yoon
4.1 Introduction *77*
4.1.1 eWLB (Embedded Wafer-Level BGA) Technology *78*
4.2 eWLB-MLP (Mold Laser Package-on-Package) Technology *80*
4.2.1 Test Vehicle Specification *80*
4.2.2 Assembly Process Flow *82*
4.2.3 Component-Level Reliability *82*
4.2.4 Board-Level Reliability *84*
4.2.5 Failure Analysis of eWLB-MLP After Board-Level Reliability Tests *86*
4.3 3D eWLB-PoP Technology *86*
4.3.1 3D eWLB-PoP Test Vehicle Specification *87*
4.3.2 Component-Level Reliability of 3D eWLB-PoP *88*

4.3.3 Experimental Thermal Characterization of 3D eWLB-PoP *88*
4.3.4 Electrical Functional Characterization of 3D eWLB-PoP *90*
4.3.5 Parasitic Electrical Simulation of 3D eWLB-PoP and fcPoP *90*
4.3.6 Board-Level Reliability of 3D eWLB-PoP *91*
4.4 3D eWLB SiP/Module *92*
4.5 Conclusions *94*
References *94*

5 NEPES' Fan-Out Packaging Technology from Single die, SiP to Panel-Level Packaging *97*
Jong Heon (Jay) Kim
5.1 Introduction *97*
5.2 Structure and Process Flow *97*
5.3 Thin Fan-Out Packaging *100*
5.4 Double-Sided Fan-Out Packaging *101*
5.5 Via Frame (VF) Fan-Out Package *101*
5.6 System-in-Package *104*
5.7 Panel-Level Package *107*
5.8 Performance and Reliability *108*
5.8.1 Thermal Performance *108*
5.8.2 Electrical Performance of Automotive Radar Package *109*
5.9 Application *110*
5.9.1 Mobile and Automotive Applications *111*
5.9.2 Sensor Products *111*
5.9.3 Optical Module *112*
5.9.4 IoT and Industrial Applications *112*
5.10 Roadmap and Remarks *115*
References *115*

6 M-Series™ Fan-Out with Adaptive Patterning™ *117*
Tim Olson and Chris Scanlan
6.1 Technology Description *117*
6.2 Basic Package Construction *119*
6.3 Manufacturing Process Flow and BOM *122*
6.4 Design Features and System Integration Capability *124*
6.5 Adaptive Patterning *128*
6.6 Manufacturing Format and Scalability *135*
6.7 Robustness and Reliability Data *137*
6.8 Electrical Test Considerations *138*
6.9 Applications and Markets *139*
Acknowledgment *140*
References *140*

7 **SWIFT® Semiconductor Packaging Technology** *141*
Ron Huemoeller and Curtis Zwenger
7.1 Technology Description *141*
7.2 Basic Package Construction *142*
7.2.1 Traditional Package-on-Package Designs *142*
7.2.2 SWIFT PoP Structure *143*
7.3 Manufacturing Process *144*
7.4 Design Features *146*
7.4.1 Form Factor *146*
7.4.2 Feature Size *147*
7.5 Manufacturing Format and Scalability *150*
7.5.1 System in Package *150*
7.5.2 Networking and High Performance Graphics (SWIFT on Substrate Applications) *151*
7.6 Package Performance *152*
7.6.1 Electrical Benefits *152*
7.6.2 Signal and Power Integrity DDR4 (AP PoP Applications) *152*
7.6.3 PCIe and Ethernet (SWIFT on Substrate Applications) *154*
7.6.4 Impedance Matching *156*
7.7 Thermal Performance *156*
7.8 Robustness and Reliability Data *159*
7.9 Applications and Markets *162*
References *167*

8 **Embedded Silicon Fan-Out (eSiFO®) Technology for Wafer-Level System Integration** *169*
Daquan Yu
8.1 Technology Description *169*
8.2 Basic Package Construction *169*
8.3 Manufacturing Process Flow *170*
8.4 Design Features *175*
8.5 System Integration Capability *175*
8.6 Manufacturing Format and Scalability *178*
8.7 Package Performance *178*
8.8 Robustness and Reliability Data *181*
8.9 Applications and Markets *183*
Acknowledgment *183*
References *183*

9 **Embedding of Active and Passive Devices by Using an Embedded Interposer: The i^2 Board Technology** *185*
Thomas Gottwald, Christian Roessle, and Alexander Neumann
9.1 Technology Description *185*
9.2 Basic Interposer Construction *191*

9.3 Manufacturing Process Flow and BOM *191*
9.4 Design Features *194*
9.5 System Integration Capability *194*
9.6 Manufacturing Format and Scalability *194*
9.7 Package Performance *195*
9.8 Robustness and Reliability Data *195*
9.9 Electrical Test Considerations *197*
9.9.1 Test Strategy *197*
9.10 Applications and Markets *197*
9.11 Summary *199*
References *200*

10 Embedding of Power Electronic Components: The Smart p^2 Pack Technology *201*
Thomas Gottwald and Christian Roessle
10.1 Introduction *201*
10.2 Technology Description p^2 Pack *202*
10.3 Basic Package Construction *203*
10.4 The p^2 Pack Technology Process Flow *204*
10.5 Smart p^2 Pack *206*
10.6 Package Performance *206*
10.6.1 Electrical Performance *206*
10.6.2 Dynamic/Switching Losses *207*
10.6.3 Inverter Efficiency *207*
10.6.4 Thermal Performance *209*
10.6.5 Robustness and Reliability Data *213*
10.7 Applications and Markets *213*
10.8 Summary *214*
Acknowledgments *215*
References *215*

11 Embedded Die in Substrate (Panel-Level) Packaging Technology *217*
Tomoko Takahashi and Akio Katsumata
11.1 Technology Description *217*
11.2 Basic Package Construction *218*
11.3 Manufacturing Process Flow and BOM *219*
11.4 Design Features *221*
11.5 System Integration Capability *223*
11.6 Package Performance *224*
11.6.1 Thermal Performance Comparison Between EDS and FBGA *224*
11.6.2 Electrical Performance Comparison Among EDS, FC-BGA, and FBGA *225*
11.6.3 Robustness and Reliability Data *227*
11.7 Diversity of EDS Technology: Module *229*

11.7.1 Technology Description *229*
11.7.2 Basic Package Construction *229*
11.7.3 Manufacturing Process Flow *229*
11.7.4 Design Features *230*
11.7.5 Robustness and Reliability Data *230*
11.7.6 System Integration Capability *231*
11.8 Diversity of EDS Technology: Power Devices *233*
11.8.1 Technology Description *233*
11.8.2 Basic Package Construction *234*
11.8.3 Manufacturing Process Flow *234*
11.8.4 Electrical Characteristics *235*
11.8.5 Size Miniaturization and Thermal Characteristics on Power Module Package *237*
11.9 Applications and Markets *239*
References *240*

12 Blade: A Chip-First Embedded Technology for Power Packaging *241*
Boris Plikat and Thorsten Scharf
12.1 Technology Description *241*
12.2 Development and Implementation *242*
12.3 Basic Package Construction *243*
12.4 Manufacturing Process Flow and BOM *246*
12.4.1 Manufacturing Equipment *246*
12.4.2 Basic BOM *247*
12.4.3 Wafer/Die/Assembly Preparation *247*
12.4.4 Die Attach and Adhesion Promotion *248*
12.4.5 PCB Processes *248*
12.4.6 Inspection and Process Controls *250*
12.5 Design Features *250*
12.5.1 Mature Design Rules and Roadmap *250*
12.6 System Integration Capability *251*
12.6.1 2D and Side-by-Side Packaging *251*
12.6.2 3D and Package on Package *251*
12.7 Manufacturing Format and Scalability *252*
12.8 Package Performance *252*
12.8.1 Electrical Performance *252*
12.8.2 Thermal Performance *253*
12.8.3 Thermomechanical/CTE, Moisture, and Warpage Issues *255*
12.9 Robustness and Reliability Data *255*
12.9.1 First Level/Component Level (CLR) *255*
12.9.2 Second-Level/Board-Level Reliability (BLR) *256*
12.10 Electrical Test Considerations *258*

12.11 Applications and Markets *258*
Acknowledgments *258*
References *259*

13 The Role of Liquid Molding Compounds in the Success of Fan-Out Wafer-Level Packaging Technology *261*
Katsushi Kan, Michiyasu Sugahara, and Markus Cichon
13.1 Introduction *261*
13.2 The Necessity of Liquid Molding Compound for FO-WLP *262*
13.3 The Required Parameters of Liquid Molding Compound for FO-WLP *264*
13.4 Design of LMC Resin Formulation *267*
13.5 Development of LMC in Connection with Latest Requirements *269*
13.6 Current LMC Representative Proprieties *269*
13.7 Conclusions *270*
Acknowledgment *270*
References *270*

14 Advanced Dielectric Materials (Polyimides and Polybenzoxazoles) for Fan-Out Wafer-Level Packaging (FO-WLP) *271*
T. Enomoto, J.I. Matthews, and T. Motobe
14.1 Introduction *271*
14.2 Brief History of PI/PBO-Based Materials in Semiconductor Applications *271*
14.3 Dielectric Challenges in FO-WLP Applications *274*
14.4 HDM Material Sets for FO-WLP *277*
14.5 PBO-Gen3 (Positive-Acting, Aqueous-Developable Material) *279*
14.6 PBO-Gen3 Process Flow *280*
14.7 PBO-Gen3 Lithography *282*
14.7.1 PBO-Gen3 Resolution *282*
14.7.2 PBO-Gen3 Thick Film Formability *283*
14.7.3 PBO-Gen3 Deep Gap Formability *284*
14.8 PBO-Gen3 Material Properties *284*
14.9 PBO-Gen3 Dielectric Reliability Testing *289*
14.9.1 PBO-Gen3 Adhesion After PCT *289*
14.9.2 PBO-Gen3 Chemical Resistance *292*
14.9.3 PBO-Gen3 bHAST *294*
14.10 PBO-Gen3 Package Reliability Performance (TCT Testing at Component and Board Level) *295*
14.11 Performance Comparison Between PBO-Gen3 and PBO-Gen2 *299*
14.12 PI-Gen2 (Negative-Acting, Solvent-Developable Material) *299*
14.13 PI-Gen2 Process Flow *299*

14.14 PI-Gen2 Lithography *303*
14.15 PI-Gen2 Material Properties *303*
14.16 PI-Gen2 Dielectric Reliability Data *306*
14.16.1 PI-Gen2 Adhesion After PCT *307*
14.16.2 PI-Gen2 Chemical Resistance *308*
14.16.3 PI-Gen2 bHAST *309*
14.17 PI-Gen2 Package Reliability Performance (Component and Board Level) *310*
14.18 Comparison Between PBO-Gen3 and PI-Gen2 *311*
14.19 Summary *313*
References *313*

15 Enabling Low Temperature Cure Dielectrics for Advanced Wafer-Level Packaging *317*
Stefan Vanclooster and Dimitri Janssen
15.1 Description of Technology *317*
15.2 Material Challenges for FO-WLP *319*
15.3 Material Overview *322*
15.4 Process Flow *328*
15.5 Material Properties *331*
15.6 Design Rules *340*
15.7 Reliability *341*
15.8 Next Steps *345*
References *346*

16 The Role of Pick and Place in Fan-Out Wafer-Level Packaging *347*
Hugo Pristauz, Alastair Attard, and Harald Meixner
16.1 Introduction *347*
16.2 Equipment Requirements for Fan-Out Bonders *349*
16.2.1 Core Capabilities of an Advanced Die Attach Equipment *349*
16.2.1.1 Die Feeding *349*
16.2.1.2 Substrate Handling *350*
16.2.1.3 Die Flip *352*
16.2.1.4 Fluxing *352*
16.2.1.5 Constant Bond Heat *353*
16.2.1.6 Pulse Heat *353*
16.2.1.7 Accuracy *354*
16.2.1.8 Clean Capability *358*
16.3 Avoiding Fan-Out Bonding Pitfalls *359*
16.3.1 Die Tilt *359*
16.3.2 Warped Carriers *360*
16.3.3 Influence of Motion Parameters *361*
16.3.4 Lack of Placement Repeatability *361*

16.3.5 Flying Die *361*
16.3.6 Process Margins *362*
16.4 Equipment Qualification for Fan-Out Pick and Place *363*
16.4.1 Step-by-Step Qualification *363*
16.4.2 Auto-Diagnostic of Core Capabilities *363*
16.5 Running a Large Area Glass-on-Glass Process *366*
16.6 Running a Glass-on-Carrier Process *367*
16.7 Running a Reference Production Lot with Test Die *368*
16.8 Conclusions *368*
References *368*

17 Process and Equipment for eWLB: Chip Embedding by Molding *371*
Edward Fürgut, Hirohito Oshimori, and Hiroaki Yamagishi
17.1 Introduction *371*
17.2 Historical Background Molding *372*
17.3 The Molded Wafer Idea: Key for the Fan-Out eWLB Technology *373*
17.3.1 Carrier System *378*
17.3.2 Molded Wafer *380*
17.3.3 Debonding *381*
17.4 The Compression Molding Process *381*
17.4.1 Molding Compound Ingredients *381*
17.4.2 Molding Compound State of Aggregation *382*
17.4.3 Processing Methods *384*
17.4.4 Molding Compound Preparation *385*
17.4.5 Compression Molding Temperature *386*
17.4.6 Cavity Filling During the Compression Molding Process *386*
17.4.7 Mold Tool Release During Compression Molding *388*
17.4.8 Transfer Molding vs. Compression Molding Capability *390*
17.5 Principle Challenges for Chip Embedding with Compression Molding *391*
17.5.1 Cavity Filling of Large Area *391*
17.5.2 Planarity of Large Area *392*
17.5.3 Dimensional Accuracy *393*
17.5.4 Molded Wafer Identical to SEMI Specification *394*
17.6 Process Development Solutions for Principle Challenges *395*
17.6.1 Cavity Filling of Large Area *395*
17.6.2 Planarity of Large Area *396*
17.6.3 Dimensional Accuracy *396*
17.7 Compression Molding Equipment for Chip Embedding *398*
17.8 Chip Embedding Features Achieved by Compression Molding *399*
17.9 Conclusions and Next Steps *399*
Acknowledgments *401*
References *402*

18 Tools for Fan-Out Wafer-Level Package Processing *403*
Nelson Fan, Eric Kuah, Eric Ng, and Otto Cheung
18.1 Turnkey Solution for Fan-Out Wafer-Level Packaging *403*
18.2 Die Placement Process and Tools for FO-WLP *404*
18.3 Encapsulation Tool for Large Format Encapsulation *408*
18.4 The Test Handling and Packing Solution for Wafer-Level Packaging and FO-WLP *415*
References *417*

19 Equipment and Process for eWLB: Required PVD/Sputter Solutions *419*
Chris Jones, Ricardo Gaio, and José Castro
19.1 Background *419*
19.2 Process Flow *422*
19.2.1 Degas *422*
19.2.2 Pre-clean *422*
19.2.3 PVD Adhesion Layer Deposition *422*
19.2.4 Cu Seed Layer Deposition *423*
19.3 Equipment Challenges for FO-WLP *423*
19.3.1 Contamination *423*
19.3.2 Increased I/O Density *425*
19.3.3 Organic Passivation Particle Management *426*
19.3.4 Contact Resistance (R_c) Management *426*
19.3.5 Wafer Warpage *427*
19.3.6 Capital Cost *427*
19.4 Equipment Developed to Overcome Challenges *428*
19.4.1 Solution for Contamination *428*
19.4.2 Atmospheric Degas vs. Vacuum Degas *429*
19.4.3 Solutions for Increased I/O Density *430*
19.4.4 Solution for Particle Management *431*
19.4.5 Solution for R_c Management *431*
19.4.6 Why Soft ICP Etch Is Best *433*
19.4.7 Solution for Wafer Warpage *434*
19.4.8 Solution for Capital Cost Reduction *435*
19.5 Additional Equipment Features *436*
19.6 Design Rules Related to the Equipment *437*
19.7 Reliability *438*
19.8 Next Steps *438*
References *440*

20 Excimer Laser Ablation for the Patterning of Ultra-fine Routings *441*
Habib Hichri, Markus Arendt, and Seongkuk Lee
20.1 Advanced Packaging Applications and Technology Trends *442*
20.2 The High Density Structuring Challenge *443*

20.3 Excimer Laser Ablation Technology *444*
20.3.1 Excimer Laser Enabled Dual Damascene RDL for Advanced Packaging Applications *448*
20.3.2 Key Advantages of Excimer Laser Enabled Dual Damascene RDL *450*
20.3.3 Electrical and Reliability Data for Excimer Laser Enabled Dual Damascene *451*
20.3.4 Process Cost Comparison to Current POR *452*
20.4 Summary and Conclusion *455*
References *456*

21 Temporary Carrier Technologies for eWLB and RDL-First Fan-Out Wafer-Level Packages *457*
Thomas Uhrmann and Boris Považay
21.1 Slide-Off Debonding for FO-WLP *459*
21.2 Laser Debonding: Universal Carrier Release Process for Fan-Out Wafer Packages *462*
21.3 Parameters Influencing DPSS Laser Debonding *463*
Acknowledgments *470*
References *470*

22 Encapsulated Wafer-Level Package Technology (eWLCSP): Robust WLCSP Reliability with Sidewall Protection *471*
S.W. Yoon
22.1 Improving the Conventional WLCSP Structure *471*
22.2 The Encapsulated WLCSP Process *474*
22.3 Advantages of the Encapsulated WLCSP, eWLCSP *477*
22.4 eWLCSP Reliability *479*
22.5 Reliability of Larger eWLCSP over 6 mm × 6 mm Package Size *481*
22.6 eWLCSP Wafer-Level Final Test *483*
22.7 Conclusions *485*
References *486*

23 Embedded Multi-die Interconnect Bridge (EMIB): A Localized, High Density, High Bandwidth Packaging Interconnect *487*
Ravi Mahajan, Robert Sankman, Kemal Aygun, Zhiguo Qian, Ashish Dhall, Jonathan Rosch, Debendra Mallik, and Islam Salama
23.1 Introduction *487*
23.2 EMIB Architecture *490*
23.3 High Level EMIB Process Flow *493*
23.4 EMIB Signaling *494*
23.5 Conclusions *498*
Acknowledgments *498*
References *498*

24 Interconnection Technology Innovations in 2.5D Integrated Electronic Systems *501*
Paragkumar A. Thadesar, Paul K. Jo, and Muhannad S. Bakir
24.1 Introduction *501*
24.2 Polymer-Enhanced TSVs *503*
24.2.1 Polymer-Clad SVs *503*
24.2.2 Polymer-Embedded Vias *506*
24.2.3 Coaxial TSVs *507*
24.3 HIST *510*
24.4 Conclusion *515*
References *515*

Index *521*

Preface

Embedded and fan-out wafer-level packaging (FO-WLP) are new electronic packaging technologies that have entered the market over the last 10 years due to a drive for low cost, high reliability, and large-scale integration of semiconductor devices for mobile, automotive, Internet of things (IoT), and medical markets. These new packaging technologies also address other electronic packaging challenges, including mechanical, materials, thermal, and electrical performance.

When designing an electronic package, there are many material challenges to consider, both in the semiconductor die and in the package. The die may have fragile low dielectric and extreme low dielectric constant interlayer dielectric (ILD) materials that must be taken into account. The stress on these layers must be minimized to prevent cracking or delamination of the passivation. One solution to avoiding these potential failure modes is to eliminate the interconnection between the device and substrate, reducing the stress on the semiconductor die. The pitch of the interconnects on the die is another consideration. For traditional packaging solutions such as wire bond and flip chip, the pitch of the wire bonds or flip-chip solder interconnects can be limited, such as in wire-bond packaging; the ability to wire bond or route the interconnection in the substrate, for example, can limit the minimum pitch. In flip-chip packaging, the ability to chip attach at fine pitch, to underfill fine-pitch interconnects, and to route the substrate at fine pitches can be a limiter. Again, it is best to eliminate the interconnections altogether to minimize the pitch. Flip-chip packaging also has the challenge of introducing low coefficient of thermal expansion (CTE) and high modulus substrate materials. These materials are required in order to minimize warpage in a package. Also, low CTE mold materials must be used to mitigate the stress introduced due to the CTE mismatch between the device, the substrate, and the interconnect material. A solution that eliminates the substrate and the interconnect, such as embedded and FO-WLP, is preferred. These material challenges have led to the development of substrate- and interconnect-free embedded packages.

The continued demand for ever thinner electronic packages has led to mechanical challenges. A simple method of thinning a package is to start with a thinner die. However, the latest wafer technologies are strain-engineered, so any bending of thin devices (<80 μm) can impact device performance. Furthermore, the handling of thin die can be difficult and costly, so it is often preferable to avoid the use of thinned die. Packaging technologies like embedded and FO-WLP enable a low package height by eliminating the substrate while still being able to use die thicker than 80 μm.

Modern electronic packages must also be highly reliable. The CTE mismatch between the materials in a package, including the mold compound, substrate, and interconnect, can induce mechanical stress. Minimizing this mismatch is one key to good package reliability; however, the elimination of interconnects and substrates can also eliminate these stresses. Warpage control is another important consideration. Warpage can be managed through the properties of the materials that make up the package, such as CTE, glass transition temperature (T_g), and modulus. The adjustment of core, prepreg, and solder resist layer thicknesses and copper density can also help manage warpage. However, elimination of the substrate altogether can eliminate the risk of warpage. Thus, these and other mechanical challenges can be addressed with embedded and FO-WLP packaging.

During the electrical design of an electronic package, signal integrity, power distribution, and any functional partitioning must be taken into account. The package should not degrade the electrical performance of the device. Thus, interconnect lengths must be short to minimize parasitics in the package, a low resistance is required to support the on-chip power network, voltage drops ($V = IR$) must be small, multiple parallel power and ground pins are necessary, and dummy metal planes are introduced for grounding. Since embedded and FO-WLP packaging can eliminate substrates and interconnects allowing for shorter packaging routing circuitry from die to solder ball, these electrical concerns are addressed.

In packaging, poor thermal paths, no airflow, closed systems, and 3D integration all exacerbate thermal issues. The thermal resistances of the epoxy mold compound, device, interconnect, substrate, and the ball grid array (BGA) for BGA and chip-scale packages (CSP) must be taken into account. The junction temperature or operating temperature of the device (T_j) must be minimized for a given power to prevent device failure. The thermal paths of embedded and FO-WLP packages are minimized by elimination of the substrate.

In embedded and FO-WLP packaging, a device is embedded into an organic laminate substrate or mold compound, respectively. There are many types of embedded and FO-WLP packages offered in the market today, such as extended wafer-level ball grid array (eWLB), redistributed chip package (RCP), M-Series, integrated fan-out (InFO), embedded chip package (ECP), and semiconductor embedded in substrate (SESUB), to name a few. In eWLB, RCP, M-Series, and InFO, the dies are molded into a wafer form, and then wafer-level packaging (WLP)-like buildup technologies are used to create the redistributed copper circuitry. These types of packages are generally referred to as fan-out wafer-level packages (FO-WLP) because the BGA is fanned out from the original device's footprint, unlike WLP where the BGA pins are constrained by the area of the die. In ECP and SESUB packages, the device is placed on a sheet of copper, and the substrate layers are built up onto the die using traditional organic laminate substrate technologies. These types of packages will be referred to here as embedded packaging.

FO-WLP and embedded packages differ from wire-bond and flip-chip packages because they do not require a wire-bond or solder bump interconnect to the die. Embedded and FO-WLP also do not require a lead frame or organic laminate substrate, because the substrate is built onto the device to create the circuitry that connects the device to the outside world. By eliminating the interconnects, these packages provide excellent electrical performance because the shorter circuitry in the embedded and fan-out wafer-level package has lower parasitics. Also, eliminating the interconnects eliminates the interconnect stress and thus avoids ILD crack and delamination issues.

FO-WLP not only eliminates the die interconnect and substrate, resulting in better electrical performance and lower cost, but also provides finer lines and spaces in the buildup circuitry, allowing for better routability. The lines and spaces are finer than typical substrate technologies because the FO-WLP process uses wafer back-end fabrication (fab)-like processes, materials, and equipment, as used in WLP. This is a significant advantage that FO-WLP has over embedded, as embedded packaging technology uses traditional organic substrate buildup processes, materials, and equipment, resulting in coarser lines and spaces. Also, since FO-WLP uses fab-like processes, the yields are much higher than those expected in embedded organic substrate laminate technologies. This is because organic laminate substrates can typically be tested before chip attach, so known good substrates can be identified. However, when building a substrate onto the device, the process must be robust enough to have >99.5% yield; otherwise both the substrate and the package are scrapped. The organic laminate substrate industry processes, materials, and equipment were not designed to provide such high yields. However, the one advantage embedded has over FO-WLP is its low cost, as it does not utilize high-end fab-like processes or materials.

Embedded and fan-out wafer-level packaging serve the same purpose as traditional electronic packaging such as flip-chip or wire-bond packaging: they provide a method to connect a device to the outside world as well as mechanical and environmental protection, and they allow the device to be handled and tested before shipping. One critical advantage embedded and fan-out wafer-level packaging have over WLP is the ability to perform testing in singulated package form. Since a WLP is a true chip-sized package made of entirely silicon, it cannot be tested using a traditional handler kit and automated tester. WLP is typically tested in wafer form with a prober, and thus the singulation process may introduce defects that cannot be screened out. Another advantage of embedded and fan-out wafer-level packaging is that the footprint of an embedded or fan-out wafer-level package can be smaller than that of a flip-chip and wire-bond package due to the finer design rules (narrower lines and spaces) provided by the fab-like processing. The thickness can also be reduced. Finally, as discussed above, embedded and fan-out wafer-level packaging offer many other advantages, such as improved electrical and thermal performance, good warpage control, and improved reliability due to lack of CTE mismatch.

This book consolidates much of the past 15 years' activity within this amazing new field of packaging technology, starting with a "History of Embedded and Fan-Out Packaging Technology" by Michael Töpper et al. of Fraunhofer IZM, followed by "FO-WLP Market and Technology Trends" by E. Jan Vardaman of TechSearch International. Sections are dedicated to chip-first FO-WLP, chip-last FO-WLP, embedded die packaging, material challenges, equipment challenges, and resulting technology fusions.

The chip-first FO-WLP section begins with a chapter on the first chip-first FO-WLP in the market, called eWLB, by Thorsten Meyer of Infineon Technologies and our coeditor Steffen Kroehnert of Amkor Technology Holding B.V. The section continues with the extension of eWLB to package-on-package (PoP) solutions by S. W. Yoon of STATS ChipPAC. Further chip-first FO-WLP chapters include nepes Corporation's fan-out packaging technology by Jay Kim and Deca Technologies' M-Series™ technology by Tim Olson et al. A new technology called chip-last, represented by Amkor Technology's silicon wafer integrated fan-out technology (SWIFT®), is then presented by Ron Huemoeller and Curtis Zwenger.

The embedded die packaging section describes the embedding of die using printed circuit board processes and materials technology represented by Schweizer Electronic (i^{2}Board and p^{2}Pack) in chapters by Thomas Gottwald et al., J-Devices (fan-out panel-level package [FO-PLP]) by Akio Katsumata et al., and Infineon Technologies (Blade) by Boris Plikat and Thorsten Scharf. This section also includes a novel idea of embedding a die in a silicon wafer by Daquan Yu of Huatian Technology Electronics Co.

The emergence of FO-WLP required innovation in both epoxy mold compound materials and low temperature cure dielectric materials. Chapters regarding these challenges are included in the materials section of the book. Katsushi Kan et al. from Nagase describe the complexity of liquid mold compounds for the FO-WLP market. Low cure dielectric material development is addressed by Ioan Matthews et al. of HD MicroSystems and Stefan Vanclooster et al. of Fujifilm Electronic Materials.

In the equipment section, pick and place equipment challenges are discussed by Hugo Pristauz et al. of BESI and Nelson Fan et al. of ASM Pacific Technology. Then, Edward Fuergut of Infineon Technologies and Hirohito Oshimori et al. of Apic Yamada Corporation describe the difficulties in creating a compression mold solution for FO-WLP. ASM Pacific Technology also describes their compression mold challenges in their chapter. Chris Jones of SPTS Technologies Ltd. and Ricardo Gaio et al. of Amkor Technology Portugal S.A. also describe the challenges of designing a physical vapor deposition (PVD) (or sputtering) machine for FO-WLP in which epoxy-molded wafers instead of silicon wafers are placed into high vacuum chambers. Additional equipment chapters describing unique solutions for embedded and FO-WLP packaging are also included. Habib Hichri et al. of SUSS MicroTec Photonic Systems, Inc. describe "excimer laser ablation for patterning of ultrafine routings," and Thomas Uhrmann and Boris Považay of EV Group give details about "temporary carrier technologies for eWLB and RDL-first fan-out wafer-level packages."

Finally, technology fusions resulting from research in the embedded and FO-WLP market are described in the last section, including S. W. Yoon's chapter on encapsulated wafer-level chip-scale package (eWLCSP) technology currently in production at STATS ChipPAC. Ravi Mahajan et al. of Intel then describe their embedded multi-die interconnect bridge (EMIB) for high-end applications. Finally, Muhannad Bakir at the Georgia Institute of Technology writes "Interconnection Technology Innovations in 2.5D Integrated Electronic Systems."

We hope you find this book on embedded and FO-WLP technologies – the first of its kind – educational and enjoy reading it as much as we enjoyed writing it.

Beth Keser
Steffen Kröhnert

List of Contributors

Markus Arendt
SUSS MicroTec Photonic Systems, Inc., Corona, CA, USA

Alastair Attard
BESI, Austria

Kemal Aygun
Intel Corporation, Chandler, AZ, USA

Muhannad S. Bakir
Electrical and Computer Engineering, Georgia Institute of Technology

Tanja Braun
Fraunhofer IZM, Berlin, Germany

José Castro
Amkor Technology Portugal S.A., Portugal

Otto Cheung
ASM Pacific Technology

Markus Cichon
Nagase (Europa) GmbH

Ashish Dhall
Intel Corporation, Chandler, AZ, USA

T. Enomoto
HDMicroSystems

Nelson Fan
ASM Pacific Technology

Edward Fürgut
Infineon Technologies AG

Ricardo Gaio
Amkor Technology Portugal S.A., Portugal

Thomas Gottwald
Schweizer Electronic AG, Schramberg, Germany

Habib Hichri
SUSS MicroTec Photonic Systems, Inc., Corona, CA, USA

Ron Huemoeller
Amkor Technology, Inc.

Dimitri Janssen
Fujifilm Electronic Materials, NV

Paul K. Jo
Electrical and Computer Engineering, Georgia Institute of Technology

Chris Jones
SPTS Technologies Ltd, Newport, UK

Katsushi Kan
Nagase ChemteX Corporation

Akio Katsumata
J-Devices

Jong Heon (Jay) Kim
nepes Corporation

Steffen Kröhnert
Amkor Technology Holding B.V., Netherlands

Eric Kuah
ASM Pacific Technology

Klaus-Dieter Lang
Fraunhofer IZM, Berlin, Germany

Seongkuk Lee
SUSS MicroTec Photonic Systems, Inc., Corona, CA, USA

Ravi Mahajan
Intel Corporation, Chandler, AZ, USA

Debendra Mallik
Intel Corporation, Chandler, AZ, USA

J.I. Matthews
HDMicroSystems

Harald Meixner
BESI, Austria

Thorsten Meyer
Infineon Technologies AG, Germany

T. Motobe
HDMicroSystems

Alexander Neumann
Schweizer Electronic AG, Schramberg, Germany

Eric Ng
ASM Pacific Technology

Tim Olson
Deca Technologies

Hirohito Oshimori
Apic Yamada Corporation

Andreas Ostmann
Fraunhofer IZM, Berlin, Germany

Boris Plikat
Infineon Technologies AG, Regensburg, Germany

Boris Považay
EV Group, St. Florian am Inn, Austria

Hugo Pristauz
BESI, Austria

Zhiguo Qian
Intel Corporation, Chandler, AZ, USA

Christian Roessle
Schweizer Electronic AG, Schramberg, Germany

Jonathan Rosch
Intel Corporation, Chandler, AZ, USA

Islam Salama
Intel Corporation, Chandler, AZ, USA

Robert Sankman
Intel Corporation, Chandler, AZ, USA

Chris Scanlan
Deca Technologies

Thorsten Scharf
Infineon Technologies AG, Regensburg, Germany

Michiyasu Sugahara
Nagase & Co., LTD

Tomoko Takahashi
J-Devices

Paragkumar A. Thadesar
Electrical and Computer Engineering, Georgia Institute of Technology

Michael Töpper
Fraunhofer IZM, Berlin, Germany

Thomas Uhrmann
EV Group, St. Florian am Inn, Austria

Stefan Vanclooster
Fujifilm Electronic Materials, NV

E. Jan Vardaman
TechSearch International, Inc., Austin, TX, USA

Hiroaki Yamagishi
Apic Yamada Corporation

S.W. Yoon
STATS ChipPAC, JCET Group

Daquan Yu
Huatian Technology (Kunshan) Electronics Co., Ltd., Economic & Technical Development Zone, Kunshan, Jiangsu, China

Curtis Zwenger
Amkor Technology, Inc.

Acknowledgments

We would like to thank all of our contributors who worked long hours to funnel their passion, experience, know-how, and years of long dedication into development and deployment of embedded and FO-WLP into chapters for this book. In addition, thanks goes to all the engineers, managers, and leaders in the electronics industry who have contributed to the success of embedded and FO-WLP technology. We would also like to thank our families for their patience and support during the writing of this book.

1

History of Embedded and Fan-Out Packaging Technology

Michael Töpper, Andreas Ostmann, Tanja Braun, and Klaus-Dieter Lang

Fraunhofer IZM, Berlin, Germany

1.1 Introduction

The fabrication of microelectronic systems purely monolithically on a wafer is limited by the need for mixed technologies and redundancy. Multi-chip modules (MCMs) provided an alternative in achieving high density interconnects (HDI) being originally developed in the 1980s for aerospace applications, where size and weight were critical requirements [1]. The core idea of MCM was therefore the reduction of the interconnection length between different electronic components like integrated circuits (ICs), passives, and others like optoelectronic components. In 1990, silicon devices of 16 mm × 16 mm were the limits of manufacturability in volume production – far from the wafer-scale integration goals necessary for large electronic systems. Therefore, multiple ICs were needed, which resulted in a hybrid manufacturing process. Standard packaging technologies have used single packaged ICs that are mounted on the printed circuit board (PCB). The electrical signals have to travel from the IC through the package and through the PCB to the next package, limiting the high-speed performance of the systems. This translates to long interconnection lengths between devices and a corresponding increase in propagation delay. MCM was the first approach to eliminate the package and to use a high density routing substrate instead of PCB. All ICs and also the substrates can be pretested to ensure a high yield module process.

The interconnection of the bare ICs to the MCM substrate can be done by direct chip connection using wire bonding (WB), tape-automated bonding (TAB), or flip-chip (FC) bonding. The basic MCM concept was first proposed by IBM in 1972 as an "active silicon chip carrier" [2]. There were three main

Advances in Embedded and Fan-Out Wafer-Level Packaging Technologies, First Edition.
Edited by Beth Keser and Steffen Kröhnert.

approaches for MCM processes developed: MCM-laminate (MCM-L), in which a multilayer prepreg laminated board was used; high density ceramic MCM (MCM-C), in which a multilayer co-fired ceramic substrate was used; and thin film deposited MCM (MCM-D), in which a thin film metal/dielectric substrate was used. The reason for building MCM-D in the late 1980s will still sound familiar today: providing off-the-chip interconnect bandwidth matching the semiconductors clocking rates, decreasing power dissipation in input and output (I/O) drivers with matched impedance, and reducing the volume and size of the circuits. The difference between then and now is a major push for low fabrication costs to satisfy the consumer markets. System in packages (SiP) can therefore be viewed as an extension of an MCM. In addition, the lack of availability of known good die (KGD) for the different applications was a major hurdle to get functional MCMs with a sufficient yield. All of these types were in principle "chip-last" technologies in which the interconnection substrate is manufactured before the chip attachment being done by FC, WB, or less used TAB. "Chip-first" is the usage of an embedding process where the chips are placed under the interconnection layer. The chips have to be placed onto or within a base substrate prior to the fabrication of the interconnect structure that is then a direct metallurgical contact eliminating FC, WB, or TAB. With such an electrical interconnect, a reference ground plane can be deposited directly onto the chip pads, which provide a very well-matched high frequency interconnect structure. The inductance and the capacitance are over 10 times lower compared with FC, TAB, and WB.

Therefore embedding-type MCMs have been developed to bypass the yield issue for monolithic wafer-scale integration. Main pioneering work was done by Wayne Johnson from Auburn University and Ray Fillion from GE Research. Johnson et al. proposed a hybrid that uses pretested ICs mounted into cavities etched in a silicon wafer [3]. A schematic of this concept is shown in Figure 1.1.

The chips are interconnected with planar thin film process steps. An additional metallization layer is used for the interconnection to the next packaging hierarchy. The cavity is etched anisotropically into a silicon wafer using standard MEMS technology, which gives 54.74° sidewall angles. A 15 μm thin Si membrane was left to facilitate the deposition of the metal routing on the top Si surface. This membrane was removed by dry etching before chip embedding. The components are inserted from the backside and glued inside the thermally oxidized Si carrier with an adhesive (epoxy resin with 75% inorganic filler). In one of these first embedding approaches, benzocyclobutene (BCB) was deposited on top of the Si wafer containing four ICs for planarizing the surface of the substrate for the subsequent routing process. The low curing temperature is of great benefit for this embedding-type approach due to the low thermal stability of the epoxy resin used for gap filling. After curing the BCB was structured by reactive ion etching (RIE). To correct for chip-to-wafer misalignment occurring during die mounting from the backside of the

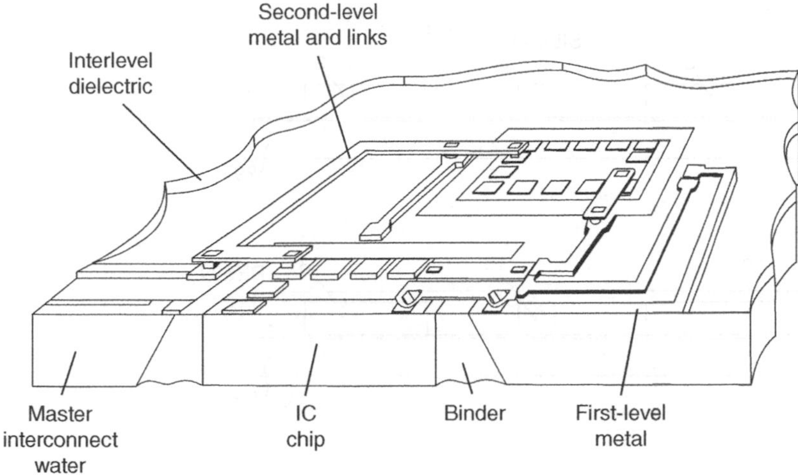

Figure 1.1 Schematic of an embedded MCM proposed by Johnson et al. [3].
Source: Reproduced with permission of IEEE.

substrate, a computer program was written that automatically calculated a corrected pattern generator file for mask making.

An even earlier approach for embedding was patented in 1968 by Ties Siebolt Te Velde and Albert Schmitz from Philips in Eindhoven in the Netherlands called semiconductor circuit having active devices embedded in flexible sheet [4]. The invention relates to a semiconductor device comprising a layer having a number of semiconductor components that are separated from one another by an electrically insulating material and are connected together on at least one side of the layer by mutually separated conductors that are located at least partly on the insulating material and adjoin surface parts of the components that are free from the insulating material. Pictures from the patent are shown in Figure 1.2.

The components were glued on a temporary support substrate preferable a flex (a). A layer of resin (polyurethane, for example) was then provided between and across the components (b). The temporary adhesive was dissolved to remove the embedded substrate (c). At that time, the focus for the semiconductor components was on diodes, transistors, resistors, and capacitors. Even an extensive search in the literature has not disclosed further information on this approach. It is not clear whether this technique has been further published or presented.

On the following several pages, early embedding technology developments will be summarized and grouped together in two categories: the first being more influenced by thin film technology used for the MCM-D and the other more laminate based. There are certainly more examples being developed in the

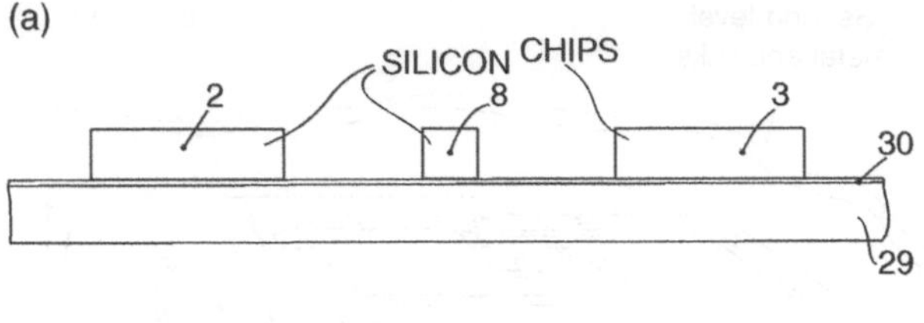

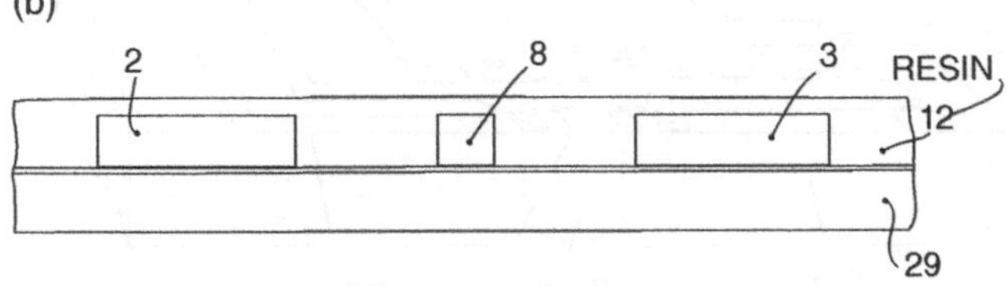

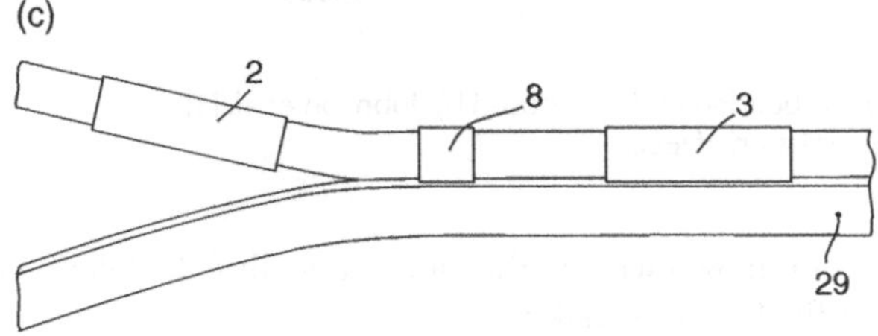

Patented May 18, 1971 3,579,056

3 Sheets–Sheet 1

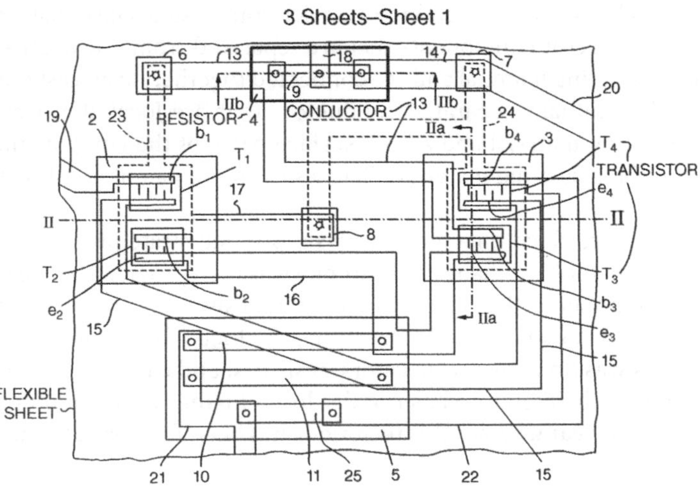

Figure 1.2 Pictures from US Patent no. 3,579,056.

last 30 years. Unfortunately for some technologies, only limited details have been published. For sure, some technologies that also have a longer history like extended wafer ball grid array (eWLB) from Infineon or redistributed chip packaging (RCP) from Freescale are discussed in separate chapters in this book.

1.2 First Embedding Technologies Based on MCM-D Concepts

1.2.1 GE High Density Interconnect

General Electric (GE) was a pioneer in the embedding technology [5–7]. In the mid-1980s, GE was the first company to commercialize the embedding concept called the GE HDI technology to address harsh military environments and/or high performance computing. The HDI MCM technology embeds bare chips into a substrate under a polymer/metal multilayer that interconnects the components directly without the need for wire bonds, TAB, or solder bumps (Figure 1.3).

GE used Kapton polyimide (PI) films for the dielectric isolation, and laser technology was used for the via formation. Texas Instruments had installed the HDI technology into a merchant market MCM foundry in Dallas. The process flow of the GE HDI embedding technology is shown in Figure 1.4.

The basis of this technology is the placement of chips of various types nearly edge to edge in cavities milled into a ceramic substrate [6]. A multilayer thin film interconnection structure is built on top of the chips and the ceramic substrate starting with a glued PI film on top of the embedded chips. The overlay film was bonded in a high temperature press under vacuum and bridges the gaps between chips and the ceramic frame. Laser ablation (5 W ultraviolet [UV] laser at 351 nm) was used to open the bond pads of the ICs. The laser system was able to accommodate the 25–75 μm misplacement of the embedded chips.

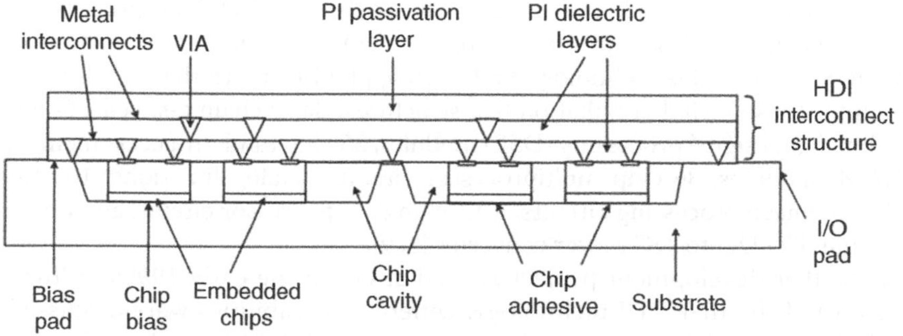

Figure 1.3 Principle of the GE HDI embedding technology.

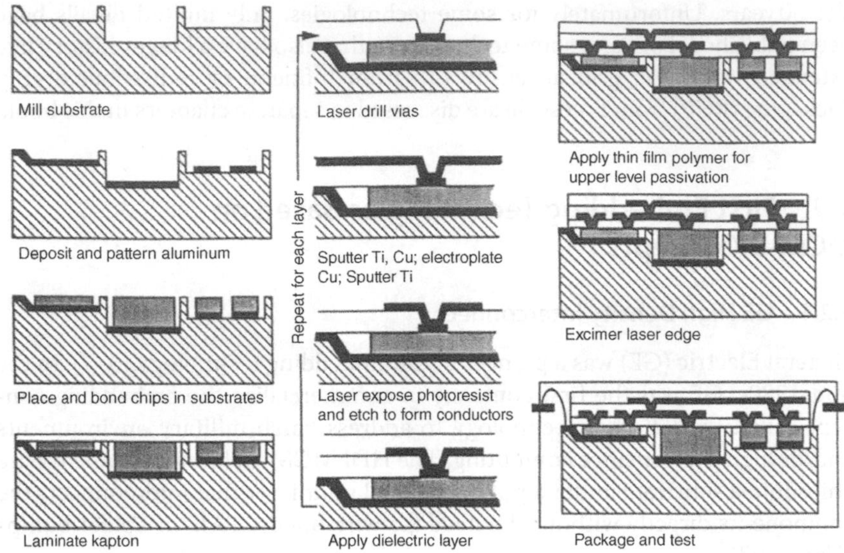

Figure 1.4 GE HDI embedding process flow [8]. *Source:* Reproduced with permission of John Wiley & Sons.

An optical microscope was attached to the laser system and was used to measure the actual chip positions including rotations. A software then adapted the via locations and interconnect (metal trace) artwork to conform to each unique substrate. Vias were formed by the laser with 25 μm width on the top and 12 μm at the bottom. In addition, the long focal depth of up to 200 μm of the laser could easily accommodate the 50 μm nonplanarity inherent with the embedded chip structure due to tolerances on the cavity depth, die thickness, and die attach bond line thickness. A standard PI/Cu thin film multilayer process of spin-coating PI and a fully subtractive metal process by sputtering (Ti/Cu) and electroplating Cu (4 μm thick) with a subsequent sputtered Ti layer to prevent Cu migration were used. The photoresist for the metal etching process was also structured by laser scanning using the data from the optical inspection. Already in 1990, this technology had been applied to more than 20 different circuits. These included radiation test structures, daisy chain test cells, 6- and 9-chip digital signal processing (DSP) modules, 25 chip random-access memory (RAM) modules, 36-chip multiprocessor circuits, analog functions, 16- and 52-chip image processing circuits, a 13-chip data processor circuit, and power modules like DC-to-DC power converter [5, 9].

A further development project started at GE in the early 1990s to lower the cost of the first HDI technology. Different possibilities were discussed, including a modified process that utilizes an FR-4 PCB as the substrate, eliminating the milling step for the ceramic substrate. FR-4 has been one of

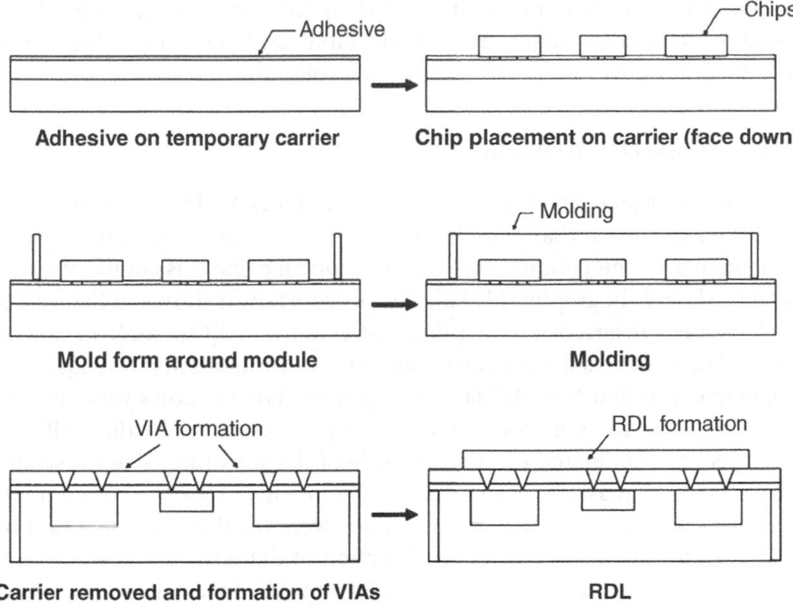

Figure 1.5 GE plastic encapsulated HDI process flow.

the standard flame-retardant (FR) (class 4) PCB materials since the 1970s. Most promising was the plastic encapsulated HDI MCM [9]. The process flow is shown in Figure 1.5.

The bare chips or passives were mounted facedown onto an adhesive-coated polymer film. An industry standard epoxy blend encapsulant was selected with a coefficient of thermal expansion (CTE) of 6 ppm K^{-1} closely matched to silicon by a 70% solid content. The material was 100% compatible with the other HDI multilayer materials. A mold form was then placed around the chips, and a molding material was deposited into the mold form. The substrate was formed out of the combination of components and cured polymer. A polymer/metal multilayer was then deposited on the flipped embedded substrate to interconnect the different embedded components. A major advantage of the plastic encapsulated process compared with the ceramic approach was the much higher planarity of the surfaces, including filling of all open gaps. The issue of high power dissipation of the chips could be eliminated by using thermal plugs made out of a high thermal conductance material such as AlN, copper, or a composite structure, which were attached directly to the exposed backside of the embedded module. Also, through package vias were under discussion in 1994 by embedding metal pin grids into the mold. The first modules were demonstrated on 150 mm square size in 1994 with the outlook on 12–18 in. in the future [9]. One commercial product was a phased array radar application [10]. EPIC was founded in 1994 by former GE employees. They developed a technology called "EPIC Chip First" and started production in 1996 primarily for government systems in low volume.

1.2.2 Fraunhofer IZM/TU Berlin

The chip-first approach developed at the Fraunhofer IZM/TU Berlin since the early 1990s was different than the HDI from GE due to the aligned bonding of the components to the ceramic opening that opened the possibility of using masks for the later lithography [11, 12]. The process flow is shown in Figure 1.6.

The chips were inserted into windows of ceramic (Al_2O_3, AlN) or silicon substrates. The windows for the active and passive components were laser cut by computer-controlled Nd:YAG laser equipment. The windows were around 100 μm larger than the components in the *x*/*y*-direction. A PI film with an adhesive was then laminated on the frontside of the substrate. The substrates were fixed to a glass mask. Using an optical alignment system, the components were inserted face-up into the windows. They were fixed in their position by the tacky PI film. The accuracy in *x*/*y*-direction of the positioning was better than 5 μm. Different from the HDI approach from GE, a mask process already used in high production for bumping and redistribution could be used to open the vias above the chip pads and to define the interconnection lines that contact the chips. No laser processing was required. The substrate was turned upside down, and a filled epoxy was dispensed into the gap between the substrate and the components. After curing the epoxy and lapping the back of the substrate, the auxiliary film was removed. The surface was cleaned with a short plasma step. The highly planar embedded substrate was now ready for the thin film wiring, starting with a layer of photosensitive polymer. Therefore, spin-on polymers rather than dry films could be used, which broadens the range of materials being used for the routing. Especially, spin-on low-k polymers were needed for RF applications [13]. A low-k polymer is a material with a dielectric constant smaller than that of silicon dioxide, enabling faster switching speeds and lower heat dissipation by reducing parasitic capacitance. An example of an embedded SRAM is shown in Figure 1.7.

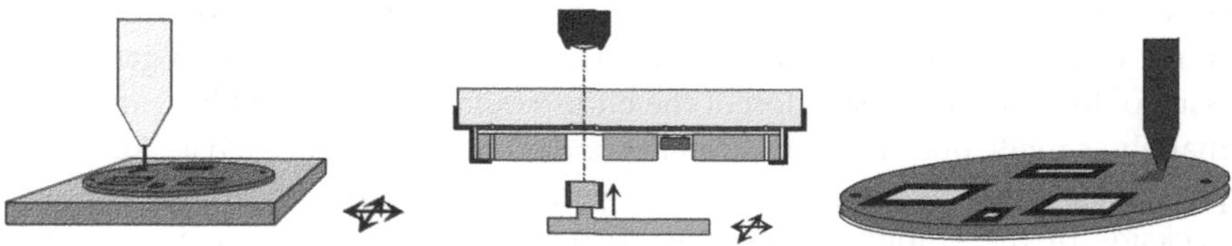

Figure 1.6 Process steps of the embedding process: laser cutting of the ceramic, chip placement using optical alignment, and filling of the gap between embedded components and substrate by an epoxy.

Figure 1.7 Embedded SRAM in AlN (SEM) and with the first photosensitive BCB and copper layer (SEM; BCB is etched for visualization).

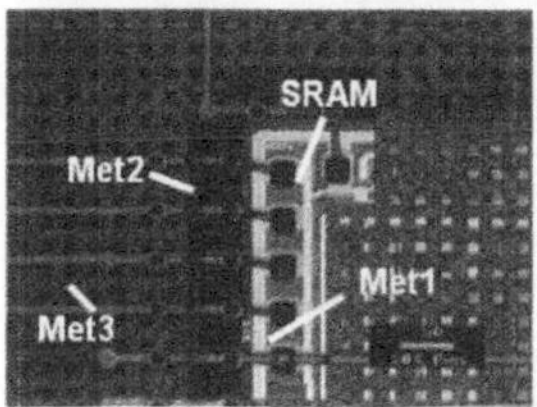

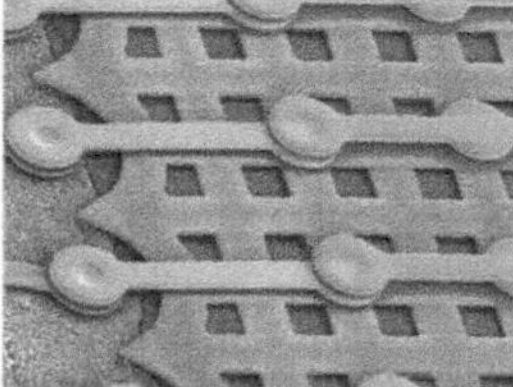
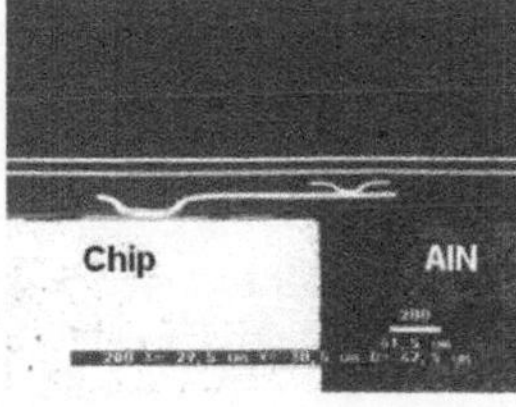

Figure 1.8 Multilayer metallization (Cu/BCB) over the SRAM (video print). Metallization over the gap (BCB was etched for SEM); cross section of the four-layer metallization interconnecting the embedded components.

To demonstrate the compatibility of the embedding technology, an MCM with eight SRAMs and four surface-mount device (SMD) capacitors (220 nF, size 1210, thickness 600 µm) was fabricated using AlN as the embedding substrate. The components were interconnected by a four-layer metallization. The dielectric polymer was mostly photosensitive BCB (10 µm thick) due to the low curing temperature of below 250 °C. A higher curing temperature would degrade the epoxy-filled gap. The copper lines were 30 µm wide and 5 µm thick. Figure 1.8 shows the first three metal layers over the chip, the metallization over the gap, and a cross section through the four-layer metallization interconnecting the embedded components.

On the embedding substrate, the redistribution is no longer limited to a fan-in, but the rerouted pattern and the final package can exceed the die size, enabling fan-out packaging structures [14]. Embedding in a carrier substrate is not restricted to only one die size, but different die from different technologies can be inserted at the same time. When a fan-out redistribution is deposited, the die itself will be completely encapsulated by ceramic and gap filler material. So the die is no longer exposed to mechanical damage risk by pick and place handling. Examples of such fan-out single chip packages using embedding technology are shown in Figure 1.9.

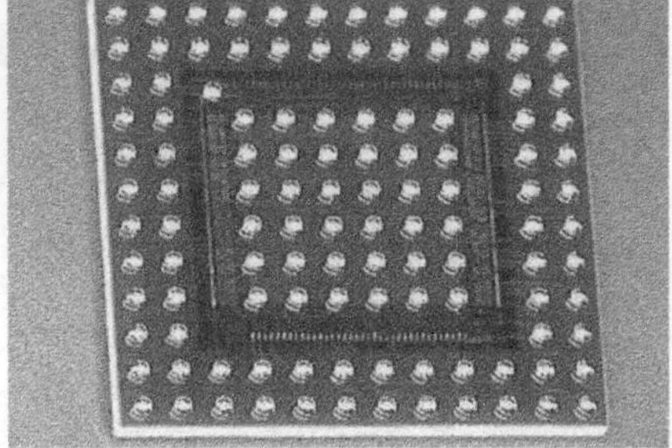

Figure 1.9 Small die with fan-out BGA pitch of 0.5 mm (left) and large die embedded with 0.4 mm BGA pitch with a package size of 9.6 mm × 9.6 mm (right).

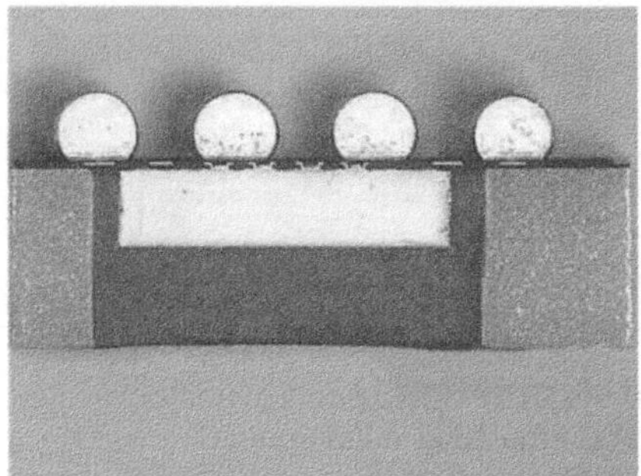

Figure 1.10 Crosscut of fan-out embedding die with a completely encapsulated die and revealing package details above embedding interface (Cu routing, Ni-UBM, PbSn solder ball).

To accommodate 300 μm solder balls, the under-bump metallurgy (UBM) dimensions were chosen as 250 μm in diameter and with a thickness of 5–10 μm. The Ni surface is protected against oxidation by a 100 nm chemical Au deposition before placement of the solder balls. Crosscuts of such fan-out packages are shown in Figure 1.10.

This planar integration of hybrid technique was also employed to achieve a stackable MCM. Vertical integrating systems of such high complexity with the shortest chip-to-chip interconnects require numerous electrical feedthroughs in the MCM substrate and vertical interconnects between mating submodules. For that purpose, prefabricated miniaturized contacting elements (thin film Cu/polymer routing) that provide the required high number of vertical electric connections were embedded in the embedded substrate materials [15]. The invention allowed vertical electric connections to achieve the high quality of usual horizontal thin film routing (Figure 1.11).

The contacting elements had been embedded in the substrate materials by the same planar junction technique. Such separately pretested submodules

Figure 1.11 Vertical interconnect element (VIE) (manufacturing, sawing, 90° turning) and SEM of such a VIE (30 µm × 50 µm).

with a dual-sided bump-array metallization had been stacked to realize the shortest interconnects between a large number of ICs [16]. To accomplish repair or system changes, removable elastomeric connectors were sandwiched between mating substrates and kept under moderate pressure. The process is shown in Figure 1.12. Compared to common 2D integration, reduced signal delay and reduced parasitic effects like resistance, capacitance, cross talk, and inductance resulted in better electrical performance. In 1997 and 1998, results indicated that the impact of small variations in geometry caused by the technological embedding process on the line parameters meets the practical requirements in terms of signal transmission. Microstrip structures for the wiring on the embedded substrate using BCB as dielectric are well suited for high frequency applications in the 20–30 GHz range [17, 18]. Small pretestable subunits, with a moderate number of integrated components and their easy exchange in the stack, should lead to a high system yield.

This embedding technology is also an excellent solution for one of the most crucial issues in power module designs, which is the assembly of large structural components – the circuit substrate and the heat sink – together. In addition, the thermal expansion mismatch between the components could be absorbed by the embedding process. A process modification was done for such power modules that involved the embedding of high power dissipating ICs in a substrate that was thinner than the chips and mechanical planarization of the substrate backside [19]. The process flow is shown in Figure 1.13. A homogeneous low thermal resistivity contact to the cooling body could be accomplished simultaneously for all embedded components due to the planar backside finish.

Using this technology, high power MCMs were fabricated with a junction to ambient thermal resistance below 0.6 $\mathrm{K\,cm\,W^{-1}}$ normalized to the chip area. This was achieved by embedding the chips into the AlN substrate, planarizing by lapping/grinding the module backside, and bonding to high

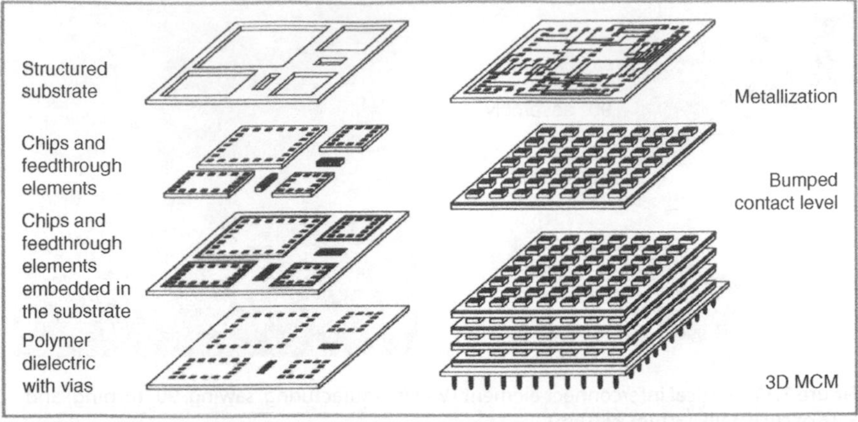

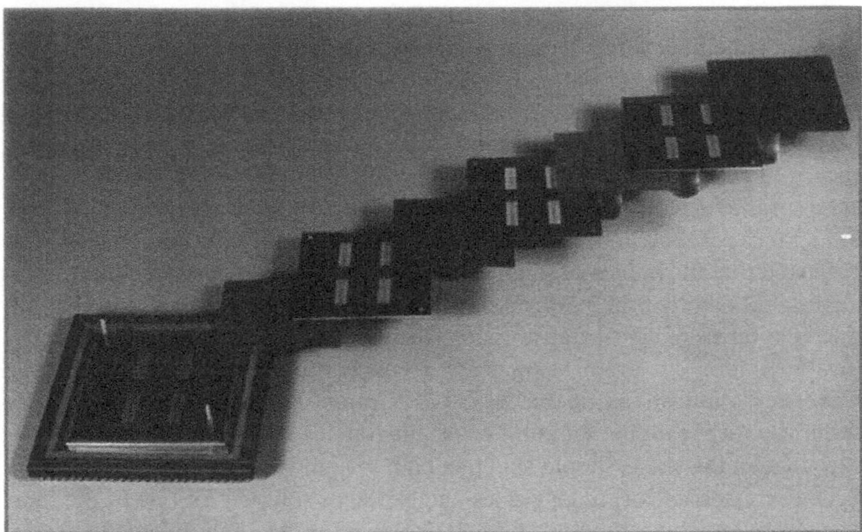

Figure 1.12 Schematic of the 3D concept of the embedding technology and view of submodules and elastomeric conductive interlayers mounted on a PGA [16].

performance microchannel heat sinks with high conductivity adhesive. Values down to 1.25 $\mathrm{Kcm\,W^{-1}}$ had been obtained by use of thermal grease materials. The low thermal interface resistance had been obtained simultaneously for all embedded dies. AlN microchannel heat sinks were fabricated by laser machining. Thermal resistances of the heat sinks below 0.6 $\mathrm{Kcm\,W^{-1}}$ (0.03 $\mathrm{K\,W^{-1}}$) were obtained at a pressure drop below 1 bar and a water flow of 60 $\mathrm{l\,h^{-1}}$. Figure 1.14 shows the thermal test of MCM with a size of 2 in. × 2 in. with six thermal test chips capable of a power dissipation up to 1000 W in a volume of 50 mm × 50 mm × 3 mm.

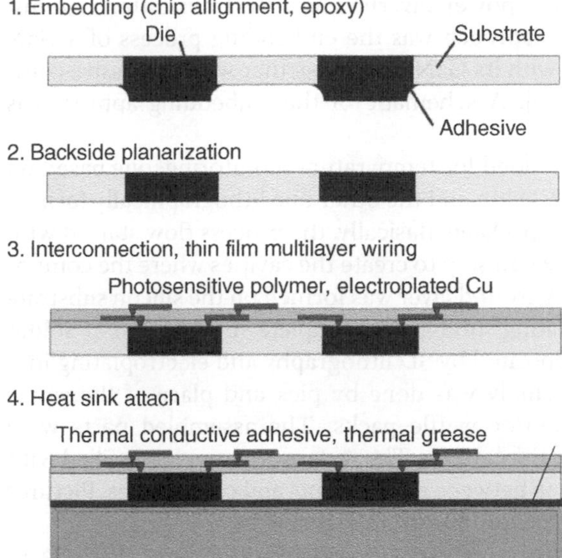

Figure 1.13 Process flow of the planar embedding technology for power modules: (1) embedding with epoxy, (2) backside planarizing, (3) thin film multilayer wiring, and (4) low thermal resistance heat sink attach.

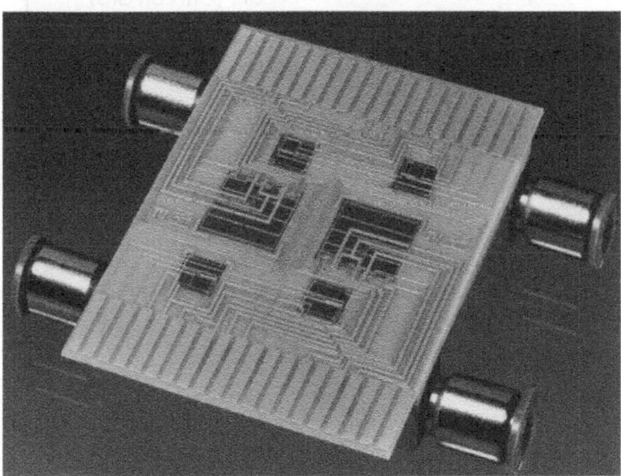

Figure 1.14 Thermal test of MCM with a size of 2 in. × 2 in. with six embedded thermal test chips and assembled on a heat sink.

A development of a fully Si-based embedded module was done in the Smart Power project from 2011 to 2016 funded by the EU. The goal of the project was to carry out the packaging and thermal management development required to achieve the efficient and cost-effective implementation of SiC- and GaN-based power modules into industrial power inverters and RF transmitter systems, respectively. One packaging approach was the embedding process of a GaN high power amplifier (HPA) with its GaN driver together with all passive components into an Si cavity [20]. A schematic of the embedding approach is shown in Figure 1.15.

Two thermal sensors were added for temperature monitoring: one based on a surface acoustic wave (SAW) chip and the other one lithographically formed directly on the silicon surface package. Basically, the process flow started with a deep reactive ion etching (DRIE) step to create the cavities where the components were later assembled. A ground layer was formed on the silicon substrate by sputtering and electroplating, lining the complete module. AuSn solder pads were structured and deposited by 3D lithography and electroplating into the cavities. Wafer-level assembly was done by pick and place of the single devices from dicing tape and/or waffle packs. The assembled parts were reflowed and soldered inside the cavities. The wafer was completely filled with a polymer layer, closing the gap between components and cavity edges. Pictures of the module at different stages are shown in Figure 1.16.

By means of successive lithography steps, a redistribution layer (RDL) layer for building component interconnections and a shield layer were deposited in

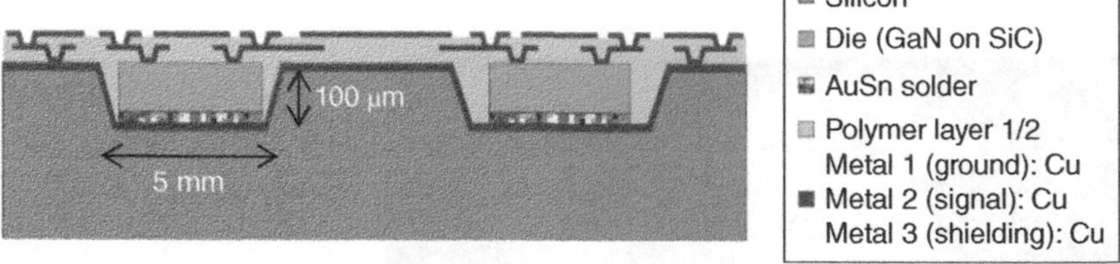

Figure 1.15 Simplified schematic of the embedding of an Si RF power module.

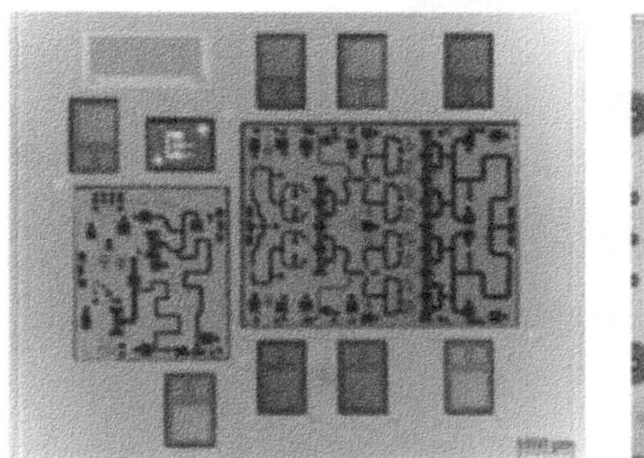

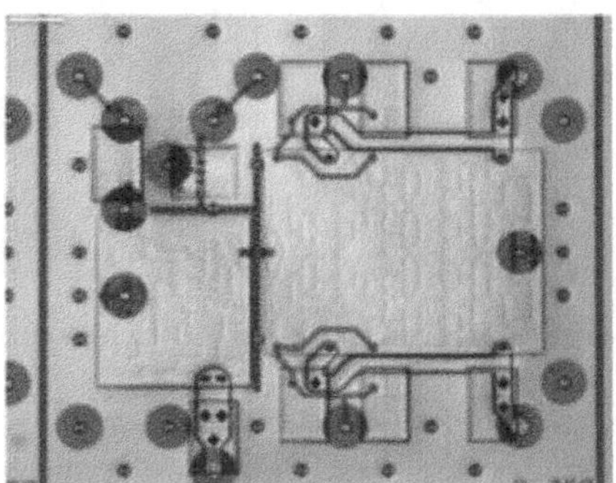

Figure 1.16 Video print of the topside of the embedded components and after multilayer RDL.

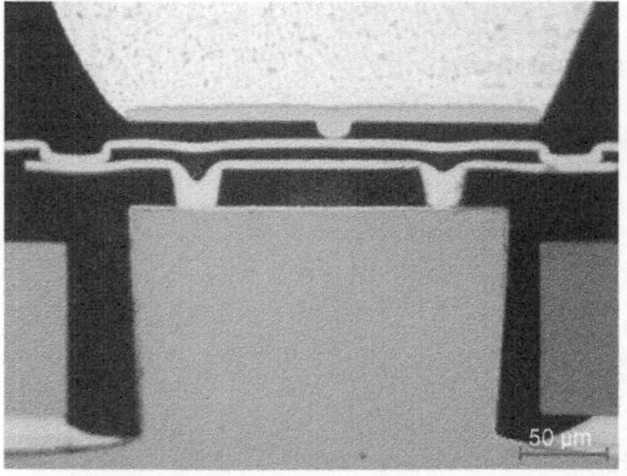

Figure 1.17 Crosscut through the embedded module with the multilayer wiring: on the left picture the embedded GaN components are seen, and on the right picture the interconnect of the RDL to the HPA Au Pad is shown in detail.

a strip line configuration. Low-k BCB polymer layers were used to isolate them. Photo-structuring in combination with excimer laser structuring the vias into the polymer was done to feed the contacts. A final passivation layer was applied and structured above the shield, and a UBM was deposited on top. Finally, a wafer-level ball mount process was used to mount 20 000 solder balls on the wafer before being diced, resulting in a hybrid component with ball grid array (BGA) connection pads, fully compatible with SMD assembly as shown in Figure 1.17.

The wafer-level processing was carried out on a 200 mm wafer. Therefore, packaged modules containing 11 active and passive components in a 10.5 mm × 9.0 mm area allowed the fabrication of up to 260 highly integrated modules per wafer.

1.2.3 Thin Chip Integration

The extreme developments in thinning techniques opened the possibility to further develop the embedding technology in 1999. The basic idea of the concept called thin chip integration (TCI) was to embed ultrathin die directly into the RDL [21]. For this technology, a base chip at wafer level was used as an active substrate for smaller and thinner die. Active components could be mounted in FC fashion on the base IC wafer using solder interconnects. To avoid the need for FC bonding, the TCI concept had been developed [22]. Key to this approach was the use of extremely thin ICs (down to 20 µm thickness and less), which were incorporated into the RDL (Figure 1.18).

The process flow for TCI modules began with the bottom wafer carrying large base chips. The completely processed device wafers for the top IC had to be mounted on a carrier substrate by a reversible adhesive bond and undergo a backside thinning process until the thinned wafers were 20–40 µm thick.

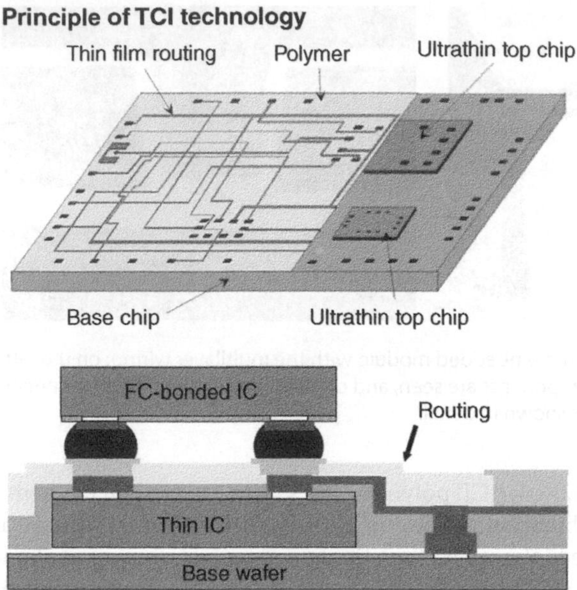

Figure 1.18 Principle of TCI.

The thinned top wafer and its carrier substrate were singulated, thus producing thin chips that could be handled by its carrier chip, just like any other standard die. A Cu routing layer was deposited on the base wafer. A combination of sputtered TiW/Cu and electroplated Cu was used, which had a low electrical resistivity. The Cu could be over 10 μm thick. The additional integration of passive components such as resistors (*R*), inductors (*L*), or capacitors (*C*) was possible. The bottom wafer was coated with a thin photosensitive BCB layer film, and the thinned top chips were placed and mounted into this adhesive layer. Next, the photosensitive thin film polymers like BCB, polybenzobisoxazole (PBO), and PI were deposited onto the surface to planarize the topography of the mounted thin chips. A high degree of planarization (DOP) was very important for this first polymer layer to overcome the height difference between the surface of the bottom chip and the thinned chips. A final metallization of Cu/Ni/Au for the FC bonding was deposited on the base wafer carrying the embedded ICs.

In a three-chip test vehicle for TCI, the ICs for the process were thinned to around 20 μm. In Figure 1.19, a picture and cross section of the TCI are shown [23].

In a public-funded project called RESTLES (Reliable System Level Integration of Stacked Chips on MEMS), the TCI had been also tested using MEMS and ASIC for a speed sensor [24]. Here a MEMS wafer was used as the base wafer. Within this project, the TCI technology was therefore named SCOM (stacked chip on MEMS) [25]. The base wafers contained inertial sensors. The wafer

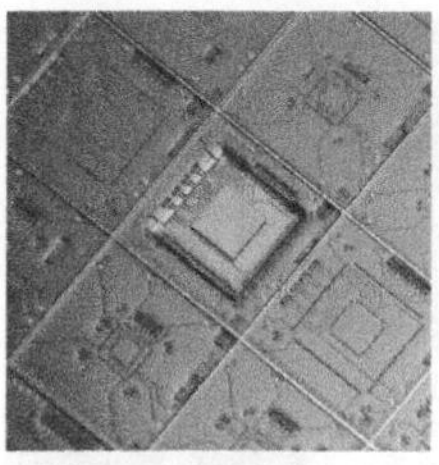
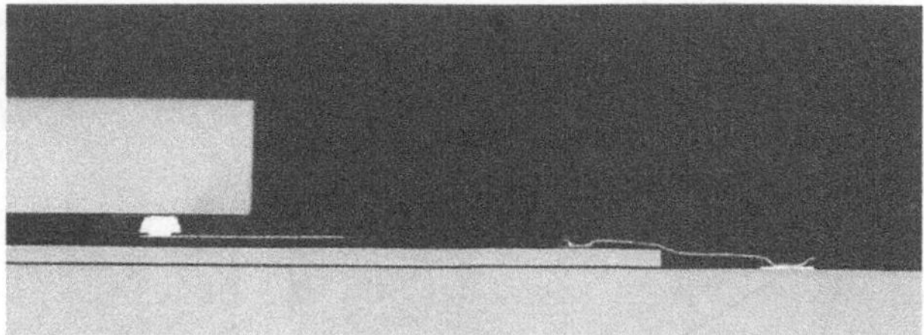

Figure 1.19 TCI test vehicle after flip-chip attach on wafer level and crosscut through the three-chip stack.

with the smaller-sized chips (e.g. ASIC chips) was thinned down on wafer level to a thickness of 10–40 μm and diced. These thinned chips were glued onto the base wafer with a BCB layer. The polymer had been deposited and structured before gluing the next chip on top. After placement of the thinned chips, the wafer was again coated with BCB to embed the chips. The next step was the buildup of metal routing using a semi-additive Cu process together with further thin film polymer layer with higher toughness like PI or PBO for stress compensation between solder balls and the later soldering process on an organic board. Then a UBM was added using electroplating. Finally, the solder ball application was done by ball drop. The wafer was then diced, and the full ASIC-MEMS package could be FC-assembled onto a PCB (Figure 1.20) [26, 27].

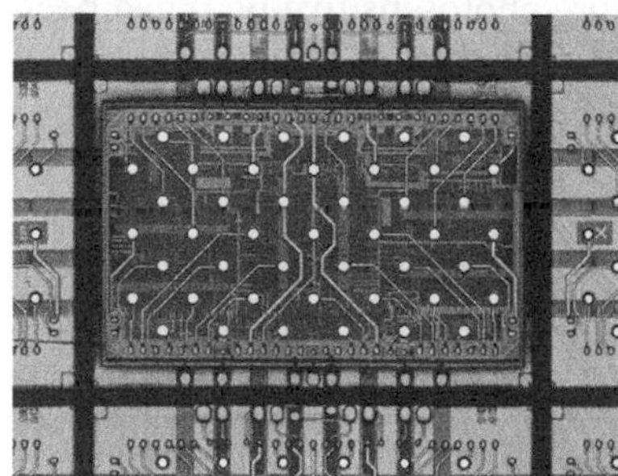
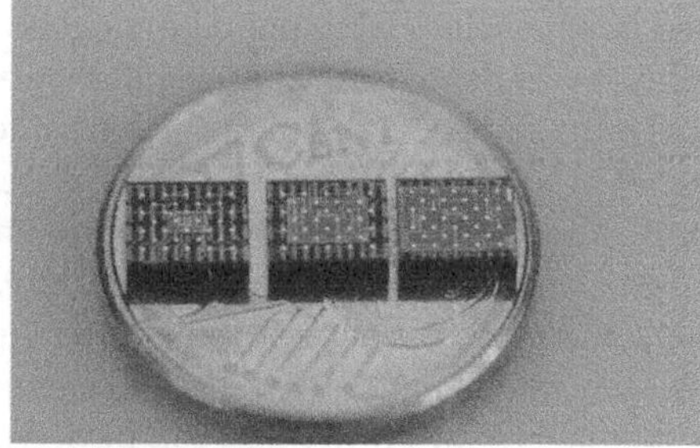
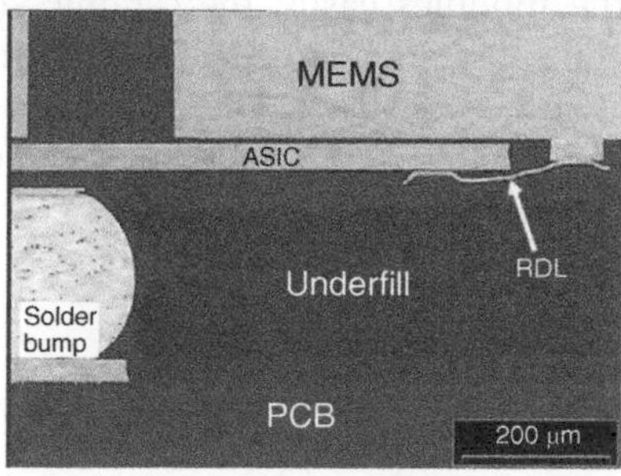

Figure 1.20 TCI for MEMS: RDL on the embedded ASIC, fully processed SiPs, crosscut.

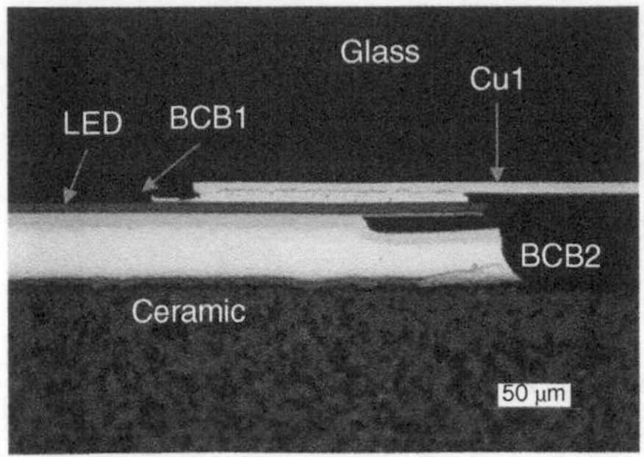

Figure 1.21 Embedded LED: cross section, complete module on AlN substrate and module operated at 7 mA.

Even after 2000 thermal cycles (−40 to +150 °C), no defect had been detected in the TCI stack for this automotive application.

A TCI test vehicle for packaging of light-emitting diodes (LEDs) was proven and published in 2013 [28]. The LED dies were fabricated by OSRAM Opto Semiconductors GmbH. The thickness of the epi-layer was 6 μm. On both sides, there was a thick gold metallization of 1–2 μm. The metallization was directly deposited onto the epi-layer. The growth substrate had been removed. Due to the very low thickness of the die, the embedding technology was perfectly applicable. The intention of the concept was to glue the LEDs with their optical active side to a glass support. Therefore photosensitive BCB had been used. Copper lines were plated on a glass wafer. To contact the corner pad, an Sn layer was deposited. Copper pillars were used to bring the corner pad contact to the same level as the LED backside contact. The LEDs were glued on a first BCB layer. Afterward, the whole wafer was reflowed to establish the solder bond. A second BCB layer was used to embed the LEDs. Copper and Sn were plated to bond the modules to a substrate. Finally each module with four LEDs was soldered to the ceramic substrate. Figure 1.21 shows a cross section after this step and a complete module operated at 7 mA. The solder joint from the copper line to the corner pad can be seen.

To test and operate the modules easily, the ceramic substrates had been glued to an insulated metal substrate board. A die attach adhesive with a thermal conductivity of $60\,W\,mK^{-1}$ was used. The pads were connected by gold WB. Using this technology, 3200 LED die could be placed on a 150 mm wafer, which is equivalent to 800 modules with 4 LEDs.

1.2.4 Irvine Sensors

Irvine Sensors was founded in 1980 to develop high density 3D stacking technologies mainly for US defense and space programs. They started a cooperation with IBM for 3D stacking using the NEO-wafer fabrication [29]. Au wire bonds were deposited on chip pads to form Au bonds for the next interconnection

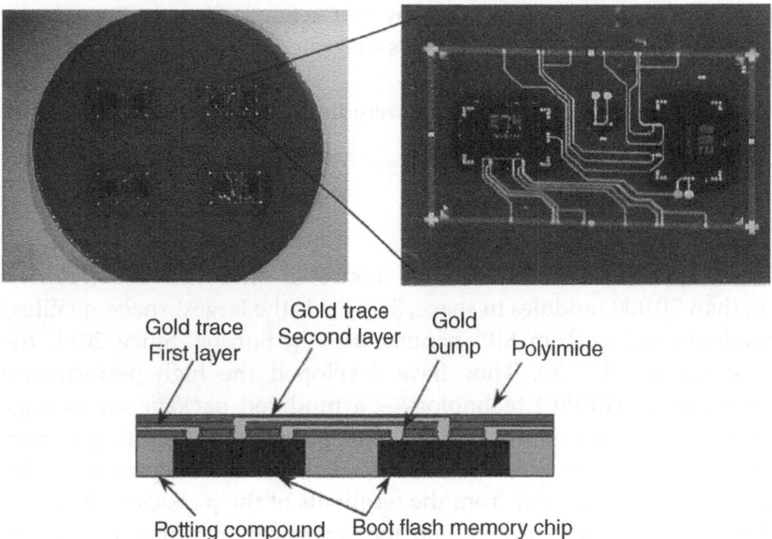

Figure 1.22 Principle and example of the NEO-wafer concept from Irvine Sensors.

level. The chips were then placed in a molding frame and were encapsulated by molding compound, forming the so-called NEO-wafer. The Au balls protruded through the mold. The RDL was formed on top of the surface using PI and Au. Finally, the NEO-wafer was thinned (around 100 μm) to open the backside of the die surface (see Figure 1.22).

This process was designed to realize 3D stacking of such NEO-wafers that could be created from individual and heterogeneous chips (see Figure 1.23). Blank silicon was added to open areas on layers where smaller dies were used

Figure 1.23 3D stacking using the NEO-wafer principle.

to enhance thermal conduction between layers if needed. After the lithography and metallization process, the NEO-wafers were diced into identical sizes that contained each layer to be stacked.

Mature stacking technology and tools were used to stack many layers and interconnect layers.

1.2.5 3D Plus

The French company 3D Plus started in 1996 as a spin-off from Thales (France). With more than 70 000 modules in space, 3D Plus is the largest space-qualified catalog products and custom SiP manufacturer in Europe. Since 2011, the company is part of HEICO. They have developed the high performance package-in-package (HiPPiP) technology – a modified package on package (PoP) [30]. The fundamental process step is the placement of plastic packages on an adhesive foil that is overmolded and thinned. After the deposition of the RDL in order to build a fan-out from the terminals of the packages, the layers with molded packages are stacked, and the vertical interconnection is performed using the laser-assisted metallization process from 3D Plus. This process was further enhanced for unpackaged components (typical thickness 100 μm) being patented as WDoD (Wirefree Die-on-Die) technology [30]. The process starts with the deposition of an adhesive on silicon or glass carrier (see Figure 1.24).

Die are assembled on the adhesive with accuracy better than ±5 μm. Compression molding of an epoxy resin over components and grinding (optional) is done as the next step. An RDL is deposited over the embedded components after de-taping (see Figure 1.25).

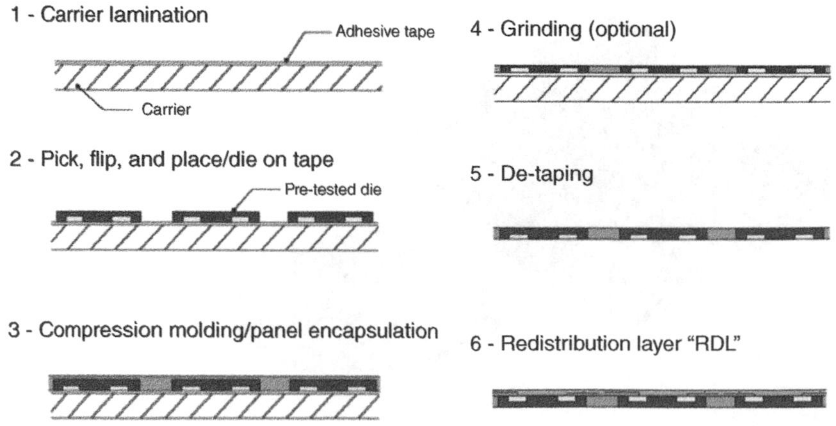

Figure 1.24 First part of the process flow of the Wirefree Die-on-Die (WDoD) technology.

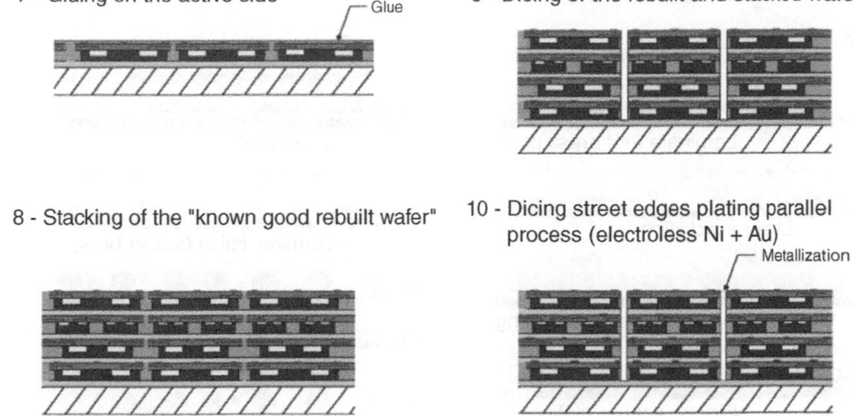

Figure 1.25 Second part of the process flow of the Wirefree Die-on-Die (WDoD) technology.

Nonconductive glue is used for stacking the tested layers. Dicing of the rebuilt and stacked layers is done as the next step, and the sidewalls are metallized. Finally, direct laser patterning with a pitch of 100 μm is performed for the electrical contact between the layers.

1.2.6 Casio and CMK

The embedding technology developed by Casio and CMK was called embedded wafer-level packaging (EWLP) [10, 31]. The process starts with standard redistributed wafer-level packages (WLPs) placed on a carrier (see Figure 1.26).

The WLPs have Cu pillars on top that are embedded by a laminate using a hot press process. A laser process is used to open the Cu pillars. A combination of electroless Cu and electroplated copper is used for the interconnection of the die. The modules are bumped after the deposition of a solder mask. One product was a miniaturized wristwatch containing a speed meter for runners.

1.2.7 Panel-Level Molding

The massive trend for higher productivity and lower cost has pushed the embedding process further to a chip-first approach using molding on large panels. At Fraunhofer IZM in Berlin, one of the first panel molding machines at 24 in. × 18 in. has been installed [32]. An example of such a panel is shown in Figure 1.27.

The large panel size equals a typical PCB format and is selected to achieve process compatibility with cost-efficient PCB processes. Such a "mold-first" process starts with die and fiducial assembly on an intermediate carrier followed by overmolding and debonding of the molded wafer/panel from the carrier. Compression molding is done at constant temperature, typically in the

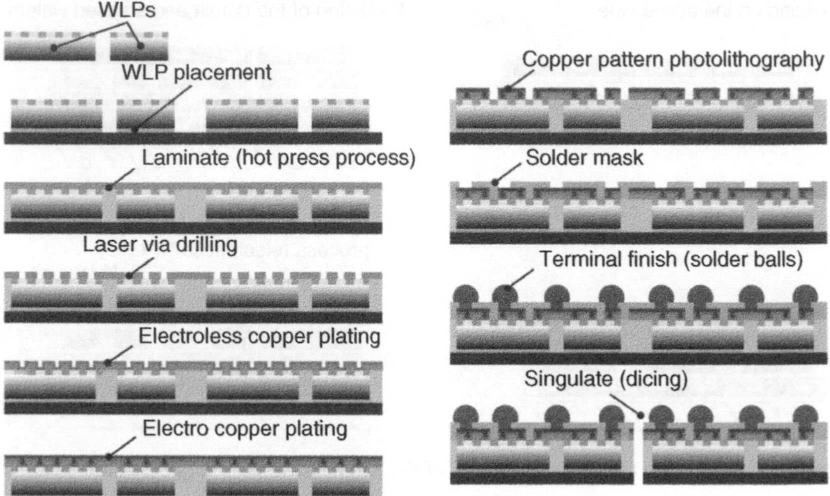

Figure 1.26 Process flow of EWLP.

Figure 1.27 Example of a 24 in. × 18 in. fully populated molded panel with embedded die.

range of 120–130 °C and under pressure and vacuum to achieve a homogeneous encapsulation without voiding or air entrapments. The molding materials are liquid, granular, or sheet compounds [33]. One key process step for homogeneous large area embedding is material application before compression molding. Where sheet compounds already deliver a uniform material layer, the application of liquid and granular compound must be optimized and adapted for a homogeneous distribution without flow marks, knit lines, and incomplete fills. Hence, dispense patterns of liquid and granular molding compounds are studied to achieve high yield and reliable mold embedding. In addition, applicable thickness ranges, total thickness variation, voids, and warpage are investigated for the different material types. From the processing point of view, the liquid materials are dispensed in the middle of the cavity and flow during closing and compression of the tooling to fill the entire cavity. However, a granular compound is distributed nearly homogeneously all over the cavity. The compound melts and the droplets have to fuse during closing and compression of the tool. Sheet compound offers easy application options, as the sheet has to be only applied on the panel and melts, flows around the die, and cures during compression molding. The molding process is shown in Figure 1.28.

A glass or metal carrier with the assembled die is shown in (a) of Figure 1.28. Granular molding compound is distributed over the panel, as shown in (b)–(d).

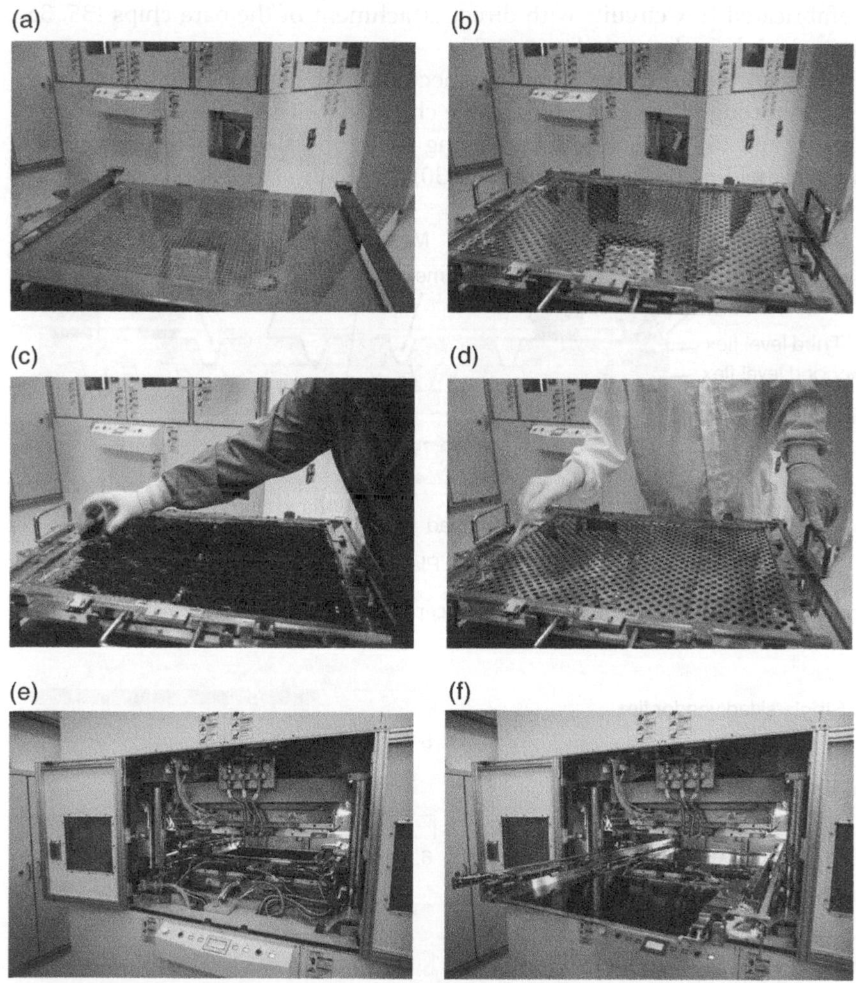

Figure 1.28 Process flow of the panel-level molding process using granular molding compounds at Fraunhofer IZM.

The panel is molded using heat and pressure (e). After this molding process, the panel is ready for the RDL process being based on thin film or PCB technology (f). The process has been also proven for multi-project runs [34].

1.3 First Embedding Technologies Based on Organic Laminates and Flex

1.3.1 GE Chip-on-Flex

A further modification of the HDI technology was done by GE using prefabricated flex circuits with direct attachment of the bare chips [35, 36]. The principle is shown in Figure 1.29.

The chips were also electrically connected to the interconnect flex by forming vias through the flex directly to the chip pads with a laser, metallization of the vias and top surface, and patterning of the metal and then followed by a transfer molding process (see Figure 1.30).

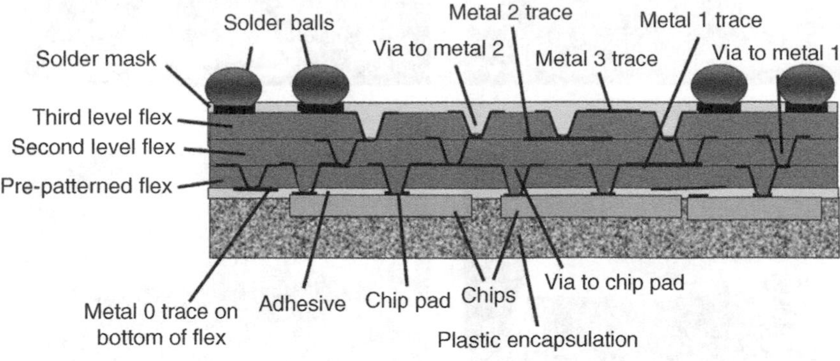

Figure 1.29 Principle of the Chip-on-Flex concept.

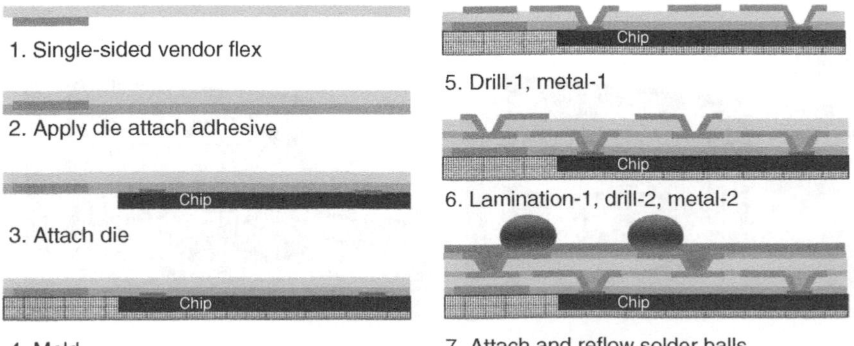

Figure 1.30 GE Chip-on-Flex process flow.

Figure 1.31 GE HDI technology: 100 W point of load power converter module.

An example of a power converter is shown in Figure 1.31.

In some publications, the name MCM-F (Multichip-on-Flex) had been used for this technology. Such a multilayer structure could be thinned down to less than 100 μm, including the Si thickness of around 50 μm, using standard wafer grinders. In addition the backside of the chips can be opened to attach cooling elements or a backside metallization.

1.4 Helsinki University of Technology and Imbera Electronics Embedded Chips

Helsinki University of Technology started in 1997 a study on the solderless interconnection and packaging technique based on embedding for active components motivated by the decreasing interconnection pitch of the die [37]. Imbera Electronics, a spin-off from the university, further developed this technology. Several generations of their integrated module board (IMB) technology to embed discrete electronic components with low- to mid-range numbers of I/Os (2–350 I/Os) inside a low cost organic PCB have been demonstrated since 2002 [38]. The technology had been proven for CMOS, GaAs, and several discrete C and R components. One focus of the development was the use of standard, widely available materials and processes. In the first IMB manufacturing processes (IMB-A and IMB-B), the components were embedded inside a PCB core layer, and the interconnections between the IC and the copper conductor were done using buildup layers and chemically plated micro-vias (see Figure 1.32).

The core layer was manufactured using normal inner layer PCB manufacturing processes like FR-4 or bismaleimide triazine (BT). The active and passive components were placed using a nonconductive adhesive. An alignment tolerance of the assembly process of ±13 μm using a chip shooter with 7000 units per hour (uph) and ±5 μm using a FC bonder was published in 2004 and 2009

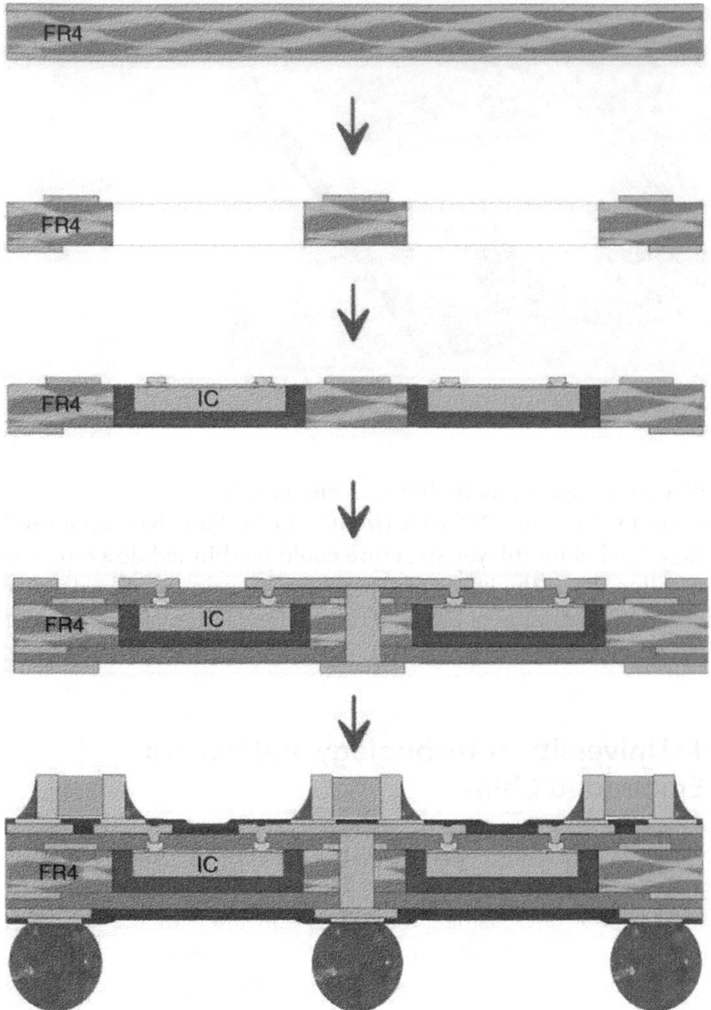

Figure 1.32 Imbera process flow for IMB.

[38, 39]. The cavities were filled with a thermosetting epoxy to create a planar surface. Resin-coated films (RCF) were pressed on both sides. UV laser drilling was used to open the vias to the components that need a minimum copper thickness of 3 μm on the I/O bond pads to withstand the process without damage. An example of the technology is shown in Figure 1.33.

Several PCB process flows could then be used to build the modules. Due to the direct metallization of the I/O pads, a low contact resistance of less than 1 mΩ is possible. Thirty micrometer lines and spaces have been demonstrated.

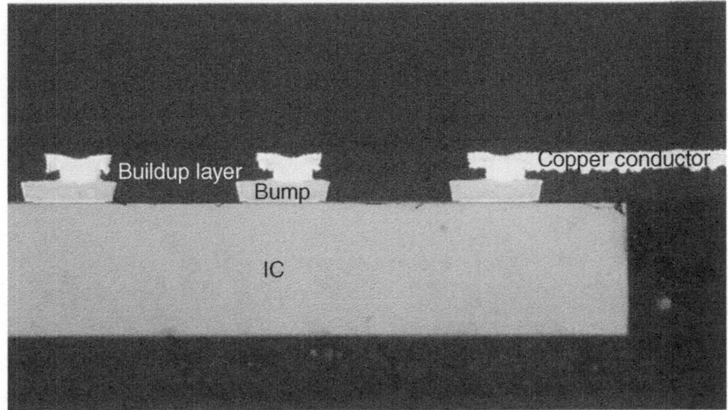

Figure 1.33 Cross section of embedded die using the Imbera process.

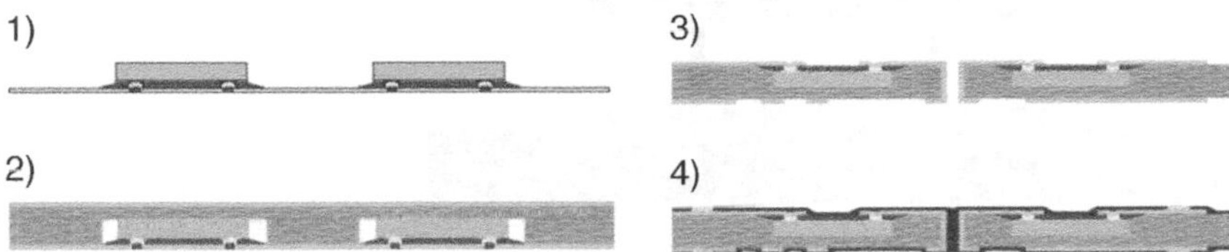

Figure 1.34 Process flow of the modified IMB process.

A slightly modified process flow was published in 2009 where the manufacturing process started with laser drilling micro-vias and alignment marks into a metal foil (see Figure 1.34, step 1) [40].

Prepreg with premanufactured openings around the components was pressed onto the foil and cured (step 2). The Cu foil and the laminate are patterned to form the electrical routing (steps 3 and 4). GE Healthcare Finland Oy, in partnership with GE Idea Works, announced in 2013 that it had completed the acquisition of Imbera Electronics Oy.

1.5 Fraunhofer IZM/TU Berlin Chip-in-Polymer (CiP)

A new concept for the integration of active components in organic substrates has been developed by Fraunhofer IZM and TU Berlin starting in the late 1990s. The chip-in-polymer (CiP) technology is based on the embedding of ultrathin chips into buildup layers of PCBs [41]. The interconnect structure, which is neither an FC nor a wire bond, is shown in Figure 1.35.

The basic idea of CiP is slotting thin semiconductor chips into standard PCB constructions. This technique can be used to fabricate 3D stacks of multiple die. The basic process flow is shown in Figure 1.36.

The process started with the electroless Ni/Cu deposition on a wafer to strengthen the original bond pad [42]. Laser drilling of micro-vias and the PCB

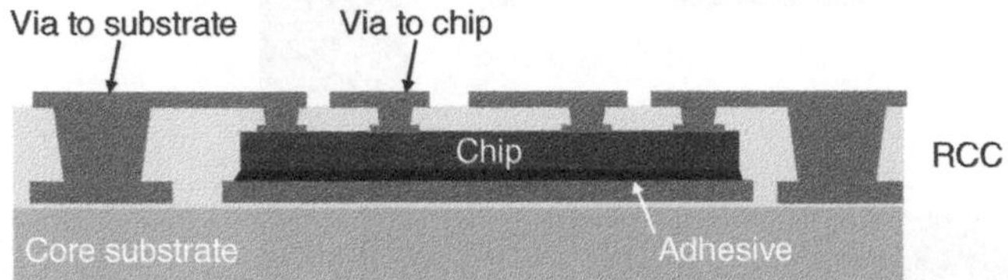

Figure 1.35 Interconnect principle of an embedded chip in a PCB buildup layer.

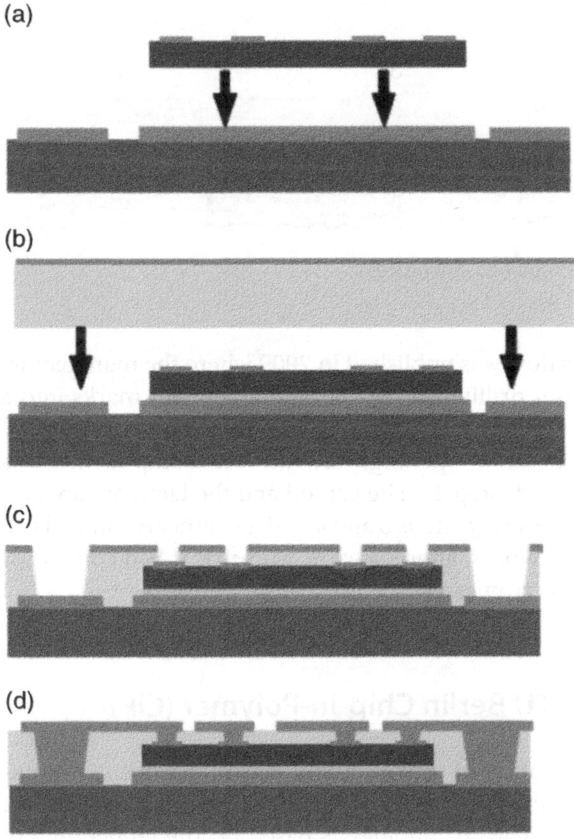

Figure 1.36 Process steps of chip embedding: (a) die bonding, (b) embedding in a polymer layer by vacuum lamination, (c) laser drilling of vias to chip and substrate, and (d) metallization of vias and Cu structuring.

metallization process are not compatible with Al or Cu contact pads of semiconductor chips. Therefore, a further layer of 5 μm Cu was applied to the bond pads of the chips to be embedded. Other metallization such as electroless Ni/Pd can be optimized for micro-via plating. Passivation layers have to be tested for their fragility as well as for their adhesion with the resin-coated copper (RCC) laminate layers. The wafers were thinned to 50 μm and diced. Die cavities were formed in FR-4 PCB. The thinned chips were placed into 40 μm deep cavities. Placement accuracy is extremely crucial for chip embedding. The process tolerances for sequential die bonding via drilling and Cu structuring have to be very low in order to achieve an acceptable yield. One of the requirements is that the machines for these three process steps should use the same alignment fiducials on the core substrate. Especially large substrates are multilayer cores with small thickness and provide very low contrast to fiducial marks. Prior to chip placement, die ejection from a dicing blue tape takes place and should be optimized according to the adhesiveness of the tape, size, and ultimate thickness of the chip. Adhesiveness of the tape is regulated by UV exposure and is reduced for component pickup and placement without endangering the chip integrity. Additional requirements for the placement accuracy are related to the PCB vision system. The PCB camera has to cover the entire area of the proposed multilayer PCB cores as well as the problems related to the low contrast fiducial marks. Furthermore, a die bonder machine should have an advanced conveyor system. Chip bonding had been developed by using printable pastes and die attach films. Screen printing allowed a precise control of volume and location of the adhesive paste over dispensing. Electrically conductive Ag-filled pastes or B-stage pastes could be used. The core substrate with the die-bonded chips was covered from both sides with an RCC layer. Temperature and pressure profiles should be adjusted carefully to promote epoxy adhesion at all interfaces and avoid chip breakage. A thickness of 15–20 μm over the chip surface was desirable for the subsequent micro-via opening and filling. The overlaying Cu layer served as the base for package routing. In the CiP technology, interconnections were achieved via micro-via laser drilling to chip pads and subsequent metallization, similar to the formation of micro-via on PCBs. For micro-via drilling, a pulsed 355 nm UV laser had been used. It could ablate Cu as well as the RCC dielectric. Accurate alignment of the micro-via drilling with respect to the underlying Cu pattern remained challenging for the yield of the interconnection. After drilling, the micro-vias were chemically cleaned and then treated by a Pd activation and electroless Cu deposition. The Cu layer was around 1 μm thick and acted as a seed layer for the consecutive Cu electroplating. A minimum thickness of 10 μm Cu was required in the micro-vias. By the use of special Cu plating chemistry, a nearly complete filling of the micro-vias could be achieved, as shown in Figure 1.37.

An example of this embedding concept was a modular micro camera (see Figure 1.38) [43]. At a size of only 16 mm × 16 mm × 12 mm, including optics

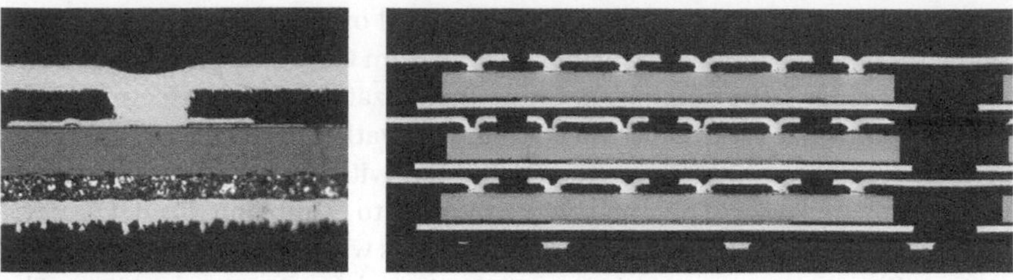

Figure 1.37 Cross section of a micro-via interconnect to an embedded chip and three-stacked embedded die.

Figure 1.38 Cross section of the micro camera module with image sensor on top and processor embedded into the PCB.

and 16 mm × 16 mm × 4.6 mm without the optics, the micro camera module is an extremely small system [43].

A total of 72 passive and 13 active components (such as oscillators, DC-to-DC converters, memory chip, and image processor) have been embedded inside the module, and the image sensor is mounted on top. A key advantage of the newly developed module is that all components are integrated directly into the PCB. By encapsulating the electronic components, the micro camera is now impervious to any kind of environmental influences. The main system advantage is that the image material is directly inside the camera, since it is equipped with an integrated processor for image processing. After the image sensor has recorded the image, the integrated processor evaluates the frame. The video itself no longer has to be sorted and analyzed by an interposing system. Instead, only the relevant signals are transmitted. The data volumes to be transmitted and processed turns out to be much less.

1.6 HiCoFlex, Chip-in-Flex, and UTCP

The HiCoFlex process (see Figure 1.39) was commercialized by the company HighTec (Switzerland), which started in 1992 as an independent company, formerly being the thin film division of the Swiss company ABB [44].

The process started with the deposition of a separation layer on a glass substrate. A multilayer PI and Cu process was added including active and passive components. After final testing and protection, the whole layer was separated from the glass substrate. The process was later transferred to 24 in. × 24 in. substrate size with 15 μm LDI technology. The technology is a kind-of RDL-first process.

Fraunhofer IZM and Technical University of Berlin (TUB) developed a Chip-in-Flex technology (FCF) [45]. By embedding chips into flexible wiring boards, the functional density of electronic systems can be dramatically increased. The benefits of flex substrates are light weight and high wiring density, which will be combined with the complexity of the active chip. However, to maintain the basic flex substrate characteristics, the buildup with an integrated chip has to be as small as possible. Chips with a thickness of only 20 μm have been used and the interconnection should not exceed a couple of microns. The technology relies on an FC-type mounting of the thin chip onto the flex substrate and lamination of the structure on both sides (see Figure 1.40). Very thin chips (range of 20 μm) with Ni/Au pads are placed on 25 μm thin PI flex substrates using anisotropic conductive adhesive.

Contacts to outer layers are realized by through-holes. Further layers can be added to the buildup. The electrical interconnections are extremely thin, and

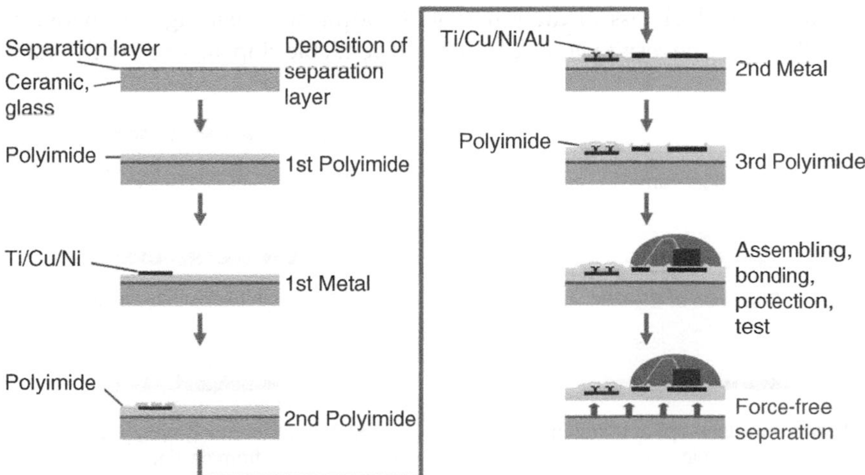

Figure 1.39 Process flow of the HiCoFlex process.

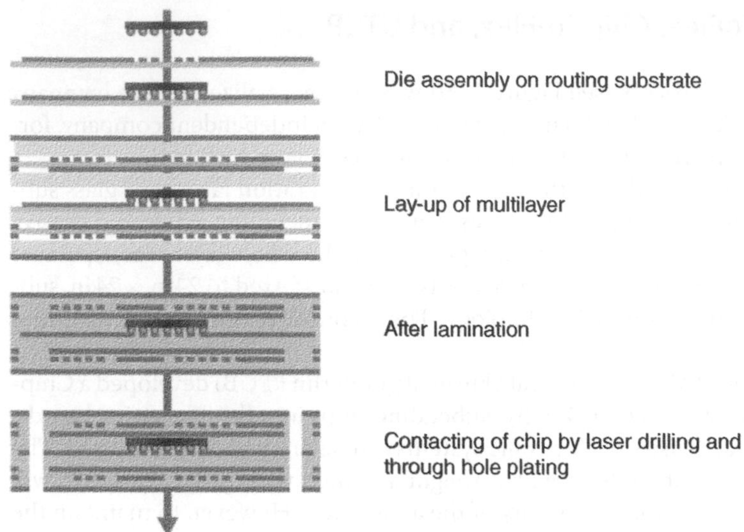

Figure 1.40 Process flow of the Chip-in-Flex technology.

the mechanical coherence of the chip to the substrate is ensured by the no-flow underfill. After testing multiple flex substrates, the flex substrates are laminated together. Interconnections of the layers are done by laser drilling combined with a metallization process.

The ultrathin chip package (UTCP) technology developed by IMEC is based on the embedding of ultrathin die in flexible substrates [46]. The process flow is shown in Figure 1.41.

Chips with thickness in the range of 20–30 μm were packaged in between two PI layers. The result was a very thin and bendable chip package, with a total

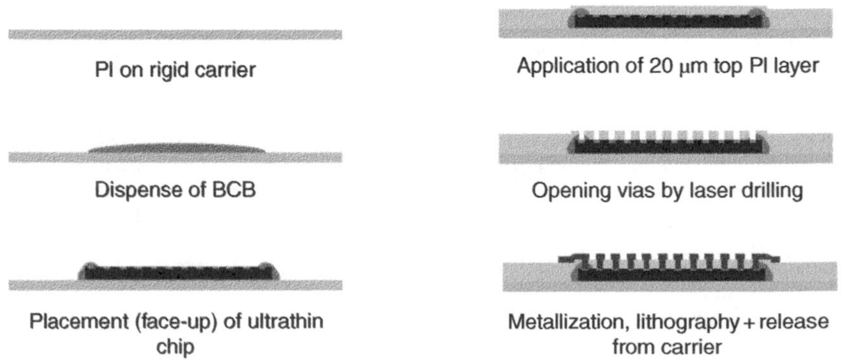

Figure 1.41 Overview of the process flow of the UTCP.

thickness of only 50–60 μm. The base substrate was a uniform spin-coated PI layer, on a rigid carrier. Ultrathin chip are placed and fixed using BCB. A second PI process embeds the chips. Vias were made by laser drilling. The metal layer provides a fan-out to the contacts of the chips. Finally the whole package is released from the rigid substrate. The potential for stacking in order to obtain 3D-type chip packages is given by this process.

1.7 Conclusion

Embedding technologies and processes have a history of nearly 40 years. A majority of the fundamental work had been published by GE, mainly by Ray Fillion. The main driver at that time was the limitation of large monolithic ICs. In addition, a reliable bumping infrastructure that is now an industrial standard was not yet installed. Also, high density organic substrates were not available. Therefore, embedding was a valid alternative to FC, WB, and TAB for MCMs, offering excellent electrical performance with dense routing. One major issue for the embedding technology was and still is the lack of repair. Components that have been embedded are very difficult to remove. This is not the case for the chip-last approaches like FC and WB. Therefore, one of the great advantages of today is the availability of KGD – one of the factors limiting the embedding process to only special application 30 years ago. These applications were mainly high-end applications that required the very high performance of electronic systems using embedding technology. Most of the examples that have been described in this chapter are from this technology area. WB was the workhorse for consumer and personal computer products. With further miniaturization and increasing performance, WB is being replaced more and more by FC bonding. High density bumping processes were established, and the wiring density of organic substrates increased steadily. Over the last 10 years, the requirements for consumer, industrial, and automotive applications are strongly increasing. The differences in electrical performance for the different electrical interconnects are not negligible anymore. For example, the bandwidth for wired and wireless links is now increasing data rates by around 10 times every five years [47]. The Internet of things (IoT) is demanding a further steep increase in bandwidth. 5G is targeting 10–100 $\mathrm{Gb\,s^{-1}}$. Below 1 GHz, the electronic performance is not that different between the chip-last interconnection technologies like FC and WB compared with the chip-first direct chip connection using embedding. For example, the chip-to-package interconnection of the embedding process from TSMC (InFO) is based on copper bump without a solder joint. This reduces the loss between chip and package. Simulation results showed that parasitic R, L, and C of the embedding interconnection are 75, 76, and 14% lower from 55 to 65 GHz than those of the chip-last technology using

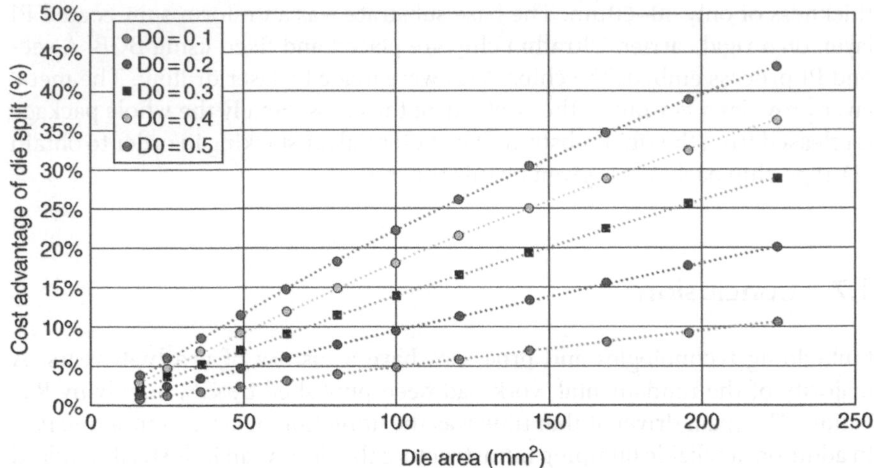

Figure 1.42 Cost advantage of die splitting for increasing die area and for different process yields [50]. *Source:* Reproduced with permission of Springer.

FC bonding [48]. Also the embedding technology from Amkor (SWIFT) has an improvement in cross talk and return loss by 6 dB at 2 GHz compared with an FC-PoP [49]. Furthermore, a current motivation for the embedding process is the demand for packaging field-programmable gate arrays (FPGA), which are ICs designed to be configured by the customer. Unfortunately, the manufacturing of large die like the FPGAs that are needed, for example, for artificial intelligence (AI) in the area of autonomous driving, is still limited by the process yield. Therefore, it may be cost effective to split a die into multiple die that are then reintegrated in a single SiP [50]. The Si yield curve is related exponentially to die area and a process-specific constant (D_0). Thus, for a given process technology, the cost advantage of a symmetrical die split into two die with equal areas of $A/2$ is then a function of the original single die area and the process yield (see Figure 1.42).

As shown in Figure 1.42, splitting a small die results in a smaller cost advantage compared with splitting a large die. However, for combining two or multiple die, a high-speed interconnection like the embedding process is required.

In general, the high functionality in a small form factor together with superior electrical performance and high thermal dissipation are the major reasons for the restrengthening of the substrate-less embedding technology. The cost of thin film routing has strongly decreased by the increasing wafer size. Also, high-speed assembly has gained a significant high precision that has to be aligned to the fine line routing process. A step to panel-level processing may further decisively lower the process cost for the high density embedding technologies.

References

1 Garrou, P. and Turlik, I. (1997). *Multichip Module Technology Handbook*. McGraw-Hill Professional.
2 Bodendorf, D.J., Olson, K.T., Trinko, J.F., and Winnard, J.R. (1972). Active silicon chip carrier. *IBM Technical Disclosure Bulleting* 15 (2): 656–657.
3 Johnson, R.W., Davidson, J.L., Jaeger, R., and Kerns, D. (1986). Silicon hybrid wafer-scale package technology. *IEEE Journal of Solid State Circuits* October: 845–851.
4 Te Velde, T.S. and Schmitz, A. (1968). Semiconductor circuit having active devices embedded in flexible sheet. US 3,579,056 A, registered on the 10 October 1968 and published 18 May 1971.
5 Fillion, R.A., Wojnarowski, R.J., and Daum, W. (1990). Bare chip test techniques for multichip modules. *Proceedings 40th Electronic Components and Technology Conference* (1990), pp. 554–558.
6 Gdula, M., Yerman, A., Krishnamurthy, V., and Fillion, R. (1993). The GE high density overlay MCM interconnect method solves high power need for GaAs system design. *Proceedings of the Fifth Annual IEEE International Conference on Wafer Scale Integration* (1993), pp. 339–345.
7 Fillion, R. (1993). A forecast on the future of hybrid wafer scale integration technology. *IEEE Transaction CHMT* 16 (7): 615–624.
8 Hannemann, R., Kraus, A., and Pecht, M. (1994). *Semiconductor Packaging: A Multidisciplinary Approach*, 156. Wiley.
9 Fillion, R., Wojnarowski, R., Gorcyzca, T., et al. (1994). Development of a plastic encapsulated multichip technology for high volume, low cost commercial electronics. *Proceedings. 44th Electronic Components and Technology Conference* (1994), pp. 805–809.
10 Vardaman, E.J. and Carpenter, K. (2007). Embedded active components and integrated passive. *TechSearch International* (October 2007).
11 Töpper, M., Wolf, J., Glaw, V. et al. (1996). MCM-D with embedded active and passive components. *Proceedings ISHM 1996*, Minneapolis, USA (October 1996), pp. 224–229.
12 Töpper, M., Buschick, K., Wolf, J. et al. (1997). Embedding technology – a chip first approach using BCB. *Advancing Microelectronics* 24 (4): 18–22.
13 Mangold, T., Wolf, J., Töpper, M. et al. (1998) A multichip module integration technology for high-speed analog and digital applications. *Proceedings ISSSE Pisa*, Pisa, Italy (September 1998), pp. 91–96.
14 Buschick, K., Glaw, V., Samulewics, K., and Karduck, C. (July 2001). Chip scale packaging of single chips for SMT assembly. *PLUS (Produktion von Leiterplatten und Systemen) Magazine* 1562–1572.
15 Buschick, K. (1995). Electric connections arranged in a high-density grid. EP 0750791 A1 WO1995025346A1, PCT/DE1995/000359.

16 Ehrmann, O., Buschick, K., Chmiel, G. et al. (1995) 3D multichip modules. *Proceedings of the International Conference on Multichip Modules (ICEMCM'95)*, CO (19–21 April 1995), pp. 358–363.
17 Wolf, J., Schmückle, F., and Heinrich, W. (1997). System integration for high frequency applications. *Proceedings IMAPS Philadelphia*, USA (1997).
18 Wolf, J., Schmückle, F., Heinrich, W. et al. (1998). System integration for high frequency applications. *Proceedings Wireless Communication Conference*, San Diego, USA (1998), pp. 37–46
19 Hahn, R., Kamp, A., Ginolas, A. et al. (1997). High power MCMs employing the planar embedding technique and microchannel water heat sinks. *IEEE Transactions on CPMT-Part A* 20 (4): 432–441.
20 Buiculescu, V., Mannier, Ch.-A., Oppermann, H. et al (2015). Micro-fabricated hybrid package optimized for RF applications. *Proceedings 16th International Symposium on RF-MEMS and RF-Microsystems*, Barcelona, Spain (2015).
21 Töpper, M., Scherpinski, K., Spörle, H.-P. et al. (2000). Novel manufacturing strategies in back end area. *Proceedings of European IEEE/SEMI Semiconductor Manufacturing Conference* (April 2000).
22 Töpper, M., Scherpinski, K., Spörle, H.-P. et al. (2000). Thin chip integration (TCI-modules) – a novel technique for manufacturing three dimensional IC-packages. *Proceedings IMAPS 2000*, Boston, USA (September 2000), pp. 208–211.
23 Fritzsch, Th., Mroßko, R., and Baumgartner, T. et al. (2009) 3-D thin chip integration technology – from technology development to application. *IEEE International Conference on 3D System Integration, 2009. 3DIC* (September 2009), pp. 1–8.
24 Töpper, M., Baumgartner, T., Klein, M. et al. (2009) Low cost wafer-level 3-D integration without TSV. *Proceedings of the ECTC 2009*, San Diego, pp. 339–344.
25 Baumgartner, T., Töpper, M., Klein, M. et al. (2009). A 3-D packaging concept for cost effective packaging of MEMS and ASIC on wafer level. *Proceedings of the EMPC*, Rimini, Italy (15–18 June 2009).
26 Baumgartner, T., Töpper, M., Klein, M., and Reichl, H. (2010). Integration of thinned silicon chips in a polymer-copper-redistribution-A-3-D stacking technology without TSVs. *The Symposium on Polymers for Microelectronics*, Winterthur, USA (May 2010).
27 Schmid, B., Gerner, M., Loreit, U. et al. (2011). Zuverlässigkeitsuntersuchungen an einem auf Waferlevel 3D umverdrahteten Raddrehzahlsensorsystem. *Proceedings MST 2011*, Darmstadt, pp. 253–256.
28 Kleff, J., Töpper, M., Dietrich, L. et al. (2013). Wafer level packaging for ultra thin (6 μm) high brightness LEDs using embedding technology. *Proceedings IEEE-ECTC 2013*, Las Vegas, USA, pp. 1219–1224.
29 Carson, K. (1996). The emergence of stacked 3D silicon and its impact on microelectronics systems integration. *Proceedings of the Eighth Annual IEEE*

International Conference on Innovative Systems in Silicon, Austin, TX (1996), pp. 1–8.

30 Couderc, P. (2012). SiPs in the medical domain and in the defense and industrial domain produced with WDoDTM technology. *IMAPS Packaging Device Conference* (March 2012).

31 Wakabayashi, T. (2003). System packaging and embedded WLP technologies for mobile products. *Extended Abstracts of the 2003 International Conference on Solid State Devices and Materials,* Tokyo (2003), pp. 386–387.

32 Braun, T., Raatz, S., Voges, S. et al. (2015). Large area compression molding for fan-out panel level packing *Proceedings ECTC,* San Diego (2015), pp. 1077–1083.

33 Braun, T., Voges, S., Töpper, M. et al. (2015). Material and process trends for moving from FOWLP to FOPLP. *Proceedings EPTC 2015,* Singapore.

34 Braun, T., Raatz, S., Maass, U. et al. (2017) Kaynak development of a multi-project fan-out wafer level packaging platform. *Proceedings ECTC,* Orlando (2017), pp. 1–7.

35 Forman, G.A., Fillion, R.A., Kolc, R.F. et al. (1995). Development of GE's plastic thin-zero outline package (TZOP) technology. *Proceedings. 45th Electronic Components and Technology Conference* (1995), pp. 664–668.

36 Fillion, R.A. (2006). *Advanced Packaging Technology for Leading Edge Microelectronics and Flexible Electronics.* Lecture at Cornell University.

37 Kujala, A., Tuominen, R., and Kivilahti, J.K. (1999). Solderless interconnection and packaging technique for active components. *Proceedings 49th ECTC* (1999), pp. 155–159.

38 Palm, P., Tuominen, R., and Kivikero, A. (2004). Integrated module board (IMB); an advanced manufacturing technology for embedding active components inside organic substrate. *Proceedings 54th ECTC* (2004), pp. 1227–1231.

39 Tuominen, R., Waris, T., and Nettovaara, J. (2009). IMB technology for embedded active and passive components in SiP, SiB and single IC package applications. *Transactions of the Japan Institute of Electronics Packaging* 2 (1): 134–138.

40 Tessier, T. G., Karila, T., Waris, T. et al. (2010). Laminate based fan-out embedded die technologies: the other option. *Conference Proceedings IWLPC 2010,* pp. 198–203.

41 Aschenbrenner, R., Ostmann, A., Neumann, A., and Reichl, H. (2004). Process flow and manufacturing concept for embedded active devices. *Proceedings EPTC 2004,* Singapore (8–10 December 2004), pp. 605–609.

42 Ostmann A., Manessis, D., Stahr, J. et al. (2008). Industrial and technical aspects of chip embedding technology. *Electronics System-Integration Technology Conference, 2008. ESTC 2008,* pp. 315–320.

43 Töpper, M., Wilke, M., and Ostmann, A. (2015). Integration technologies for image sensors. *Imaging Conference Semicon Europa 2015,* Dresden.

44 Godin, L. (2011). Miniaturization of medical devices thanks to flexible substrates. *1st Advanced Technology Workshop on Microelectronics, Systems and Packaging for Medical Applications*, France (November 2011).

45 Löher, Th., Pahl, B., Huang, M. et al. (2004). An approach in microbonding for ultra fine pitch applications. *Proceedings Technology and Metallurgy, The 7th VLSI Packaging Workshop of Japan*, Kyoto (30 November–2 December 2004).

46 Christiaens, W., Vaandevelde, B., Bosman, E., and Vanfleteren, J. (2006). UTCP: 60 μm Thick Bendable Chip Package. *Proceedings IWLPC 2006*, pp. 114–120.

47 Nam, B.-G. (2017). Digital Architectures and Systems Sub-Committee Chair ISSCC Trends, San Francisco, USA (February 2017).

48 Hsu, Ch.-W., Tsai, Ch.-H., Hsieh, J.-Sh. et al. (2017). High performance chip-partitioned millimeter wave passive devices on smooth and fine pitch InFO RDL. *Proceedings IEEE ECTC 2017*, Orlando, USA, pp. 254–259.

49 Zwenger, C. (2017). Advancements in high fan-out and heterogeneous integration. *Proceedings Semicon Taiwan 2017*.

50 Radojcic, R. (2017). *More-than-Moore 2.5D and 3D SiP Integration*, 5–67. Springer International Publishing ISBN 978-3-319-52547-1.

2

FO-WLP Market and Technology Trends

E. Jan Vardaman

TechSearch International, Inc., Austin, TX, USA

2.1 Introduction

The quest for thin, low profile packaging solutions for mobile devices and cost reduction with improved performance continues to drive the development of new packages. The development of the fan-out wafer-level package (FO-WLP) is the latest industry trend. There are an increasing number of suppliers for FO-WLP and a growing number of applications. This chapter examines the market trends and drivers for package adoption and the technology trends for FO-WLP.

2.2 FO-WLP: A Disruptive Technology

An FO-WLP is a substrate-less package that typically uses a carrier and mold compound to create a format on which single or multiple dies are placed and a redistribution layer (RDL) is used to "fan out" the inputs and outputs (I/Os) to create a package. When Infineon developed its FO-WLP, the embedded wafer-level ball grid array (eWLB), one of the drivers was to reduce package cost by removing costly parts of the package including the substrate and the flip-chip bump. The absence of a substrate makes FO-WLP a disruptive technology. Thin film metallization is used for interconnect instead of bumps or wires. Infineon's FO-WLP has the active side of the die placed facedown in the process and so is often referred to as a face-down FO-WLP. In the case of a face-up process, the die has a thick Cu post. Figure 2.1 shows a face-up and a face-down option.

Advances in Embedded and Fan-Out Wafer-Level Packaging Technologies, First Edition.
Edited by Beth Keser and Steffen Kröhnert.

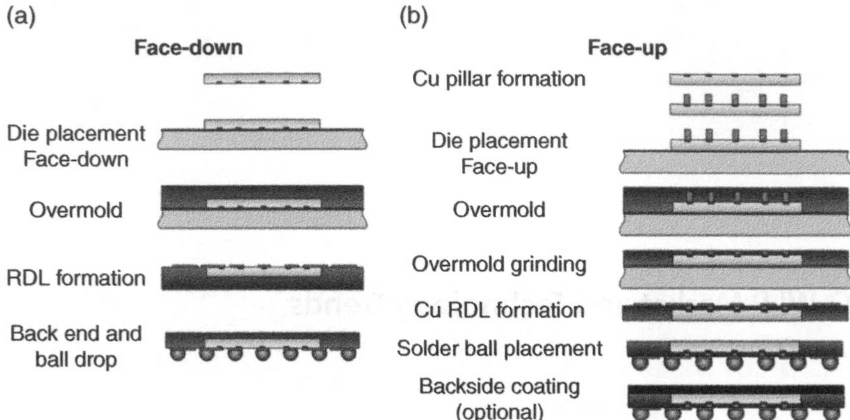

Figure 2.1 (a) Active die face-down FO-WLP die placement. (b) Active die face-up FO-WLP die placement. *Source:* Courtesy of ASE.

The use of a thin film process instead of a laminate substrate makes it possible to achieve much finer feature sizes. Laminate substrates are in high volume production with 15 μm lines and spaces, with some leading-edge production of 10 μm lines and 9 μm spaces. Achieving higher routing densities such as less than 5 μm lines and spaces may be costly.

2.3 Embedded Die Packaging

FO-WLP is a form of embedded die. An embedded component is defined as an active or passive device that is placed or formed on an inner layer of an organic circuit board, module, or chip package, such that it is buried inside the completed structure, rather than on the top or bottom surface. Adoption of embedded active components is being driven by demands for smaller form factors and improved performance. Some companies embed active components on an inner layer of the IC package or module substrate with a lamination process. Other companies embed RF components such as Bluetooth devices in the product board with a lamination process. The embedded component can take the form of bare die or a wafer-level package.

One of the early examples of embedded components is General Electric (GE) with its chips first buildup technology. The GE technology was originally targeted at applications such as high-end military and space applications including complex multiprocessor modules, memory modules, 3D structures, and CSPs. The key feature of this technology is the direct metallization connection between the chip and the package, accomplished through laser drilling of vias directly through the polyimide film to the chip I/O pads followed by sputtering of a seed layer metal and semi-additive buildup plating of copper conductors [1]. GE subsequently acquired the assets and technology from Imbera Electronics Oy and has developed a process to embed die in flex circuit for power device applications. The technology has been licensed to Shinko Electric and Flexceed in Japan.

Austria Technologie & Systemtechnik (AT&S) is located in Austria and started building and testing embedded component packaging (ECP®) technology in 2009. This technology is used for power modules and other applications. Applications at AT&S include automotive and mobile products. AT&S provides the embedding for Texas Instrument's (TI) MicroSiP TPS8267, a 650 mA DC/DC converter. Products use the ECP chip-first process flow.

TDK Corporation developed the embedded die solution called semiconductor embedded in substrate (SESUB®). The technology is in production for power management integrated circuits (PMICs). Battery charger products have also been demonstrated with the technology. TDK also introduced an ultracompact Bluetooth low-energy module designed for Bluetooth 4.0 low energy specifications targeted for wearable electronics products. The package is 4.6 mm × 5.6 mm × 1.0 mm. The Bluetooth IC die is embedded into the thin substrate; the peripheral circuitry including a quartz resonator, bandpass filter, and capacitors is integrated on the top. The module is 65% smaller than the alternative with individual discrete components.

Casio Micronics and CMK Corporation have developed an embedded wafer-level packaging (EWLP) technology. Target applications include stacked memory modules, memory cards, Bluetooth™ modules, and camera modules. Casio shipped several products using the technology. The main product was a mobile TV tuner.

DNP is located in Japan and began volume production of embedded passive devices (EPD) in 2006 and embedded active devices (EAD) in 2007. The devices are used in a range of configurations from packages to PCBs, including package-on-package (PoP) substrates, camera modules, and motherboards for mobile phone applications.

Fujikura developed a multilayer flex circuit fabrication process that allowed WLPs to be embedded in the flexible substrate layers. This freed the surface space for other components to be mounted, and the module can be reduced in size. Fujikura shipped samples designed for camera and Bluetooth modules used in mobile phones.

Driven by the need for miniaturization in mobile applications, DNP developed a B^2 it PCB with EPD and EAD. The use of embedded components offers a smaller area and improved electrical performance, and the technology was used to embed passive devices.

Samsung Electro-Mechanics (SEMCO) developed the Smart Functional Circuits® for the integration of embedded actives and passives in PCBs and modules.

Taiyo Yuden has been manufacturing EAD substrates with its embedded organic module involved nanotechnology (EOMIN®) process since 2006. Most of the EAD are power management die, signal control die, and sensors. Applications include power management, signal processing, and sensing. Both active and passives can be embedded.

TI has been shipping its TPS8267x family of step-down converters since 2010. These products are offered in an 8-pin MicroSiP™ package. The active component is a switching regulator in TI's PicoStar™ format that is embedded in an FR-4 laminate-type substrate. I/O capacitors and an inductor are mounted on the top surface of the substrate. This format enables a very low profile package.

3D Plus developed a stacked module technology that is also considered an embedded component package [2]. This module, called Walpack, was developed with partners STMicroelectronics, CEA-Leti, Thales, Schlumberger, Cards/Axalto, Cybernetics, and IBS in a four-year European program that operated from 2001 to 2004.

2.4 FO-WLP Advantages

Advantages of FO-WLP include:

- Form factor, low profile package, with small gap between the die and the package edge (≤50 μm).
- Support of increased I/O density while at the same time allowing the use of a WLP with die shrinks as companies move to advanced semiconductor technology nodes.
- Ability to support fine features roadmaps (≤10 μm lines and spaces).
- Split die or multi-die packaging/system in package (SiP).
- Excellent electrical performance (low parasitic inductance, low insertion loss, and good signal integrity) because of the short signal path from the chip to package and to the board.
- Excellent thermal performance, especially when thermal balls are placed directly below the silicon die and the direct thermal path to the PCB due to the absence of a substrate.
- Sidewall protection on all slides resulting from the use of molding resin is advantageous when compared with fan-in wafer-level packaging (WLP), which has unprotected silicon sidewalls.

2.5 FO-WLP Versions

Infineon developed the eWLB process in which dies are placed facedown on a metal carrier and molded into a reconstituted wafer. After removal from the carrier, a single or double RDL is used to fan out the I/Os. Solder balls are attached to the active side of the die, and then the reconstituted wafer is laser-marked and singulated. The face-down fan-out process is illustrated in Figure 2.1.

ASE, NANIUM, and STATS ChipPAC licensed Infineon's eWLB technology and have production lines. JCET, based in China, purchased STATS ChipPAC in 2015. Amkor purchased NANIUM in May 2017. SPIL is also establishing a production line with a chip-first face-down FO-WLP process. SPIL's encapsulated mold process uses a laser via formation and Cu post plating through package interconnection to fabricate a PoP. The die also has a copper pillar (this differs from the original eWLB process).

Freescale developed the redistributed chip package (RCP) process, which is a chip-first face-down process similar to eWLB in structure. Freescale Semiconductor (now NXP and soon to be part of Qualcomm) sold its 300 mm production equipment to nepes Corporation in 2009, but maintained a 200 mm pilot development line in Tempe, Arizona, until February of 2017. The nepes Corporation 300 mm production line is located in Korea. The process uses glass and ceramic carriers, and the devices are mounted facedown before molding (see Figure 2.2).

TSMC's FO-WLP version is called integrated fan-out WLP (InFO-WLP). In TSMC's InFO process, a small Cu pillar interconnect is electroplated on the application processor. The die is placed faceup on the tape, followed by molding to create the reconstituted wafer. After grinding to reveal the top of the Cu pillar, the RDL is formed on the polished surface [3, 4]. A face-up process is show in Figure 2.3.

Deca Technologies, based in the Philippines, has also developed a face-up structure. M-Series is a fully molded FO-WLP structure wherein mold compound protects the frontside of the die, in addition to providing fan-out area at the periphery (see Figure 2.4). Covering the frontside of the die also prevents mold flash at the die edge boundary and die-to-mold compound coplanarity issues found in face-down FO-WLP solutions. These issues can cause issues for photolithography and solder bump reliability [5]. In the manufacturing process, copper pillar interconnects are fabricated on the native wafer, and the devices are singulated. The devices are attached faceup to a carrier with a temporary adhesive and overmolded. The temporary adhesive is removed, and the front of the panel is ground to reveal the copper studs.

The next step uses Deca's unique adaptive patterning process (see Figure 2.5). A high-speed optical scanner measures the true position of the copper studs on each die on the reconstituted panel. A portion of the RDL design in close proximity to the die bond pads is recalculated using proprietary design software, generating a unique layout that accommodates die shift. Adaptive alignment shifts and rotates the first via and RDL to match the die location. A unique via and RDL pattern is dynamically applied to each panel using a maskless lithography system. The under-bump metallization (UBM) pattern is kept

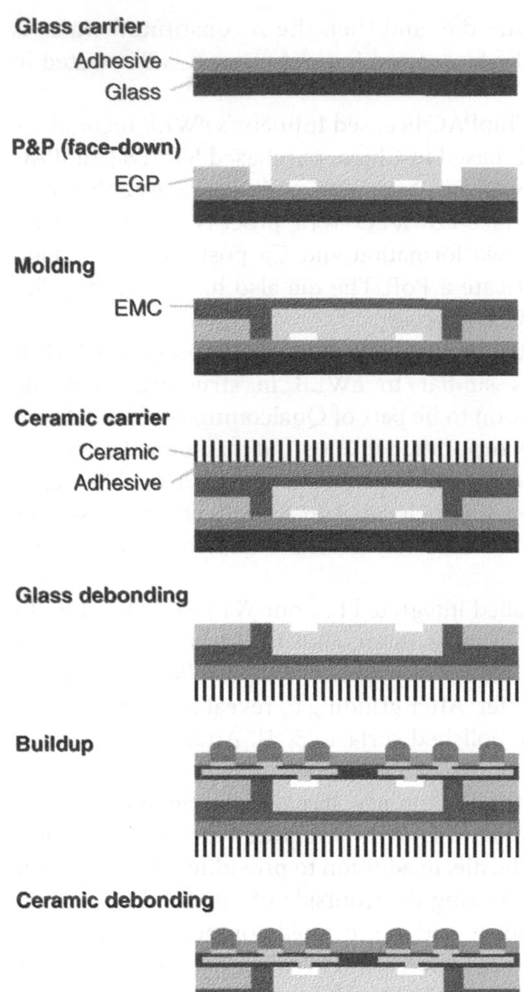

Figure 2.2 NXP's RCP FO-WLP process flow. *Source:* Courtesy of NXP.

fixed with respect to the package edge to maintain compliance with the package outline specifications [6].

Amkor has introduced a chip-last technology as an alternative to an FO-WLP called silicon wafer integrated fan-out technology (SWIFT®). It provides increased I/O and circuit density with a reduced footprint that can be used for single- or multi-die applications. In the process, RDLs are formed on a carrier (see Figure 2.6). Flip-chip bumped dies are attached to the high density structure with a minimum of 2 μm lines and spaces. Cu pillar pitch down to 30 μm is supported. A mold compound is used to encapsulate the structure and wafer backgrinding is used to thin the package [7]. Structures as thin as 500 μm with

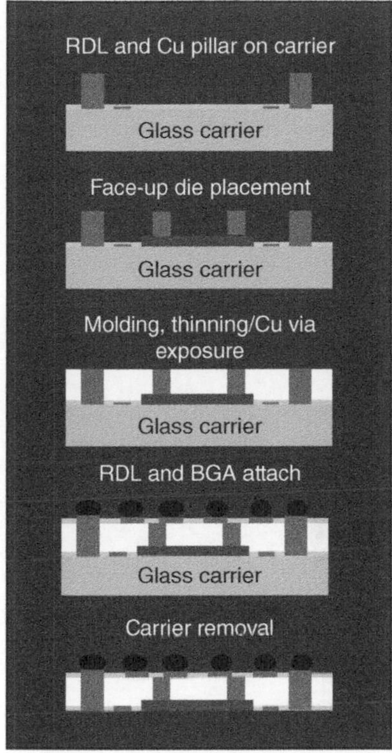

Figure 2.3 Face-up FO-WLP process flow.

up to three RDLs can be created using the backgrinding process. The carrier can be inspected using automated optical inspection (AOI) to ensure it is a known good substrate and the die can be probed so that known good die (KGD) can be placed on good sites. With the Amkor process it is possible to create the tall Cu pillars for connection, or a through-mold via (TMV) process with solder balls can be used.

JCAP, a subsidiary of JCET based in China, has developed the encapsulated chip package (ECP). It can be used as an FO-WLP for single-die or multi-die packaging to provide a thin package. It can be used to provide six-sided encapsulation and protection (four sidewalls, top, and bottom) for fan-in WLP. A silicon carrier or backing is used in the ECP process (see Figure 2.7). This helps control warpage to <0.5 mm. A process without silicon backing is under development, and a version with EMI shielding is also being developed. The

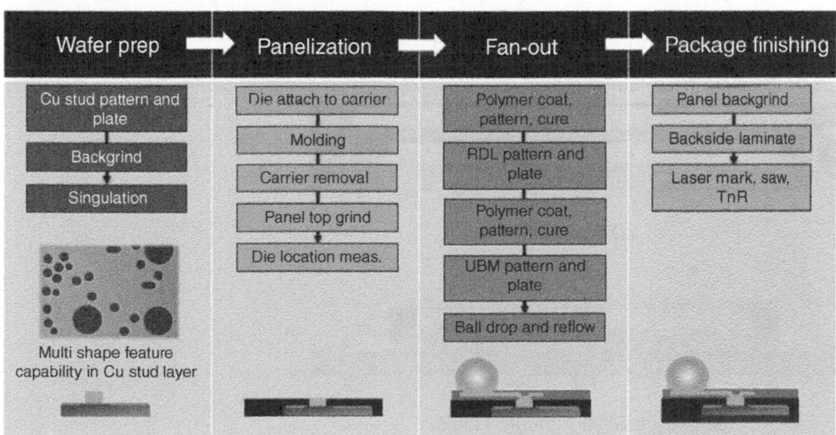

Figure 2.4 Deca's M-Series® face-up FO-WLP process flow. *Source:* Courtesy of Deca Technologies.

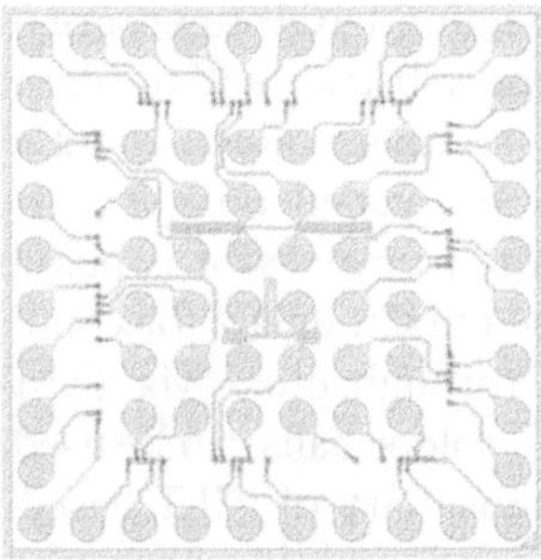

Adaptive routing
• Dynamically adapt RDL routing to accurately align to true die position

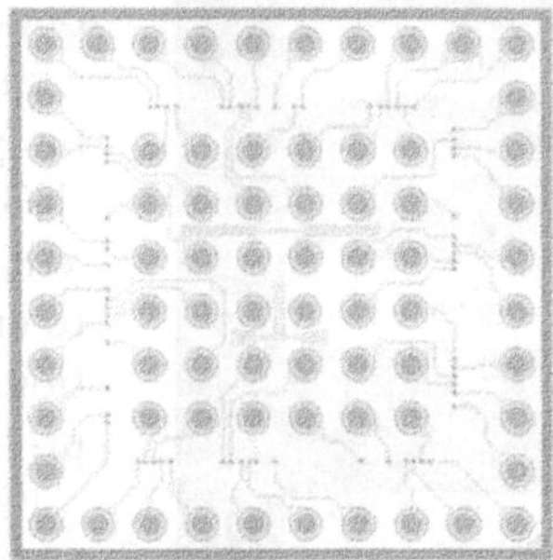

Adaptive alignment
• Align the entire RDL layer to true die position within the unit

Figure 2.5 Deca's adaptive patterning process. *Source:* Courtesy of Deca Technologies.

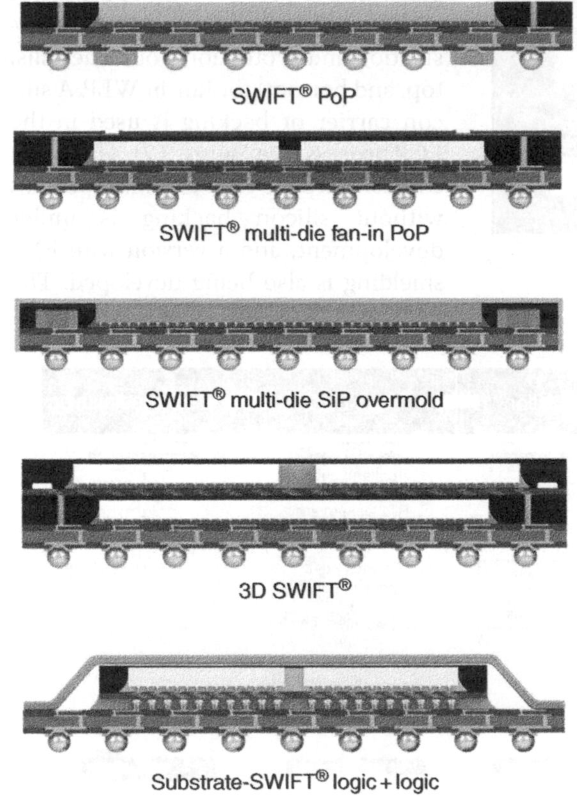

Figure 2.6 Amkor's SWIFT package structure proposals. *Source:* Courtesy of Amkor Technology.

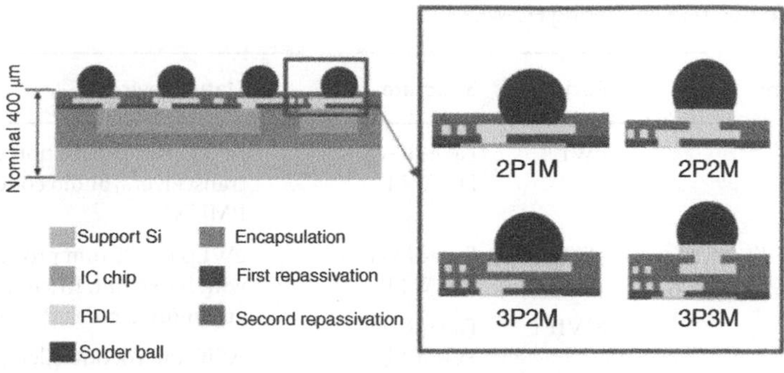

Figure 2.7 JCAP's ECP package and structure proposal. *Source:* Courtesy of JCAP.

electrical performance is similar to WLP. ECP is a chip-first face-up process. A dry film laminate process is used to encapsulate the chip. The dry film uses a < 2 μm filler size. A Cu/Ni/Au bump is plated or a preformed solder ball is attached to form the package. Laser scribe plus dicing is used, but plasma dicing is under investigation [8].

Huatian Technology (Kunshan) Electronics has developed a wafer-level embedded silicon fan-out (eSiFO) in which KGDs are embedded in cavities formed on a silicon wafer (see Figure 2.8). The microscale gap between the die and the cavity is filled with epoxy resin, and RDL is fabricated on the top [9]. There is no molding, temporary bonding, or debonding. There is no warpage

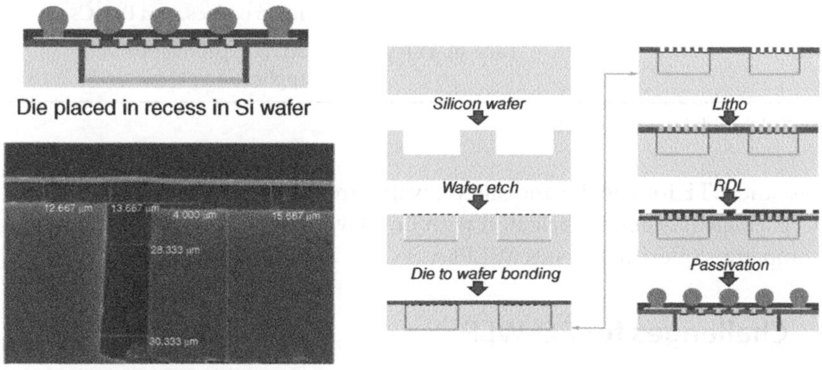

Figure 2.8 Huatian's eSiFO package and structure proposal. *Source:* Courtesy of Huatian Technology.

Table 2.1 Companies offering FO-WLP.

Company	Package	Structure	Status/devices
ASE	eWLB	Face-down FO-WLP	In 300 mm production with RF, transceivers, audio codecs, and PMICs
Amkor Technology (including NANIUM)	eWLB	Face-down FO-WLP	eWLB in 300 mm production with baseband, RF, and automotive radar modules
	SWIFT	Face-down FO-WLP	ASIC microcontrollers, medical, sensors, including passives demonstrated
Deca Technologies	M-Series™	Face-up FO-WLP (panel compatible)	300 mm and panel in qualification
Huatian Technology	eSiFO	Face-up FO-WLP	Qualified
Infineon	eWLB	Face-down FO-WLP	200 mm production with automotive radar modules
JCAP	ECP	Face-up FO-WLP	Production for analog devices
NXP (formerly Freescale Semiconductor)	RCP	Face-down FO-WLP	
nepes Corporation	RCP	Face-down FO-WLP	In 300 mm production with automotive radar, connectivity modules
SPIL	TPI-FO	Face-down FO-WLP	300 mm qualificationbaseband, PMIC, RF
STATS ChipPAC	eWLB	Face-down FO-WLP	300 mm production with baseband and RF, transceivers, audio codecs, and PMICs
TSMC	InFO-WLP	Face-up FO-WLP	300 mm production with application processor, RF

Source: TechSearch International, Inc.

because the CTE for the die and silicon wafer are the same. A 3.3 mm × 3.3 mm eSiFO package has been fabricated with one-layer RDL. Table 2.1 summarizes the companies offering fan-out WLPs.

2.6 Challenges for FO-WLP

FO-WLP is based on the use of a reconstituted wafer where singulated dies are placed on a carrier and spaced precisely from one another to facilitate the subsequent RDL fan-out process. Warpage is an issue with FO-WLP, and several organizations have documented methods to reduce warpage. Research from IME shows solutions to reduce wafer warpage include the ratio of die to molding thickness, overmold effect, molding compound material properties, carrier wafer material, dielectric material, and the copper RDL area percentage. The thickness ratio of the die to mold, molding compound and carrier wafer materials, dielectric material, and copper RDL design are some of the most important factors [10].

Die shift in the molding process is also an issue with FO-WLP, which adversely impacts yield. Deca Technologies has developed an FO-WLP process that deals with the issue of die shift. In the manufacturing process, copper pillars are electroplated on the native wafer, and the devices are singulated. The devices are attached faceup to a carrier with a temporary adhesive and overmolded. The temporary adhesive is removed, and the front of the panel is ground to reveal the copper studs. The next step uses Deca's unique adaptive patterning process. A high-speed optical scanner measures the true position of the copper studs on each die on the reconstituted panel. A portion of the RDL design in close proximity to the die bond pads is recalculated using proprietary design software, generating a unique layout that accommodates die shift. Adaptive alignment shifts and rotates the first via and RDL to match the die location. A unique via and RDL pattern is dynamically applied to each panel using a mask-less lithography system. The UBM pattern is kept fixed with respect to the package edge to maintain compliance with the package outline specifications [11].

2.7 Drivers for FO-WLP

Many companies prefer the small form factor, low profile features of WLPs, but as companies move to the next semiconductor nodes, the face of the die is too small to route all the I/O without using fine-pitch small solder balls (<200 μm) that require high density boards to route high pin count parts with fine pitch. In addition, board reliability issues have been noted with fine-pitch solder balls on WLPs. This is especially true for audio codec, PMIC, and RF IC parts. Figure 2.9 shows audio codec and RF parts from Qualcomm packaged in FO-WLP.

In the application processor space, demands for thinner packages have focused on options such as the continued use of flip chip with a laminate substrate or embedded die. 3D IC with through-silicon vias (TSVs) to connect memory and logic has been ruled out because of business issues such as the supply chain, cost, and the absence of a commercially available thermal solution to manage the heat dissipated by the processor. Thermal problems include heat from the processor, resulting in failures in the memory as hot spots from the logic die cause the temperature to exceed the operating temperature of the memory specification.

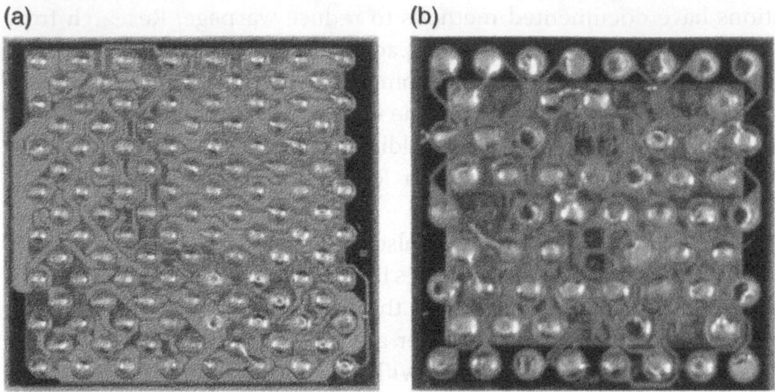

Figure 2.9 Qualcomm's FO-WLP in production. (a) Audio codec 4.25 mm × 3.90 mm package size. (b) RF transceiver 3.3 mm × 3.3 mm package.

Figure 2.10 TSMC's InFO-PoP technology used in Apple's A10 process for the iPhone7. *Source:* Courtesy of TSMC and TechInsights.

TSMC's FO-WLP version called InFO-WLP has been adopted for Apple's A10 processor in the iPhone 7 as the bottom package for the PoP (see Figure 2.10). In TSMC's InFO process, a small Cu pillar is formed on the application processor. The die is placed faceup on the tape, followed by molding to create the reconstituted wafer. After polishing to reveal the top of the Cu pillar, the RDL is formed on the polished surface. Adoption of InFO for the application processor is driven by form factor, as well as electrical and thermal performance (at the board level). InFO-PoP is reported to have seven times lower power distribution network (PDN) impedance (up to 5 GHz) than FC-PoP because three metal layers are used in the RDL routing rather than four metal layers in the laminate substrate. TSMC also reports that when InFO-PoP also provides improvement in power noise reduction and signal integrity improvement [12].

Qualcomm and Spreadtrum selected eWLB because it provides a low profile package with good electrical performance. Qualcomm has used eWLB for PMICs, RF transceivers, and audio codecs.

FO-WLP has been selected for automotive radar modules from Infineon, NXP, and others. Some of these modules contain multiple dies. Both the eWLB and the RCP process are used for these products. The 300 mm RCP production line is located at nepes Corporation in South Korea. Infineon has its own 200 mm line in Regensburg. Infineon has also introduced an eWLB package for 60, 70, and 80 GHz LTE backhaul transceiver applications. NANIUM has also packaged radar modules in eWLB for customers. FO-WLP was selected for automotive radar modules for the following reasons:

- Improved system performance due to lower electrical parasitics (*R*, *L*, and *C*), attenuation, and insertion losses.
- Controlled impedance achievable with embedded ground plane or multilayer RDL.
- Excellent RF isolation.
- Ability to achieve AEC-Q100 Grade 1 reliability performance (operation from −40 to 125 °C) at moisture sensitivity of MSL 1.
- Enablement of heterogeneous integration.
- Volume shrink of up to 90%.
- Potentially lower cost due to the absence of a substrate.

Both single-die and multi-die packages have been adopted. The eWLB process and the RCP process were developed by Freescale Semiconductor and licensed to nepes Corporation.

2.7.1 Markets and Applications for FO-WLP

The highest volume application for FO-WLPs is smartphones – smartphone models from Apple, Huawei, Xiaomi, and Vivo. FO-WLP volumes for automotive radar modules are increasing, and connectivity modules have emerged.

FO-WLP growth is driven by its adoption for a variety of applications and device types. These include application processors, RF transceivers and switches, PMICs, and audio codec, connectivity modules, and radar modules for automotive safety systems.

Both single-die and multi-die versions of FO-WLP are in production. NANIUM has also introduced 3D stacked multi-die solutions with the integration of surface-mount devices (SMDs). These applications are targeted at Internet of things (IoT) and Internet of everything (IoE), and wearables [13]. Examples of fan-out WLPs in production are provided in Table 2.2.

Several companies have introduced versions of FO-WLP targeted at high performance computing, including networking, data center, server, and

Table 2.2 Fan-out WLP examples.

Company, device type (semiconductor node)	Package size (mm)	Number of balls	WLP process	Lines and spaces (μm)	Ball pitch (mm)
Apple, A10 processor	14.3 × 15.6 × 0.37	1407	InFO	10/10	0.4/0.35
Infineon, 77 GHz mid-range radar	6 × 6	112	eWLB	20/20	0.5
NXP (former Freescale), short-range radar	6 × 6 × 1	90	RCP	—	0.5
NXP, 77 GHz radar front-end IC	13.5 × 11.5	272	eWLB (dual chip)	15/15	0.5
Spreadtrum, TD-SCDMA modem plus RF transceiver	9 × 9	230	eWLB (2 die side-by-side)	10/10	0.4
Qorvo, Switch	1 × 1	2	eWLB	15/15	—
Qualcomm audio codec WCD9335	4.25 × 3.95 × 0.6	113	eWLB	15/15	0.5
Qualcomm PMIC PM8956	5.4 × 5.4 × 0.49	164	eWLB	15/15	0.4
Qualcomm RF transceiver WTR2965	3.3 × 3.3 × 0.56	61	eWLB	15/15	0.4

Source: TechSearch International, Inc.

artificial intelligence applications. FO-WLP on substrate versions would replace high density silicon interposers or high density organic interposers. Large 24 mm × 26 mm fan-out parts are mounted on laminate BGA substrates ranging from 42.5 mm × 42.5 mm to 45 mm × 45 mm. NANIUM has demonstrated reliability for parts as large as 25 mm × 23 mm [14]. TSMC and ASE have introduced chip-first versions, and Amkor has introduced a chip-last version of SWIFT on a substrate. ASE's fan-out chip-on-substrate (FOCoS) process has been in production for HiSilicon since 2016 (see Figure 2.11). TSMC is expected to move into production in 2018 with its InFO_oS [15].

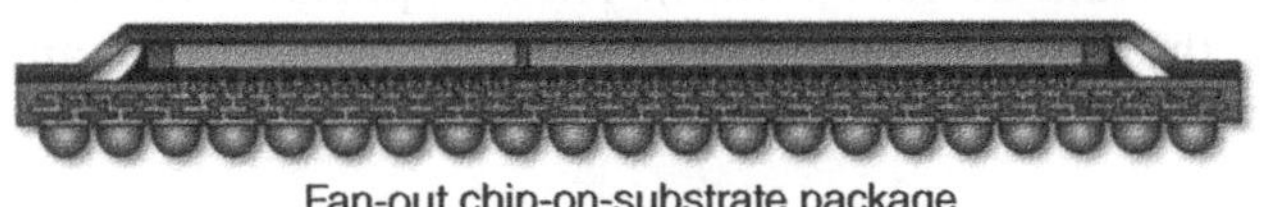

Fan-out chip-on-substrate package

Figure 2.11 ASE FOCoS process. *Source:* Courtesy of ASE.

FO-WLP on substrate meets the needs of heterogeneous integration. Advantages include:

- High density interconnect for multiple dies.
- Thin package for high I/O counts.
- A narrow gap between dies (<150 μm) provides good electrical performance.
- Package substrate helps to control warpage.

2.8 Strong Demand for FO-WLP

Strong demand for FO-WLP is projected for both high density and low density packages. More than 1.2 billion FO-WLPs were shipped in 2016. A greater than 33% CAGR is projected for FO-WLP from 2016 to 2021. Figure 2.12 shows the growth for FO-WLP.

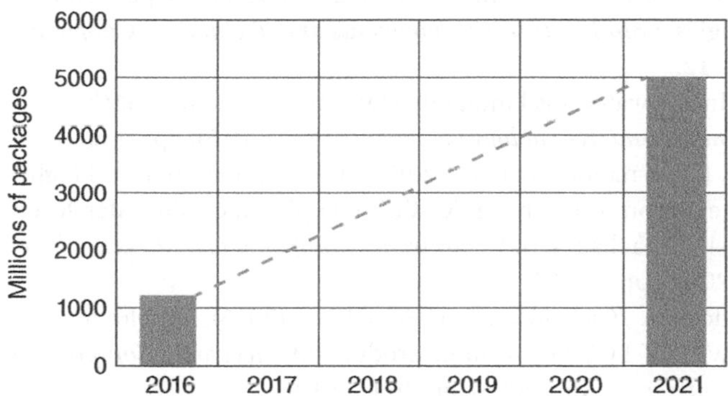

Figure 2.12 Forecasted growth of FO-WLP.

References

1 Fillion, R.A. (2006). Embedded chip build-up for high-end processors. *ICEP 2006 Proceedings* (April 2006), pp. 65–70.

2 Val, C. (2005). Ultra low profile 3-D cube with wafer level packaging technique. *International Wafer-Level Packaging Conference* (3–4 November 2005), pp. 134–137.

3 Tseng, C.-F., Liu, C.-S., Wu, C.-H., and Yu, D. (2016). InFO (wafer level integrated fan-out) technology. *IEEE 66th Electronic Components and Technology Conference* (May 2016), pp. 1–6.

4 Wang, C.-T. and Douglas Y. (2016). Signal and power integrity analysis on integrated fan-out PoP (InFO_PoP) technology for next generation mobile applications. *IEEE 66th Electronic Components and Technology Conference* (May 2016), pp. 380–385.

5 Rogers, B., Melgo, M., Almonte, M., et al. (2014). Enhancing WLCSP reliability through build-up structure improvements and new solder alloys. *IWLPC* (11–13 November 2014).
6 Bishop, C. (2016). Adaptive patterning design methodologies. *IEEE 66th Electronic Components and Technology Conference* (May 2016), pp. 7–12.
7 Huemoeller, R. and Zwenger, C. (2015). Silicon wafer integrated fan-out technology. *Chip Scale Review* March/April: 34–37.
8 TechSearch International, Inc. (September 2016). *Advanced Packaging Update.* TechSearch International, Inc.
9 Yu, D., Huang, Z., Xiao, Z. et al. (2017). Embedded Si fan out: a low cost wafer level packaging technology without molding and de-bonding processes. *67th Electronic Components and Technology Conference* (May 2017).
10 Che, F.X. Ho, D., Ding, M.Z., and Minwoo, D.R. (2016). Study on process induced wafer level warpage of fan-out wafer level packaging. *66th Electronic Components and Technology Conference* (May 2016), pp. 1439–1885.
11 Bishop, C., Rogers, B., Scanlan, C., and Olson, T. (2016). Adaptive patterning design methodologies. *66th Electronic Components and Technology Conference* (May 2016), pp. 7–12.
12 Tseng, C. (2016). InFO (wafer level integrated fan-out) technology. *66th Electronic Components and Technology Conference* (May 2016), pp. 1–6.
13 Cardoso, A., Dias, L., Fernandes, E. et al. (2017). "Development of novel high density system integration solutions in FOWLP-complex and thin wafer-level SiP and wafer-level 3D packages. *67th Electronic Components and Technology Conference* (May 2017), pp. 14–21.
14 Chatinho, V., Cardoso, A., Campos, J., and Geraldes, J. (2015). Development of very large fan-in WLP/WLCSP for volume production. *Electronic Components and Technology Conference* (May 2015), pp. 1096–1099.
15 TechSearch International (June 2017). *Advanced Packaging Update.* TechSearch International.

3

Embedded Wafer-Level Ball Grid Array (eWLB) Packaging Technology Platform

Thorsten Meyer[1] *and Steffen Kröhnert*[2]

[1] *Infineon Technologies AG, Germany*
[2] *Amkor Technology Holding B.V., Netherlands*

3.1 Technology Description

Wafer-level packaging (WLP) is one of the fastest-growing packaging technologies on the market today. The performance and dimension shrink requirements of today's advanced products fit WLP technology better than other packaging platforms for many applications and markets. WLP is packaging on the wafer as received from the wafer fabrication (FAB) facility. In comparison with other packaging technologies, the wafer is not separated into individual die before packaging. Solder balls are attached to the active side of the wafer after under-bump metallization (UBM) is applied. For WLP, two options are known as shown in Figure 3.1, WLP with and without copper redistribution layers (RDL).

No redistribution is required if the die pads are located at the desired positions of the BGA balls and the die size is smaller than 3 mm × 3 mm – only a UBM and a solder ball are applied. For die with peripheral pad-out, like typical for wire-bondable designs, RDL is required. RDL is typically applied with thin film technology, well known from the front-end processing. A passivation layer (dielectric layer) is normally applied by spin coating, one or more RDL are manufactured by sputtering and electroplating (semi-additive process), and a final passivation layer again by spin coating.

Finally a UBM may be applied before attachment of the solder ball. Front-end-based process steps such as spin coating, lithography, sputtering, and electrolytic plating are the major contributors to this packaging technology. Figure 3.2 shows a sketch of a typical WLP with two RDL (RDL1 and RDL2).

A major restriction of WLP is the real estate available for solder balls on the die. All solder balls must fit on the die area. For large pitches or small die, the

Advances in Embedded and Fan-Out Wafer-Level Packaging Technologies, First Edition.
Edited by Beth Keser and Steffen Kröhnert.

Without redistribution

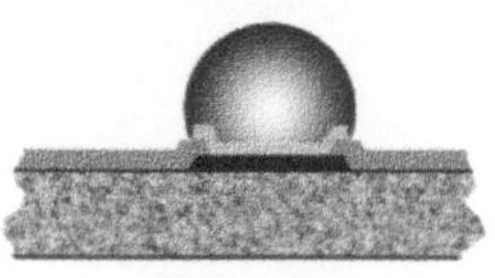

• Low I/O count (<30)
• No redistribution
• Applications: passives, etc.

With redistribution

• Low to medium I/O count (<60)
• Redistribution needed
• Applications: bluetooth, GPS, transceiver, etc.

Figure 3.1 Principle of WLP with and without redistribution layer (RDL).

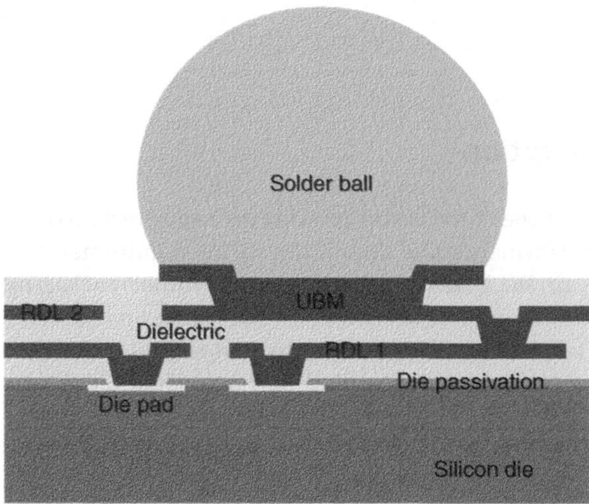

Figure 3.2 RDL stack with UBM for WLP.

available space may not be sufficient to position all balls on the die surface, so the use of WLP would not be possible. In order to overcome this ball count and pitch dilemma for WLP, a fan-out WLP (FO-WLP) began development in 2005. The version of FO-WLP technology reviewed in this chapter is called eWLB for embedded wafer-level ball grid array.

FO-WLP is the extension of WLP and provides higher input and output (I/O) capability. An area around the die is created prior to the application of the RDL and serves as a fan-out area for these additional I/O, which typically would not fit on a WLP. All performance and dimension advantages of WLP are enriched by the fan-out region due to the addition of I/O and, as later discussed, system integration capability. Figure 3.3 indicates the driving factors for FO-WLP.

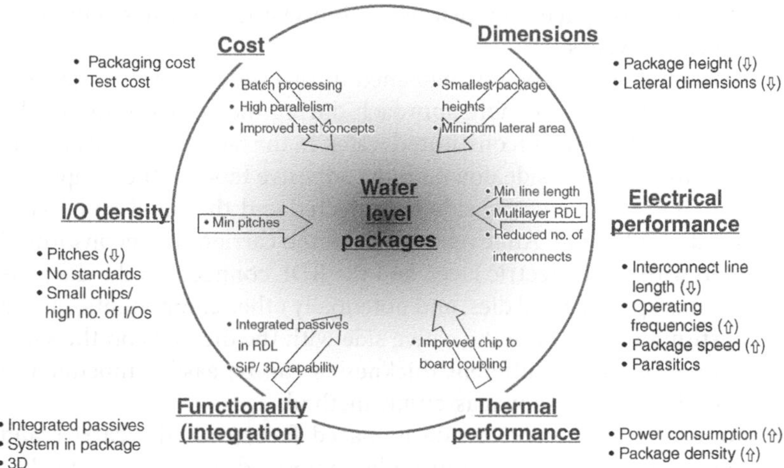

Figure 3.3 Driving factors for FO-WLP development.

FO-WLP has been developed primarily for the mobile application market. Infineon was the first company to introduce an FO-WLP technology called eWLB to the market in 2009. Today, eWLB is in high volume manufacturing at multiple contract assembly suppliers, and many other FO-WLP variants are offered by other suppliers including foundries and integrated device manufacturers (IDM) in addition to assembly suppliers and are reviewed in this book. Many of them, including those invented by consortiums and universities and also included in this book, are still in the development phase, but some of them recently also started high volume production. The most prominent technology is the integrated fan-out (InFO) package from silicon foundry TSMC. It was introduced in 2015 and is a high volume product in the mobile arena now.

FO-WLP can be classified as chip-first and chip-last technologies as well as chip face-up or chip face-down assembly approach during the wafer reconstitution process. Chip-first starts with the placement of the die on a temporary mold carrier laminated with an adhesive tape. Afterward, the die on the mold carrier with adhesive tape is embedded with mold compound using a wafer molding process, and after carrier removal, the RDL and solder balls are applied.

In the chip-last FO-WLP technology, a thin, typically high density RDL is deposited on a temporary carrier first, and then the die are attached to this layer using typical flip-chip attach and solder reflow technology. Finally, the mold compound is applied for encapsulation of the die. Chip-last FO-WLP is in development and not currently offered in high volume manufacturing. All technologies in production at the time of publication are chip-first. This is because the first and most prominent member of the FO-WLP technologies (eWLB) was first to market 10 years ago. The ability to create the thin film and fine line and space dimensions of chip-last fan-out has only been developed in recent years.

Chip-first technology can be distinguished in two different categories: die face-down or die face-up assembly approach during the construction of the molded wafer, also called the reconstituted wafer. In the face-down variant, the die is placed with the active side down on the adhesive tape on the temporary mold carrier. This way, the active side is protected and the mold compound embeds five sides of the die. After removal from the carrier, the reconstituted wafer is coated with the dielectric layer, and the RDL connects to the die pads directly. With this approach, all dies, and potentially other components, placed on the reconstituted wafer have the active side with the die pads on the same plane. Also, multiple die of different thicknesses can be easily embedded in mold compound in the face-down assembly method.

In face-up assembly, the active side is placed faceup on the carrier and molded, embedding the active side in mold compound. In order to do this, application of copper pillars is required at the wafer level prior to wafer singulation. For example, copper pillars are typically electroplated copper cylinders on the wafer surface with a height of 20–60 μm and a diameter of 60 μm. After Cu pillar application and wafer saw, the die are placed faceup on the adhesive tape on the mold carrier. This way, the mold compound embeds all five sides of the die. In comparison with chip-first face-down FO-WLP, the active side of the die is now embedded and the die backside is not. A molded wafer grinding step is required to expose the copper pillar and make it accessible for contact by RDL. Both variants are in high volume production: eWLB for chip-first face-down FO-WLP and InFO for chip-first face-up FO-WLP. There are multiple FO-WLP products on the market. The first product being introduced was a baseband die (see Figure 3.4).

Figure 3.4 X-GOLD 616 baseband product in eWLB package technology.

The baseband X-GOLD 61x-series were monolithically integrated, low power 65 nm baseband devices that included all the digital, analog, and power management functions. Following products in the market packaged in FO-WLP – face-down and face-up assembly approaches – have been RF, audio codec, power management, filters and switches, and combinations in FO-WLP-based wafer-level system in package (WL-SiP) up to application processors (AP) for the mobile communication market. FO-WLP started to enter other markets such as automotive, medical and health care, bio-devices, Internet of things (IoT), and security. Also, microelectromechanical systems (MEMS) have been packaged using FO-WLP technologies.

3.2 Basic Package Construction

Wafer-level packages are produced at wafer level, meaning in wafer form, utilizing wafer FAB equipment typically used in front-end process technology and now also in the back-end ones. The term front end is used to describe the area and process of processing wafers, e.g. the formation of transistors directly in the silicon. The term back end is used for area and process blocks from wafer separation through packaging until component test. The main innovation in the step from classical WLP to FO-WLP was the introduction of an artificial molded wafer to generate the fan-out area. This process is called reconstitution.

Figure 3.5 shows the basic construction of an FO-WLP. The die is embedded in the mold compound on five sides. RDL, embedded in isolating dielectric layers, redirect the die pads to the interconnect positions for the solder ball. These balls can be placed in the fan-in zone on the die or in the fan-out zone on the mold compound.

FO-WLP can be used to meet many packaging challenges of today. The reconstituted wafer diameter is independent from the size of the incoming wafer. The reconstituted wafer allows for heterogeneous integration of silicon,

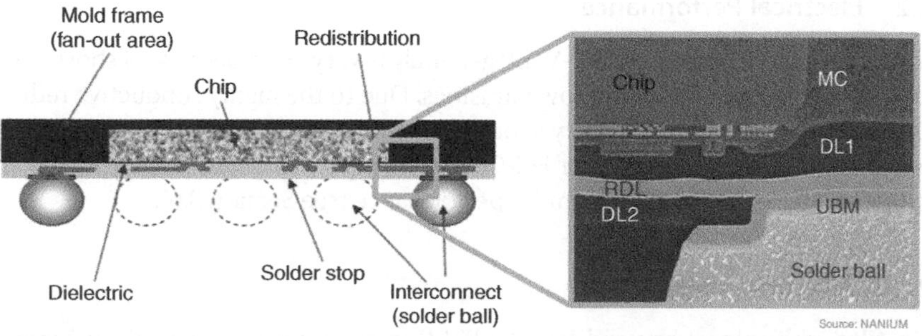

Figure 3.5 Basic construction of FO-WLP.

GaAs, or SiGe die as well as glass parts, discrete passives, or other packaged parts, like prefabricated vias or prefabricated routing layers. The fan-out area is adaptable to the area needs, defined by the assembly distance between the die in the reconstitution process.

The challenge of z-height reduction in mobile applications can be addressed. Package thicknesses below 0.3 mm are possible, because a laminate substrate is not needed for rerouting the signals and the reconstituted wafer can be ground after application of the RDL. The redistribution with shortest interconnects provides best electrical performance and highest interconnect density capability, compared with classical packages. Finally, FO-WLP is best suited for integration purposes, multi-die packages, passive integration (in the RDL or as discrete components), and also stacking approaches. In general, the advantages of FO-WLP are listed below [1].

3.2.1 Cost Performance

FO-WLP achieves an excellent packaging cost position, especially for small packages. Since it is a fairly new packaging technology on the market, additional opportunities for cost reduction are still available (such as substrate size increase, material cost improvements, equipment efficiencies, and yield improvements). WLP cost is sensitive to the package size, because the process steps mainly are parallel process steps. The cost for one process step is divided by the number of good die on the wafer. FO-WLP also uses wafer-level processes, but cost is dependent on the package size or number of good packages per reconstituted wafer rather than the number of good die per wafer. FO-WLP is cost competitive to flip-chip chip-scale packages (FC-CSP) and wire-bond chip-scale packages (WB-CSP) up to a certain package size, because of the parallel process steps and the absence of a substrate and an additional interconnect element. However above a certain package size, FO-WLP is more expensive than WB-CSP and FC-CSP.

3.2.2 Electrical Performance

The electrical performance of FO-WLP generally is very good because of short, low resistance connections and very low parasitics. Due to the highly conductive redistribution and the possibility of very short interconnections, FO-WLP is superior to classical packages like WB-CSP or FC-CSP with respect to electrical performance. Also, FC-CSP cannot provide a similar performance (see Section 3.6).

3.2.3 Interconnect Density

Thin film technology, as used for FO-WLP interconnects, is based on highly accurate and reproducible process steps. Due to this high accuracy, very high package interconnect density can be achieved. FO-WLP can address even the smallest pad pitches of today's most advanced front-end technology nodes. Due to the thin film redistribution technology, small lines and spaces can be realized. 20 μm lines and spaces are standard today, and the leading-edge products are already providing 5 μm line width and space. A limitation of FO-WLP is the shift of the die during reconstitution, which limits the capability to fit the redistribution to the die. Flip-chip interconnects, which typically are soldered to the substrate, benefit from a self-centering effect of the solder interconnect.

3.2.4 Thermal Performance

Due to the thin stack-up of FO-WLP, there is good thermal dissipation into the board. The thermal dissipation is characterized by the sum of the thermal resistances of the individual contributors. Due to the absence of a laminate substrate and very thin layers of the RDL, the thermal resistivity of FO-WLP is low. The heat flow into the customer board therefore has less resistance than is the case for other comparable package technologies, like WB-CSP or FC-CSP.

3.2.5 Board-Level Reliability

FO-WLP has reliability limitations similar to WLP at board level. The coefficient of thermal expansion (CTE) mismatch between printed a circuit board (PCB) of ~16 ppm $°C^{-1}$ and a silicon die of ~3 ppm $°C^{-1}$ has to be buffered by the solder ball interconnects between the package and PCB. The temperature cycling on board (TCoB) reliability requirement is common in the electronics industry. FC-CSP and WB-CSP do have the advantage of having the substrate as a CTE buffer between the silicon and the PCB. This buffering reduces the stress in the interconnects (e.g. solder balls), which is mainly caused by the mismatch of the different coefficients of expansion in a final configuration. Therefore the reliability of traditional substrate laminate packages like FC-CSP and WB-CSP is higher than the reliability of FO-WLP and WLP.

It is also very important to have high package coplanarity and low package warpage at the reflow temperature when the package is soldered to the PCB in order to achieve a high PCB assembly yield, especially for package-on-package (PoP) applications. Low warpage change up to reflow temperature is a key feature of FO-WLP (see Section 3.6).

3.2.6 System Integration

FO-WLP is best suited for system integration, and multiple different options are available for the integration of passives, actives, or special features into an FO-WLP. Since multiple active and passive components can be placed very close together during the reconstitution, integration with very short interconnects is possible. Heterogeneous integration – the assembly and packaging of multiple separately manufactured components onto a single package – is possible, and the RDL can easily be used for the implementation of passives like coils, baluns, or similar. Multiple options for connections in the z-direction can also be realized, which allows the realization of stacked components. This includes methods like the filling of laser-drilled vias with conductive material or the embedding of prefabricated via blocks. For the first mentioned option, vias are drilled in the mold compound by a laser. These holes are then filled with metal, typically by electroless plating or electroplating. The second option is the integration of known good vias in prefabricated via blocks using PCB technology or through-silicon via (TSV) blocks, which are embedded next to active or passive components. These via blocks can be tested for functionality before embedding it into the mold compound.

The integration capability of FO-WLP is superior to any other packaging technology, while the highest electrical performance and smallest dimensions can be maintained. FO-WLP competes mainly with FC-CSP due to the similar field of application and performance. A comparison between both packages is shown in Figure 3.6.

FO-WLP is typically used in applications with a small package body size and a pin count (I/O count) below 600 pins. Since FO-WLP is produced on the wafer level with mainly parallel process steps, the package size has a direct impact on the cost of the package. The sweet spot for FO-WLP therefore is below 600 pins [1]. Automotive, mobile, and IoT applications are sweet spots of the technology. Figure 3.7 shows this product package position for WLP and FC-CSP.

Part	Task	Functional layers	
		Flip-chip	eWLB
Sketch		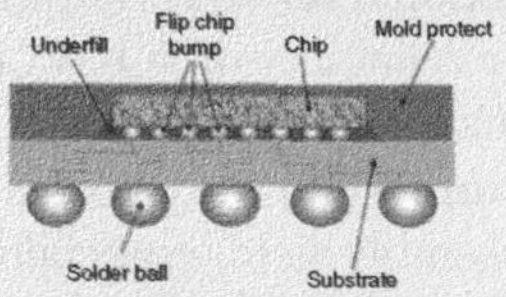 Underfill, Flip chip bump, Chip, Mold protect, Solder ball, Substrate	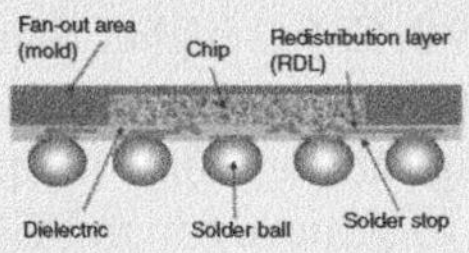 Fan-out area (mold), Chip, Redistribution layer (RDL), Dielectric, Solder ball, Solder stop
First level interconnect	Interconnection from chip to package	• Flip-chip UBM w/ bump • Substrate	• Dielectric • Redistribution • Solder stop
Second level interconnect	Interconnection from package to customer board	• Substrate • BGA solder ball	• Redistribution • BGA solder ball
Encapsulation	Mechanical protection of silicon die	• Underfill/ molding/ molded underfill	• Molding

Figure 3.6 Comparison between FC-CSP and FO-WLP (example: eWLB).

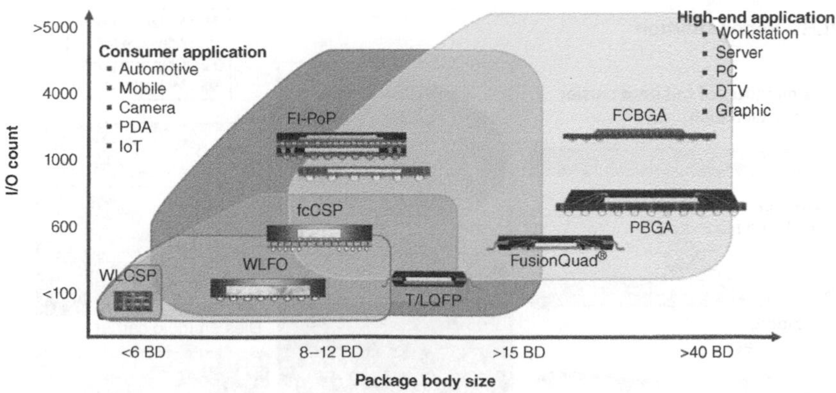

Figure 3.7 Product package position for WLP and FC-CSP [1]. *Source:* Courtesy of Amkor Technology, Inc.

3.3 Manufacturing Process Flow and BOM

FO-WLP is generated in four major process blocks: reconstitution, which itself includes four major process steps; RDL; solder ball application and package singulation; and electrical test. In reconstitution, a tested wafer is sawn into separate components, while in parallel a double-sided adhesive foil is attached to a carrier, typically by lamination. The adhesive foil is releasable and loses its stickiness by addition of energy (e.g. by heating). Then, tested known good die are picked from the singulated silicon wafer and are placed on sticky tape on the carrier. The die spacing defines the fan-out area of the package. It can be small or large, depending on the final package size required. Next, mold compound in liquid, granular, or sheet format is applied onto the die. By using a compression molding process, the die are embedded into mold compound, filling the gaps between the die and forming the artificial wafer. The artificial wafer consists of silicon die and mold compound and is round like a semiconductor wafer with a notch or flat as rotation indication. The wafer is cured. Finally, the foil is debonded from the carrier. Figure 3.8 shows the reconstitution process flow of eWLB as an example. Other technologies have similar process flows.

The typical process blocks of FO-WLP technologies are shown in Figure 3.9. After the reconstitution, the artificial wafer is processed like WLP. A dielectric is applied by spin coating. Due to the limited temperature stability of the mold compound, a low temperature cure material is used (see chapters later in this book on low temperature cure dielectric materials). Imaging vias in the dielectric can be done by laser direct imaging (LDI) or lithography. In eWLB, lithographic stepper technology is used for the exposure of the dielectric layer. The vias are the openings in the dielectric above the pads of the die as well as in the dicing street.

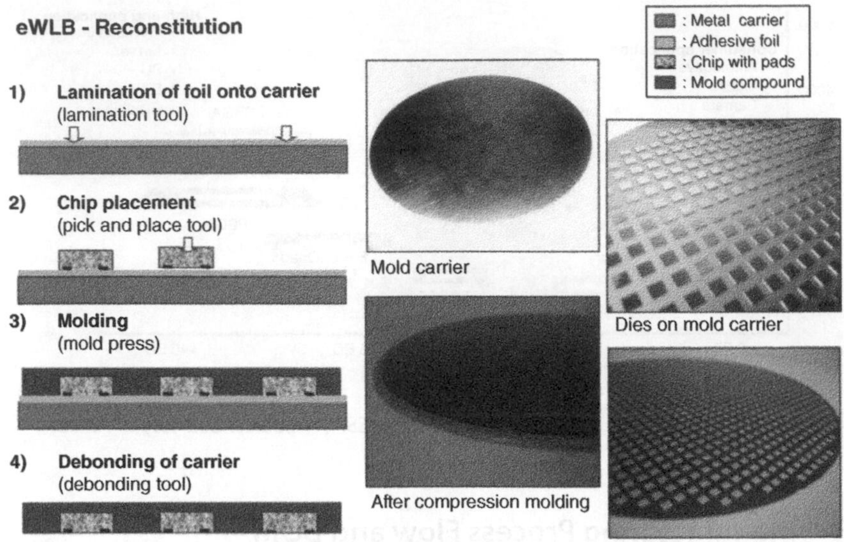

Figure 3.8 Reconstitution of artificial wafer [2]. *Source:* Reproduced with permission of IEEE.

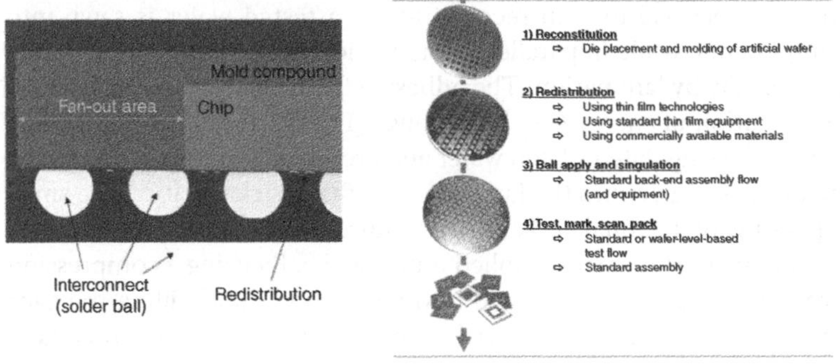

Figure 3.9 eWLB cross section and process overview [2]. *Source:* Reproduced with permission of IEEE.

The RDL is applied using thin film technology. An adhesion promoter layer such as titanium or titanium–tungsten is applied using sputtering. Next, a copper seed layer is sputtered. A plating resist is applied by spin coating and is photoimaged to create the pattern for the RDL circuitry. In eWLB, a stepper technology is used for the exposure of the plating resist. After development of the resist, the copper redistribution is now plated on the seed layer in the openings of the plating resist. Thicknesses between 4 and 8 μm of copper layer are typical. Next, the plating resist is removed and the two seed layers are etched away one after the other.

The last dielectric layer covers the redistribution and defines the landing pads for solder balls. Then the solder ball application is done at wafer level. Flux is applied onto the landing pads, typically in a printing process with a stencil, and then preformed solder balls, which are available in many different diameters, are rolled into openings of a stencil above the positions of printed flux. The solder balls stick in the flux and the wafer is transported into a reflow furnace. After melting, the solder balls are connected to the redistribution.

Finally, wafer thinning by standard wafer grinding and polishing is completed, and then laser marking on the die backside, reconstituted wafer dicing, and final test of the components. Figure 3.10 shows a cross section of the final package.

The bill of materials (BOM) for FO-WLP contains the following basic materials:

Material	Function	Typical material class
Mold compound	Fan-out area, die protection	Liquid epoxy mold compound, granular epoxy mold compound
Dielectric layer	Isolation, mechanical buffering	Polyimide, epoxy, silicone
Redistribution	Electrical contact, ball pad, passive structures	Electroplated copper
Solder stop	Isolation, ball pad definition	Polyimide, epoxy, silicone
Under-bump metallization	Solder ball landing pad	Electroplated copper or copper/nickel/gold
Solder ball	Customer interface, second-level interconnect element	Lead-free SAC (tin, silver, copper) with different contents (e.g. SAC105, SAC305, etc.)

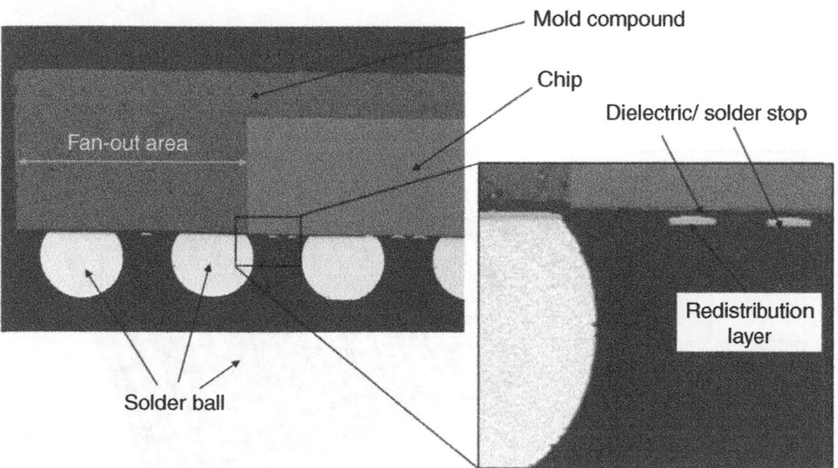

Figure 3.10 Cross section of eWLB.

3.4 System Integration Capability

FO-WLP is best suited for system integration: both horizontal multi-die packages as well as for vertical stacked integration of active and passive components. SiP applications are realized by placing two, three, or more active die in one package, for example, baseband, power management unit, and an RF module. Also, the integration of passive devices like surface-mount devices (SMD) and integrated passive devices (IPD) saves significant space on the application board and improves electrical system performance. Multiple components can be placed side-by-side above the die, or face-to-face on the active die area without increasing z-height. Furthermore coils, capacitors, resistors, and especially antennas can be directly formed using the RDL. Coils and antennas in particular can be realized in the fan-out area with an excellent quality factor since they are isolated over mold compound instead of over the die.

Side-by-side integration is already well known in the industry for classical package technologies. FO-WLP can provide performance and dimensional advantages due to the fine width and space of the redistribution lines and resulting high packaging density. Multiple die can easily be placed into the mold compound, next to each other and very close. Die distances at and below 50 μm have been achieved to date. Die of different front-end nodes can be used, as well as active or passive components. Figure 3.11 shows a two-die test vehicle with a die-to-die distance of 300 μm. The redistribution connects the two die as well as connecting to the solder balls. This way, multiple connections with very tight pitches can be realized.

The example above is only one flavor of realized side-by-side FO-WLP. In order to achieve a higher interconnect density, different package types have

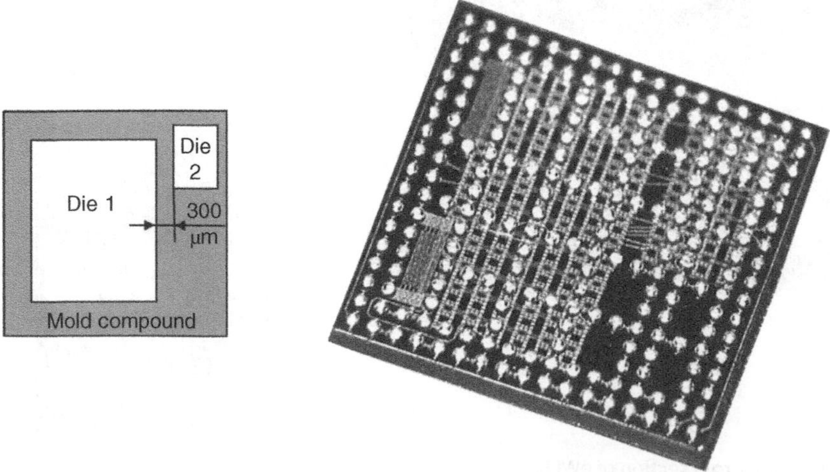

Figure 3.11 Side-by-side approach of FO-WLP.

Figure 3.12 Side-by-side approach of SWIFT [3]. *Source:* Courtesy of Amkor Technology, Inc.

been developed. Amkor's silicon wafer integrated fan-out technology (SWIFT) is one example of a chip-first, RDL-first fan-out multi-die package (see Figure 3.12) [3]. With its fine feature photolithography and thin film dielectrics, SWIFT bridges the gap between TSV and traditional wafer-level FO-WLP. Since very tight pitches can be realized, SWIFT offers strong improvements in form factor, signal integrity, power distribution, and thermal performance compared with classical packages. SWIFT incorporates an RDL-first process, which means that the RDL are generated before the silicon is attached.

FO-WLP is also suitable to be used for PoP stacking. Typically, a memory package is placed on the backside of a bottom package, which often contains a processor or another logic die. Top-to-bottom connections can be realized by laser-drilled through-mold vias or by placing prefabricated contact bars in the fan-out area of the bottom die. Also, by implementing a backside RDL with landing pads and connection to the frontside, an area array can be achieved. The very stable coplanarity over temperature is another big advantage of this technology for package stacking.

The most prominent example for a PoP package is TSMC's InFO package. InFO is a chip-first face-up FO-WLP using copper pillars to provide 3D interconnections to connect the bottom RDL circuitry to the topside of the package, so a DRAM package can be attached on top of the applications processor for a mobile phone. Figure 3.14 shows another example of system integration in FO-WLP. Oscillators as well as discrete passive components of different dimensions (e.g. 0201 or 01005) are embedded in the mold compound next to each other. Also, components from tape and reel can be picked and placed onto the mold carrier next to silicon die, as shown in Figure 3.13.

The integration of MEMS or nanoelectromechanical systems (NEMS) components often requires special packaging solutions. MEMS structures often are sensitive to mechanical stress in production or later in application and also often require mechanical decoupling, shielding, or hermetic housings. FO-WLP offers special capabilities for those applications. For example, a pressure sensor with a mechanically sensitive membrane can be embedded in mold compound close to an ASIC, but may not be connected to the board by solder balls, as indicated in the sketch in Figure 3.14. This way, the sensitive sensor is decoupled from the mechanical mismatch of CTE between the die and package and board. ASIC and MEMS are integrated, which requires codesign, a design focusing not only on the single die but also on the design of the complete system. Products like this actually are under development, not in high volume production.

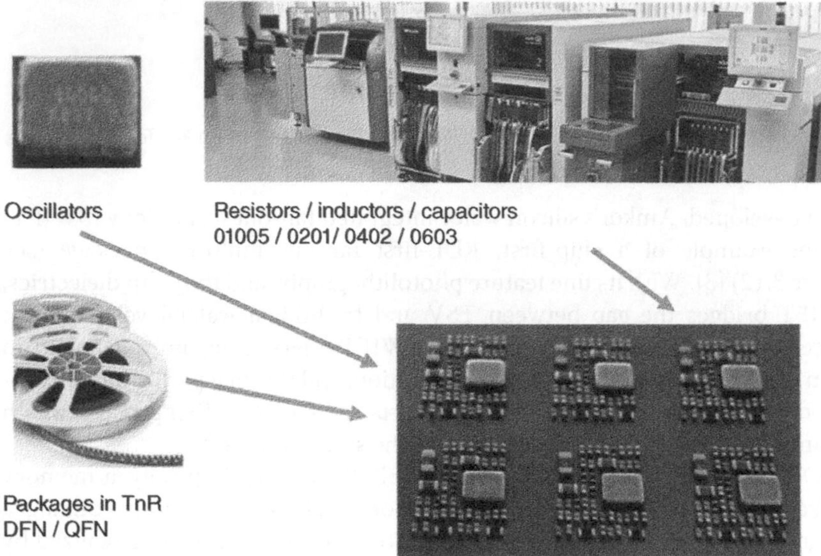

Figure 3.13 Heterogeneous system integration [3]. *Source:* Reproduced with permission of Amkor Technology, Inc.

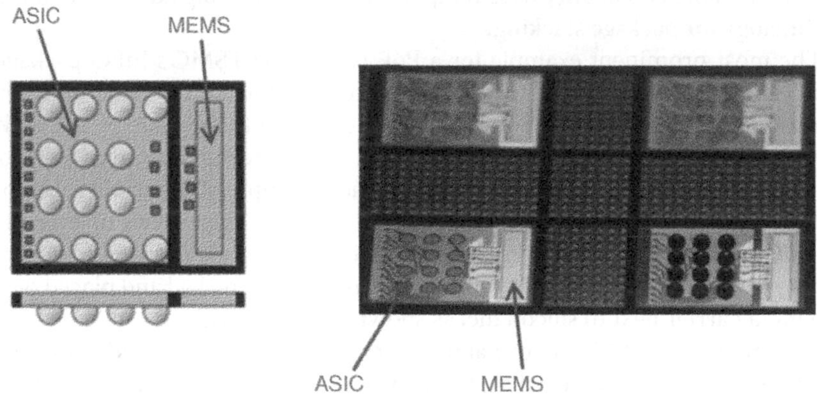

Figure 3.14 Heterogeneous system integration of ASIC and MEMS.

3.5 Manufacturing Format and Scalability

FO-WLP in high volume manufacturing today is typically produced in 300 mm wafer format. Some users still produce in 200 mm wafer format due to availability of the production line, but the low economy of scale of this smaller size is not cost competitive. One supplier of FO-WLP has moved to 330 mm round

Table 3.1 Formats and area scaling factors for different substrate formats.

Format	Dimensions	Area scaling factor
Round	300 mm Ø	1×
	$300 \times 300\,mm^2$	1.64×
Square	$500 \times 500\,mm^2$	4.54×
	$600 \times 600\,mm^2$	6.5×

Source: Reproduced with permission of IEEE.

format after adaptation of toolings and handling systems. The additional economy of scale lowers the production cost.

A strong trend of the recent past is the move to panel format for low-end FO-WLP. Multiple companies are moving to rectangular substrate formats in order to increase the economy of scale further and therefore reduce packaging cost. Table 3.1 shows the area scaling factors for different dimensions. So far, no standards have been set for this production technology. The first panel technologies in high volume manufacturing are expected in 2018.

3.6 Package Performance

As indicated earlier, FO-WLP offers many advantages in comparison with traditional WB-CSP or FC-CSP. In this section, the warpage behavior and the electrical and thermal performance of different package platforms will be compared. The lack of warpage change over temperature of FO-WLP is a distinct advantage. While the warpage of the FO-WLP changes little with temperature, strong and very strong warpage change over temperature is documented for embedded die and for FC-CSP [4]. This can be a strong advantage for testing, assembly, reliability, or 3D technologies, for example, PoP. For high yield package assembly or PoP stacking, warpage of the package is critical. If the warpage is too large, open solder joints may occur between the bottom package and the board or between the two stacked packages. Therefore, not only warpage at room temperature is a concern, but also the warpage at solder reflow temperatures (up to 260 °C) should be considered. As a result, warpage control at both temperature extremes is critical for package assembly [4].

From an electrical point of view, FO-WLP shows lower RLC package parasitics (resistance, inductance, and capacitance) compared with WB-CSP and FC-CSP. This is mainly because FO-WLP has shorter interconnects due to its substrate-less design. In Figure 3.15, package resistance and inductance of the interconnects from die to package and the package construction are compared

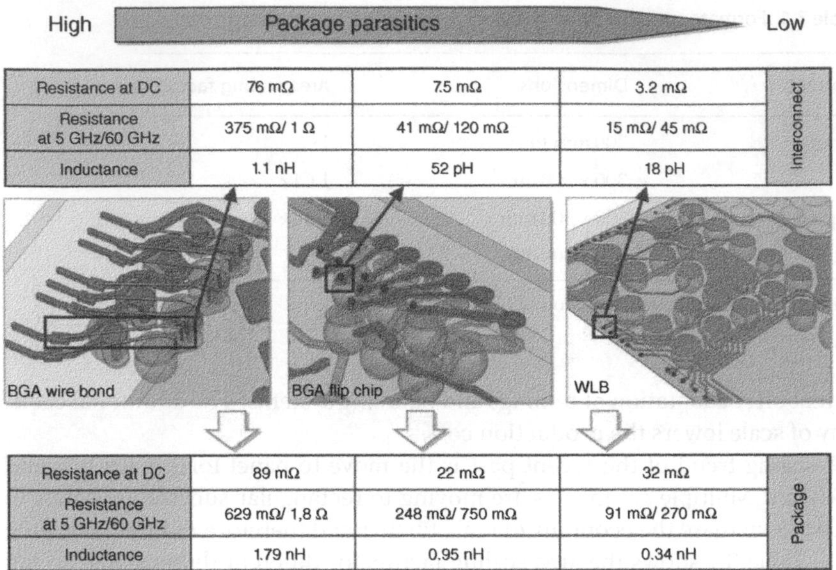

High
Package parasitics
Low
Resistance at DC
76 mΩ
7.5 mΩ
3.2 mΩ
Resistance at 5 GHz/60 GHz
375 mΩ/ 1 Ω
41 mΩ/ 120 mΩ
15 mΩ/ 45 mΩ
Inductance
1.1 nH
52 pH
18 pH
Interconnect
BGA wire bond
BGA flip chip
WLB
Resistance at DC
89 mΩ
22 mΩ
32 mΩ
Resistance at 5 GHz/60 GHz
629 mΩ/ 1,8 Ω
248 mΩ/ 750 mΩ
91 mΩ/ 270 mΩ
Inductance
1.79 nH
0.95 nH
0.34 nH
Package

Figure 3.15 Electrical performance comparison (resistance, inductance) [5].

for the different packaging platforms in simulation. Resistance and inductance values have been extracted from the simulation. For the comparison, packages with the same functionality have been chosen according to existing package design rules. The simulation of the interconnect compares aluminum wire bond with a solder-bumped flip-chip interconnect on a two-layer substrate with a one-layer RDL for the FO-WLP.

The direct current (DC) resistance and the electrical resistance at 5 GHz of the wire-bonded package is a factor of 10 higher than the resistance of the flip-chip interconnect with solder bump with a diameter of 150 μm. The electrical resistance of the FO-WLP interconnect is even reduced by a factor of 2 compared with the flip-chip interconnection. For the inductance, of the three interconnects, a similar picture is seen. It is reduced for the flip-chip interconnect compared with the wire-bond variant, but again improved by the FO-WLP due to the laminate substrate-less design. Translated to a potential package solution, this leads to lowest electrical resistance and inductance for the FO-WLP.

Comparing the thermal performance between WB-CSP, FC-CSP, and FO-WLP leads to a similar picture as the electrical performance comparison (see Figure 3.16). Simulations of the thermal resistivity (Rth) consider constant power dissipation for all packages and a constant die size for all platforms. In Figure 3.16, a comparison of Rth is shown. A wire-bond BGA with two-layer substrate, a flip-chip package with two-layer substrate, and an FO-WLP

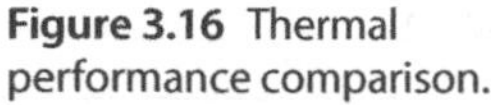
Figure 3.16 Thermal performance comparison.

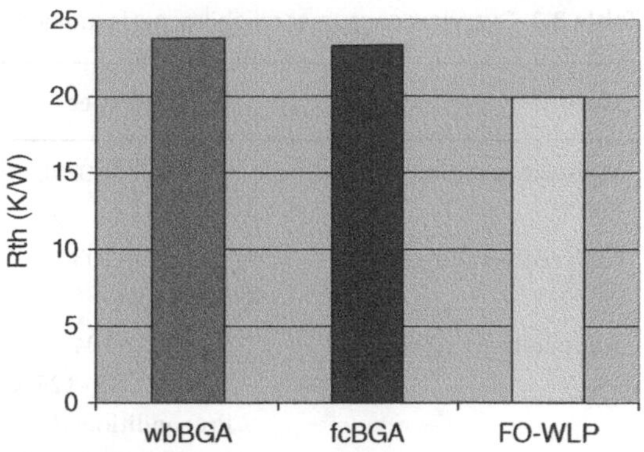

with one-layer redistribution are compared in simulation. The simulation model contains constant die sizes, and package size is chosen according to standard package design rules. FO-WLP shows better thermal performance compared with WB-CSP and FC-CSP because of the thinner package, the absence of an isolating substrate, and less interfaces between the die and the heat sink (board).

3.7 Robustness and Reliability Data

The excellent electrical and thermal performance of FO-WLP is caused by short and direct interconnects, the absence of a board with additional thermal resistance and electrical performance reduction, and a reduced number of interconnects. This causes a performance advantage, but it causes challenges as well. The package is directly connected to a PCB, without buffering layers. Therefore, the mismatch of CTE between the silicon die, the mold body, and the PCB has to be compensated by the solder balls and a thin dielectric layer. The major hurdle for WLP is the TCoB, where the component is attached to a standardized PCB and exposed to thermal cycling (e.g. −40 to 125 °C).

Package boundary conditions like die and package size, package thickness, number of solder balls, solder ball material and size, and many more influence the reliability results. Table 3.2 shows an overview of typical reliability tests for consumer products. In this case, the following boundary test vehicle conditions have been used:

Package size	9.25 mm × 8.80 mm
Die size	5.28 mm × 5.62 mm
Ball pitch	0.5 mm BGA pitch
Construction	Single die, one-layer-Cu-RDL, polyimide dielectric, no UBM, 122 solder balls

Table 3.2 Overview consumer reliability performance of FO-WLP.

Stress test	Standard/spec	Pass criteria
Moisture sensitivity level (MSL)	EIA/J-STD-020C (Level 1)	MSL 1
High temperature storage (HTS)	JESD22-A103 (Ta: 150 °C)	1000 h
Temperature cycling (TC)	JESD22-A104 (Cond B: 55–125 °C)	1000×
	Preconditioned (Level 1; Tr: 260 °C)	1500×
Unbiased HAST (uHAST)	JESD22-A118 (Cond A: 130 °C/85% RH)	96 h
	Preconditioned (Level 1; Tr: 260 °C)	188 h
Temperature humidity bias (THB)	JESD22-A101 (85 °C/85% RH, VCC)	1000 h
Temperature cycling on board (TCoB JEDEC)	IPC 97-01 (−40 °C/+125 °C, 1 cy h^{-1})	500× 1000×
Temperature cycling on board (TCoB NOKIA)	NOKIA spec. (−40 °C/+125 °C, 2 cy h^{-1})	FF > 500 cycles
Drop test	JESD22-B111	<10% fails at 20 drops

The described FO-WLP passed all required tests. This allows a move toward automotive applications with higher reliability requirements.

FO-WLP has entered the automotive market for radar applications. The package is best suited for high frequency applications due to the thin film RDL with very reproducible properties and low parasitics. Also, the solder ball interconnects can be placed on the fan-out area, an important feature for electrical performance at high frequencies. Figure 3.17 shows the 77 GHz radar component in eWLB package technology. The die in the center of the package contains only four thermal balls, and all other interconnects are placed over the fan-out area. It can be seen that a redundant ball concept is used and multiple solder balls from center to edge are connected by the RDL. This way, all solder balls carry the same signal. During temperature cycling, stress will occur in the solder balls due to the mismatch of CTE of the package (3–9 ppm $°C^{-1}$) and the PCB (14–20 ppm $°C^{-1}$). The highest stress will occur in the solder balls with the largest distance to the center of the package or distance from neutral point (DNP). This stress will cause the solder balls with largest DNP to crack first, but redundant balls are continuing to transfer the signals. With the

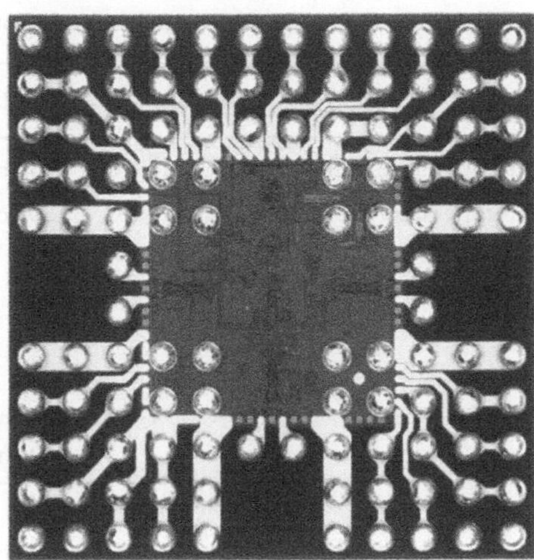

Figure 3.17 eWLB 77 GHz radar package.

redundant ball concept, connected solder balls can take over the transmission of the signal with no impact on the electrical functionality due to the cracked solder ball. This way, the redundant ball concept improves the TCoB performance strongly.

FO-WLP passed automotive standard AEC-Q100 Grade 1 for advanced driver assistance systems (ADAS). Table 3.3 gives an overview of the reliability test results.

3.8 Electrical Test Considerations

The proof of electrical functionality is an important milestone for any package platform. CSP packages are tested after singulation into discrete packages. This flow is indicated in Figure 3.18 as Flow 1. A wafer from the front end is tested on wafer level and diced after test. Functional die are then assembled in the package. After package singulation, the package test is performed. Defects caused by dicing will be testable in this flow. For FO-WLP, a reconstituted wafer test is possible as well, as indicated in Flow 2 of Figure 3.18. However, FO-WLP can also be tested after package singulation like CSP (Flow 1) as shown in Flow 3.

Flow 2 is a wafer-based test flow. The front-end wafer is tested and diced. The wafer-level testing is important to find known good die for reconstitution. This way, the reconstituted wafer starts with typically 100% yield. After the reconstitution and the application of dielectric, RDL, and solder balls, the reconstituted wafer can be tested using a wafer prober. After this, package singulation is performed. Due to the singulation after testing, potential

Table 3.3 Overview automotive reliability performance of FO-WLP.

Test	Conditions	Criteria	Result
Temperature cycle	−65 °C/+150 °C	1000 cycles	Pass continued up to 1500 cycles
Temperature cycle on board	−40 °C/+125 °C	1300 cycles	Pass 1st fail at 2215 cycles/first 6 fails up to 2500 cycles
Drop test	JEDEC JESD22-B111		Pass 1st fail at 120 drops
uHAST	130 °C/85%	192 h	Pass
HTS	150 °C/175 °C	2000 h 1000 h	Pass

defects due to the singulation process are not detected. Since package singulation is done through mold compound and not through silicon, this is seen as low critical. In order to avoid discussion of yield loss due to dicing defects, frame prober testing can be carried out. It follows the same steps as Flow 2, but package singulation is done prior to the final testing. This is then done with a frame prober on the dicing tape/frame, and it also indicates potential defects from the singulation process.

3.9 Applications and Markets

End customer acceptance of new package types and multi-sourcing strategies impede the introduction of new packaging technologies like FO-WLP. Most important, packaging cost for large FO-WLP is still high. Packaging cost reduction is a major target for FO-WLP technologies at the low end. Therefore, the switch from round wafer format to rectangular panel format using cost-effective PCB process technologies will be a next step in this area. Even with a PCB-technology-related lower pattern density and a possible need for additional RDL, the lower process costs and the much higher process economy of scale will improve the cost per package significantly.

Continuous shrinking of the silicon nodes leads either to reduced die sizes or to higher I/O counts due to increased functionality. In both cases, a need to switch from the die-size-limited WLP to the much more flexible FO-WLP technology arises. Therefore, the future focus for high-end FO-WLP will move from the single-die package, using one RDL, to highly integrated SiP, using multiple routing layers on the package frontside and for package stacking technology (PoP), also additional routing layers on the package backside. This will speed up the integration of sensors and MEMS close to the AP or at least to bundled modules.

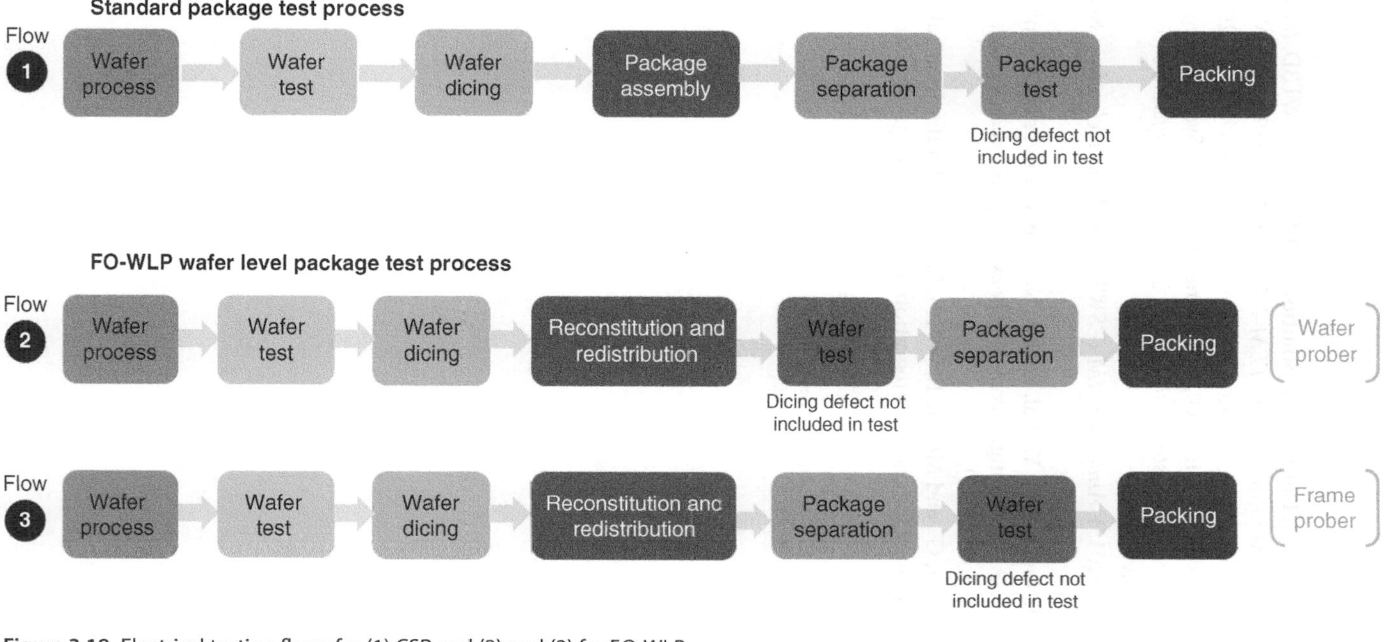

Figure 3.18 Electrical testing flows for (1) CSP and (2) and (3) for FO-WLP.

References

1 Kroehnert, S. (2017). Fan-Out SiP Solutions – WLSiP and WL3D. *Inaugural Conference and Exhibition on SYSTEM-IN-PACKAGE (SiP) TECHNOLOGY*, Doubletree Sonoma Wine Country, Sonoma, California, USA (27–29 June).

2 Meyer, Th., Pressel, K., Ofner, G., and Römer, B. (2010). System integration with eWLB. *Electronic System-Integration Technology Conference (ESTC)*, Berlin, Germany (2010).

3 Huemöller, R. (2015). *Amkor's SLIM & SWIFT Package Technology*. Semicon Taiwan.

4 Prashant, M., Liu, K., Jin, Y. et al. (2010). Next generation eWLB (embedded wafer level BGA) packaging. *Electronics Packaging Technology Conference (EPTC)*, Singapore (2010).

5 Pressel, K., Beer, G., and Meyer, T., (2010). Embedded wafer level ball grid array (eWLB) technology for system integration. *IEEE CPMT Symposium*, Japan.

4

Ultrathin 3D FO-WLP eWLB-PoP (Embedded Wafer-Level Ball Grid Array-Package-on-Package) Technology

S.W. Yoon

STATS ChipPAC, JCET Group

4.1 Introduction

Newly emerging applications in the consumer and mobile products, the growing impact of the Internet of things (IoT) and wearable electronics (WE), and the complexities in sustaining Moore's law have been driving many new trends and innovations in advanced packaging technology. The semiconductor industry now has to focus on density scaling and system level integration to meet the ever-increasing electronic system demands for performance and functionality as well as the reduction of form factor, power consumption, and cost.

This paradigm shift from chip scaling to system level scaling is and will continue to reinvent microelectronics system packaging, drive increased system bandwidth and performance, and help sustain Moore's law. Demand for maximum functional integration in the smallest and thinnest package will continue to grow with an order-of-magnitude requirement for lower manufacturing cost and reasonable cycle time. The challenge of the semiconductor industry is to develop a disruptive packaging technology capable of achieving these goals in timely manners.

To meet the above said challenges, eWLB, shown in Figure 4.1a, is a continually evolving technology platform that offers additional space for routing higher inputs and outputs (I/O) chips on top of the silicon (Si) chip area, which is not possible in conventional wafer-level chip-scale packaging (WLCSP) or wafer-level bump (WLB) [1]. It also offers comparatively better electrical, thermal, and reliability performance at a reduced cost with the possibility to address Moore's law (decreasing technology nodes with extreme low-k dielectrics in system on a chip [SoC]) and more than Moore (heterogeneous integration of chips with different wafer technology as an SiP solution in multi-die or 3D eWLB approaches) as shown Figure 4.1b.

Advances in Embedded and Fan-Out Wafer-Level Packaging Technologies, First Edition.
Edited by Beth Keser and Steffen Kröhnert.

(a)

(b)

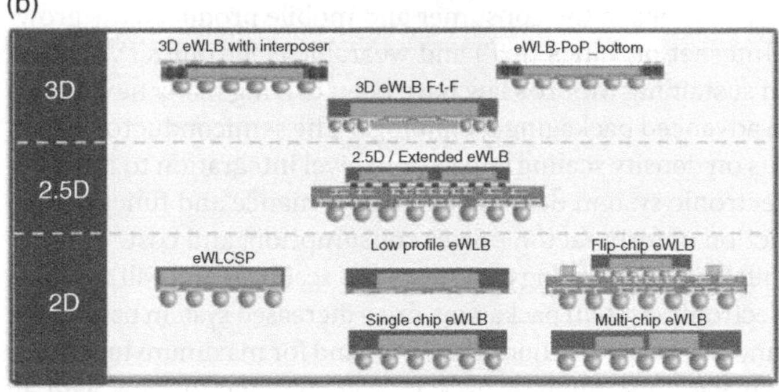

Figure 4.1 (a) 300 mm eWLB carrier and eWLB packages and (b) evolution of eWLB technology from 2D to 2.5D/3D packaging solution.

eWLB technology uses a combination of front- and back-end manufacturing techniques with parallel processing of all the chips on a wafer, which can greatly reduce manufacturing costs. Its benefits include a smaller package footprint compared with conventional lead frame or laminate packages, medium to high I/O count, maximum connection density, and desirable electrical and thermal performance. It also offers a high performance, power-efficient solution for the wireless market [2].

4.1.1 eWLB (Embedded Wafer-Level BGA) Technology

eWLB technology is addressing a wide range of factors. At one end of the spectrum is the packaging cost along with testing costs. Alongside, there are physical constraints such as footprint and height. Other parameters that were considered during the development phase included I/O density, a particular challenge for small chips with a high pin count, the need to accommodate SiP approaches, and thermal issues related to power consumption and the device's electrical performance (including electrical parasitic and operating frequency) [3]. Two choices presented themselves: fan-in WLCSP or fan-out wafer-level packaging (FO-WLP). With fan-in WLCSP, the I/O density is limited to the die size. In FO-WLP or eWLB, the interconnection system is processed directly on the wafer, and the I/O density is unconstrained by die size, making it compatible with motherboard technology pitch requirements.

WLCSP was introduced in the late 1990s as a semiconductor package wherein all manufacturing operations were done in wafer form with dielectrics, thin film metals, and solder bumps directly on the surface of the die with no additional packaging [1]. Unlike conventional WLCSP, the first step in eWLB manufacturing is to thin and singulate the incoming silicon wafer. Although this is commonly done for other semiconductor package formats, it has not been practiced for conventional WLCSP. Following singulation, the diced silicon wafers are then reconstituted into a standardized wafer (or panel) shape for the subsequent process steps as shown in Figure 4.2.

The reconstitution process as shown on the left in Figure 4.2 includes four main steps:

1) The reconstitution process starts by laminating an adhesive foil onto a carrier.
2) The singulated die are accurately placed face down onto the carrier with a pick and place tool.
3) A compression molding process is used to encapsulate the die with mold compound while the active face of the die is protected.

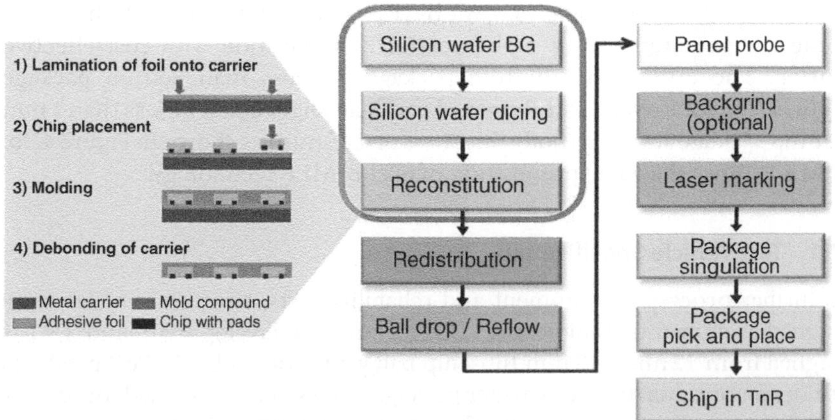

Figure 4.2 The eWLB assembly process flow.

4) After curing the mold compound, the carrier and foil are removed with a debonding process, resulting in a reconstituted wafer where the mold compound surrounds all exposed silicon die surfaces.

The eWLB process is unique in that the reconstituted wafer does not require a carrier during the subsequent wafer-level packaging processes. The implementation of this process flow into 300 mm diameter reconstituted wafers has been described in detail in previous publications [2].

4.2 eWLB-MLP (Mold Laser Package-on-Package) Technology

The continued demand for a higher level of integration has led to the industry's adoption of 3D packaging technologies and, in particular, the package-on-package (PoP) configurations. This technology allows for vertical integration of the memory package and the logic package into one stacked package. The top package is primarily a memory module including some combination of flash and SDRAM, while the bottom package typically contains the logic die, which is a baseband or an application processor. Top and bottom packages are connected via the pads that are located on the topside of the bottom PoP package, and these pads are used to connect the top PoP (memory module) ball grid array (BGA) solder balls to the bottom PoP package. There are various PoP package types including bare die PoP, embedded solder on pad (eSOP) PoP, and laser-via PoP that have proliferated to meet the increasing market demand.

eWLB-MLP technology is a new combination of FO-WLP (eWLB) and molded laser PoP, which is used in conventional flip-chip substrate packaging. eWLB-MLP has unique advantages due to the adoption of merits of each of the two technologies, bringing significant advantages in profile and cost compared with current PoP technologies. eWLB-MLP is designed to meet the lower profile PoP requirement for mobile or tablet application with cost-effective solution. The eWLB-MLP bottom package has less than 500 μm package height, so the total eWLB-MLP stacked package height could be less than 1 mm after top package stacking (body thickness of 0.45 mm) as shown in Figure 4.3b. Table 4.1 shows the value proposition of eWLB-MLP technology.

4.2.1 Test Vehicle Specification

For further process development and reliability tests, two test vehicles were designed as shown in Table 4.2. TV2 is for process development, and TV1 is designed from 12 mm × 12 mm flip-chip ball grid array (fcBGA) PoP products with design optimization for smaller package body size. TV1 is used for further reliability tests with ball shear and OS (open-short) tests (Figure 4.4).

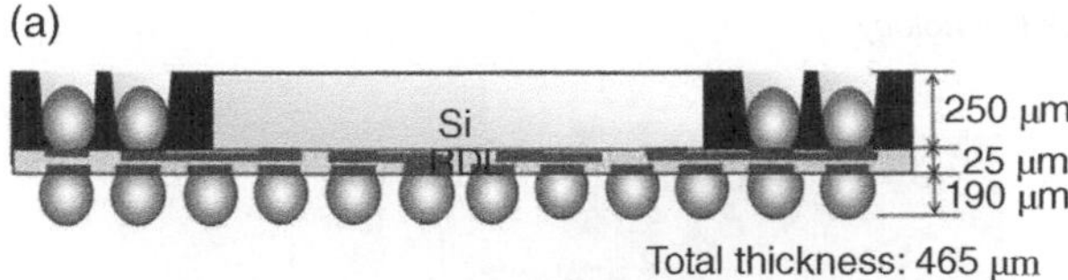

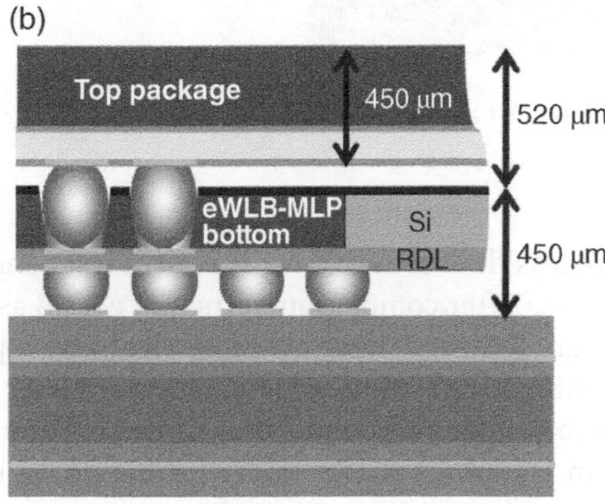

Figure 4.3 Schematics of package structure of eWLB-MLP. (a) eWLB-MLP bottom package and (b) example of stacked eWLB-MLP with top package of 450 μm body thickness (total thickness is less than 1.0 mm).

Table 4.1 Value proposition of eWLB-MLP.

1) Thin POP solution (PoP height is <1 mm)
2) Low warpage during solder reflow cycles
3) Larger Si die cavity
4) Flexibility in memory interface
5) High routing density: line and space widths (L/S) = 10/10 (μm)
6) Compatible with ELK (extreme low-k dielectric devices)
7) Good thermal performance (Θ_{JA} 18–22[°C W^{-1}], Θ_{JB} 3–7[°C W^{-1}] for 12 mm × 12 mm eWLB-MLP)

Table 4.2 eWLB-MLP test vehicles specification.

	TV1	TV2
Package body size (mm)	10 × 10	12 × 12
Die size (mm^2)	64	36
Top ball pitch (mm)	0.4	0.4
Bottom ball pitch (mm)	0.4	0.4
Ball size (mm)	0.25	0.25
Body thickness (mm)	0.25	0.25

Figure 4.4 Micrograph of eWLB-MLPs: (a) TV1 and (b) TV2.

4.2.2 Assembly Process Flow

Figure 4.5 shows a schematic process flow of eWLB-MLP. First, a test vehicle is assembled using the eWLB process. After completion of the full eWLB assembly process, MLP process steps are followed: laser ablation, cleaning, top ball attachment, reflow and flux cleaning, singulation, and final testing. Laser ablation was carried out to form vias for solder interconnects as shown in Figure 4.6. The SEM cross-sectional view in Figure 4.7 shows solder formation on laser-ablated via holes.

The tapered via shape with a lager top diameter and smaller bottom diameter helps to achieve stable solder ball loading during solder filling and stable solder heights for uniform PoP SMT stacking. A residue-free Cu surface of the PoP interface lands is critical for solder wetting and PoP stacking electrical continuity. There was no visible mold compound residue or contamination on the Cu interface lands after optimization of the pad cleaning process after laser ablation.

4.2.3 Component-Level Reliability

Table 4.3 shows package-level reliability results for next-generation 3D eWLB package. They passed Joint Electron Device Engineering Council (JEDEC)

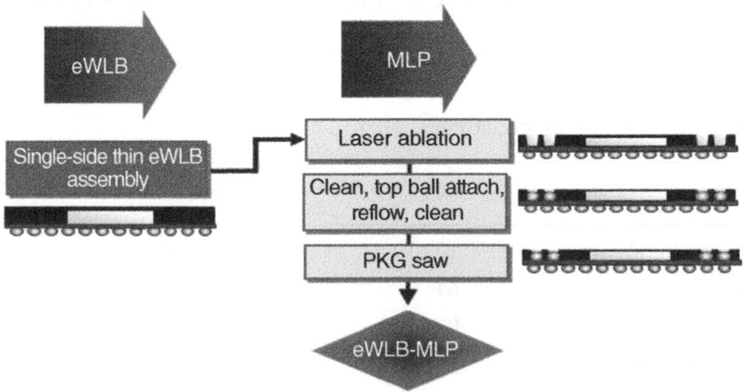

Figure 4.5 Assembly process flow of eWLB-MLP.

Figure 4.6 Micrograph of top view of eWLB-MLP, after laser ablation process to form via holes in eWLB.

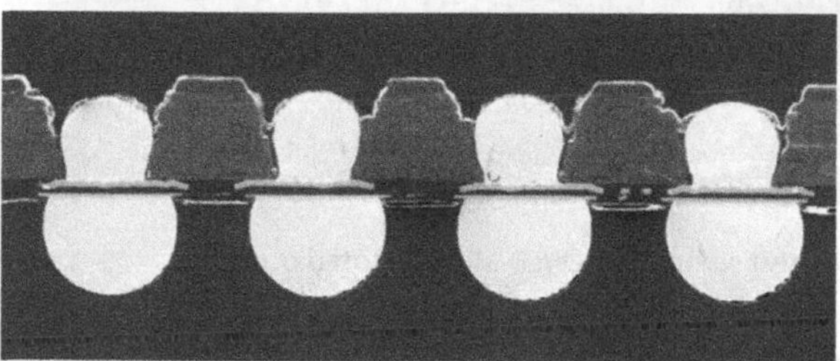

Figure 4.7 SEM micrograph of a cross-sectional view of eWLB-MLP, after solder filling of laser-ablated via holes.

Table 4.3 Package-level reliability results of eWLB-MLP packages.

Reliability test	JEDEC	Test condition	Readout	Results
Unbiased HAST (w/MSL)	JESD22-A118	130 °C, 85%RH	168 h	Pass
Temperature cycling (TC-B, w/MSL)	JESD22-A104	−55 °C/125 °C; 2 cycle h^{-1}	1000×	Pass
High temp. storage (HTS)	JESD22-A103	150 °C	1000 h	Pass

standard package reliability tests such as moisture sensitivity level (MSL) 1 with Pb-free solder conditions. Test vehicle (TV1) has 10 mm × 10 mm eWLB-MLP. It successfully passed all industry standard package-level reliability with ball shear and OS tests.

4.2.4 Board-Level Reliability

For board-level reliability tests, eWLB-MLP (stacked with top package) was assembled and mounted on PCB as shown in Figure 4.8. For PoP assembly, 500 μm high eWLB (body thickness is 250 μm) top packages were assembled, and the total eWLB-MLP stacked package was 750–770 μm high, as shown in Figure 4.9. These samples were tested in JEDEC TCoB and drop reliability test conditions.

Figures 4.10 and 4.11 show Weibull plots of 10 mm × 10 mm eWLB-MLP board-level reliability of JEDEC TCoB and drop tests. The test vehicles were

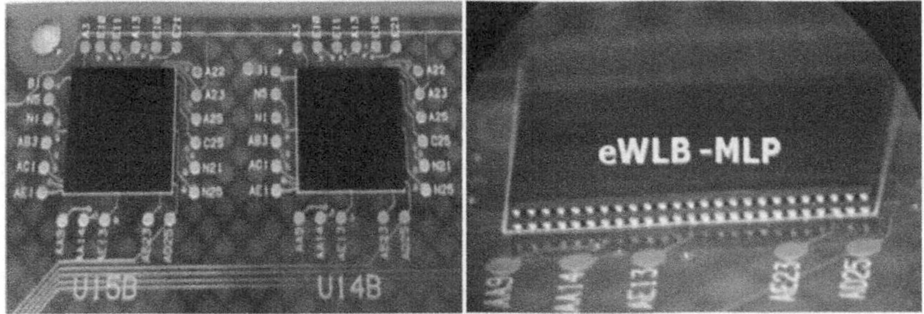

Figure 4.8 Board-level reliability test samples of eWLB-MLP mounted on PCB.

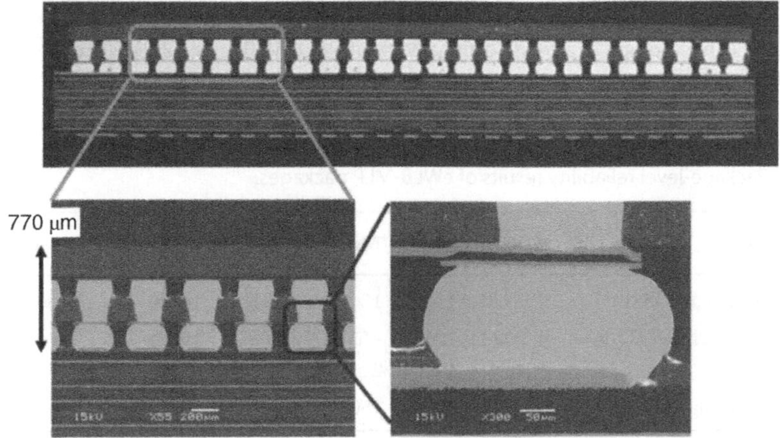

Figure 4.9 SEM micrographs of eWLB after stacking 500 μm high eWLB top package. Total package height is 0.77 mm after mounting on PCB.

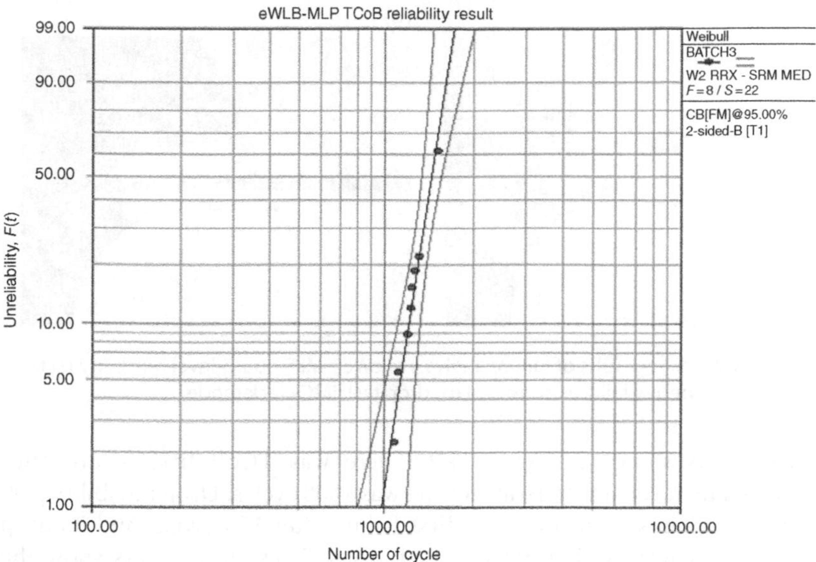

Figure 4.10 Weibull plot of TCoB reliability of 10 mm × 10 mm eWLB-MLP (−40/125 °C, 2 cycles per hour).

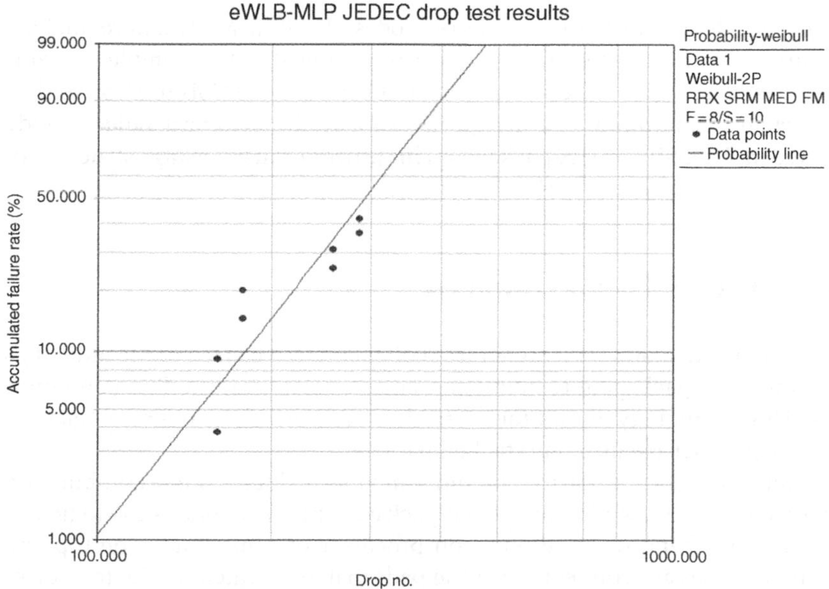

Figure 4.11 Weibull plot of drop reliability of 10 mm × 10 mm eWLB-MLP.

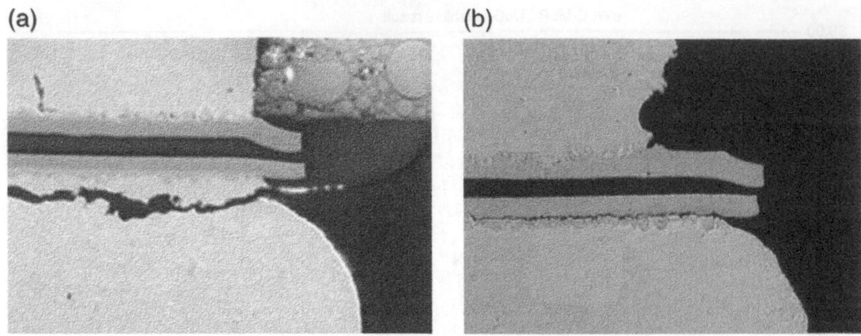

Figure 4.12 SEM micrographs of failure analysis of eWLB-MLP board-level reliability tests. (a) TCoB: solder cohesive failure mode and (b) drop test: IMC crack mode.

PoP stacked as shown in Figure 4.9. There was TCoB first failure after 1000 cycles, and its characteristic lifetime was 1500 cycles. Drop reliability performance was robust and showed a first failure after 150 drops, and its drop reliability characteristic lifetime was 320 drops. These test results show the robustness of board-level reliability of eWLB-MLP.

4.2.5 Failure Analysis of eWLB-MLP After Board-Level Reliability Tests

In order to identify board-level failure modes of eWLB-MLP, failure analysis was carried out by cross-sectional observation after standard sample preparation procedure. Figure 4.12 shows solder cohesive (for TCoB test) or solder-intermetallic compound fracture (drop test) as the standard failure mode reported in WLCSP or BGA packages. This proves robust package structure of eWLB-MLP with thinner package solution.

4.3 3D eWLB-PoP Technology

The continued demand for a higher level of integration has led to the industry's adoption of 3D packaging technologies and, in particular, the PoP configurations. This technology allows for vertical integration of the memory package and the logic package into one stacked package.

The top package is primarily a memory module including some combination of flash and DRAM, while the bottom package typically contains the logic die, which is a baseband or an application processor of some kind. The top and bottom packages are connected via the pads that are located on the topside of the bottom PoP package, and these pads are used to connect the top PoP (memory module) BGA solder balls to the bottom PoP package. There are various

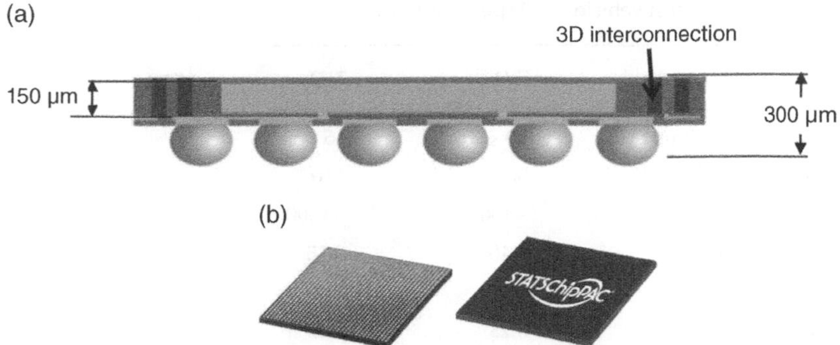

Figure 4.13 (a) Schematics of package structure of 3D ultrathin eWLB-PoP bottom and (b) 3D eWLB-PoP stacked (with top memory package) to a total of 0.8 mm in height.

Table 4.4 Value proposition of eWLB-PoP.

1) PoP packages larger than 15 mm × 15 mm have been enabled using eWLB HVM processes
2) Ultrathin PoP solutions of 0.8 mm total stacked package (300 μm total bottom package thickness with embedded high density vias)
3) Further 0.6 mm PoP total stacked height, with top memory package made thinner in eWLB technology
4) Package meets all component-level and board-level reliability tests per JEDEC standard
5) Enhanced thermal and electrical performance with shorter interconnection length compared with flip-chip or WB solutions
6) Well-controlled warpage for thinner package height
7) Top ball pitch is scalable down to 0.2 mm (~1000 I/O in 16 mm × 16 mm PKG). Prestacked assembly option available for top package

PoP package types including bare die PoP, eSOP PoP, and laser-via PoP that have proliferated to meet the increasing market demand (see Figure 4.13) [4].

3D eWLB-PoP offers significant advantages in thin profiles and lower cost compared with current PoP technologies, particularly for mobile or tablet applications. The 3D eWLB-PoP bottom has a 300 μm package height, enabling the total stacked PoP height to be less than 0.8 mm after top memory package stacking (body thickness of 0.40 mm). Table 4.4 shows the value proposition of 3D eWLB-PoP technology.

4.3.1 3D eWLB-PoP Test Vehicle Specification

For further process development and reliability tests, two test vehicles were designed as shown in Table 4.5. Both packages were used for further component and board-level reliability tests with ball shear and OS tests. TV2 was

Table 4.5 eWLB-PoP test vehicles (TV) specifications.

	TV1	TV2
Package body size (mm)	10×10	15×15
Die size (mm^2)	50	110
Ball IO	~400	~1000
Top ball pitch (mm)	0.40	0.35
Bottom ball pitch (mm)	0.40	0.40
Ball size (mm)	0.25	0.25
Body thickness (mm)	0.20	0.20

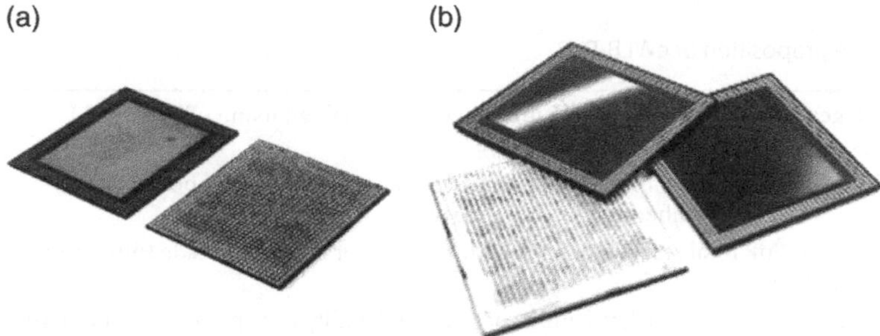

Figure 4.14 Micrograph of 3D eWLB-PoP: (a) TV1 and (b) TV2.

used for thermal characterization with thermal die assembly. In addition, the 28 nm fab-node functional devices of TV1 and TV2 (see Figure 4.14) were assembled for electrical functional characterization and compared to a flip-chip PoP (fcPoP) [5].

4.3.2 Component-Level Reliability of 3D eWLB-PoP

Table 4.6 shows the component-level reliability results of 3D eWLB-PoP. The TVs passed JEDEC standard package reliability tests such as MSL 3 with Pb-free solder conditions. The test vehicles (TV1/2) were 10 mm × 10 mm and 15 mm × 15 mm 3D eWLB-PoP. Both successfully passed all industry standard package-level reliability with ball shear test and OS test.

4.3.3 Experimental Thermal Characterization of 3D eWLB-PoP

For thermal characterization, the test vehicle was prepared with thermal die. The test vehicle specification was the same as for TV2 in Table 4.5. In this study, the die thickness effect was studied with three different die thicknesses:

Table 4.6 Package-level reliability results of 3D eWLB-PoP.

Test	Test condition	Test conditions	Readout	Test results
Unbiased HAST (w/MSL3)	JESD22-A118	130 °C/85% RH	168 h	Passed
TC, temp. cycling	JESD22-A104	Ta = −55/+125 °C 1000 cycles	1000×	Passed
HTSL, high temp. storage life	JESD22-A103	Ta = 150 °C 1000 h	1000 h	Passed

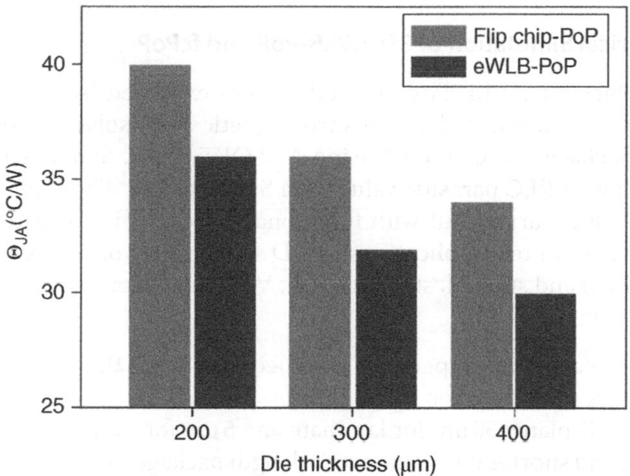

Figure 4.15 Experimental thermal characterization data for 3D eWLB-PoP with different die thicknesses compared with fcPoP.

200, 300, and 400 μm. The same die sizes for fcPoP were prepared for the comparison study.

All test vehicles had thermal die with a transistor and heating circuit block as well as a temperature sensor to more easily detect the temperature at the hot spot of the die with applied power. After SMT on a thermal test PCB, 2.0 W of power was applied and measured junction temperature with various die thickness. As shown in Figure 4.15, eWLB-PoP has an 8–10% thermal performance improvement for the same die thickness compared with fcPoP. eWLB can use thicker die than fcPoP for embedding, achieving a >20% improvement in thermal performance with the same package height of eWLB-PoP compared with fcPoP.

4.3.4 Electrical Functional Characterization of 3D eWLB-PoP

TV1 and TV2 were assembled with live devices from 28 nm low power foundry technology. After assembling the 3D eWLB-PoP, final functional test, bench test, and system level test (SLT) were performed with existing test infrastructure including test hardware of fcPoP. A room temperature test and a hot test at 110 °C were carried out. Both TV1 and TV2 passed SLT testing and all stress tests (MSL3, TC, HTS), as shown in Table 4.6.

Test data shows that 3D eWLB-PoP performance is equivalent or slightly improved compared with fcPoP solutions. Multiple retests did not result in cracking, demonstrating the mechanical robustness of low profile 3D eWLB-PoP.

4.3.5 Parasitic Electrical Simulation of 3D eWLB-PoP and fcPoP

The RLC parasitic values for eWLB-PoP and fcPoP were extracted by computer simulation using a commercial 2D electromagnetic field solver. The S parameter of each package was extracted using ANSOFT HFSS. Simulated results were compared with RLC parasitic values and S parameters. The simulation modeling design was carried out with functional devices to investigate package-level performance in real applications. In 3D simulation works, a few critical pins were selected and studied, such as clock, VDD, and data pins.

In this simulation work:

1) High-speed bus with each trace impedance matched (e.g. to 50 Ω).
2) Trace lengths were typically between 2 and 3 mm.
3) Trace distance to GND plane: 50 μm for laminate and 5 μm for eWLB.
4) eWLB is a very thin and shorter interconnection length package. As a result, the cross talk in eWLB is typically much lower, by more than 10 dB.

Table 4.7 illustrates the simulated results that were compared with RLC parasitic values and S parameters. For eWLB, it was reported as a >60%

Table 4.7 Electrical parasitic values of *R*, *L* of eWLB-PoP and fcPoP at 1 GHz.

	Inductance, *L* (nH)			Resistance, *R* (mΩ)		
Net	**Flip chip**	**eWLB**	**Δ (%)**	**Flip chip**	**eWLB**	**Δ (%)**
1	1.77	0.43	−76	240	67	−72
2	2.03	0.24	−88	308	42	−86
3	1.51	0.57	−62	348	112	−68
4	1.08	0.25	−77	268	66	−75

reduction of inductance and resistance compared with flip chip, mainly due to its shorter interconnection length with thin film RDL without bump or organic substrate.

4.3.6 Board-Level Reliability of 3D eWLB-PoP

For board-level reliability tests, eWLB-PoP was prestacked with top memory and mounted on the PCB. For PoP assembly, 0.4 mm body thickness FBGA top packages were assembled separately with a standard wire-bonding process and finally prestacked on an eWLB-PoP bottom package. The total eWLB-PoP stacked package height was less than 0.8 mm after SMT on PCB (see Figure 4.16). These samples were tested in JEDEC TCoB and drop reliability test conditions.

Table 4.8 shows 3D eWLB-PoP board-level reliability of JEDEC TCoB and drop test results of test vehicles 1 and 2 (Table 4.5 and Figure 4.14). The first

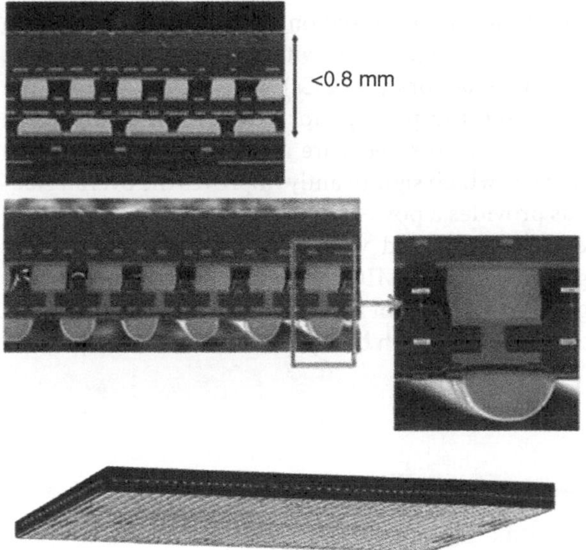

Figure 4.16 Micrographs of a cross section of a 3D eWLB-PoP stacked after top package attachment.

Table 4.8 Board-level reliability test results of 3D eWLB-PoP.

Test	Test condition	Test conditions	Test results
TCoB	JESD22-A103	−40/125 °C, 500 cycles	Passed
Drop test	JESD22-B111	1500G, 100 drops	Passed

TCoB failure was after 1000 cycles. Drop reliability performance was robust and showed no failure after 300 drops. These test results show the robust board-level reliability of 3D eWLB-PoP.

4.4 3D eWLB SiP/Module

FO-WLP in a 3D configuration has received considerable customer interest for memory and advanced application processors by virtue of the higher routing density and form factor reduction. The requirement for SiP integration is also a growing trend for advanced application processors, MEMS, and sensors in IoT and WE as way to cost-effectively achieve advanced silicon die partitioning for increased performance and integration in a reduced form factor [6, 7].

Figure 4.17 shows a 3D eWLB SiP module that has a number of discrete components in the top package and is prestacked on the bottom eWLB-PoP to form a 3D SiP module with a thin package profile with a total height of 1.0 mm. Twelve discrete inductors and capacitors (MLCCs) were removed from the motherboard and relocated in the top package for a reduction in the space required on the motherboard. These discretes are also more power effective when they are close to the device, which significantly improves the overall electrical performance as well as provides a power-saving advantage.

Functional test samples were prepared with a mobile processor and a power management integrated circuit (PMIC) as shown in Figures 4.18 and 4.19, respectively. For Figure 4.18, a 15 mm × 15 mm eWLB-PoP was assembled as described, and a thin substrate with bump was attached on top of the

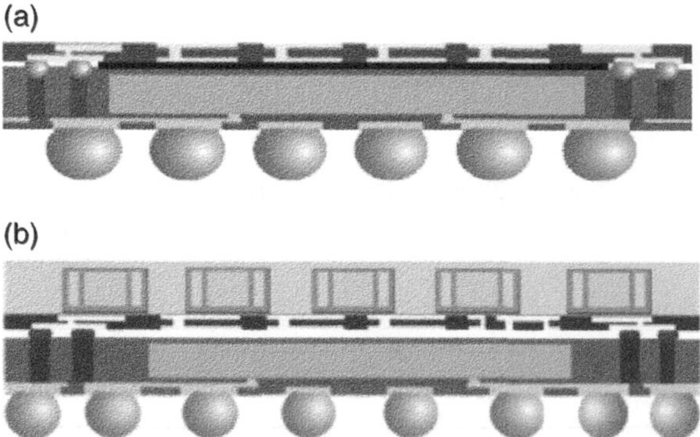

Figure 4.17 Schematics of 3D eWLB SiP (a) with interposer and (b) discretes on interposer or top package of discretes.

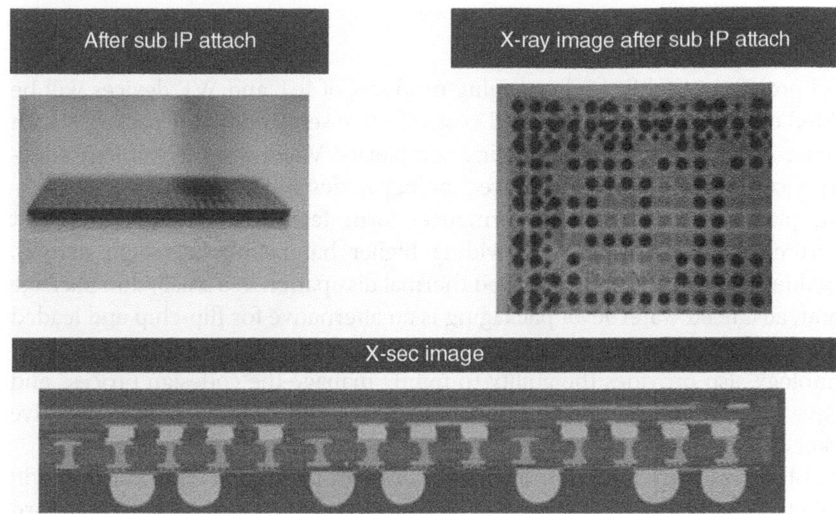

Figure 4.18 Micrographs of 3D eWLB SiP with interposer.

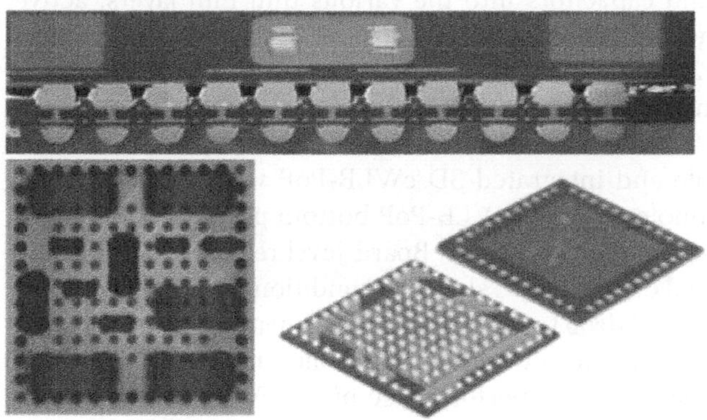

Figure 4.19 Micrographs of a 3D SiP eWLB-PoP with discretes on an interposer or a top package of discretes.

eWLB-PoP [8]. The total height was less than 0.5 mm. In addition, the Figure 4.17b concept was demonstrated as shown in Figure 4.19. It was a 6 mm × 6 mm package size with a 4 mm × 4 mm Si die and 12 discretes on top. This 3D eWLB SiP demonstrated more attractive power efficiency performance compared with conventional packaging, and it is representative of a significantly smaller packaging solution that is well suited for IoT or WE devices.

4.5 Conclusions

Rapid growth of mobile and emerging products of IoT and WE devices will be enabled only by more compact and cost-effective semiconductor packages with increased performance and packaging complexity. Wafer-level technology effectively accommodates new foundry technology nodes and provides a strong packaging platform to address performance, form factor, integration, and cost requirements. In addition to providing higher bandwidth, ultrahigh density, embedded capabilities, and improved thermal dissipation in a small, thin package format, advanced wafer-level packaging is an alternative for flip-chip and leaded packages and is becoming an attractive choice in the evolving market. FO-WLP technology also provides the ability to tightly manage the codesign process and achieve silicon design optimization, which is critical in ultra-cost-sensitive markets.

eWLB technology is an important wafer-level packaging solution that will enable the next generation of mobile applications. The advantages of standard fan-in WLCSPs, such as low assembly cost, minimum dimensions and height, and excellent electrical and thermal performance, are equally true for eWLB. The differentiating factors with eWLB are the ability to integrate passives like inductors, resistors, and capacitors into the various thin film layers, active/passive devices into the mold compound, and achieve 3D vertical interconnections for new SiP and 2.5D and 3D packaging solutions. Advanced eWLB technology will play an important role in the new wave of IoT and WE devices today and in the near future.

Advanced low profile and integrated 3D eWLB-PoP was developed using eWLB (FO-WLP) technology. A 3D eWLB-PoP bottom passed JEDEC standard component-level reliability conditions. Board-level reliability tests of pre-stacked PoP were carried out in JEDEC standard conditions and showed robust reliability in both TCoB and drop tests. Thermal and functional electrical characterizations were also carried out with thermal die and 28 nm functional devices and showed the enhanced performance of 3D eWLB-PoP compared with conventional fcPoP.

Advanced 3D eWLB-PoP and eWLB SiP technology provides more smaller form factor performance value add and is proving to be a new 3D or SiP packaging platform that can expand its application range to various types of emerging mobile, IoT, and WE applications, including sensors and MEMS or automotive applications.

References

1 Brunnbauer, M., Fürgut, E., Beer, G., and Meyer, T. (2006). Embedded wafer level ball grid array (eWLB). *Proceedings of 8th Electronic Packaging Technology Conference*, Singapore (10–12 December 2009).

2 Pitcher, G. (2009). Good things in small packages. *Newelectronics* (23 June 2009), pp. 18–19.
3 Yoon, S.W., Caparas, J.A., Lin, Y., and Marimuthu, P.C. (2012). Advanced low profile PoP solution with embedded wafer level MLP (eWLB-MLP) technology. *Proceedings of Electronic Component Technology Conference*, USA.
4 Eslampour, H., Lee, S., Park, S. et al. (2010). Comparison of advanced PoP package configurations. *Proceedings of ECTC 2010*, Reno, NV.
5 Chen, K., Chua, L., Choi, W.K. et al. (2017). 28nm CPI (chip/package interactions) in large size eWLB (embedded wafer level BGA) fan-out wafer level packages. *Proceedings of Electronic Component Technology Conference*, USA.
6 Choi, W.K., Na, D.J., Yong, A. et al. (2014). Ultra fine pitch RDL development in multi-layer eWLB (embedded wafer level BGA) packages. *Proceedings of Electronic Component Technology Conference*, USA.
7 Yoon, S.W., Petrov, B., and Liu, K. (2015). Advanced wafer-level technology: enabling innovations in mobile, IoT and wearable electronics. *Chip Scale Review* (May/June 2015), pp. 54–57.
8 Lin, Y., Kang, C., Chua, L. et al. (2016). Advanced 3D eWLB-PoP (embedded wafer level ball grid array – package on package) technology. *Proceedings of Electronic Component Technology Conference*,USA.

5

NEPES' Fan-Out Packaging Technology from Single die, SiP to Panel-Level Packaging

Jong Heon (Jay) Kim

nepes Corporation

5.1 Introduction

nepes Corporation's fan-out wafer-level packaging (FO-WLP) technology was first introduced in the year 2010 on a 300 mm platform, which is based on the redistributed chip package (RCP) licensed from NXP (formerly Freescale Semiconductor) using a chip-first, face-down concept, which is similar to most conventional FO-WLP platforms in the industry [1–5].

Further development has been made, enhancing the process robustness and high volume mass (HVM) production capability for automotive products, mobile applications, and system solutions with multiple devices embedded and integrated. nepes' FO-WLP system-in-package (SiP) solution offers 40–90% volumetric shrink from existing modules with flexible product design to end user [6]. Numerous active or passive components could be embedded and connected in 2D or 3D via connections to the backside of an FO-WLP package for package-on-package (PoP) structures typically designed for communication modules and system control applications.

5.2 Structure and Process Flow

As shown in Figures 5.1 and 5.2, nepes' FO-WLP is based on chip-first technology. Embedded ground planes (EGPs) (Cu material) are placed onto adhesive tapes laminated on substrates in advance. This EGP, however, is optional depending on the necessity and the package design. Incoming wafers, regardless of size, 6, 8, 12, or even 18 in. (in the future), are subjected

Advances in Embedded and Fan-Out Wafer-Level Packaging Technologies, First Edition.
Edited by Beth Keser and Steffen Kröhnert.

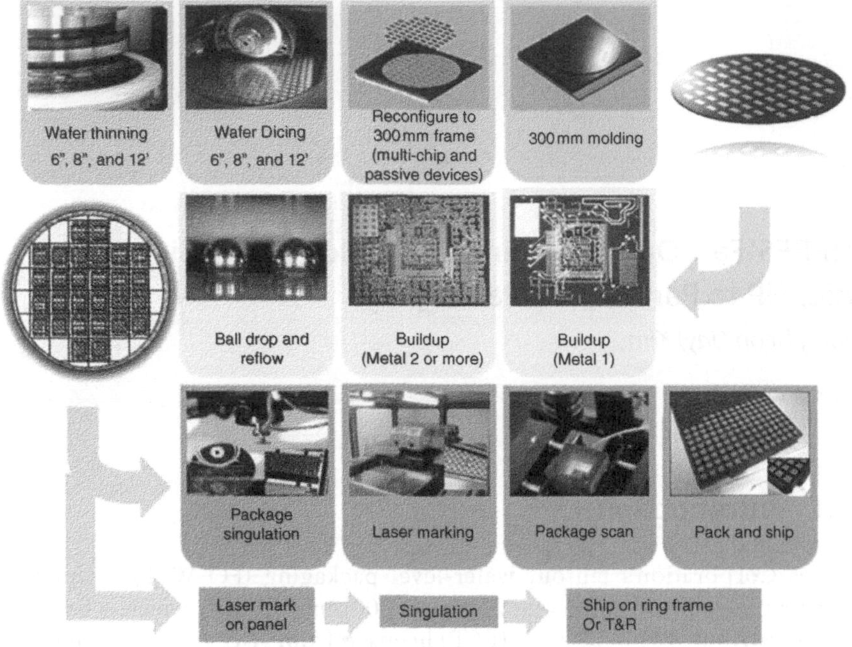

Figure 5.1 Fan-out wafer-level packaging process flow.

to backgrinding and dicing to segregate individual die. The die are subsequently picked and placed onto the substrate. Optional passives, integrated passive devices (IPD), and discrete components could be placed as well. Epoxy compounds are then used to encapsulate the EGP and die in the form of 300 mm wafers by either printing or compression molding. This process is known as panelization. The molded panel is attached to a carrier, which is reusable for subsequent process. Redistribution layers (RDLs) comprising polymeric dielectrics and metallization are then built onto the wafers using tools similar to back-end silicon wafer processing. The number of layers of metallization is dictated by device complexity and governing FO-WLP design rules. Upon RDL completion, ball grid array (BGA) balls are placed, or bumps are formed to be used for the next level of interconnection. Finally, the wafers are sawn into individual packages, laser-marked for product traceability, and shipped to the customer [7].

The value proposition of nepes' FO-WLP includes PoP larger than 14 mm × 17 mm enabled in HVM, four metal layer capability with ~100 μm die to die spacing realizing a volumetric shrink of 40–90% for SiP, heterogeneous integration of several active die and more than a hundred passives with robust design and process control, a thin package profile of ~250 μm including solder

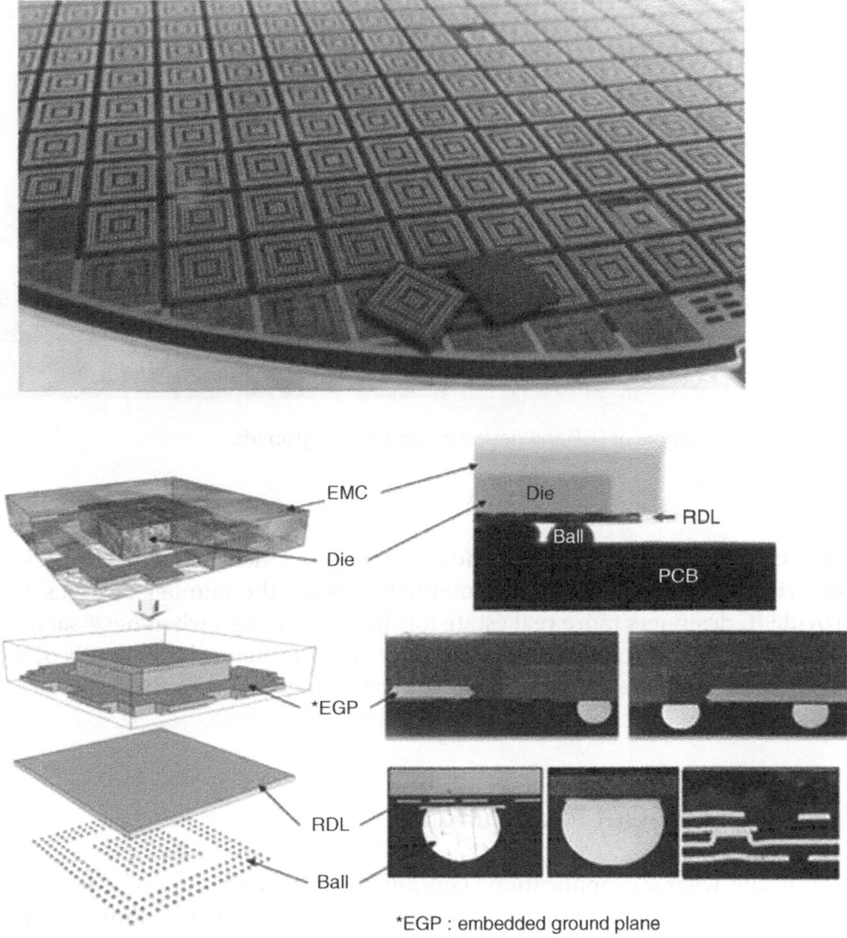

Figure 5.2 Schematic and picture of FO-WLP.

ball for PoP, low contact resistance between silicon die and RDL, enhanced thermal and electrical performance with EGP, and a via frame structure available for the top package in 3D integration.

In addition, one key distinctive feature in nepes' FO-WLP is embedded metal frame, known as EGP. This frame was designed initially to localize die drift during the encapsulant cure step, which controls the die drift within the pocket of the EGP locally. As the EGP is copper material, it allows for better heat dissipation when connected to a chip or exposed to the environment through the package backside after backgrinding. Furthermore, EGP could also enhance package electrical performance through various designs and shapes. As shown in Figure 5.3, EGP can be partitioned to perform ground, VCC (voltage

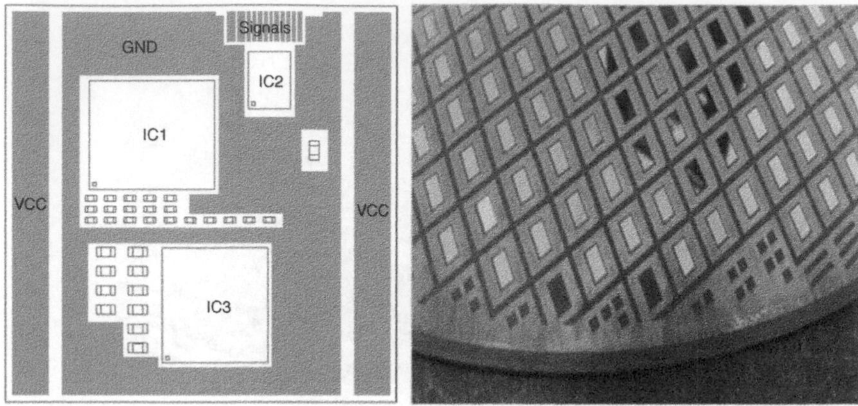

Figure 5.3 An example of EGP and design for signals and grounds.

connected between ground and the collector), and electromagnetic interference (EMI) shielding that could potentially reduce the number of RDLs and provide IC designers more real estate for their chip design when necessary. Of course, EGP as a metal frame also reinforces the mechanical robustness of the package, assisting the handling of the thin panel together with the process carrier attached on the backside of the panel while processing.

5.3 Thin Fan-Out Packaging

Mobile and wearable applications continue to drive reduction of Z height of the assembly due to more dense and complex integration within limited space. Another added advantage and perhaps a more critical one of thin profile packaging is the enhanced thermal performance. nepes Corporation's continual development of thinner packages is one key focus of the packaging roadmap. Figure 5.4 shows a two-chip with two-RDL FO-WLP device produced in nepes. Package dimension is 10 mm × 10 mm × 0.28 mm with more than 500 balls with PoP structure incorporated using a via frame.

Figure 5.4 Thin profile FO-WLP (two die, PoP supportive).

Figure 5.5 FO-WLP with double-sided RDL.

5.4 Double-Sided Fan-Out Packaging

For further flexibility to adapt various formats of top package, backside (top) RDL, which is called double-sided RDL, has been introduced (Figure 5.5). The bottom package has a 0.4 mm thin profile including solder ball, where 10 μm/10 μm line and space RDL has been applied. This technology will allow the main chip in the bottom package to stay faceup so that the top device and bottom package could communicate more efficiently electrically or optically. Table 5.1 shows a package reliability test of the fan-out package with double-sided RDL and an embedded vertical structure. The test vehicle (bottom) was 6 mm × 6 mm × 0.35 mm and passed Joint Electron Device Engineering Council (JEDEC) standard package reliability tests such as MSL2.

5.5 Via Frame (VF) Fan-Out Package

Via frame fan-out package (VF-FOP) is shown in Figure 5.6 as a solution developed for 2D and 3D SiP and module packages using a printed circuit board (PCB)-based via frames structure. This technology enables chip face-up packages for sensor application and package stacking for 3D integration as well.

In principle, the via frame is an interconnection media designed with through holes filled with conductive material where one side has pads to connect to chips while the other side has metal bumps, which will be connected to other packages through PoP. Via frames can be designed in various forms matching the ball layout and size and number of balls of the upper package.

As shown in Figure 5.7, the process to provide through vias within the FO-WLP base includes using a laminate-based via unit with top and bottom electrodes as a critical element in SiP and 3D interconnections. Materials and processes for VF-FOP were adopted from the PCB industry but with added proprietary features for a fan-out PoP structure. As shown in the panelization process (the left of Figure 5.7), conductive balls or bumps were adopted on the other side of the via frame, which needs precise temperature control throughout the entire fan-out package process. These via frame units are embedded during the chip attachment process. FO-WLP panels having an epoxy-based substrate with conductive through vias (via frame) will need to be coplanar with both top and bottom surfaces of the package as well as solderable

Table 5.1 The package-level reliability of FO-WLP with double-sided RDL.

Test mode	Test condition	Sample size	Sampling plan	Results (pass/fail)	Result	Ref. document
Precon	Bake: 24 h@125(−0, +5) °C	90 ea	Visual inspection	0/90	Passed	JESD22-A113F:2008
T/C	500 cycle/−55(+0, −10)↔125(+15, −0) °C	55 ea	:All	0/55	Passed	JESD22-A104D:2009
PCT	96 h/121 °C/100%	77 ea	CSAM :11 ea/item	0/77	Passed	
uHAST	96 h/130 °C ± 2 °C, 85 ± 5% RH/230 kPa	55 ea	Cross section	0/55	Passed	JESD22-A118A:2011
HTS	1000 h/150 °C	77 ea	:2 ea/item	0/77	Passed	

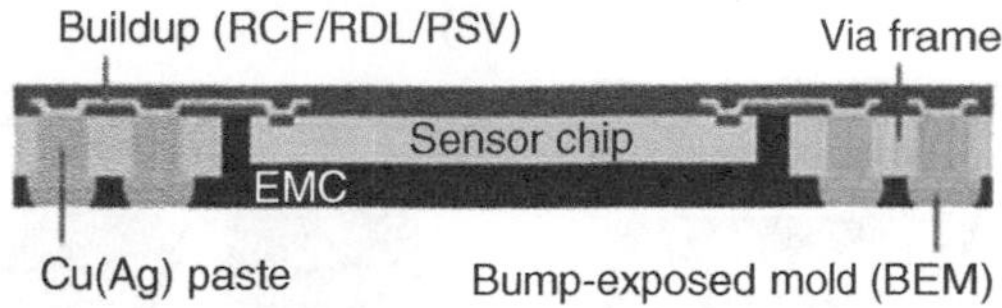

Figure 5.6 Typical structure of VF-FOP.

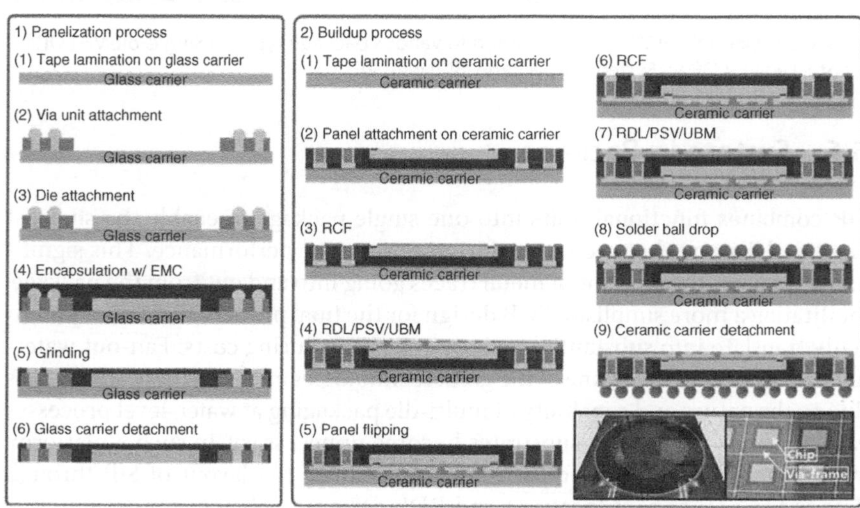

Figure 5.7 Process flow of VF-FOP.

surfaces (on package top). As panels contain semiconductor devices that might have different thicknesses, panels would require delicate control during backgrinding to expose the electrode on one side of the via frame [7]. After panelization, embedding chip and via frame, the post process is almost identical to a typical FO-WLP or wafer-level package (WLP) process like dielectric and metal redistribution.

3D stacking process for some package configuration using VF-FOP was also developed as shown in Figure 5.8. Having metal balls on one side of the VF-FOP, a more advanced and complex SiP could be manufactured with bottom package for PoP (Figure 5.8c). Such SiP has multiple active dies and up to a hundred passive components with a via frame surrounding all devices that are interconnected to one another. The via frame also acts as a 3D connection to the package top (with backside). Via frame pads are subsequently exposed through epoxy mold compound using a panel backgrind process step. Figure 5.8 shows several types of VF-FOP for different product application. As shown in Figure 5.8a, VF-FOP is able to face chip-up (opposite to solder ball side) with certain selected chip surface area to be exposed or protected by very thin dielectric layer.

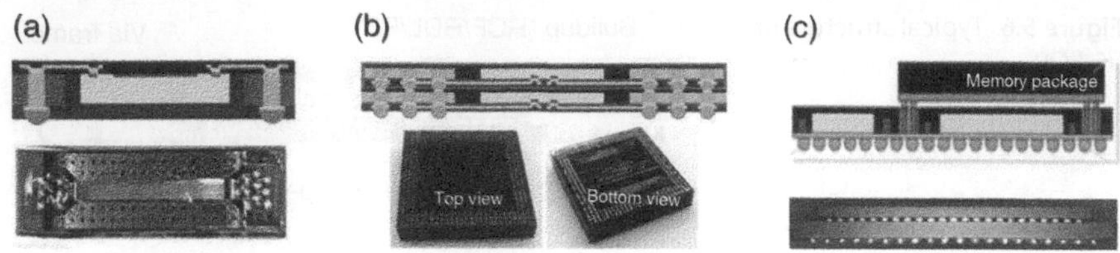

Figure 5.8 Example of VF-FOP application to various package type. (a) Single-die VF-FOP. (b) Stacked VF-FOP. (c) System-in-package (SiP) with VF.

5.6 System-in-Package

SiP combines functional units into one single package to enable the shortest electrical distance between parts for more superior performance. This significantly reduces the amount of metal traces going into and out from the package, facilitating a more simplistic PCB design for the final product that could potentially translate into substantial savings in manufacturing costs. Fan-out wafer-level SiP (FOWL-SiP) is one of the great technologies enabling these advantages due to the nature and capability of multi-die packaging at wafer-level processes where bumps, wires, and substrates become unnecessary. In such a platform, system designers only need to tune and optimize the layout of SiP through device and component locations and RDL designs. This reduces design cycle time significantly in the development stage so that time to market-in will be much shorter than others. However, in order to adequately support this, FOWL-SiP needs to have the capability to build multilayer RDL to minimize or completely eliminate the use of additional substrates. It is also essential to understand the behaviors of various components during fan-out processing through electrical and reliability data.

PoP takes this integration a step further, placing one package on top of another, allowing greater integration complexities and interconnect densities. PoP also enables procurement flexibility, lower cost of ownership, better total system and solution costs, and faster time to market. Scalability is another PoP key advantage because PoP body sizes comply with JEDEC standards, so customers can maintain control of memory procurement, enabling multiple suppliers, supplier qualifications, and certifications with minimal inventory vulnerability. Customers can choose from a wide range and combinations of memory such as flash, PSRAM, SDRAM, or DDR. With the understanding of this requirement, it is possible to expand the applications of FOWL-SiP, PoP solutions such that they can truly serve as subsystems.

The FOWL-SiP process flow is almost identical to single-die package except for multiple embedded devices and repeated RDL processes. However, in order to truly realize SiP in FO-WLP platforms, it is crucial to understand different components' (passive, active) drift behaviors and also the criticality of robust

RDL designs to pass stringent environmental reliability tests [1, 7]. The angle of the RDL crossing from component to component, device to EMC interface at die edge to control surface topography, surface condition controls of devices to prevent the resin bleed during the molding process, adhesion control between multi-RDL to multi-dielectric through PVD, and photo process parameter controls are just few of the critical factors for high volume manufacturing. On top of that, package backside quality also needs to be well controlled for subsequent assembly processing of PoP [8].

Typical processes involve chip bonding, via frame bonding, and passive components bonding (not necessarily in the same order) onto temporary glass substrates followed by molding (either print mold or compression mold). This is followed by iterations of dielectric and RDL processes before finishing with ball drop and panel backside grind to expose terminals for PoP assembly. Figure 5.9 includes examples of images from panel through final assembly. One could notice very dense and complicated RDL structures (dielectric removed on purpose) in the center picture of Figure 5.9. The package (Table 5.2) passed commercial and industrial grade package reliability with less than 50 μm package-level warpage over the entire reflow profile range. It has three-die comprising processor, memory, and power chips with 109 passive components [9, 10]. As illustrated in Figure 5.10, all of the active devices and components are integrated and embedded in one single package, representing a true modular level package. Robust process controls and capabilities mentioned earlier are the key prerequisites in order to enable such highly integrated SiP eliminating the use of substrate modules.

Multi-RDL capability is also utilized to build inductor coils on the package surface as shown in Figure 5.11. Three inductors (2 of 180 nH, 1 of 48 nH) were built on 9 mm × 9 mm FO-WLP utilizing two layers of Cu spiral trace, and the inductance is achieved well with ±5% tolerance range

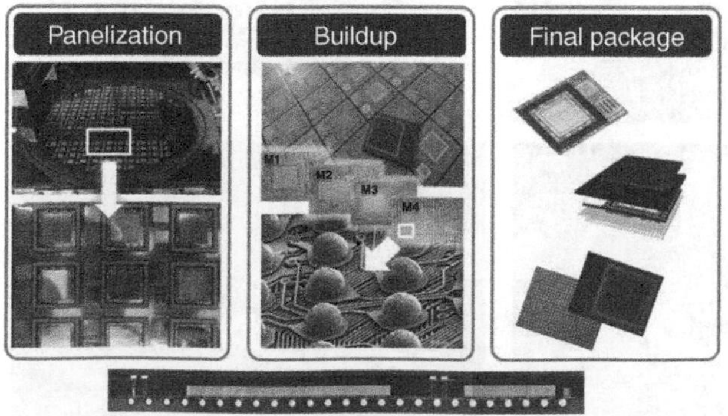

Figure 5.9 Process images of FOWL-SiP.

Table 5.2 Package structure details of FOWL-SiP and package reliability test results.

Item	Description	Specification	Remark
POD (bottom)	PKG dimension (mm^3)	14 × 17 × 1.0	Top package information • 12 × 12 mm^2
	Pitch/IO	0.65 mm/500 pins	• Pitch/IO: 0.4 mm/216 pins
Key feature	Though via pitch	0.4 mm/0.25 mm (dia)	For PoP interconnection
	Topography	Max 10 μm	Component terminal thickness control
	Component (inch)	Min 01005	Min 0.4 × 0.2 mm^2
	Contact resistance	<10 mΩ	For PMIC die
	L/S	15/20 μm, 4 metals	
	Warpage (in package)	<50 μm	Room to peak 260 °C
Condition		**Sample size (units)**	**Remarks**
MSL 3 Precon		240	Passed
Temp. cycle (−40 to125 °C)		80	500 cycles passed
Temp. cycle (0–100 °C)		80	1000 cycles passed
THB (85 °C, 85% RH)		80	504 h passed
HTS (150 °C)		80	504 h passed

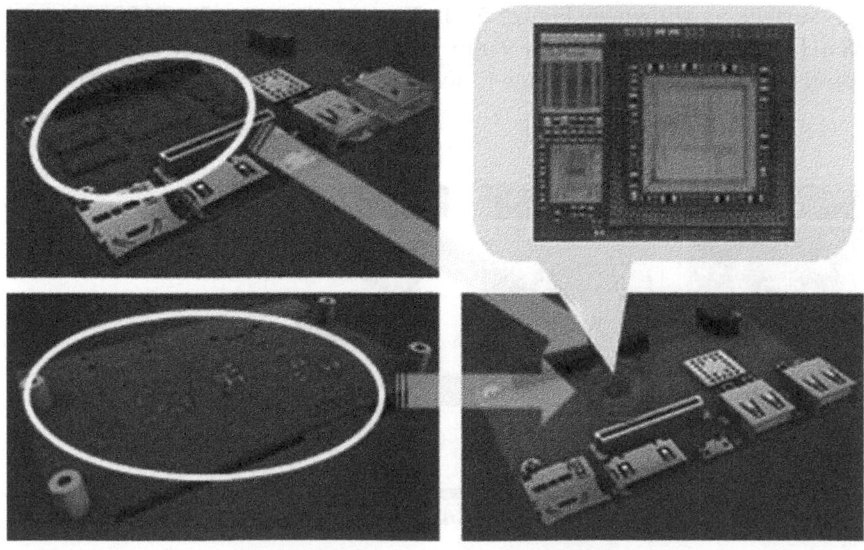

Figure 5.10 Image of an actual module product built with FO-WLP technology.

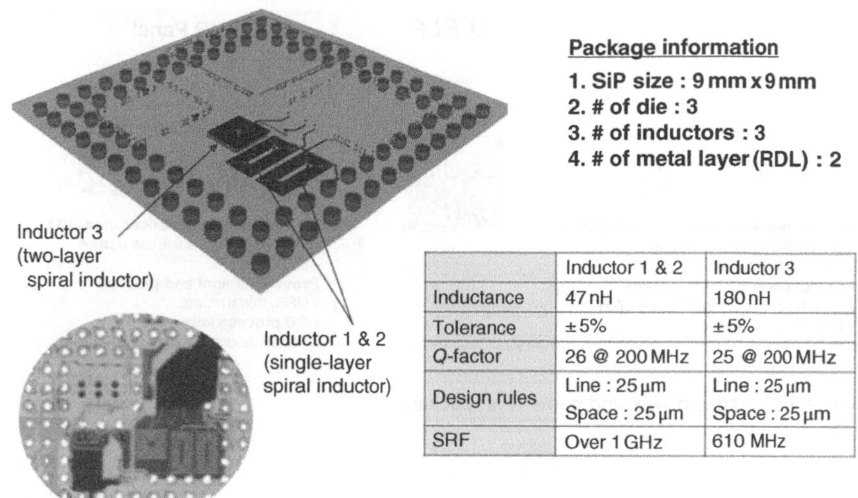

	Inductor 1 & 2	Inductor 3
Inductance	47 nH	180 nH
Tolerance	±5%	±5%
Q-factor	26 @ 200 MHz	25 @ 200 MHz
Design rules	Line : 25 μm Space : 25 μm	Line : 25 μm Space : 25 μm
SRF	Over 1 GHz	610 MHz

Figure 5.11 Inductor on multi-die fan-out WLP with multi-RDL.

for the target value. The quality factor (*Q* factor) is higher than 25 at 200 MHz, implying the inductor is well designed.

5.7 Panel-Level Package

FO-WLP built in large-scale panels, called panel-level package (PLP), is designed for direct cost reduction from 300 mm scaled FO-WLP. PLP maximizes the benefits of fan-out packaging, which resolves the input and output (I/O) limitation of WLP and offers SiP capability with shorter system development cycle time, flexibility of system design due to RDL routing with no substrate, and simple logistic flow in the entire supply chain at large scale. Key processes and technologies of PLP are leveraged from FO-WLP. The key challenges in fan-out packaging need to be known first before scaling up. On the other hand, the panel format of PLP requires a different approach of handling, materials, and equipment. Considering all this and the capital expense, starting PLP development is very challenging. The prime motivation for starting PLP development at nepes Corporation is the broad technology experience in WLP, FO-WLP, and more importantly also in large panel manufacturing of liquid crystal display (LCD) product. With years of experience in both FO-WLP and LCD processing including touch screen panel for mobile product using fourth-generation LCD equipment, 600 mm × 600 mm PLP technology has been developed and demonstrated in a mass production platform (Figure 5.12).

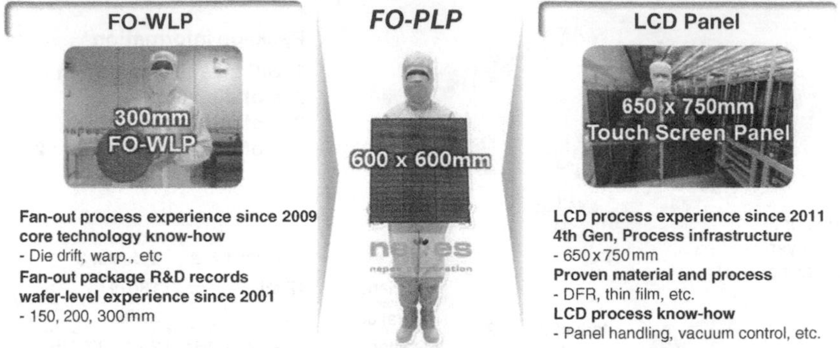

Figure 5.12 Fan-out WLP and panel-level package.

5.8 Performance and Reliability

5.8.1 Thermal Performance

The effect of EGP demonstrated by the simulated thermal performance of a package structure is shown in Figure 5.13 using the FloTHERM analysis tool. Conditions include JEDEC still air enclosure of X : 304.8 mm, Y : 304.8 mm, Z : 304.8 mm with test board of 114.3 mm × 76.2 mm × 1.6 mm, which has four layers with Cu coverage of Trace 1 (20%)/Trace 2 (90%)/Trace 3 (90%)/Trace 4 (20%). The test is based on JEDEC standards JESD51-2/JESD51-8/JESD51-14.

Simulation results showed 25% better thermal performance in package with EGP compared with no EGP structure as summarized in Figure 5.14. The hot spot zone is reduced as heat is more efficiently transferred externally since the EGP carries heat more from the chip through RDL than the no EGP structure. This implies nepes' fan-out package (RCP) has better heat dissipation ability than a typical FO-WLP.

❖ **Modeling structure**

Item		Specification
Package	Dimension	6.0 mm × 6.0 mm × 0.4 mmt
	Ball pitch/dia.	500 µm / ϕ300 µm
	Ball I/O	90 ea
Die dimension		2.3 mm × 2.8 mm × 0.125 mmt
Buildup layer thickness		35 µm
Cooling	TIM thickness	50 µm
	H/S thickness	300 µm

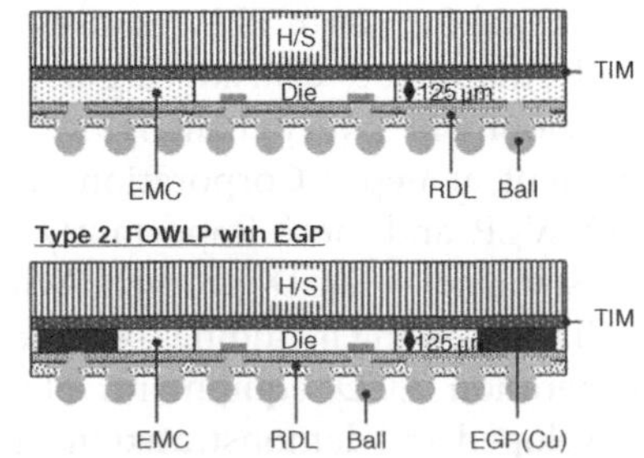

Figure 5.13 Thermal simulation of fan-out WLP on the effect of EGP.

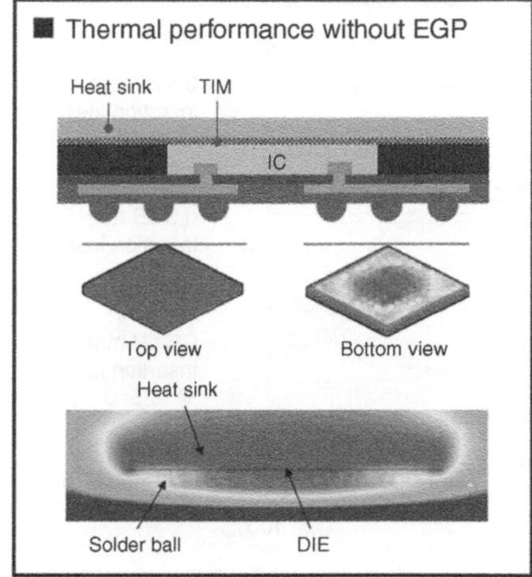

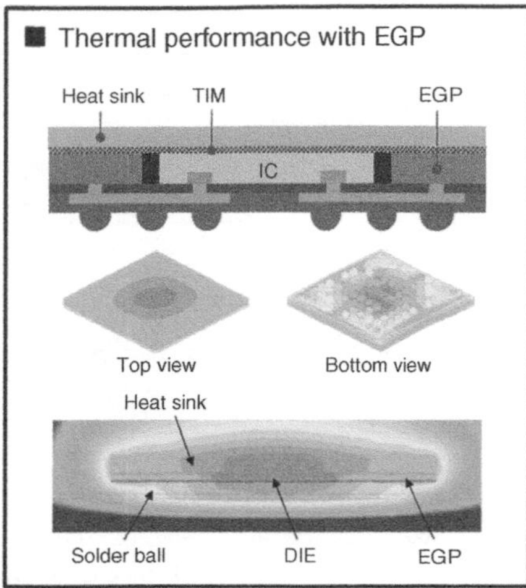

Type	Theta Ja	Theta Jb	Theta Jc	Remark
Type 1	46.0 K W^{-1}	18.37 K W^{-1}	7.13 K W^{-1}	TIM material : polymer
Type 2	34.4 K W^{-1}	13.87 K W^{-1}	6.80 K W^{-1}	TIM material : polymer

cf. Cold plate temp. 20 °C (@Jb & Jc)

Figure 5.14 Comparison of thermal performance between no EGP (type 1) and EGP (type 2). (Absolute value of simulation results may vary depending on the condition or package and chip dimension.)

5.8.2 Electrical Performance of Automotive Radar Package

The requirements for packaged automotive radar solutions include excellent RF isolation, controlled impedance, low insertion loss, low attenuation, good thermal dissipation up to 2 W in an ambient temperature of 125 °C, and fulfilling stringent AEQ-100 G1 reliability criteria. Using the advantages of EGP and robustness of nepes' FO-WLP, this platform was selected for 77 GHz Radar Package used for automotive solution. The package has an EGP size of 6 mm × 6 mm and is shown in Figure 5.15.

The package has been tested over a range of frequencies and temperatures for voltage-controlled oscillators (VCO), transceiver, and receiver as shown in Figures 5.16 and 5.17. It shows extremely low loss of <1 dB, and insertion loss degradation is small across a wide temperature range.

Figure 5.15 Radar sensor fan-out package designed for automotive application.

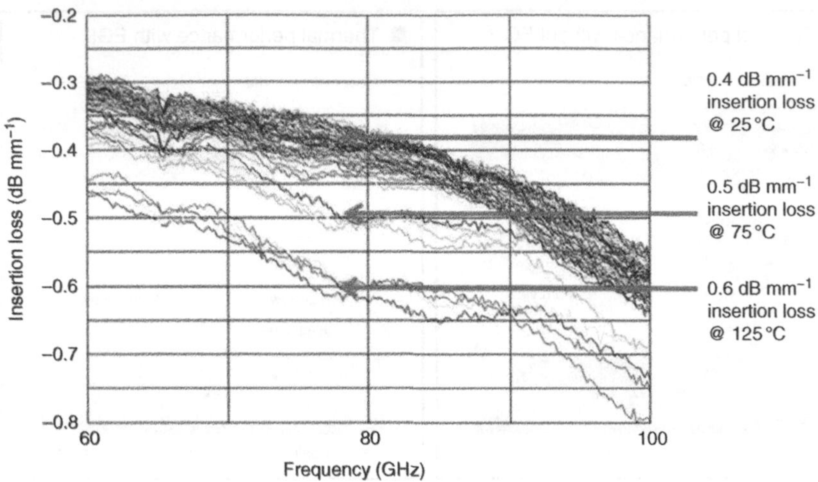

Figure 5.16 Insertion loss test result at various temperatures.

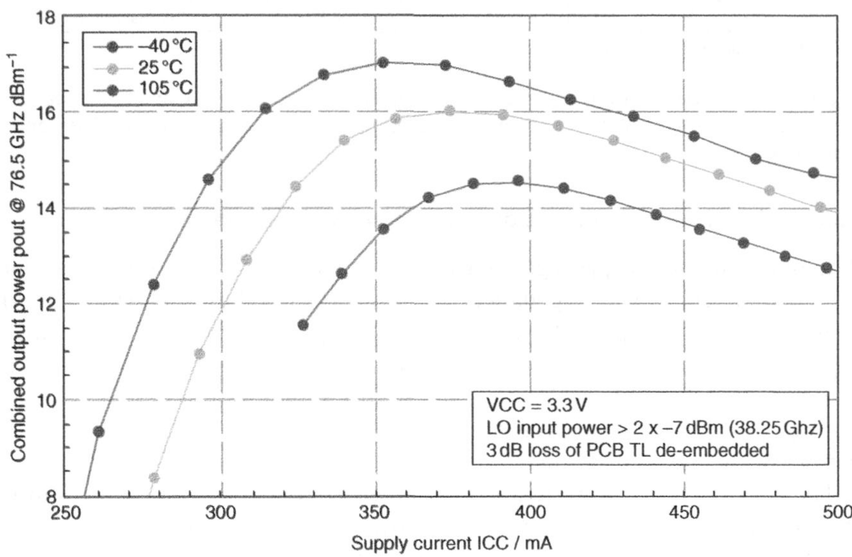

Figure 5.17 Insertion loss test result at 77 GHz.

5.9 Application

New and emerging applications in the consumer and mobile space, the growing impact of the automotive, Internet of things (IoT) and wearable electronics (WE), and the complexities in sustaining Moore's Law have been driving many new trends and innovations in advanced packaging technology. Demand for

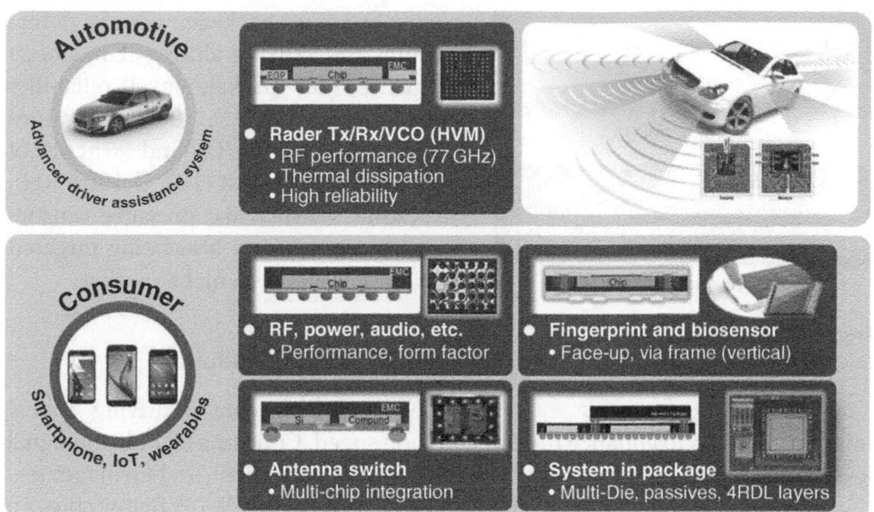

Figure 5.18 The FO-WLP types and application products produced by nepes Corporation.

functional integration in the small and thin package will continue to grow with the requirements for lower cost and power consumption. To develop a disruptive packaging technology capable of achieving these goals is challenging, and FO-WLP is playing important roles in many application areas as shown below. Figure 5.18 shows various application products of FO-WLP from mobile consumer to automotive.

5.9.1 Mobile and Automotive Applications

Automotive IC's are traditionally wire-bond packages. Due to the increasing complexity and higher performance requirements of automotive applications, the packaging industry is moving toward high performance packages like FO-WLP, which provides a smaller form factor and much less interconnection parasitics that are very critical for high frequency applications. Other advantages of wafer-level processing are smaller tolerances, which enable better assembly yield results and a lower cost. The driving forces in the mobile space are always form factor and I/O constraints due to die shrink or high pin count, especially for power management, RF devices, etc.

5.9.2 Sensor Products

Figure 5.19 shows a double-sided RDL with VF-FOP for a fingerprint sensor (FPS) package with a thin profile of less than 0.2 mm thickness. FO-WLP greatly enhances the detectability or sensing ability of sensor devices since the distance from the chip surface to the external surface is shorter than with a

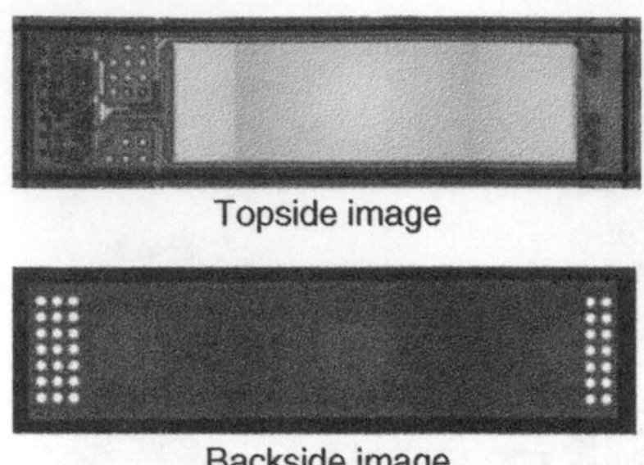

Figure 5.19 Via frame FO-WLP for fingerprint sensor device application.

wire-bonding package, which determines the sensing ability. It has been qualified for package-level reliability conditions of MSL2, 700 temperature cycles (−55 to 125 °C), and 96 hours of pressure cooker test (121 °C/100% RH).

Other sensors like pressure sensors and biosensors are also being targeted and implanted as a niche market.

5.9.3 Optical Module

The era of big data is driving optical interconnections to enhance the transmission speed. Concerns include the signal loss at wire bonding, complicated assembly of fiber alignment, and form factor.

These concerns can be relaxed with key benefits of FO-WLP. Below shows a simple concept of an optical module for high-definition multimedia interface (HDMI) application that may need a smaller version of module for mini or micro HDMI (see Figure 5.20). The package size is small with optical IC and driver IC as well. This technology and concept provide not only a small form factor but also some special features on the package top, which allows easy fiber alignment as well. It leads compact-sized package and also easy assembly providing the module manufacturer with better productivity at lower cost. The package size is 3.50 mm × 3.00 mm and 0.315 mm thick. The chip sizes are 0.27 mm × 1.0 mm, 1.3 mm × 0.9 mm. The structure is face-up with two RDL layers and via frame embedded for vertical interconnection.

5.9.4 IoT and Industrial Applications

An Arduino (Orange Board™) was successfully optimized and redesigned in VF-FOP. Die and components have been assembled on the modules embedded in FO-WLP. This is based on one VF-FOP size of 7.35 mm × 7.35 mm excluding the connector. Figure 5.21 shows the package structure and size

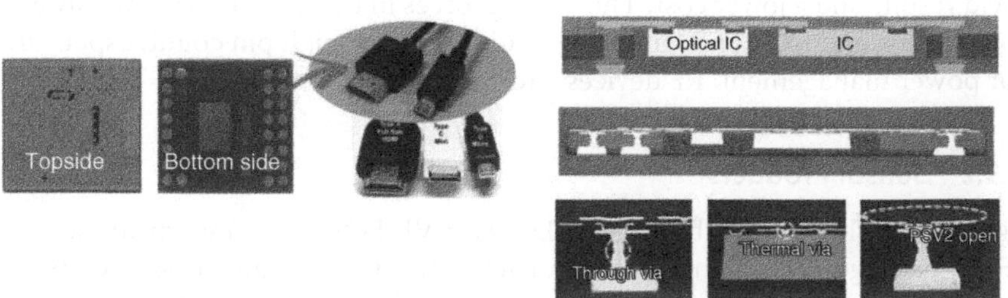

Figure 5.20 Via frame FO-WLP for optical module.

comparison of this new redesigned package against the original Arduino. This small dot-sized Arduino (called DotDuino) is 100% compatible with original Arduino module but with a 90% size reduction. After prototyping this new product, the inventors can commercialize their idea very quickly and efficiently for this new smaller module to the market on various applications as showcased in Figure 5.22. Nepes' FO-WLP roadmap is show in Figure 5.23.

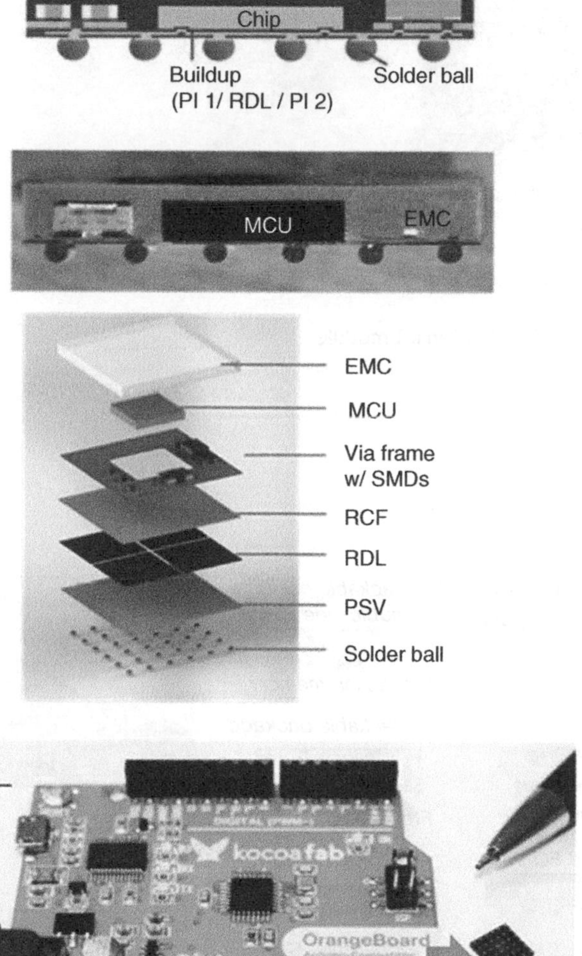

Figure 5.21 Package structure of DotDuino and comparison with Arduino board.

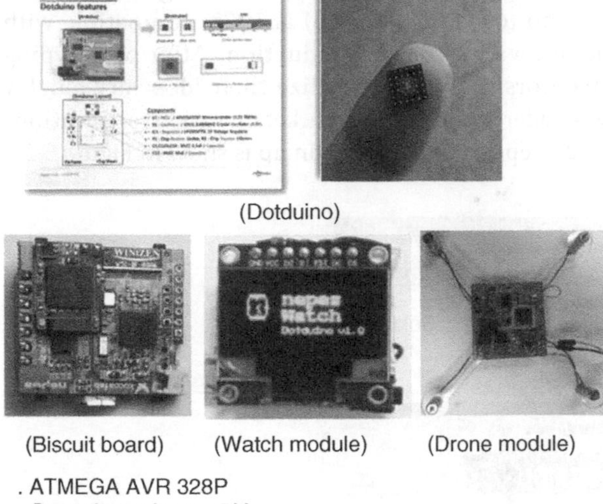

Figure 5.22 Application of FOWL-SiP in IoT module.

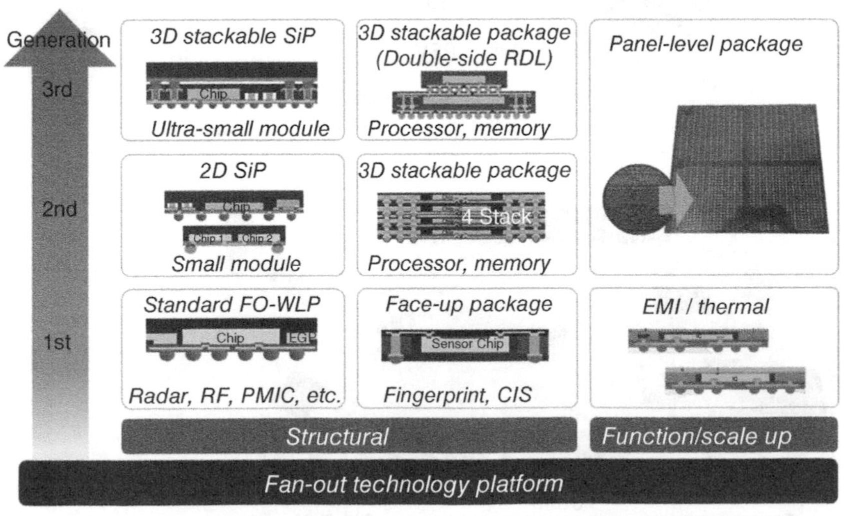

Figure 5.23 FO-WLP lineup and roadmap.

5.10 Roadmap and Remarks

Development focus is driven by markets where form factors and cost are important while new features such as thermal, speed, and integration capability are equally important. As introduced, large-scale PLP has been implemented in high volume manufacturing mode and is in the initial production stage. Special features and materials are applied and implemented to enhance the thermal and EMI shield performance as well. Further reduction in package thickness to less than 100 μm will be implemented.

References

1 Keser, B., Amrine, C., Duong, T. et al. (2007). *Electronic Components and Technology Conference, 2007. ECTC '07. Proceedings. 57th.*

2 *Solid State Technology* (2010). *Fujitsu renews packaging license with Tessera on original terms* (22 September 2010).

3 Techsearch *Advanced Packaging Update-Market and Technology Trends*, vol. 2-0617. Austin, TX: Techsearch.

4 Yole Development (2016). *Fan-Out Packaging: Technologies and Market Trends*. Yole Development.

5 Watkins, J. (February 2015). *Advanced IC Packaging Technologies, Materials, and Markets*, 210–216. New Venture Research Corp.

6 Kim, Jong Heon (Jay) (2015). Fan out WLP technology as 2D, 3D System in Packaging (SiP) solution. *International Wafer Level Packaging Conference (IWLPC)*, San Jose, CA, USA (2015).

7 Kang, I.S. and Kim, Jong Heon (Jay) (2011). 3D SiP solutions with wafer level package technology. *7th International Conference and Exhibition on Device Packaging (DPC)*, Arizona, USA (8–10 March 2011).

8 Lin, Y., Kang, C., Chua, L. et al. (2016). Advanced 3D eWLB-PoP (embedded wafer level ball grid array – package on package) technology. *2016 IEEE 66th Electronic Components and Technology Conference Proceedings*, Las Vegas, Nevada (31 May–1 June 2016).

9 Kim, Jong Heon (Jay) (2016). Fan out wafer level package for high performance and integration application. *The 15th International Symposium on Microelectronics and Packaging*, Suwon, South Korea (2016).

10 Freescale (2015). *Single Chip System Modules – A Disruptive New Set of Products Offered by Freescale*. Austin: Freescale Technology Forum.

6

M-Series™ Fan-Out with Adaptive Patterning™

Tim Olson and Chris Scanlan

Deca Technologies

6.1 Technology Description

M-Series is a chip-first, face-up fan-out wafer-level packaging (FO-WLP) technology with a unique structure wherein the semiconductor device active surface and vertical sidewalls are fully encapsulated within the epoxy molding compound (EMC) with device interconnect enabled by Cu studs through the EMC layer as shown schematically in Figure 6.1a and in actual cross section [1] in Figure 6.1b

The EMC layer between the active region of the chip and the end electronic appliance printed circuit board (PCB) provides an intermediate stress buffer that has proven to significantly extend board-level reliability (BLR) as compared with conventional wafer-level chip-scale packaging (WLCSP) with only polyimide or polybenzoxazole (PBO).

Figure 6.2 shows a JEDEC condition G BLR temperature cycle comparison of a 6 mm × 6 mm semiconductor device in a conventional WLCSP versus the same chip in a 6.25 mm × 6.25 mm M-Series fan-out package with SAC405 solder balls. All samples utilized 0.35 mm pitch with approximately 8% depopulation on the WLCSP devices and approximately 15% on M-Series. The additional LF35 (Sn/Ag1.2/Cu0.5/Ni0.05) solder ball data was included since it is an alternative solder used on WLCSP in drop or shock-sensitive applications. LF35 has not been tested on the M-Series structure.

Adaptive patterning was developed in conjunction with M-Series to overcome the inherent inaccuracies associated with creating a composite wafer or panel consisting of individual devices embedded within EMC or alternate materials. Adaptive patterning includes measurement of each die location within the panel, creation of a complete panel design that perfectly fits the as-built fan-out

Advances in Embedded and Fan-Out Wafer-Level Packaging Technologies, First Edition.
Edited by Beth Keser and Steffen Kröhnert.

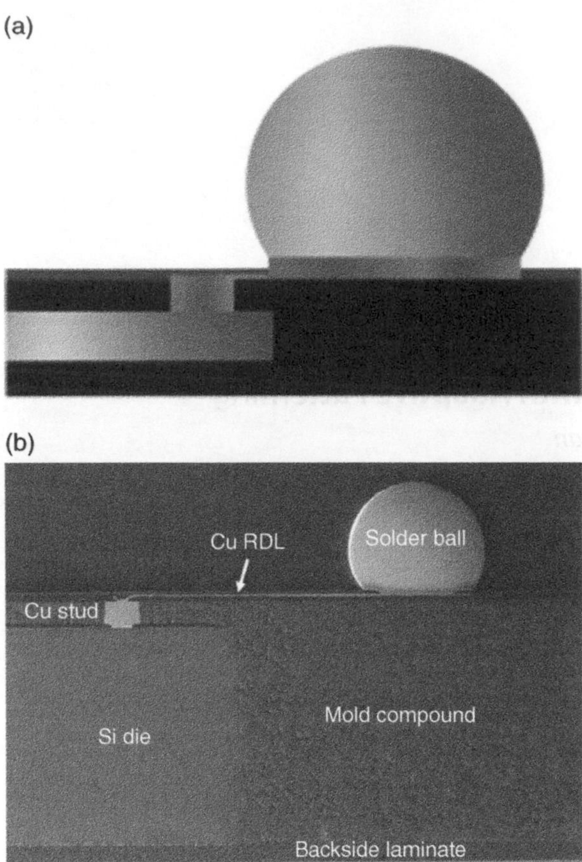

Figure 6.1 (a) Schematic of M-Series FO-WLP. (b) Actual cross section of M-Series FO-WLP.

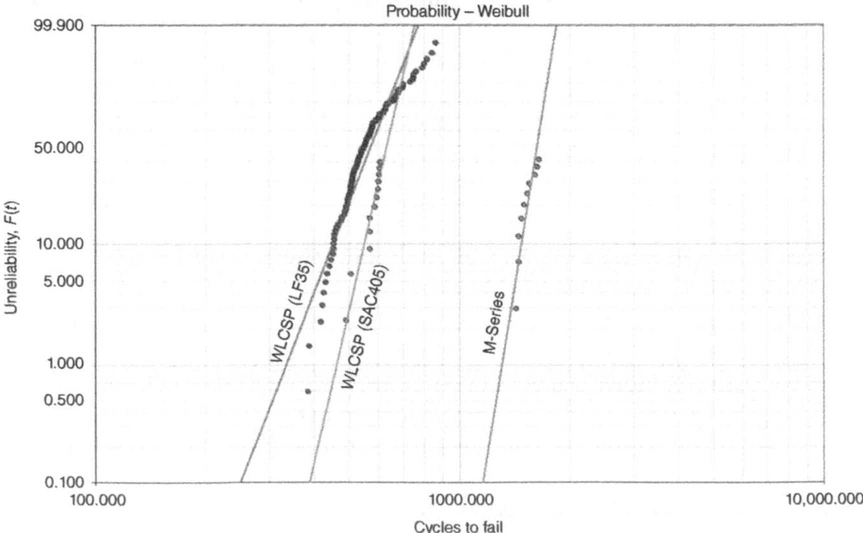

Figure 6.2 BLR comparison of M-Series versus WLCSP.

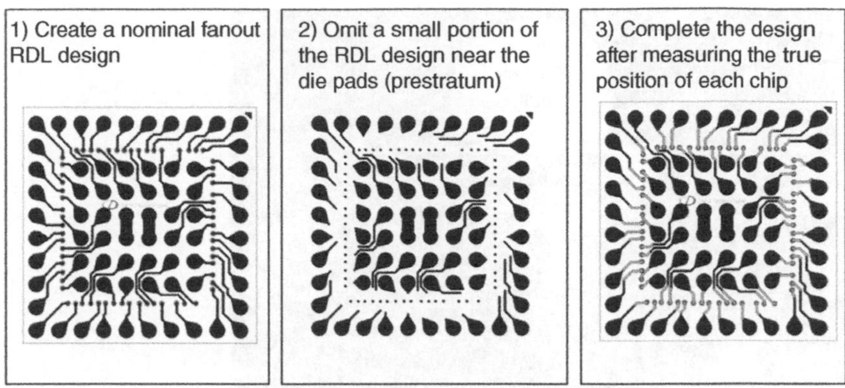
1) Create a nominal fanout RDL design
2) Omit a small portion of the RDL design near the die pads (prestratum)
3) Complete the design after measuring the true position of each chip

Figure 6.3 Adaptive patterning methodology using adaptive routing.

panel, and application of the unique design to each panel. Through adaptive patterning, simple devices can be constructed cost-effectively, while complex and multicomponent devices can be assured of high yields in manufacturing. Figure 6.3 provides a high level overview of the adaptive patterning methodology [2]; a detailed description follows in Section 6.5.

M-Series has been qualified by multiple semiconductor companies with silicon devices fabricated in 150–350 nm geometries on 200 mm wafers as well as 14–65 nm technologies on 300 mm for leading cellular phone handset producers.

6.2 Basic Package Construction

M-Series is fundamentally an embedded chip fully molded fan-out technology that blends the benefits of classic lead frame or laminate packaging in terms of providing complete molded protection to the active semiconductor device while utilizing the massively parallel processing capability of wafer- or panel-level processing to create the interconnect layers. The basic structure with a single layer of RDL (Cu redistribution layer) is shown in Figure 6.4.

As compared with traditional fan-in (WLCSP) and conventional fan-out [3, 4], M-Series provides several important reliability and manufacturability benefits. A primary reliability benefit includes lowering of e-chip-board interaction (CBI) stresses applied to the device [5] as a result of the molded stress buffer layer between the active region of the semiconductor device and the PCB acting in a similar manner to a laminate substrate in a classic flip-chip ball grid array (BGA) package. The M-Series molded stress buffer layer has already proven to provide superior BLR performance exceeding a 200% improvement as compared with traditional fan-in WLCSP technologies as

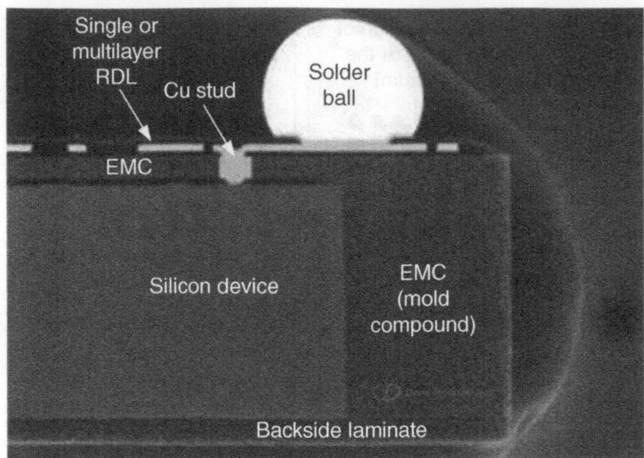

Figure 6.4 Basic package construction of M-Series.

shown in Figure 6.2. Since conventional fan-out technologies utilize organic spin-on polymers as the only stress buffer between the active device and the PCB, similar to fan-in WLCSP, e-CBI stresses are expected to be significantly higher as compared to M-Series.

A second key benefit of M-Series versus conventional fan-out structures is elimination of EMC from the physical vapor deposition (PVD) process while completing first-level interconnect to the aluminum bond pads wherein the process window is typically limited to ensure low contact resistance (R_c). With M-Series, the Cu stud-to-Al bond pad connection is performed on the native silicon wafer, allowing a much wider process window to ensure low R_c prior to dicing and chip attach of the devices into a reconstituted panel. Panel-level interconnects, typically a TiCu PVD seed followed by electroplated Cu, are formed to the Cu surface of the stud, which is opened up during the topside planarization. The Cu-to-Cu interconnect scheme in panel form enables a more robust PVD process with a much wider operating window.

A third benefit of M-Series versus conventional fan-out structures is the embedding of sawn device edges within the EM preventing residual metallic buildup from laser grooving or dicing potentially protruding through the organic spin-on polymer dielectric over the surface of the silicon to EMC interface. An example of metallic buildup protruding through the dielectric on a conventional fan-out structure causing electrical shorting failures at the device edge is shown in Figure 6.5 (courtesy of Advanced Semiconductor Engineering [ASE]).

A fourth benefit of M-Series versus conventional fan-out is the planarized structure that eliminates the discontinuity from the active device surface to the

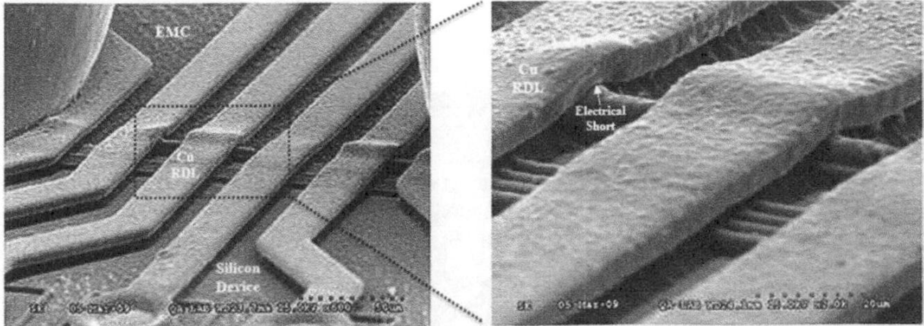

Figure 6.5 Conventional fan-out structure with electrical shorting failure. *Source:* courtesy of ASE.

EMC surface, resulting in nonplanarity of the first dielectric layer. In a structure where two semiconductor devices are placed in close proximity to each other, the resulting topography from silicon surface to EMC surface in conventional fan-out can be significant as shown in Figure 6.6.

The resulting 5.5 μm of dielectric nonplanarity as shown in Figure 6.6 severely limits the ability of scaling conventional fan-out to fine lines and spaces (2 μm and below) given the need for tightly controlled depth of focus (DOF) in the lithography process.

In contrast, the planarized nature of M-Series as shown in Figure 6.7 allows for straightforward scaling to 2 μm and below since planarization is built into the process flow. For extremely fine future submicron lines, a potential planarization enhancement with more advanced polishing technology may be implemented to reduce the current surface R_a (average surface roughness) if required.

M-Series is ideally suited for fan-in WLCSP applications requiring enhanced BLR capabilities as well as for fan-out applications where the package size is

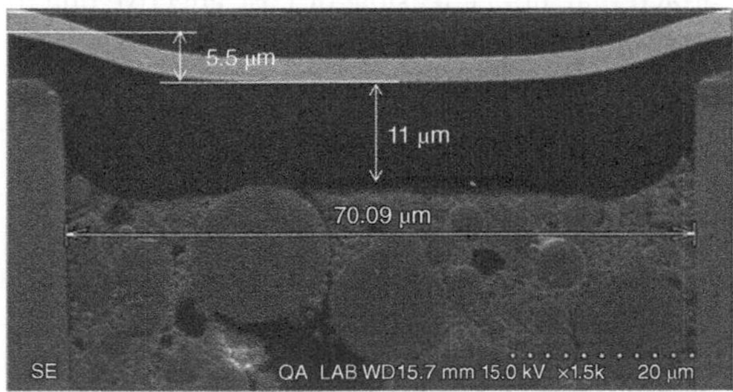

Figure 6.6 Nonplanarity of silicon to EMC surface in conventional fan-out. *Source:* courtesy of ASE.

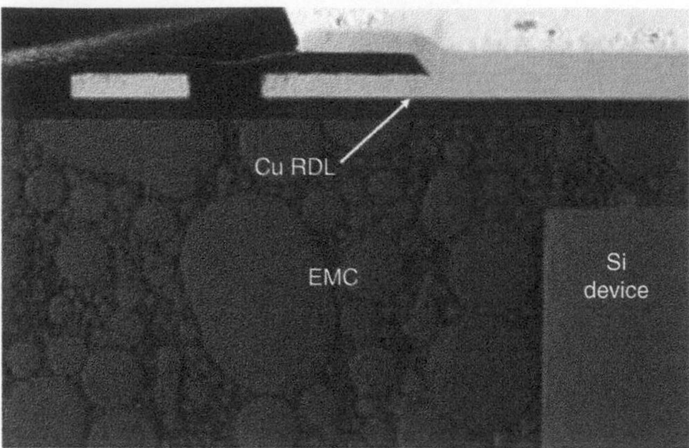

Figure 6.7 Planar surface over silicon device to EMC interface on M-Series.

within approximately two times the area of the chip, essentially a fan-out ratio (package area to die area) of 2.0 or less. While technically capable for much larger fan-out ratios, classic packaging technologies based on laminate or lead frame substrates will likely provide a more cost-effective solution within the coming few years. Over time, it is forecasted that large panel M-Series fan-out will be able to compete on cost with much larger fan-out ratios.

The limits of package size for M-Series have not yet been fully explored as of this writing. However, the fundamental structure and materials set should be extendable for multi-die and component modules approaching 2500 mm^2 such as integration of advanced graphics processors with supporting high-speed, high bandwidth memory chips surrounding their periphery.

M-Series applications as intermediate pitch expanders for large die sizes in advanced silicon nodes are also foreseen wherein device bond pad pitch is so dense that direct flip-chip attach may not be feasible or cost-effective due to limitations in escape routing within organic substrates. In this application, the M-Series package becomes a pitch-expanding pseudo-die that would include micro bumps, enabling flip-chip mounting to a laminate substrate to complete the package-level assembly.

6.3 Manufacturing Process Flow and BOM

M-Series utilizes a semiconductor wafer manufacturing environment including advanced lithography, PVD, electroplating, etching, cleaning, and visual inspection methodologies consistent with fabricating wafer-level packaging

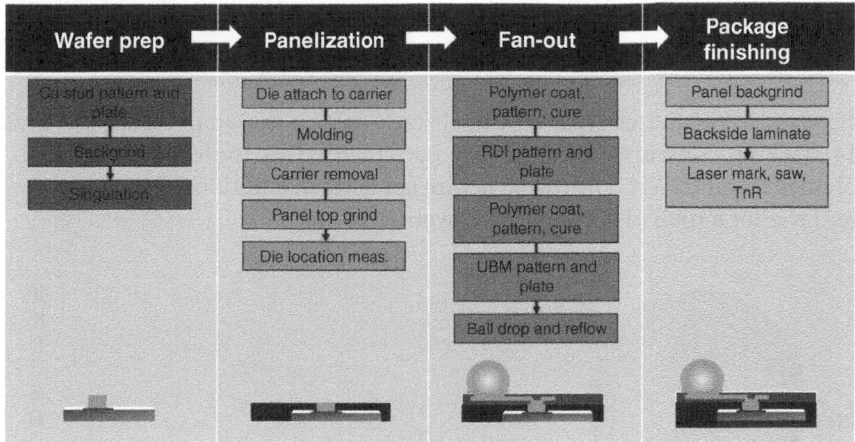

Figure 6.8 M-Series process flow for a fully molded FO-WLP.

as well as the final back-end-of-line (BEOL) interconnect layers within semiconductor device fabs.

The basic process flow for M-Series shown in Figure 6.8 begins with the "wafer prep" segment, which includes the fabrication of Cu studs on the native semiconductor wafer by sputter deposition of seed layers, patterning of a thick photoresist, electroplating Cu studs, performing strip, and etch of the photoresist and seed layers followed by backgrind and singulation to provide a thinned and diced wafer ready for pick and place of each device.

In the second segment, called panelization, the singulated devices are attached faceup to a reusable temporary carrier at the desired fanned-out pitch. The carrier and chips are then overmolded to create a plastic panel, currently in the form of a SEMI standard 300 mm wafer with development underway on a large panel format to be discussed later. The plastic panel is then debonded from the carrier and post-mold cured. Top grind is performed on the panel to reveal the Cu studs and create a highly planar surface. Finally, an optical scanner is used to measure the precise *x*,*y* location and theta rotation of each device on the panel to enable the adaptive patterning technique described in Section 6.5.

Following panelization, wafer-level fan-out processing is performed on the molded panel. This includes patterning and curing one or more polymer layers as well as patterning and electroplating one or more Cu RDL or under-bump metallization (UBM) layers. Ball drop and reflow are then performed.

For package finishing, panel backgrind is performed for final thinning and a backside laminate is applied, followed by laser mark, package saw, and tape and reel.

6.4 Design Features and System Integration Capability

The standard M-Series cross-sectional stack-ups are shown in Figure 6.9. The M4 stack-up contains a total of four patterned layers (polymer 1, Cu RDL, polymer 2, and UBM). The M6 structure contains an additional RDL and polymer layer for a total of six patterned layers.

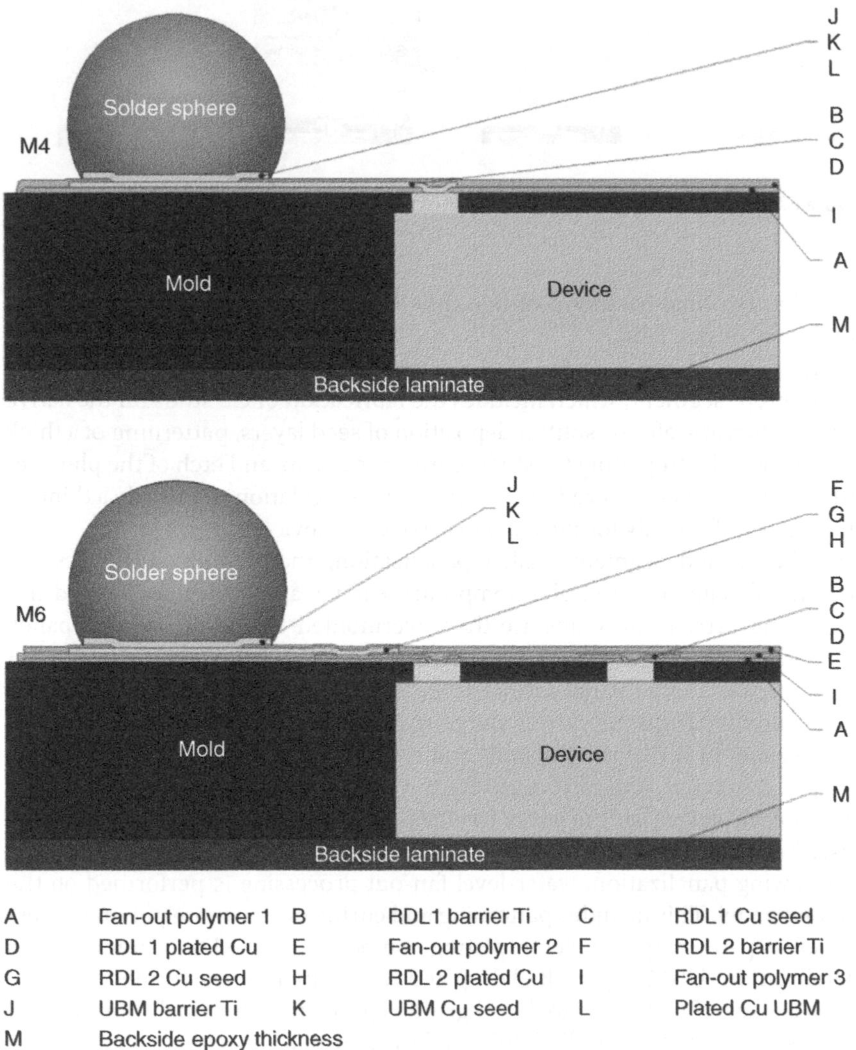

Figure 6.9 Standard M-Series cross-sectional stack-ups.

Several different via stack-up options are possible within the M-Series fan-out buildup structure. In many designs a staggered via structure is employed as showed in Figure 6.10. In this case, the vias connecting Cu RDL and UBM layers are offset with respect to each other.

It is possible to position the UBM and BGA ball directly over a Cu stud. Figure 6.11 shows this embodiment.

Via-in-pad structures with the via connecting the UBM capture pad to the first RDL layer (in the case of the M6 stack-up) or the Cu stud (in the case of the M4 stack-up) are also possible as shown in Figure 6.12.

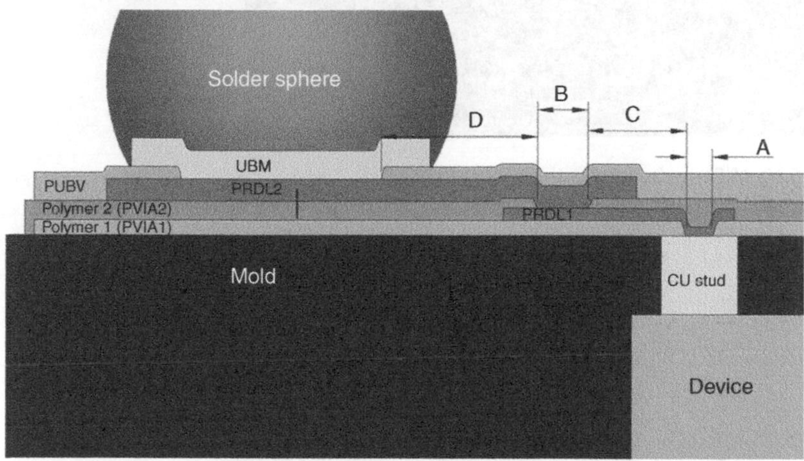

Figure 6.10 M-Series buildup with staggered vias.

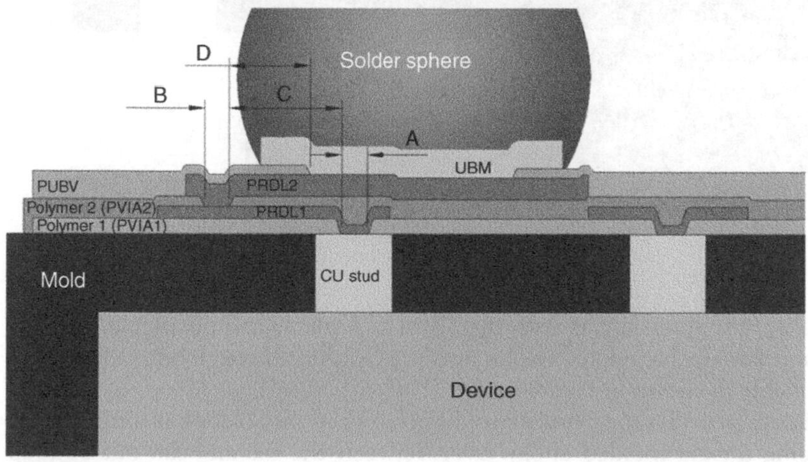

Figure 6.11 M-Series buildup with UBM and BGA ball directly over a Cu stud.

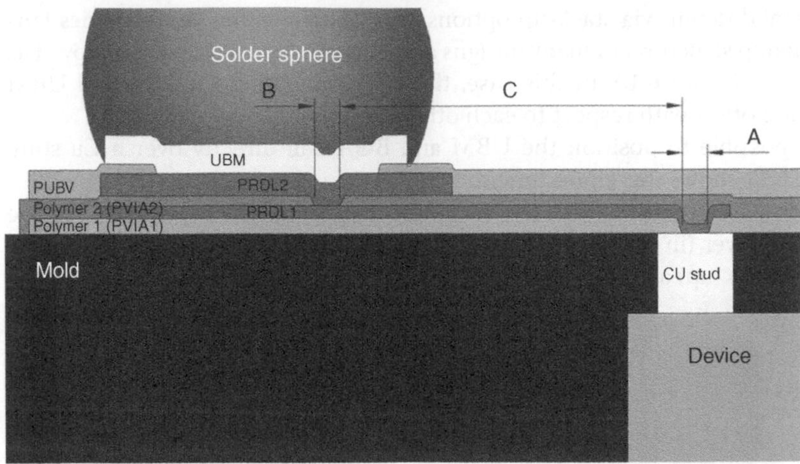

Figure 6.12 M-Series buildup with UBM via in RDL2 via.

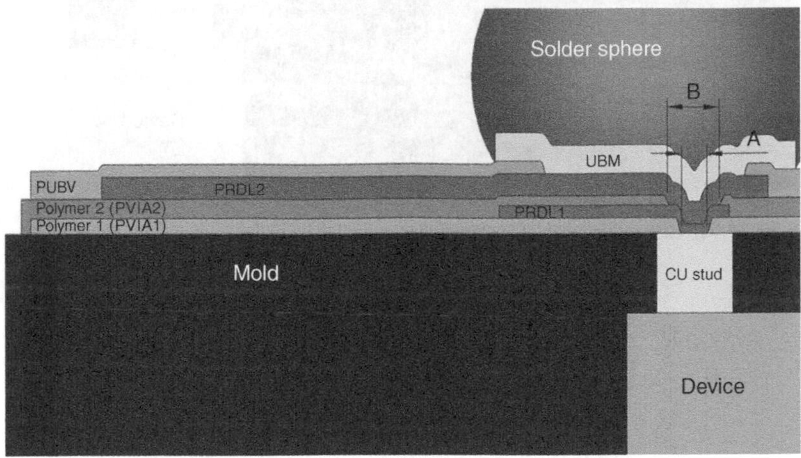

Figure 6.13 M-Series buildup with fully stacked vias.

Finally, fully stacked vias with the UBM and solder ball positioned directly over and directly connected to one or more Cu studs through both RDL layers are possible as shown in Figure 6.13.

M-Series provides mechanical isolation between the solder ball and Si die by providing a rigid molded epoxy layer between the die and fan-out buildup

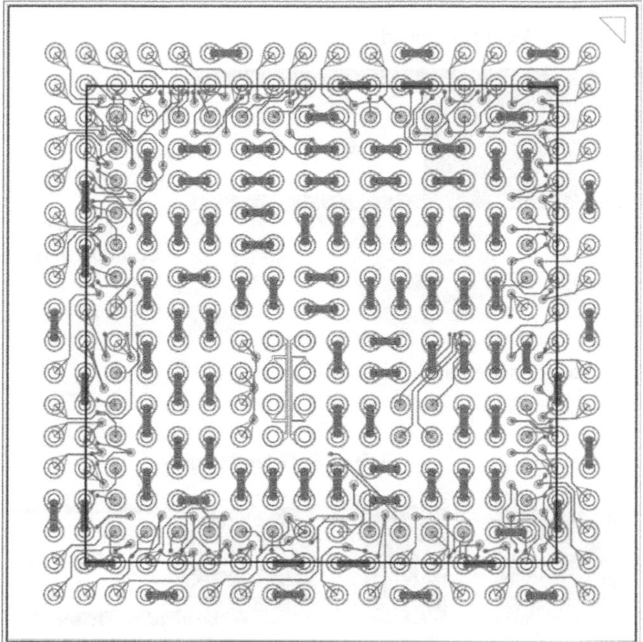

Figure 6.14 M-Series BGA view showing BGA balls over die shadow shown as black line.

structure. As a result, there is no design rule restriction with respect to placement of BGA balls in the "die shadow" region of the package. BGA pads can be placed directly over the die edge as shown in Figure 6.14.

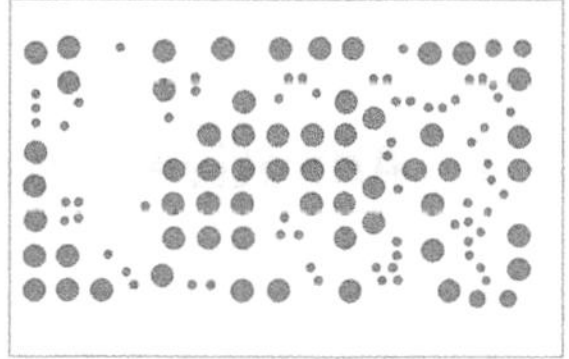

Figure 6.15 M-Series BGA view showing BGA balls over die shadow in light.

Since the M-Series structure uses a plated Cu interconnect without solder to connect the die bond pads to the fan-out routing layer, there is a lot of flexibility on designing the Cu stud layer. Different interconnect sizes and shapes can be used within the same design as shown in the die layout drawing in Figure 6.15.

3D package designs are also possible using the M-Series structure. Common 3D structures include peripheral package on package (PoP) as shown in Figure 6.16 and fan-in PoP as shown in Figure 6.17.

A summary of M-Series critical design rules, roadmap design rules, and design features is given in Table 6.1.

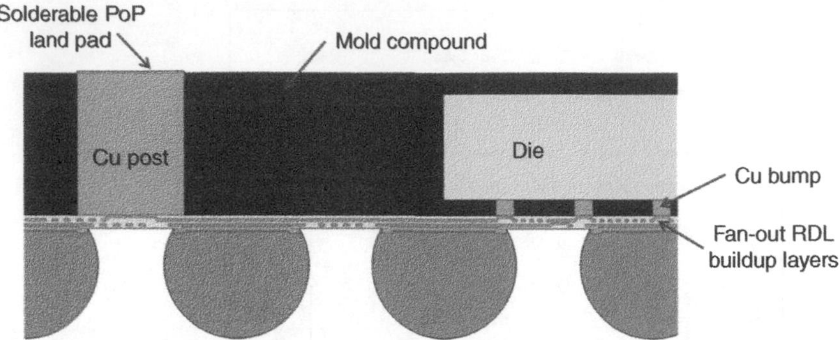

Figure 6.16 M-Series peripheral PoP schematic.

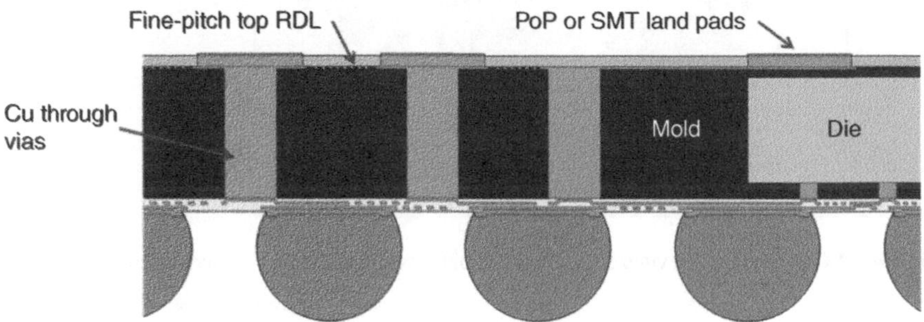

Figure 6.17 M-Series fan-in PoP schematic.

6.5 Adaptive Patterning

One of the main challenges that has prevented the widespread adoption of wafer-level fan-out technology is yield loss due to imperfect alignment of fixed lithography masks to shifting devices during manufacturing. Die shift is an accumulation of die position error from chip-attach equipment tolerances, as discussed in the previous section, and movement during molding as shown schematically in Figure 6.18. A portion of the movement due to molding can be predicted and compensated for during die placement. However, equipment and process tolerances give each die a random and unpredictable offset and rotation. Although higher throughput chip-attach equipment can reduce manufacturing costs, the increased throughput typically results in decreased placement accuracy. Increased die shift is a major contributor to yield loss, and the wider variance limits design rules. Previously, this presented an engineering trade-off. For M-Series, Deca Technologies invented a technology called adaptive patterning [6] to overcome the die-shift problem without

Table 6.1 M-Series design rules.

M-Series design feature roadmap		
Design feature	**2018**	**In development**
Cu stud layer		
Maximum stud thickness (μm)	35	50
Minimum stud thickness (μm)	25	10
Minimum stud diameter (μm)	30	10
Minimum stud spacing (μm)	12	5
Minimum stud pitch (μm)	42	15
First fan-out polymer via layer		
Minimum via diameter (μm)	15	5
Minimum Cu stud enclosure (μm)	7.5	3
Minimum PI thickness – first layer (μm)	6	3
Maximum PI thickness (μm)	12	12
Fan-out RDL layers		
Minimum trace width (μm)	8	2
Minimum trace space (μm)	8	2
Maximum RDL thickness (μm)	11	11
Minimum via 1 enclosure (μm)	7.5	3
RDL layers	2	4
Final fan-out polymer via layer (under UBM)		
Maximum PI thickness (μm)	12	12
Adaptive via truncation (%)	25	50
Via 2 – RDL enclosure (μm)	15	5
UBM layer		
Maximum UBM thickness (μm)	9	20
Minimum BGA pitch (μm)	350	250
Package outline		
Minimum die thickness (200 mm) (μm)	170	100
Max fan-out ratio	2.5	3.5
Minimum die to package edge (μm)	60	30
Max die edge length (mm)	8	12
Max package edge length (mm)	10	18
Minimum BSL thickness (μm)	25	15

(*Continued*)

Table 6.1 (Continued)

M-Series design feature roadmap		
Design feature	**2018**	**In development**
Total package thickness (mm)	0.45	0.25
Land side caps	—	✓
3D features		
Minimum TMI[a] pitch (μm)	—	100
Minimum TMI diameter (μm)	—	50
Minimum TMI spacing (μm)	—	50
TMI[a] termination (peripheral PoP)	—	SnAg
TMI recess (peripheral PoP) (μm)	—	40
Topside routing layer count	—	2
Topside routing minimum feature size (μm)	—	5
Topside land termination (fan-in PoP and SiP)	—	NiAu
Embedded passives	—	✓
EMI shielding	—	✓
Die count	3	6+
1.5 D routing (routing in Cu stud layer)		
Minimum trace width (μm)	25	15
Minimum trace space (μm)	25	15
Maximum routing density (% of package area)	40	60
Minimum PM0 enclosure (μm)	8.5	5

[a] TMI, through-mold interconnect.

compromising chip-attach throughput or design rules and to further enable high yield multi-die system-in-package (SiP) devices.

Rather than attempting to minimize die placement tolerance and compromising design rule density, adaptive patterning takes a novel approach: adjust the lithography pattern in response to the die placement [1]. First, the actual position of each die after pick and place and molding is measured using a high throughput optical inspection system. This produces a dataset of the final XY translation and rotation for each die. This dataset is processed by a custom software system to produce a unique lithography pattern for each package that accounts for the as-chip-attached and as-molded die shift. Finally, for

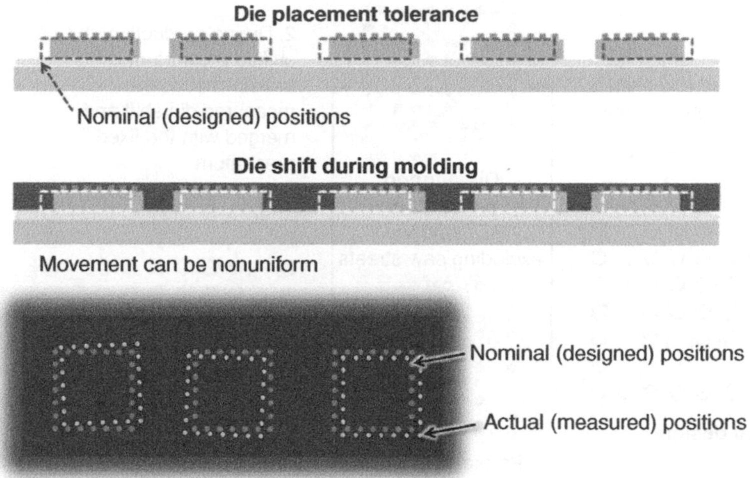

Figure 6.18 Illustration of die shift from pick and place tolerances and molding.

each buildup layer, a custom pattern generated for each FO-WLP device is automatically applied within the photolithography system. By uniquely generating a lithography pattern for each package, the adverse effects of unavoidable random die shift can be avoided. There are several techniques that can be used to generate the per-package lithography patterns.

The first technique, called adaptive alignment, dynamically translates and rotates elements of each package layer to match the measured die shift. In this technique and others, the design must be prepared by splitting it into portions, called prestratums, as shown in Figure 6.19. Each prestratum contains elements of the design, such as vias and RDL traces, that can be separately translated and rotated per package. In adaptive alignment, one prestratum, consisting of the first via layer openings and the copper RDL pattern, is rotated and translated to exactly match the measured die position. In this way, the first via layer openings are precisely aligned to the die contacts. Another prestratum is untranslated and unrotated and thus remains aligned to the package. The package-aligned prestratum contains the under-bump via (UBV) openings, the UBM pads, and the saw street openings for all polymer layers. In this way, the BGA grid pitch and alignment are unaffected.

Die shift is accounted for in the interface between the RDL and the UBV layer. Effectively, the RDL capture pad is allowed to move underneath the UBV opening within an allowable range determined by the design rules. In this case, the critical rule is the enclosure of the UBV by the RDL, or the distance between the edge of the UBV and the surrounding RDL capture pad. As shown in

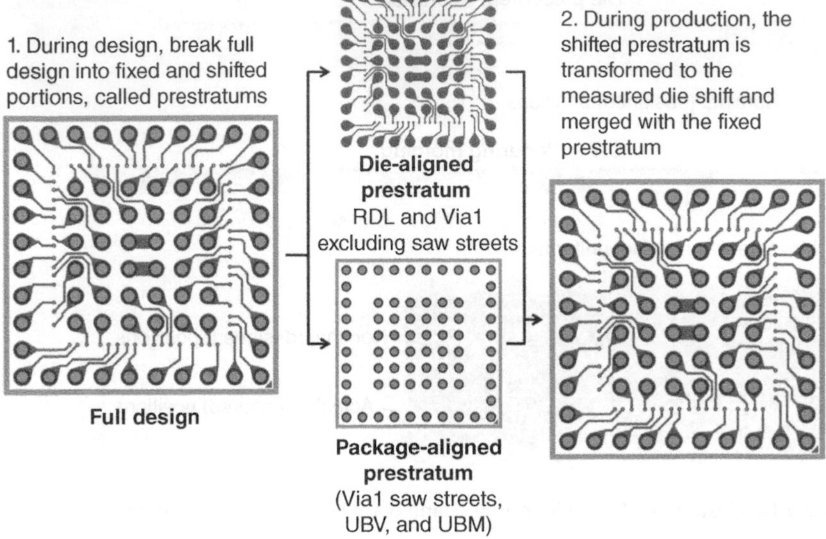

Figure 6.19 The generation of unique per-package lithography patterns by adaptive alignment.

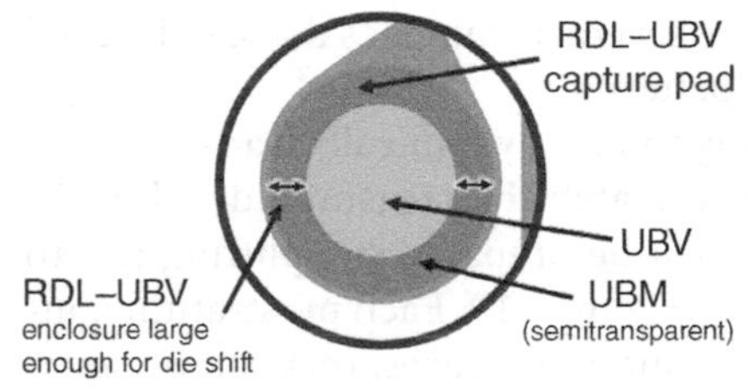

Figure 6.20 For adaptive alignment, the PRDL-UBV interface must be designed with sufficient enclosure for the expected range of die shifts.

Figure 6.20, the RDL to UBV pad stack must be designed with sufficient enclosure to ensure that the UBV overlaps the RDL capture pad over the entire expected range of die shifts. Typically, the UBV diameter is reduced by the expected die shift on each side.

In some cases, such as high current devices, decreasing the UBV area can be undesirable, and increasing the RDL capture pad diameter is not always an option. For these cases, Deca has developed another technique, called adaptive via truncation (AVT), wherein the UBV is minimally changed to account for die shift. Rather than designing the pad stack with extra enclosure, AVT dynamically truncates the UBV area to meet the enclosure design rule. With this technique, rather than reducing the UBV opening diameter during design, the UBV area is reduced dynamically and only by the necessary amount for each specific die shift. After truncation, shown in Figure 6.21, the majority of

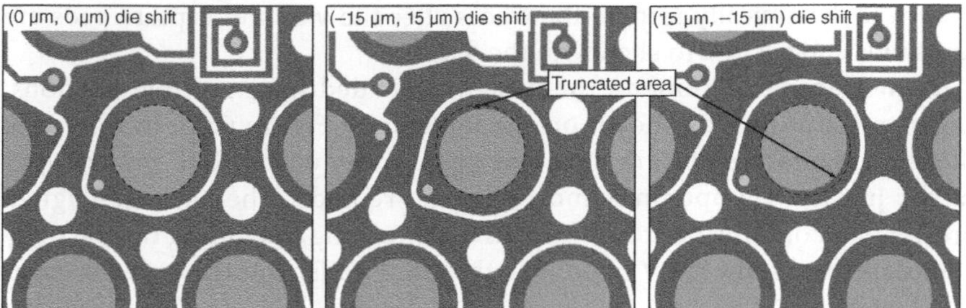

Figure 6.21 Adaptive via truncation maximizes UBV area with minimal changes to the design. The left figure shows the nominal design with zero die shift. The middle figure highlights the truncated UBV area with a left and upward die shift; the right figure illustrates truncation for a right and downward die shift.

via area is preserved, as only a small crescent-shaped portion is removed for even the worst-case die shift.

Another technique, called adaptive routing, compensates for die shift by dynamically rerouting portions of RDL traces on each package. To prepare a design for adaptive routing, small portions are removed from the RDL traces that connect to the capture pads for the first via layer as shown in Figure 6.22. Their removal leaves sufficient space for the adaptive patterning software

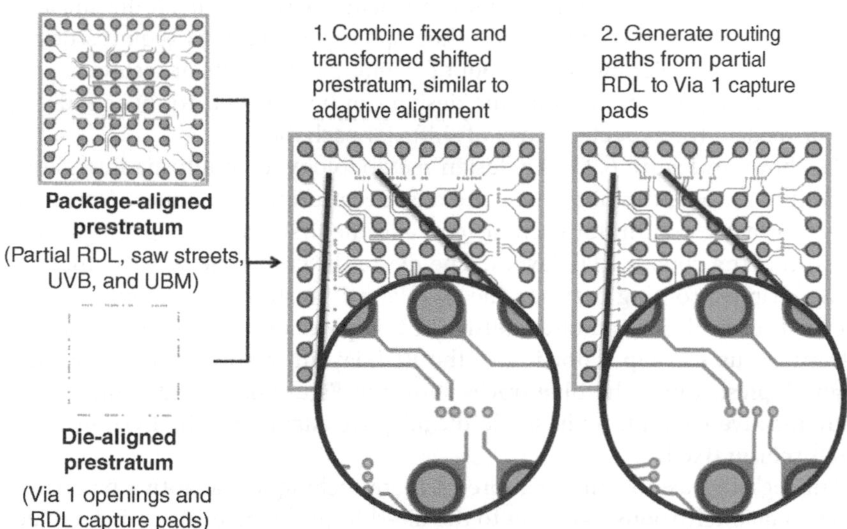

Figure 6.22 Adaptive routing combines the die-aligned and package-aligned prestratums with dynamically generated routing from the partial RDL pattern to the via opening capture pads; a small section of each trace is removed to leave room for the first via layer openings and capture pads to shift with the die.

system to dynamically reroute the traces after the vias and capture pads are translated and rotated to align with the die.

As in adaptive alignment, the design is divided into multiple portions, called prestratums, as shown on the left side of Figure 6.22. In this case, the die-aligned prestratum does not contain the entire RDL pattern, but rather just the via openings and RDL capture pads. The package-aligned prestratum contains the rest of the RDL patterning, including the partially complete traces, the UBV openings, the UBM pads, and the polymer layer saw street.

During manufacturing, the die-aligned and package-aligned prestratums are merged using the measured die positions. As in adaptive alignment, the BGA grid and pitch remain unaffected. Adaptive routing compensates for die shift by dynamically generating new RDL trace patterns that connect the partial RDL pattern to the translated and rotated capture pads for the first via layer. As shown on the right in Figure 6.22, the software system generates short RDL trace portions that meet design rules and can account for large die shifts. The capability of adaptive routing to compensate for die shift is limited only by the amount of clearance available for the first via capture pads to shift and for the subsequent rerouting.

Adaptive patterning technology has also been extended to multi-die SiP designs, providing an improved capability for high yield integration. For traditional embedded fan-out technologies, integrating multiple die poses a yield risk because each die has a random independent shift, but all die must be within the tolerance of one fixed lithographic pattern. This requires design rules with wider tolerances to account for multiple die shifts or results in yield loss. With adaptive patterning, the design can be adjusted to independently accommodate the expected range of shifts for each die.

As shown in Figure 6.23, the design methodology for multi-die integration combines the adaptive alignment and adaptive routing techniques. A separate die-aligned prestratum for each die contains the first via layer openings and a portion of the copper RDL layout. The die-to-die connections are completed using adaptive routing between the die-aligned prestratums, accommodating the shift of both die simultaneously. In this particular example, it was also advantageous to keep a portion of the RDL layout in the fixed, or package-aligned, prestratum (the blue traces shown in Figure 6.21) and connect to it with adaptive routing. As in single-die adaptive alignment, the BGA grid and pitch remain fixed.

Through the use of one or more of these techniques, adaptive patterning offers a uniquely robust solution to the die-shift problem, overcoming a major yield challenge in traditional embedded fan-out packaging technologies. In addition, the technology enables cost-effective production using high throughput chip-attach machines that can make FO-WLP economical even for smaller die sizes.

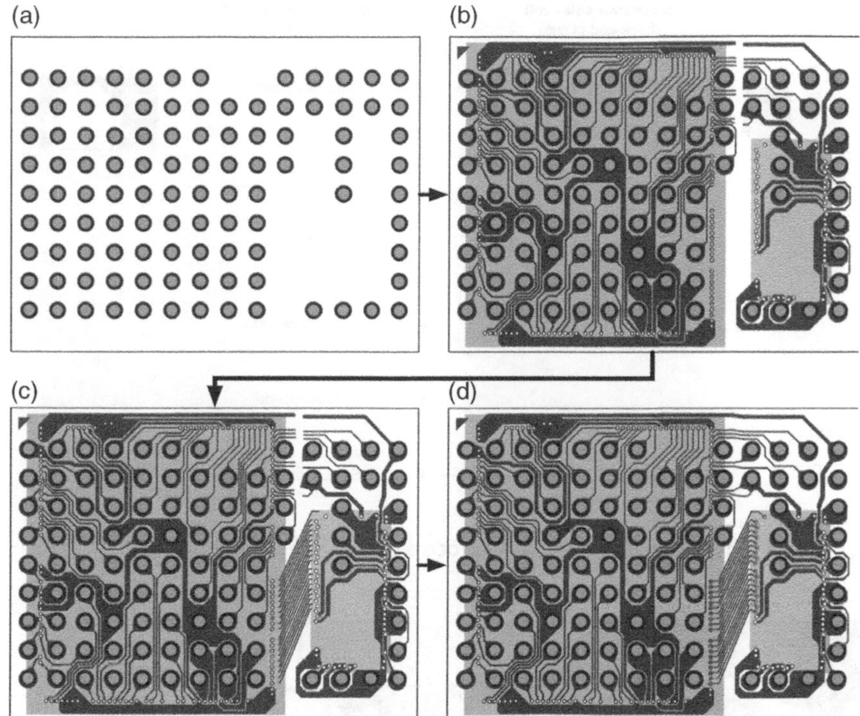

Figure 6.23 Adaptive alignment can be combined with adaptive routing for multi-die designs. Shown in step A, the UBM and UBV pad stacks are fixed with respect to the package edge. Then, in step B, a prestratum including a portion of the RDL and via 1 openings is aligned to each die using adaptive alignment. In many designs, it is beneficial to keep a portion of the RDL fixed with respect to the package edge, such as the partial traces between the die in step C. Finally, in step D, adaptive routing is used to complete die-to-die connections.

6.6 Manufacturing Format and Scalability

A large panel format M-Series has been the goal since its inception. With Deca's foundation of largely non-semiconductor equipment, creating equipment and processes for a large panel format was foreseen as relatively straightforward. M-Series utilizes several processes derived from solar wafer fabrication including PVD seed, Cu electroplating, photoresist strip, seed layer etch, and high pressure cleans. Figure 6.24 provides a high level mapping of solar cell wafer processing versus Deca WLCSP processes, which are utilized in both conventional fan-in WLCSP and M-Series production.

In 2012, Deca established a 600 mm × 600 mm square format for M-Series and began development of the materials, equipment, and processes to support

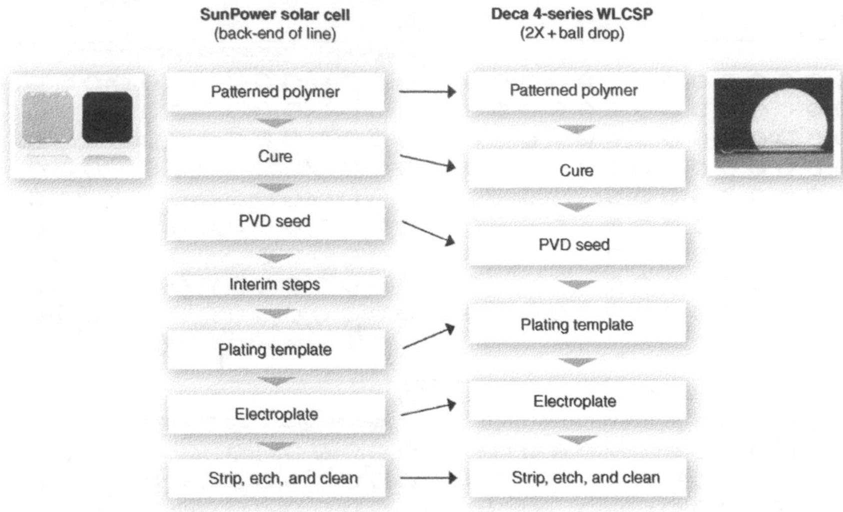

Figure 6.24 Similarities of solar cell wafer process versus Deca's WLCSP process.

Figure 6.25 The 600 mm × 600 mm panel configured for quartering to 300 mm squares.

future high volume production. A 600 mm square was chosen for the capability to cut the panels into 300 mm × 300 mm square segments following wafer processing for the initial production implementation. The resulting 300 mm square subpanels were planned to allow utilization of existing 300 mm round wafer probe electrical test assets with minor equipment upgrades. A photo of an actual 600 mm panel post chip attach that enables 300 mm quartering post fan-out processing is shown in Figure 6.25.

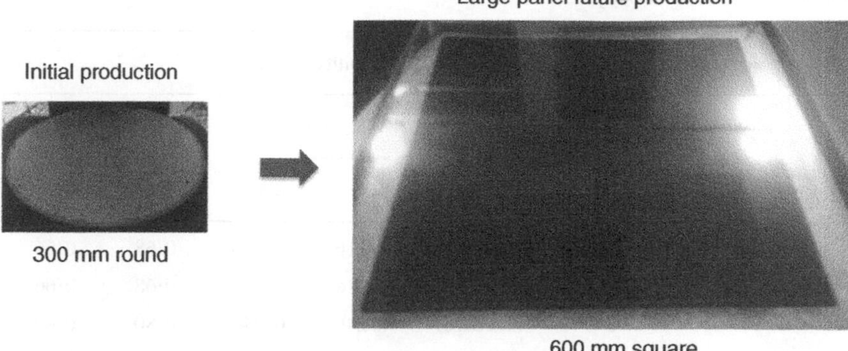

Figure 6.26 M-Series manufacturing formats.

600 mm allows the further flexibility of segmenting the panel into 200 mm squares following wafer processing for utilization of existing 300 mm round BEOL equipment in the early development phase. The 200 mm subpanels are within the maximum square dimension that can be mounted on standard 300 mm wafer saw rings and fall below the 212 mm maximum dimension that can spin within a 300 mm diameter. Figure 6.26 provides images of an actual 300 mm round initial production panel as well as an early large panel prototype through top grind.

ASE and Deca are cooperating to establish 600 mm production capability with plans to implement capacity in both companies. Current engineering work has demonstrated that the 300 mm processes, structure, and direct material set scale successfully to 600 mm, allowing flexibility to produce identical finished products for end customers from either 300 mm round or 600 mm square format.

It is contemplated that the 600 mm format will be carried through the entire manufacturing process to package singulation once final test and back-end process and equipment capabilities have been developed.

6.7 Robustness and Reliability Data

M-Series has been fully qualified according to JEDEC and OEM reliability standards for mobile phone applications. As of March 2018, M-Series packages up to 8 mm × 8 mm body size and fan-out ratio as low as 1.05 have been qualified through MSL 1 component level and BLR requirements. Table 6.2 shows results for package sizes ranging from 5 to 8 mm per side across four different wafer foundries and three device technology nodes. Since M-Series was designed to meet MSL 1 preconditioning requirements, more extensive package-level testing was completed through larger sample sizes as compared with MSL 3.

Table 6.2 M-Series reliability results.

M-Series package-level reliability results							
				MSL 1		MSL 3	
Device	**Si node**	**Fab**	**Test type**	**TCG**	**uHAST**	**TCG**	**uHAST**
A	150 nm	1	Electrical	0/216	0/220	0/68	0/66
			CSAM	0/216	0/220	0/68	0/66
		2	Electrical	0/240	0/240	0/80	0/80
			CSAM	0/240	0/240	0/80	0/80
B	40 nm	3	Electrical	0/400	0/400	0/240	0/240
			CSAM	0/400	0/400	0/240	0/240
C	14 nm	4	Electrical	0/240	0/240	0/80	0/80
			CSAM	0/240	0/240	0/80	0/80
Test conditions		MSL 1 and MSL 3 – JEDEC/IPC joint industry standard J-STD-020A TCG = 1000 cycles air to air temperature cycling from −40 to +125 °C uHAST = 96 h at 130 °CC in 85% RH, 33.3 psia Electrical = open, short, or leakage failures CSAM = visible delamination as detected with 20 μm resolution					

The significant improvement in BLR temperature cycle capability of M-Series versus conventional WLCSP and by similarity conventional fan-out constructed only with spin-on dielectric stress buffer layers was shown in Figure 6.2.

6.8 Electrical Test Considerations

As mentioned in Section 6.6, M-Series is planned to be produced in both 300 mm round and 600 mm square formats in high volume production. With the M-Series structure, 300 mm round warpage is maintained within the specification limits of ±3 mm. Multiple existing commercially available 300 mm test probe handler suppliers can accommodate this level of warpage, allowing M-Series to be tested or probed as a 300 mm wafer.

Figure 6.27 300 mm round to 600 mm by 600 mm square panel.

Quartering of 600 mm square panels post fan-out processing into 300 mm squares allows the potential for upgrading 300 mm round probe handlers to process the panels as shown schematically in Figure 6.27.

Given the fully molded structure of M-Series, singulated component final test is straightforward. The molded stress buffer layer over the active device region, molded sidewall protection, and epoxy-covered backside of the device allow it to also be handled with existing commercially available test handlers designed for singulated laminate BGAs.

6.9 Applications and Markets

The first applications to adopt M-Series are chipsets for mobiles phones and other miniaturized electronic systems such as wearable electronics. Within the mobile phone chipset, M-Series has been qualified on devices such as power management ICs (PMICs), codec ICs, and RF transceivers. M-Series in a PoP structure is planned to be employed as the bottom package for application processors. Multi-chip M-Series will be used for a variety of SiP applications including IoT modules containing RF functionality, sensors, and processing. A roadmap capability of M-Series is fine-pitch (< 2 μm line and space) high density multilayer interconnect enabling a cost-effective alternative to silicon interposers. Applications for high density M-Series include heterogeneous systems on a chip, split-die architectures, and graphics or microprocessor integration with high bandwidth memory.

Acknowledgment

The authors would like to express their sincere appreciation to Craig Bishop as well as Boyd Rogers and Cliff Sandstrom who were instrumental in creating adaptive patterning and the overall M-Series technology, respectively. We also appreciate Craig's contribution of the adaptive patterning section of this chapter.

References

1 Rogers, B., Olson, T., and Scanlan, C. (2013). Implementation of a fully molded fan-out packaging technology. *IWLPC Proceedings* (November 2013).
2 Bishop, C., Olson, T., and Scanlan, C. (2016). Adaptive patterning design methodologies. *2016 IEEE 66th Electronic Components and Technology Conference (ECTC)*, Las Vegas, NV (2016), pp. 7–12.
3 Brunnbauer, M., Fürgut E., Beer, G., and Meyer, T. (2006). Embedded wafer level ball grid array (eWLB). *Electronics Packaging Technology Conference 8th Proceedings* (December 2006).
4 Keser, B., Amrine, C., and Leal G. (2007). The redistributed chip package: a breakthrough for advanced packaging. *2007 Electronic Components and Technology Conference*, pp. 286–291.
5 Zhao, W., Nakamoto, M., Dhandapani, K. et al. (2017). Electrical chip-board interaction (e-CBI) of wafer level packaging technology. *IMAPS Advancing Microelectronics* (November/December 2017).
6 Scanlan C., Rogers, B., and Olson, T. (2012). Adaptive patterning for panelized packaging. *IWLPC Proceedings* (November 2012).

7

SWIFT® Semiconductor Packaging Technology

Ron Huemoeller and Curtis Zwenger

Amkor Technology, Inc.

7.1 Technology Description

The continued scaling of transistor geometries for semiconductor devices has been placing an increased demand on the next-level interconnect technologies. Heterogeneous integration of memory and logic devices is increasingly becoming the norm for next-generation mobile, high performance graphics, and network applications. This integration requires advanced packaging technologies with capabilities for very high signal routing densities, efficient power distribution, and superior signal integrity. In addition, 3D package integration is often required, especially for mobile applications. The 3D aspect places an increased emphasis on the package technology's z-height reduction and thermal performance capabilities.

The scalability of traditional organic and inorganic substrate technologies is limited. Organic laminate technologies are currently constrained to 10 μm/10 μm line/space (L/S) for trace circuitry. Conventional fan-out wafer-level packaging (FO-WLP), also called wafer-level fan-out (WLFO), can achieve 5 μm/5 μm L/S but is limited to one or two metal layers due to the topology challenges in photoimaging a molded wafer. Although silicon (Si) interposers can easily provide less than 2 μm/2 μm L/S interconnect densities, the inherent cost impact and supply chain limitations of through-silicon via (TSV) technology often make it an unrealistic option. Consequently, there is an interconnect gap in the 2–5 μm L/S range.

A new, innovative chip-last high density fan-out (HD-FO) structure called SWIFT (for Silicon Wafer Integrated Fan-out Technology) packaging incorporates conventional FO-WLP processes with leading-edge, thin film patterning techniques to bridge the gap between TSV and traditional FO-WLP packages [1].

Advances in Embedded and Fan-Out Wafer-Level Packaging Technologies, First Edition.
Edited by Beth Keser and Steffen Kröhnert.

The SWIFT methodology is designed to provide increased I/O and circuit density within a reduced footprint and profile for single- and multi-die applications. The improved design capabilities of the technology are due, in part, to the fine feature capabilities associated with this new, innovative wafer-level packaging (WLP) technique. The fine features allow much more aggressive design rules to be applied compared with competing FO-WLP and laminate-based technologies. The redistribution layer (RDL)-first, also called chip-last, process flow allows the SWIFT substrate to be built first, much like a traditional laminate substrate, which has distinct cycle time and yield advantages compared with chip-first packaging technologies. In addition, the unique characteristics of the SWIFT process enable the creation of innovative 3D structures that address the need for IC integration in emerging mobile applications.

7.2 Basic Package Construction

7.2.1 Traditional Package-on-Package Designs

For smartphones and wearable devices, 3D package-on-package (PoP) structures have become the standard for application processor (AP) and DRAM integration. Laminate organic substrate technology is typically used for these designs. The two most common PoP structures used in today's advanced mobile devices are illustrated in Figure 7.1: (a) exposed-die through-mold via PoP (TMV® PoP) and (b) interposer PoP. TMV PoP incorporates through-mold via technology to create the vertical signal path between the low power double data rate (LPDDR) top memory package and the underlying AP logic package. The memory package is directly attached to the TMVs that are located around the perimeter of the package. Consequently, custom memory must be used. Alternatively, interposer PoP incorporates a laminate substrate on the top of the AP (bottom) package. Although this adds thickness to the bottom package, it allows standard pitch memory packages to be used, which opens the supply chain for memory sourcing.

Figure 7.1 Traditional PoP structures. (a) Exposed-die TMV PoP. (b) Interposer PoP.

These laminate-based PoP structures have limitations, which affect their ability to extend into high-speed and high bandwidth memory applications for advance mobile devices that include:

- *Feature size:* Laminate-based organic substrates are currently limited to ~15 μm minimum L/S and 15 μm thick copper trace feature sizes. The laser-drilled via diameter is also limited. These limitations restrict the designer's ability to optimize power distribution and control the impedance of critical signals.
- *Z-height:* The organic laminate substrate's core and prepreg materials add thickness to the interconnection structure, which has an adverse effect on the power distribution network (PDN) impedance.
- *Memory interface pitch:* The current limit for TMV PoP and interposer PoP is 0.35 and 0.27 mm minimum memory pitch, respectively. This is due to the physical space required for the vertical interconnect media. Consequently, the memory I/O is limited based on the overall package body size.

The limitations of organic laminate-based PoP structures make their application more difficult as memory bus speeds extend into the DDR4/5, PCIe, and Ethernet speed ranges. To address these challenges, an existing interposer PoP device was converted to SWIFT packaging. The physical attributes were compared, and the electrical performance was analyzed via simulation modeling. The following sections discuss the SWIFT structure and the physical and electrical comparison study in detail.

7.2.2 SWIFT PoP Structure

A typical SWIFT PoP structure is illustrated in Figure 7.2. SWIFT technology is based on wafer-level processing, incorporating the use of a carrier and thin film photolithography to pattern the fine metal and dielectric features in a multilayer RDL structure. SWIFT is a "chip-last" process that has the advantages associated with conventional organic substrates regarding yield and cycle time. Once the chip-last HD-FO package's RDL is fabricated on the carrier, the structure is inspected, known bad RDL sites are identified, and the SWIFT "substrate" is placed in inventory. This allows for a shorter overall cycle time

Figure 7.2 SWIFT PoP structure. (Note: Copper pillars on left and right sides of lower substrate.)

and ensures a known good die (KGD) is dedicated to a known good substrate. This is not possible with "chip-first" wafer fan-out technologies where the RDL process is responsible for the majority of the yield impact. The chip-last HD-FO technology approach also has superior flexibility and scalability, since it can support many package variants, such as 3D PoP, system in package (SiP), multiple die, passive components, and variable component thicknesses. The benefits of SWIFT technology are further detailed in the following sections.

SWIFT package technology is designed to overcome many of the issues associated with conventional FO-WLP technology. In addition, it provides increased I/O and circuit density within a reduced footprint and profile for single- and multi-die applications. The improved design capability of the SWIFT technology is due, in part, to the fine feature capabilities associated with this new, innovative WLP technique. Since they do not use sophisticated IC processes, laminate-based IC assembly techniques cannot match a wafer-based approach. The wafer-based spin-on thin film dielectrics and photoresist enable topologies that are much flatter and provide better photo-resolution. With SWIFT technology's RDL-first, chip-last approach on a carrier, the topology has a much flatter, pristine glass surface. In contrast, other FO-WLP technologies are all chip-first and then overmolded, so they are much more difficult to deal with due to warpage. This can allow much more aggressive design rules to be applied compared with competing FO-WLP and laminate-based IC assembly techniques. In addition, the unique characteristics of the SWIFT process enable the creation of innovative two-dimensional (2D) and 3D structures that address the need for IC integration in emerging mobile and networking applications.

Although SWIFT appears to be a typical fine-pitch flip-chip (FC) construction, it incorporates some unique features not associated with conventional IC packages. These unique SWIFT features include:

- Polymer-based dielectrics.
- Multi-die (in theory, as many dies as possible within the field size limits of today's steppers) and large ($\gg$17 mm × 17 mm) die capability.
- Large ($\gg$17 mm) package body capability.
- Interconnect density down to 2 μm L/S (critical for system-on-chip [SoC] partitioning applications).
- Copper (Cu) pillar die interconnects down to 30 μm pitch.
- 3D/PoP capability utilizing TMV or tall Cu pillars.

7.3 Manufacturing Process

SWIFT's attributes are realized by applying a unique process flow that incorporates both FC assembly and wafer-level processing techniques. The SWIFT process flow is shown in Figure 7.3.

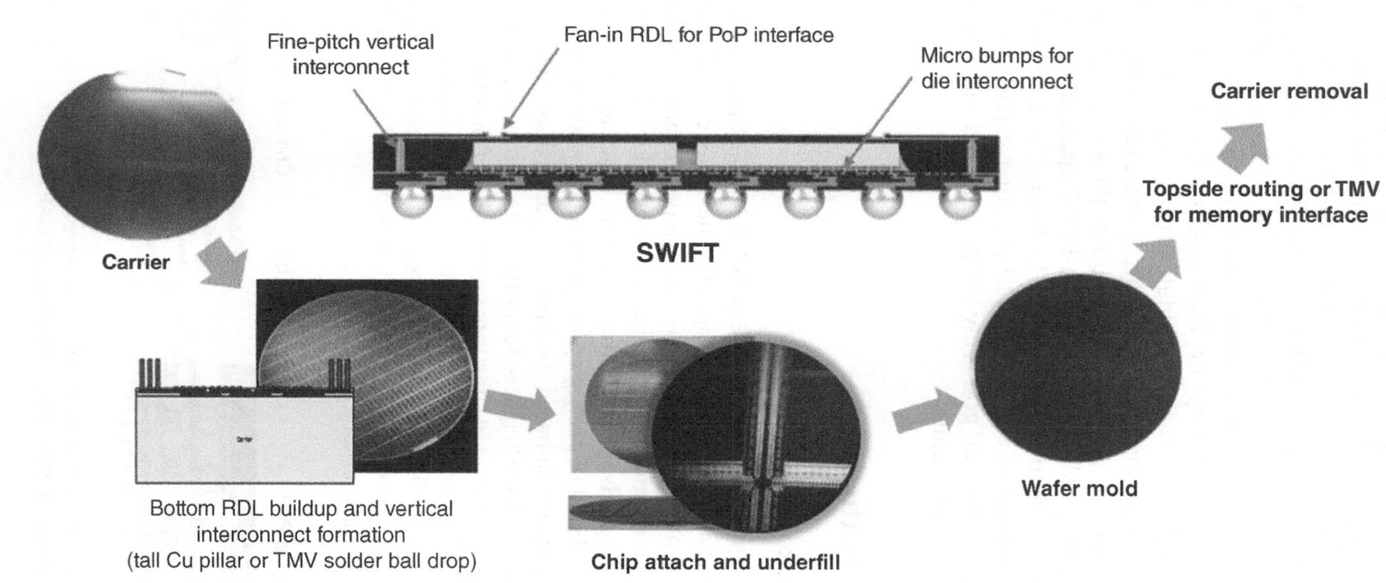
Fine-pitch vertical interconnect
Fan-in RDL for PoP interface
Micro bumps for die interconnect
Carrier removal
Topside routing or TMV for memory interface
Carrier
SWIFT
Wafer mold
Bottom RDL buildup and vertical interconnect formation (tall Cu pillar or TMV solder ball drop)
Chip attach and underfill

Figure 7.3 SWIFT process flow.

A carrier is used as the platform on which to build up the RDL using conventional WLP technology [2]. With the RDL-first, chip-last approach providing a very flat surface, fine line and space routing can be applied for high density interconnect applications. The SWIFT processes' small feature size capability also provides the opportunity for reducing the package footprint. Vertical interconnects can be formed using tall Cu pillars (manufactured using a thick photoresist and pattern plating) or TMV solder balls. The FC die is then attached to the high density RDL buildup structure and encapsulated with epoxy-based molding compound. Wafer backgrind may be applied to create very thin structures. For 3D PoP constructions, the solder balls are exposed using TMV technology. Alternatively, for fan-in PoP applications, topside routing is applied using traditional wafer RDL buildup techniques with thin film dielectrics and photoresist and pattern plating. After carrier removal, solder balls are attached to the ball grid array (BGA) pads and the molded wafer is singulated into individual units.

7.4 Design Features

7.4.1 Form Factor

One of the significant benefits of using WLP is the form factor reduction including the package *z*-height. Package thickness has critical effects on the signal and power integrity since the conductive path changes as the substrate height and 3D interconnect height change. In addition, the thermal performance of the package also benefits from a thinned package since the resistive path is decreased. HD-FO packaging is approximately 40% thinner compared with competing laminate-based technologies and thus shows substantial improvements [3]. Package height reduction results from replacing the modified semi-additive process (MSAP) substrate with the thin film RDL and interposer stack-up for both the bottom and top substrates. Figure 7.4 shows a comparison of a FC-PoP vs. a HD-FO design cross section where a total package thickness of 450 μm is achieved with HD-FO design vs. a FC-PoP with a thickness of 630 μm (excluding the memory package).

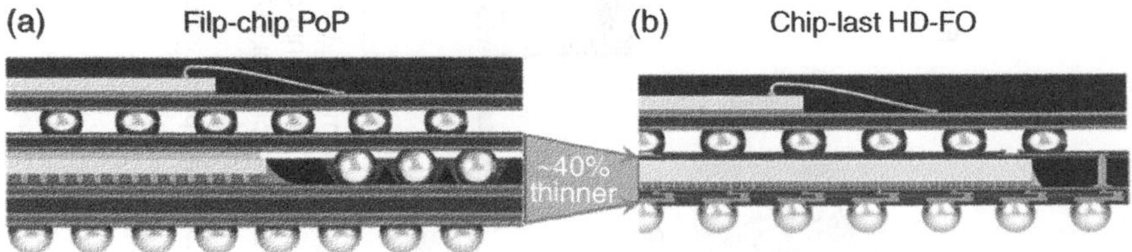

Figure 7.4 Package height comparison with cross-sectional view of an FC-PoP and the chip-last HD-FO package. (a) Standard approach with MSAP substrate. (b) SWIFT design.

7.4.2 Feature Size

The chip-last HD-FO design builds on conventional FO-WLP technology and provides improvements to the shortcomings of the limited line width, spacing, and 3D capabilities. With the unique structure of the chip-last HD-FO package and the use of key assembly processes such as stepper photoimaging, this technology enables line width and spacing down to 2 μm [4]. This increases the I/O count and circuit density within a reduced area for single- and multi-die applications. These fine feature capabilities eliminate the need to use 2.5D TSV connections required for SoC partitioning and networking applications. Multilayer (up to four layers) RDLs can be built with fine (≥2 μm) line width and spacing and polymer-based dielectrics, allowing a single body construction vs. standard FC technologies.

Although the overall package structure resembles a traditional FC stack, the use of fine-pitch micro-bump die interconnects with 30 μm pitch capability based on RDL and reduced layer count in chip-last HD-FO packaging offers key improvements. Figure 7.5 shows an image of the RDL capability and fine-pitch micro-bump die interconnects.

The flexibility of the chip-last HD-FO package structure also offers benefits for creating 3D assemblies. Tall Cu pillars can be used to create high density vertical integrations that enable 3D PoP-like structures where the advanced memory chips can be mounted on top of the package. Figure 7.6 shows the tall Cu pillars (>200 μm pillar height) and a cross-sectional picture of a 3D PoP chip-last HD-FO design with dual die structure with top RDL.

Table 7.1 compares some of the key attributes of the SWIFT process versus traditional FO-WLP technologies. Although the SWIFT approach requires Cu pillar processing at the silicon wafer level, it enables the die to be attached to a pre-inspected known good RDL structure with very fine-pitch interconnect features. This, in turn, helps ensure KGD are not subjected to any yield loss

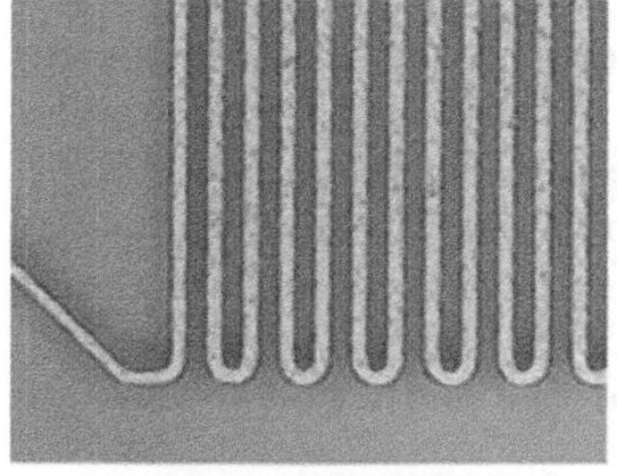

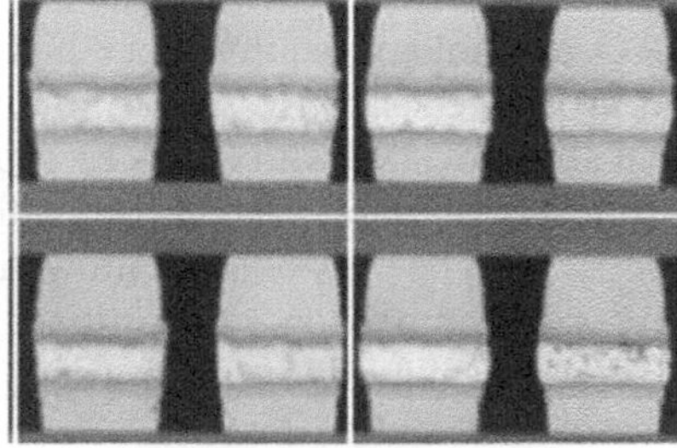

Figure 7.5 Key enabling chip-last HD-FO technologies.

Figure 7.6 Chip-last HD-FO package with tall Cu pillars and top RDL.

Table 7.1 SWIFT vs. FO-WLP process.

Key attributes	SWIFT	FO-WLP
Die wafer processing	Cu pillar	None required
Die dedication	Die last (on known good RDL)	Die-first
Die attach	High accuracy FC bond on known good RDL site	High accuracy D/A (slow)
Patterning	Photo stepper	Photo mask (align/ stepper)
Line/space	2–10 μm	6–15 μm
# RDLs	1–3	1–2

associated with the RDL buildup process. In comparison, for FO-WLP, KGD are dedicated prior to RDL creation, which increases the risk for die yield loss due to the inherent defect density of the FO-WLP process. In addition, traditional WLFO technology requires very high accuracy die-attach equipment and careful process characterization to minimize die shift and molded wafer warpage that are associated with FO-WLP processing. Finally, since the SWIFT RDL structure is built up on a flat carrier with minimal warpage, the topology of the dielectrics can be controlled to allow fine resolution circuit formation in multiple layers. In comparison, conventional FO-WLP has limitations on RDL count and L/S capability due to the wafer warpage and topology variation intrinsic to molded wafer processing.

SWIFT package and technology roadmaps are shown in Figure 7.7 and Table 7.2, respectively. Material and tooling primarily limit the development of all the roadmap items simultaneously and set the expectations for each phase timing.

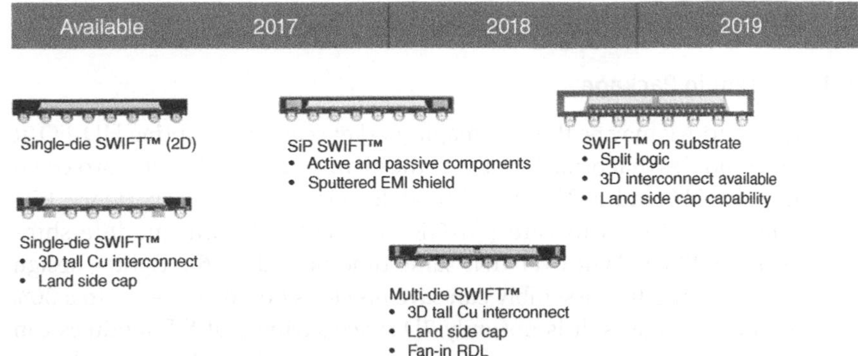

Figure 7.7 SWIFT package roadmap.

Table 7.2 SWIFT technology roadmap.

SWIFT roadmap	Intern qual	2017	2018	2019	2020
Process flow	RDL-first/chip-last				
μBump					
• Pitch (μm, min)	30	←	20		
• Die (μm, min)	15	←	10		
• Ht. (μm, min–max)	20–40	20–40	10–40		
• Bump metal stack	Cu/SnAg, Cu/Ni/SnAg	←	+ Cu		
RDL (redistribution)					
• Line/space (μm, min)	2/2	←	←		
• Layer count (max)	4–6	4–6	6		
• Via opening (μm, min)	10	←	6		
• Re-passivation material	LT PI	←			
Package					
• Pitch (μm, min)	15	25	←		
• Die (μm, min)	0.30	←	0.25		
• Ht. (μm, min–max)	0.30	←	0.25		
Memory interface					
• TMV pitch (μm, min)	0.30	0.25	0.20		
• Tall pillar pitch (μm, min)	0.30	←	0.10		
• Tall pillar Ht. (μm, max)	0.26	←	0.25		

7.5 Manufacturing Format and Scalability

7.5.1 System in Package

While high-end AP may be the first major market segment adopting HD-FO, it may be SiP modules that bring HD-FO to the mainstream market. It is projected that between 2016 and 2022, RF SiP module shipments will experience 15% compound annual growth rate (CAGR), but FO-WLP SiP module shipments will see 23% CAGR over that same time period [5, 6]. Tighter design rules allowed by module assembly can save precious board area – up to a 30% reduction in some cases. It is not only 2D space saving that SiP modules can offer in HD-FO format but height reduction as well in 3D space due to eliminating the substrate and replacing with the thin film structure.

In a recent design study, approximately 80 mm^2 of board area containing 50+ components and multiple WLPs from the motherboard was condensed to a module with an area less than 50 mm^2. This was accomplished by embedding a thinned die in the bottom portion of the chip-last HD-FO package as well as the thinnest components and then utilizing the tall Cu pillars to connect the bottom substrate to the top of the mold surface. The increase in cost for processing is offset by a substantial package size reduction and board area savings in the end application due to the increased 3D integration possibilities. The surface contains an RDL to provide landing pads for additional components as shown in Figure 7.8. While there is significant area optimization in this configuration, it should be considered volume optimization, since it takes advantage of every mm^3 allowed by the system designer.

SWIFT packaging is also capable of conformal shielding, which is increasingly requested for laminate-based SiP modules. In this case, the ground plane must extend to the edge of the "substrate" or, in the case of HD-FO packaging, into the saw street. The major concern here is delamination of the spin-on dielectric film and the plated Cu traces, in addition to being able to obtain a reliable connection between the conformal metal film and the ground plane in the buildup structure as seen in Figure 7.9.

Figure 7.8 Representative cross section of embedded WL-SiP.

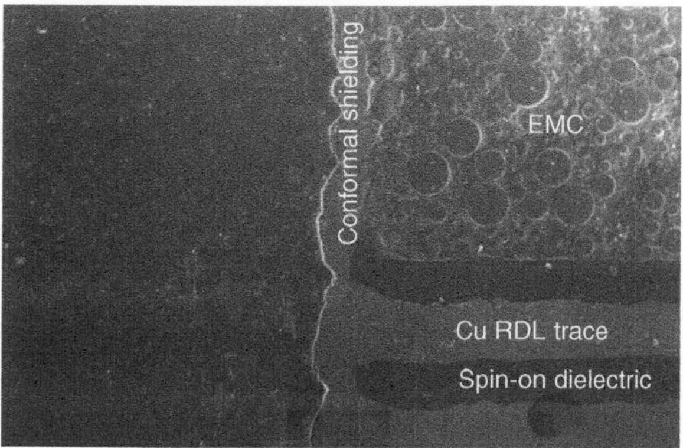

Figure 7.9 SEM photo of shielding Cu trace interface. (Note: the Cu RDL trace is 3 μm.)

7.5.2 Networking and High Performance Graphics (SWIFT on Substrate Applications)

In addition to the AP and SiP markets, SWIFT packaging has applications in the networking and high performance graphics segments. This is especially true for the high-speed market that does not want to make the jump/transition to silicon photonics or 2.5D silicon interposer. However, high-speed performance is not the only benefit of HD-FO packaging. Depending on the use case and mixing with conventional laminate substrate technology, it can reduce the cost of the overall implementation. As advanced silicon nodes shrink the die size, the interconnect density only increases, forcing the use of very expensive substrates in a laminate solution or very expensive main boards in a pure FO-WLP solution. If the two are mixed, the overall costs can decrease. Instead of fanning out a small die to a large package size, the small die is fanned out to an intermediate size, where the interconnect pitch is increased, allowing for a less advanced substrate in the final (large) package size.

The combination of both substrate and SWIFT technology can also provide the ability to have deconstructed (i.e. split) logic die and/or memory with high density interconnects between all devices. The SWIFT "interposer" extends to the outside perimeter of the interconnected logic and/or memory devices. The SWIFT structure is then attached and fanned out on a lower cost organic substrate. An example of a substrate SWIFT package is shown in Figure 7.10. This type of construction can be considered as a viable alternative to 2.5D TSV interposers.

Figure 7.10 SWIFT on substrate package with deconstructed logic.

7.6 Package Performance

7.6.1 Electrical Benefits

In this section, the chip-last HD-FO technology process is compared with both exposed-die PoP and fan-in PoP for signal integrity, power integrity, and impedance matching. A 3L chip-last HD-FO design with tall Cu pillars is compared with 3L exposed-die PoP, and a 3L chip-last HD-FO design with one top RDL is compared with 3L + 2L fan-in PoP. All test devices are 12 mm × 12 mm with a chip height of 100 μm, package height of 400 μm, mold cap thickness of 200 μm, and ball pitch of 400 μm. Figure 7.11 shows the schematic cross-sectional view of the four package types being compared.

Figure 7.11 Cross section of four test packages.

7.6.2 Signal and Power Integrity DDR4 (AP PoP Applications)

Since DDR4 memory operates with a timing margin of only 313 ps and the maximum ripple of the PDN is ±60 mV, optimizing signal and power integrity is even more critical than with DDR3. There were even those that believed DDR5 would not happen because the timing budget would be consumed in the package due to PDN noise and package variability. While JEDEC is expected to release the DDR5 specification in 2017, DDR5 production is not expected to ramp up until 2020.

In comparing a chip-last HD-FO design with its most likely competitors at DDR4 speeds (4 Gbps), it has a 15–25% improvement in eye amplitude and 19–22% improvement in eye height. The jitter is 62–66% less and rise/fall time is decreased by 26–31%. This can be seen in Figure 7.12, Tables 7.3 and 7.4.

The improvement in signal integrity is achieved by reducing the Cu trace to 5 μm wide and 3 μm tall, compared with the laminate substrate where the minimum trace width is 10–15 μm with 15 μm thickness. These improvements continue as the frequency increases. When increased to 6 Gbps, the margins widen up to a 44% improvement in eye amplitude and up to a 73% improvement in eye height. The differences in eye width and decrease in jitter and rise/fall time are similar to 4 Gbps, which can be seen in Figure 7.13, Tables 7.5 and 7.6.

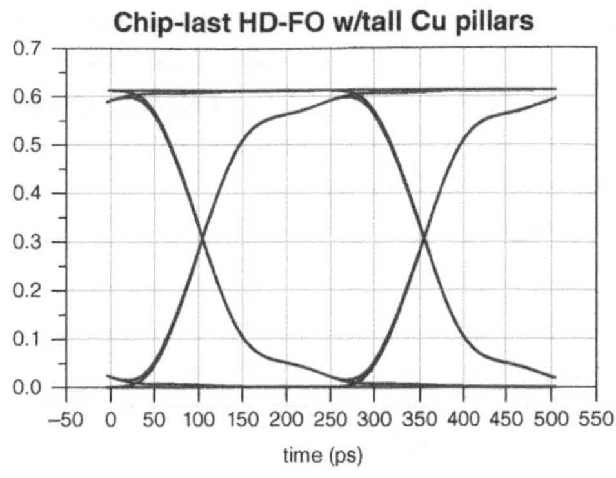

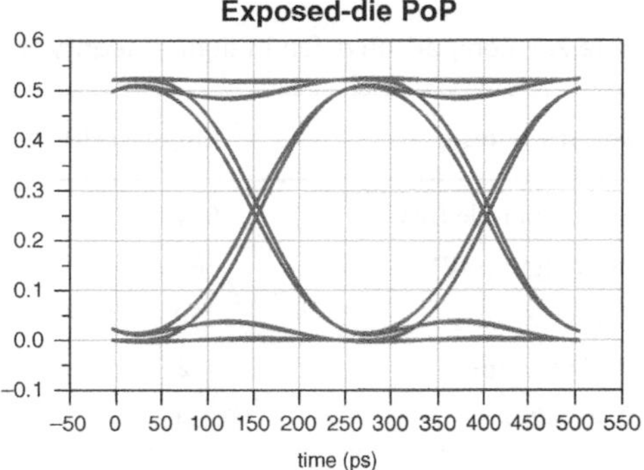

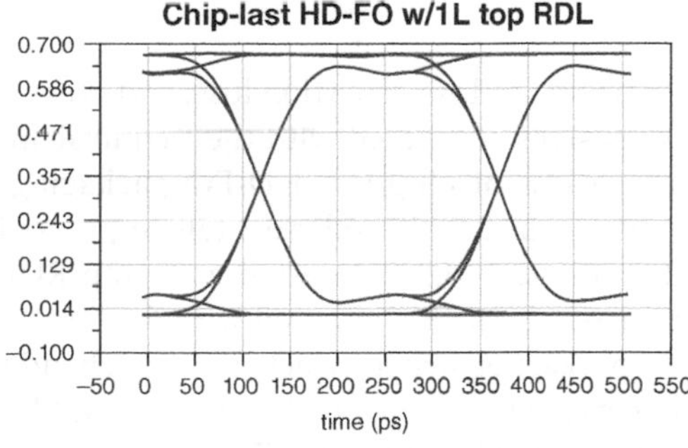

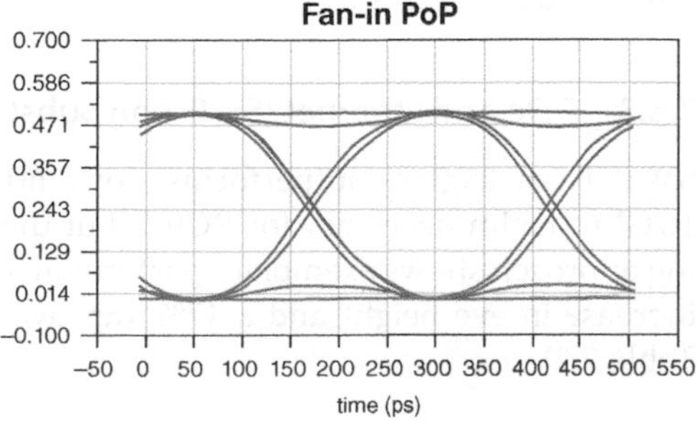

Figure 7.12 Comparison of eye diagrams at 4 Gbps.

Table 7.3 Comparison of fan-out signal integrity at 4 Gbps.

Key attributes	Chip-last HD-FO w/CuP	Exposed-die PoP
Eye amplitude (mV)	575	499
Eye height (mV)	535	450
Eye width (ps)	249	240
P–P jitter (ps)	2.2	6.4
Rise/fall time (ps)	68	92

Table 7.4 Comparison of fan-in signal integrity at 4 Gbps.

Key attributes	Chip-last HD-FO w/1L top RDL	Fan-in PoP
Eye amplitude (mV)	591	472
Eye height (mV)	562	461
Eye width (ps)	250	243
P–P jitter (ps)	2.3	6.1
Rise/fall time (ps)	50	72

The return loss, insertion loss, and cross talk all improve with chip-last HD-FO designs compared with their equivalent laminate packages. At 4 GHz, the return loss of chip-last HD-FO packaging compared with exposed-die PoP improves by 2.5 dB, and cross talk improves by 18 dB. At 2 GHz, the return loss of a chip-last HD-FO design w/1L top RDL improves by 6 dB, and cross talk improves by 6 dB compared with fan-in PoP. In both cases, the chip-last HD-FO packaging structure acts as a wideband low-pass filter, whereas both laminate packages are narrowband low-pass filters with steep roll-offs. This data is illustrated in Figures 7.14–7.16.

7.6.3 PCIe and Ethernet (SWIFT on Substrate Applications)

SWIFT packaging outperforms conventional laminate substrates in the 16 Gbps realm necessary for PCIe4, but the SWIFT chip-last HD-FO packaging approach shows exemplary performance at 28 Gbps where there are a 126% increase in eye height and a 47% improve in eye width (see Figure 7.17 and Table 7.7).

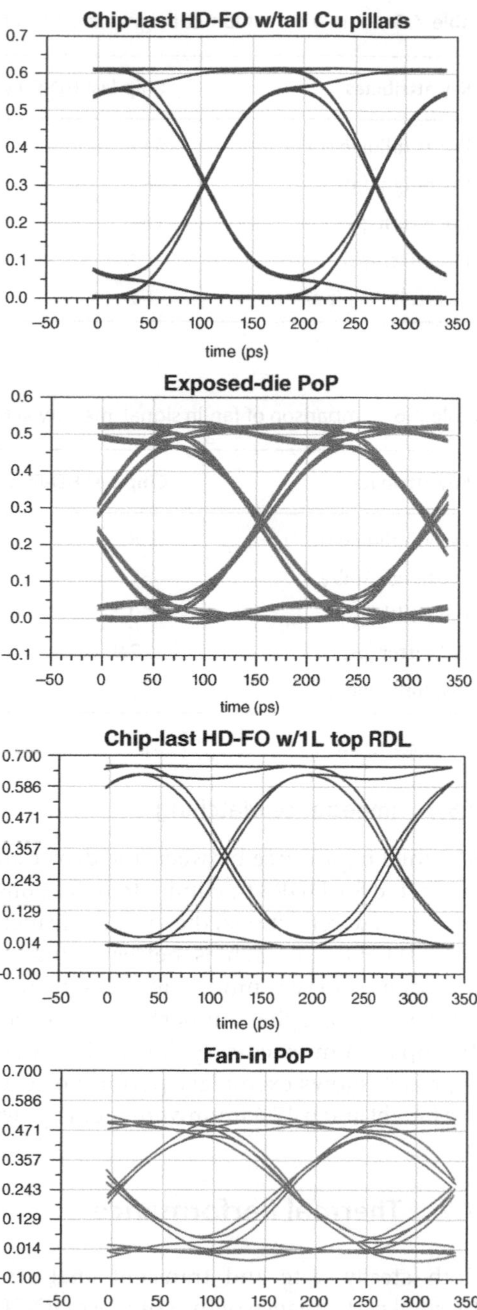

Figure 7.13 Comparison of eye diagrams at 6 Gbps.

Table 7.5 Comparison of fan-out signal integrity at 6 Gbps.

Key attributes	Chip-last HD-FO w/CuP	Exposed-die PoP
Eye amplitude (mV)	548	451
Eye height (mV)	481	339
Eye width (ps)	164	158
P–P jitter (ps)	3.7	9.8
Rise/fall time (ps)	64	75

Table 7.6 Comparison of fan-in signal integrity at 6 Gbps.

Key attributes	Chip-last HD-FO w/1L top RDL	Fan-in PoP
Eye amplitude (mV)	638	444
Eye height (mV)	551	318
Eye width (ps)	159	145
P–P jitter (ps)	7.9	14.1
Rise/fall time (ps)	63	75

7.6.4 Impedance Matching

Matched impedance between the driver and the transmission line is strongly suggested for DDR4 to eliminate reflections that return to the driver. By eliminating these reflections, the eye has cleaner edges and is more open. The chip-last HD-FO approach is better suited to a matched impedance since its impedance curve is more stable as shown in Figure 7.18.

SWIFT packaging outperforms conventional laminate substrates in the 16 Gbps realm necessary for PCIe4, but the chip-last HD-FO packaging approach shows exemplary performance at 28 Gbps, with a 126% increase in eye height and a 47% improvement in eye width (see Figure 7.19 and Table 7.7).

7.7 Thermal Performance

Both steady-state and transient analyses were performed to evaluate the thermal performance of the chip-last HD-FO package since the thermal dissipation is one of the main definitions of the package characteristics. The junction temperature influences both the thermal and electrical performance of the

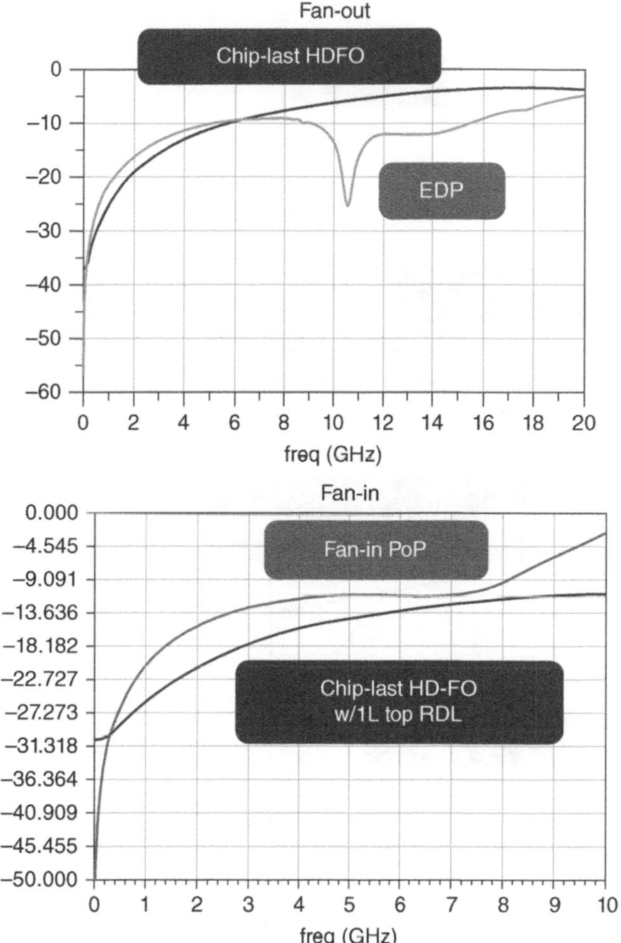

Figure 7.14 Return loss comparison.

package since the leakage current would directly increase with increased junction temperature. All test devices used 12 mm × 12 mm packages as described in the electrical performance section. Refer to Figure 7.11.

The thermal resistance (θ_{JA}) obtained via thermal simulation was ~6% less than the FC-PoP. Similarly, the maximum die temperature was ~5% less for chip-last HD-FO packages. Figure 7.20 shows the related thermal simulation data. While they were not simulated, θ_{JB} and θ_{JC} are expected to show similar improvements.

In addition, the transient analysis is performed to compare the time to reach the allowable max die temperature when a 9 W duty cycle is applied. The

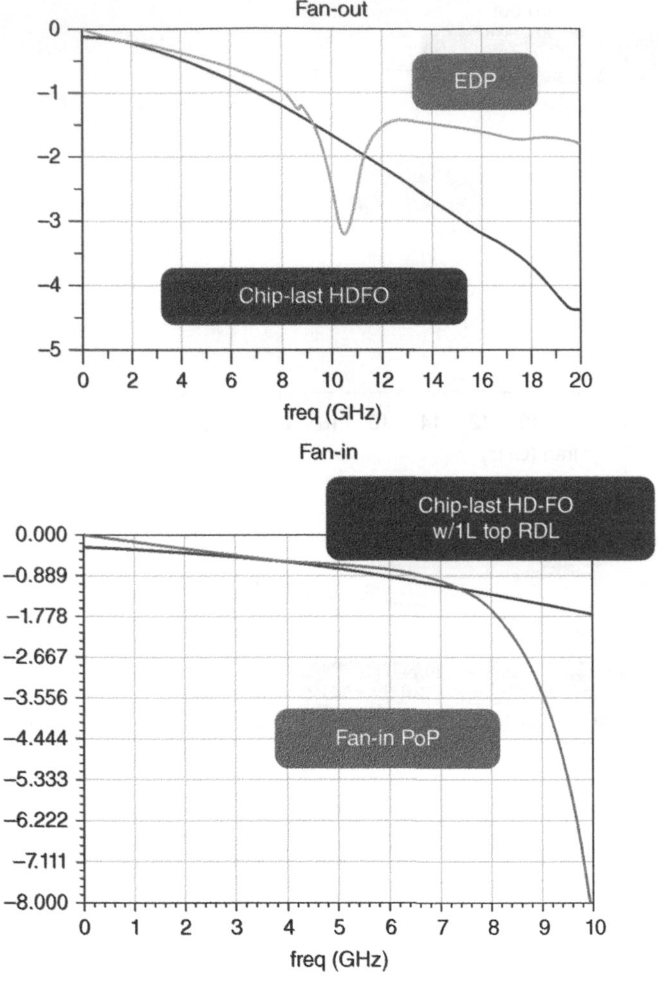

Figure 7.15 Insertion loss comparison.

transient characterization is particularly important because when the die reaches its maximum allowable temperature, the die must throttle itself to prevent permanent damage.

In this case, the assumed allowable maximum die temperature is assumed to be 105 °C, and chip-last HD-FO packaging extended the time by 9 and 16 seconds compared with the conventional FC-PoP. Therefore, the mobile phone can stay at its maximum speed for a longer period before throttling back. Figure 7.21 and Table 7.8 show the related thermal simulation data.

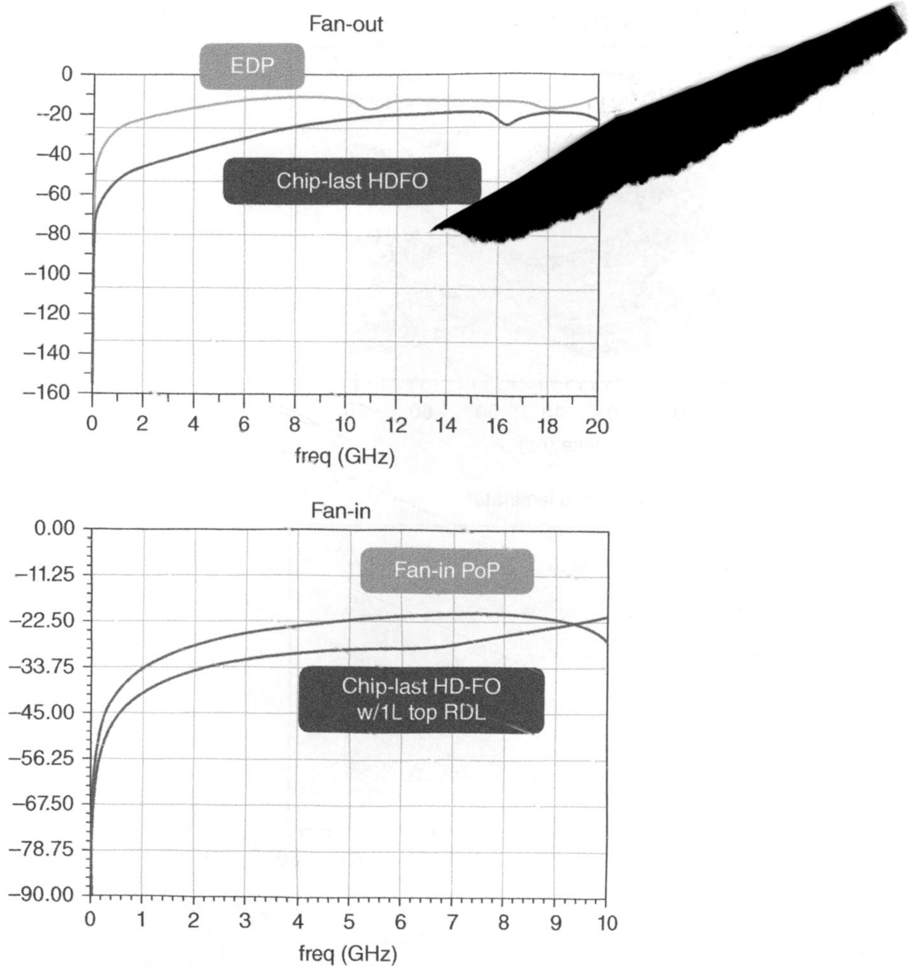

Figure 7.16 Cross-talk comparison.

7.8 Robustness and Reliability Data

To assess the component-level reliability (CLR) and board-level reliability (BLR), a 15 mm × 15 mm body package is used with three RDLs with a memory stacked in a PoP configuration. Figure 7.22 shows the package with and without memory attached.

The chip-last HD-FO package TV passed MSL 3/260 °C (per JEDEC J-STD-020D.1) package-level reliability tests including unbiased highly accelerated stress tes (uHAST) (96 hours, 130 °C/85% RH) (JESD22-A118A) and temperature cyc condition B (TC-B) (1000 cycles, –40 °C/150 °C) (JESD22-A104D). BLR tests w

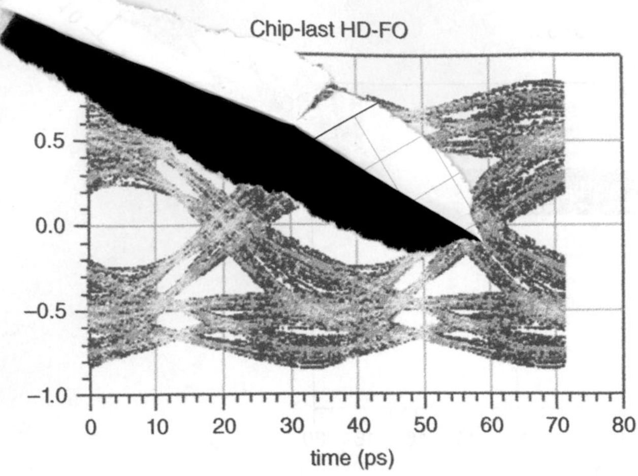

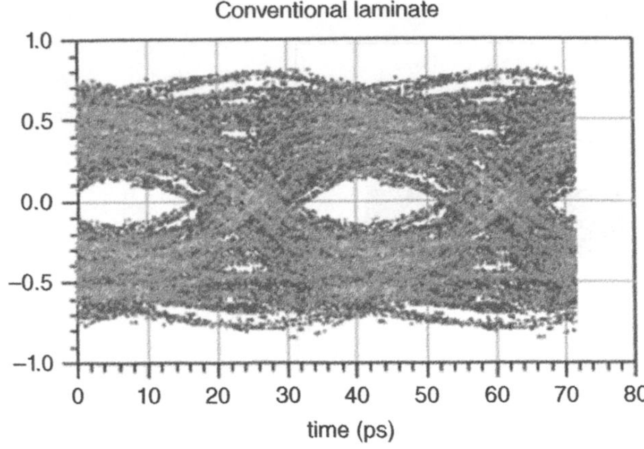

Figure 7.17 Comparison of eye diagrams at 28 Gbps.

Table 7.7 Comparison of signal integrity at 28 Gbps.

Key attributes	Chip-last HD-FO	Conventional laminate
Eye amplitude (mV)	998	993
Eye height (mV)	372	164
Eye width (ps)	25.7	17.5
P–P jitter (ps)	10.1	22
Rise/fall time (ps)	21	18

passed with temperature cycle (TC) 2000 cycles (JEDEC condition G) and completed 1000 drops with first failure obtained at 442 drops. Test results can be seen in Table 7.9. Figure 7.23 also shows the drop test Weibull analysis.

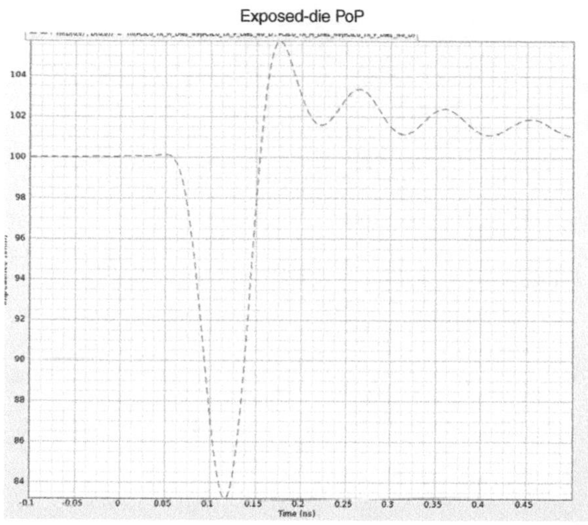

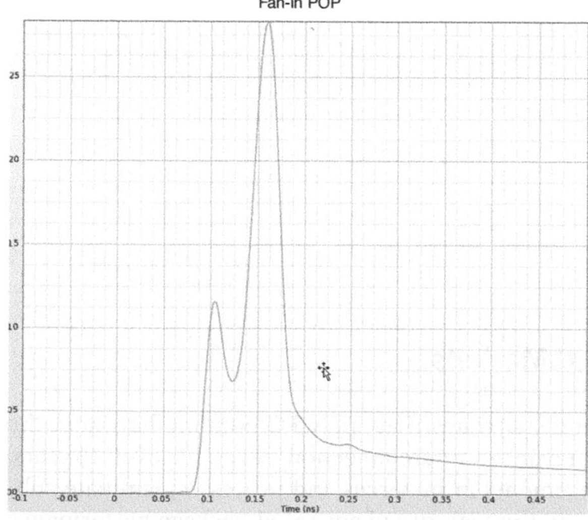

Figure 7.18 Time-domain reflectometry comparison.

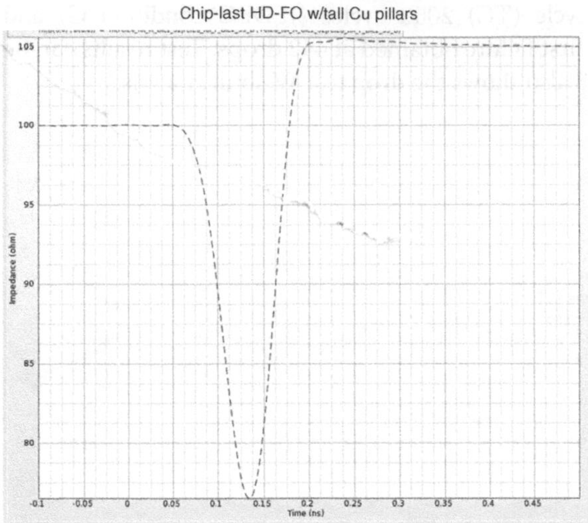

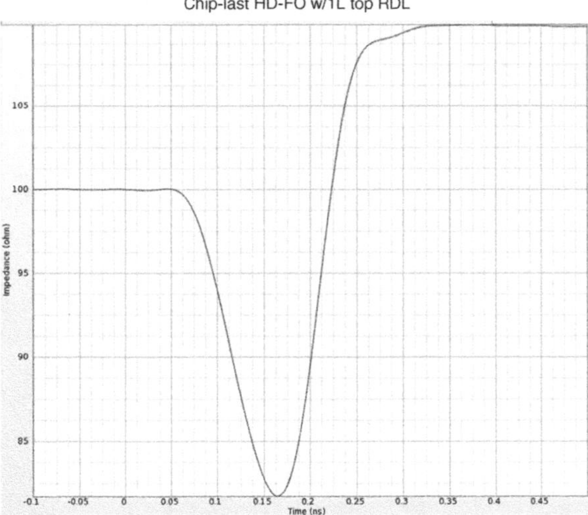

Figure 7.18 (Continued)

7.9 Applications and Markets

Traditional organic laminate substrates that apply FC bonding have met the semiconductor industry's advanced interconnection needs for over 15 years. With the continued advancements in materials and processes, laminate substrates are expected to satisfy the majority of advanced package performance and cost requirements for years to come. However, the feature size limitations

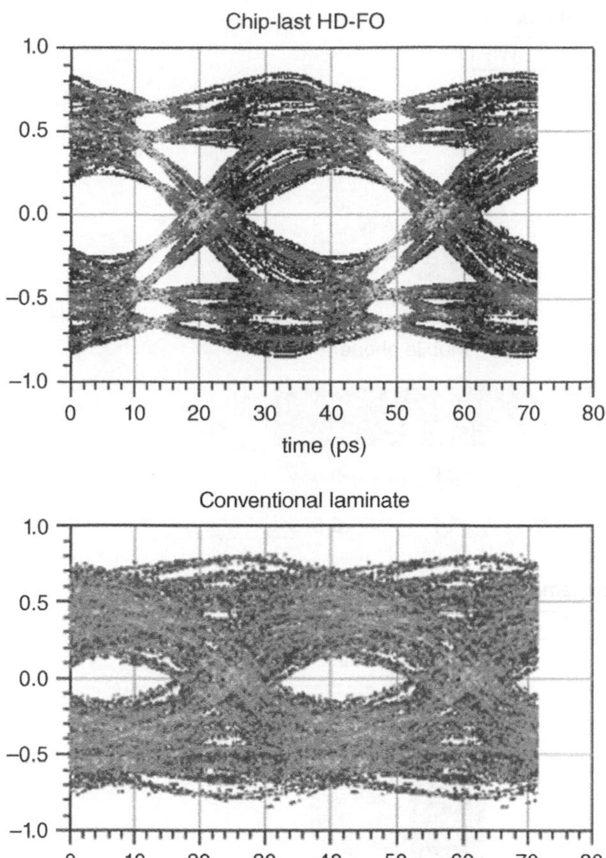

Figure 7.19 Comparison of eye diagrams at 28 Gbps.

and the electrical and thermal performance constraints will continue to restrict laminate substrates from meeting the integration requirements for next-generation mobile, high performance graphics, and networking applications.

Emerging silicon-based interconnection technologies such as TSV have shown promise in this area. By leveraging the back-end-of-line (BEOL) damascene processes of the wafer fab, multilayer submicron signal trace densities can be achieved. However, the supply chain limitations and intrinsic cost implications have limited the proliferation of 3D IC technology. In particular, for silicon interposers, there can be an undesirable effect to z-height and electrical performance due to the inherent thickness and parasitics of the silicon interposer.

Figure 7.24 illustrates the interconnection technologies that support the predominant semiconductor assembly platforms. There is a pronounced gap

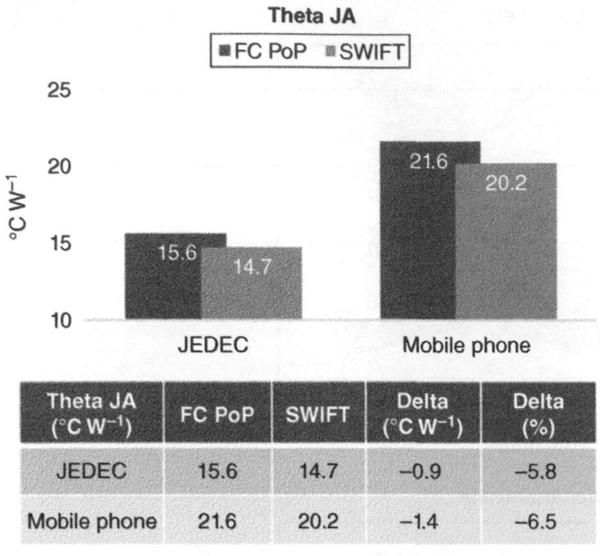

Theta JA (°C W^{-1})	FC PoP	SWIFT	Delta (°C W^{-1})	Delta (%)
JEDEC	15.6	14.7	−0.9	−5.8
Mobile phone	21.6	20.2	−1.4	−6.5

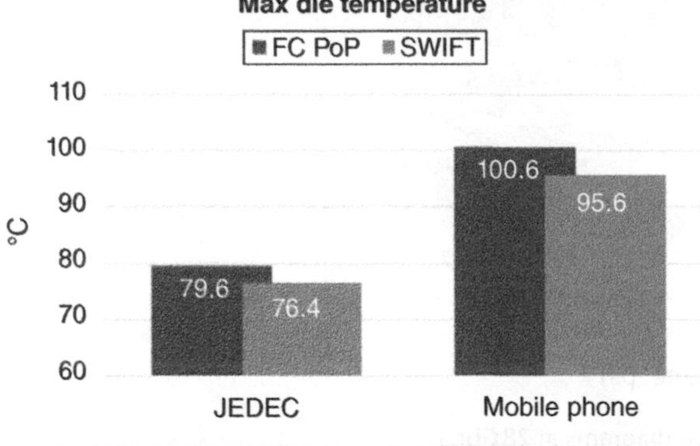

Max die temp (°C)	FC PoP	SWIFT	Delta (°C)	Delta (%)
JEDEC	79.6	76.4	−3.2	−4.0
Mobile phone	100.6	95.6	−5.0	−5.0

Figure 7.20 Chip-last HD-FO packaging vs. FC-PoP thermal simulation data – steady-state analysis.

for advanced fan-out applications that require disparate die integration on a single semiconductor package platform.

To close the platform technology gap, a new innovative chip-last wafer fan-out structure, called SWIFT technology, has been developed. This novel

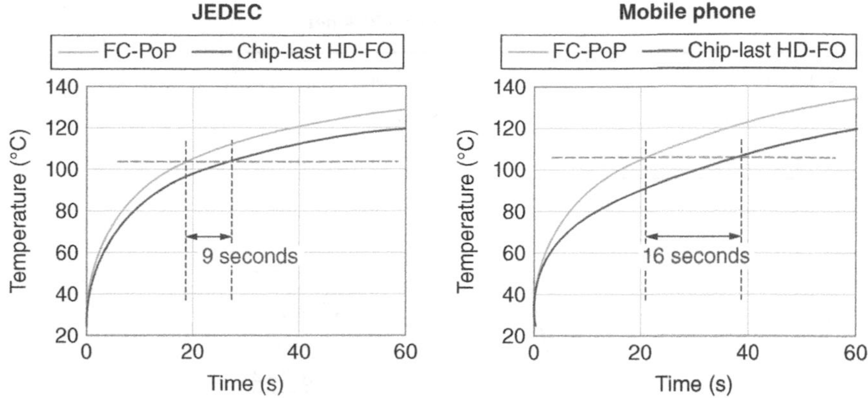

Figure 7.21 Chip-last HD-FO package vs. FC-PoP thermal simulation – transient analysis.

Table 7.8 Time for die temperature to reach 105 °C (9 W power dissipation).

	JEDEC	Mobile phone
FC-PoP	19.5 s	20.5 s
Chip-last HD-FO	28.2 s	36.8 s
Difference	8.7 s	16.3 s
	45%	80%

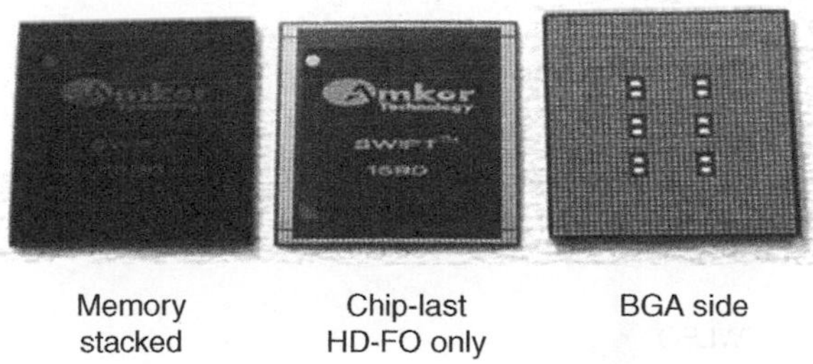

Figure 7.22 15 mm body package for reliability data collection.

packaging approach incorporates the fine feature size capabilities of WLP and the flexible platform attributes of FC assembles. Target applications include AP, baseband (logic + memory), and power management units for advanced mobile applications and high-speed switches for network servers.

Table 7.9 Drop test results.

	Corner BGA net
First failure	442 drops
Mean life	692 drops
63.2% life	754 drops

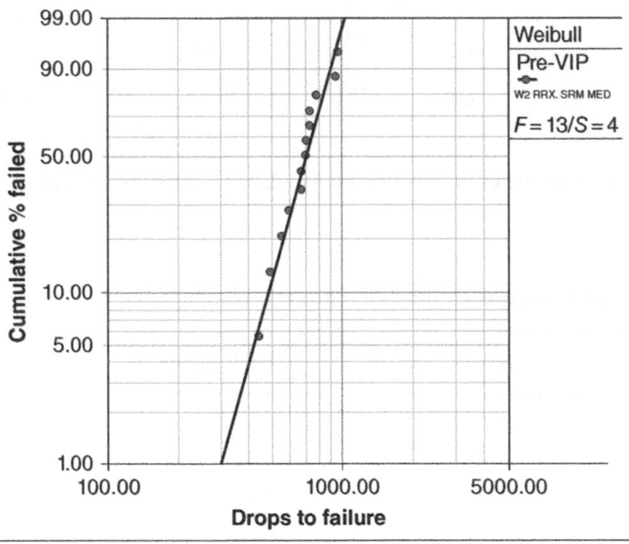

Figure 7.23 Drop test Weibull results.

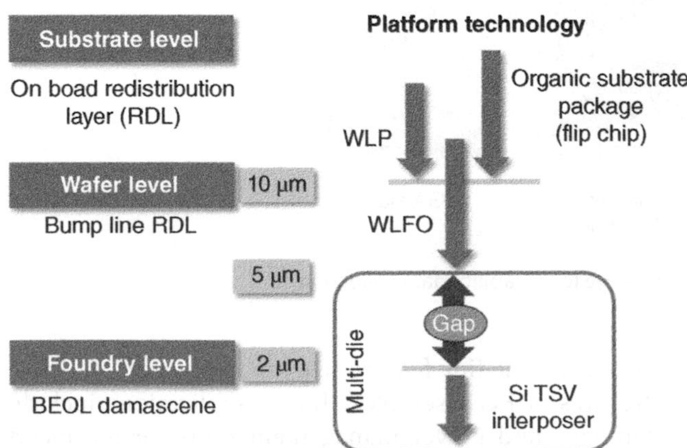

Figure 7.24 IC package integration roadmap.

References

1 Huemoeller, R. and Zwenger, C. (2015). Silicon wafer integrated fan-out technology. *Chip Scale Review* (March/April 2015).
2 Zwenger, C. and Huemoeller, R. (2015). Silicon wafer integrated fan-out technology. *IMAPS DPC Proceedings* (March 2015).
3 Baloglu, B., Scott, G., and Zwenger, C. (2016). Silicon wafer integrated fan-out technology. *IWLPC Proceedings* (October 2016).
4 Best, K., Steve Gardner, S., and Donaher, C. (2016). Photolithography alignment mark transfer system for low cost advanced packaging and bonded wafer applications. *International Symposium on Microelectronics: Fall 2016* 2016 (1): 000315–000320.
5 Yole Development (November 2017). *Advanced RF SiP for Cellphones.* Yole Development.
6 Yole Development (June 2017). *Status of the Advanced Packaging Industry 2017.* Yole Development.

8

Embedded Silicon Fan-Out (eSiFO®) Technology for Wafer-Level System Integration

Daquan Yu

Huatian Technology (Kunshan) Electronics Co.,Ltd., Economic & Technical Development Zone, Kunshan, Jiangsu, China

8.1 Technology Description

The demand for miniaturized package size, higher performance and integration density, lower power consumption, and lower manufacturing cost drives the development of various new packaging technologies. Among those new packaging technologies, the fan-out wafer-level package (FO-WLP) has emerged as a successful technology in providing the solution to fulfill the abovementioned requirements. FO-WLP has also become a key-enabling technology for multi-chip and 3D system integration [1]. As one type of FO-WLP, embedded silicon fan-out (eSiFO) technology, in which silicon instead of molding compound was used as the fan-out area, was proposed in 2015 [2]. It is a chip-first and face-up process, eliminating molding, temporary bonding, and debonding process and has been in volume production not only for single-die package but also for multi-die system-in-package (SiP) applications.

8.2 Basic Package Construction

Figure 8.1 shows the structure of an eSiFO package. In the package, known good die (KGD) were embedded in a silicon carrier, and the microgaps between the die and silicon carrier were filled by polymer. The die and silicon carrier reconstruct a surface for the routing of redistribution layers (RDL) and fabrication of solder balls. The main difference between eSiFO® and the typical FO-WLP is that there is no epoxy molding compound (EMC) in eSiFO package.

Advances in Embedded and Fan-Out Wafer-Level Packaging Technologies, First Edition.
Edited by Beth Keser and Steffen Kröhnert.

Figure 8.1 Schematic view of the embedded silicon fan-out (eSiFO) structure.

There are a number of advantages of the eSiFO package. The wafer warpage is very small during the manufacturing process since a silicon wafer is used as the reconstruct substrate. Therefore, wafer handling challenges and misalignment in lithography during the manufacturing process can be avoided [3]. The process is simple since there is no molding, temporary bonding, and debonding requirement. As a mature wafer-level process, fine-pitch high density RDL manufacturing is more easily built on a silicon wafer compared with an EMC wafer [1]. In addition, a small form factor can be achieved especially for SiP since multiple dies can be put close together and connected via fine-pitch RDL. Furthermore, an ultrathin package of 150 μm can be achieved by thinning the wafer at the end of the packaging process. The ultrathin package is very important to achieving thinner smartphones and wearable electronics.

8.3 Manufacturing Process Flow

Typically, there are five major steps for FO-WLP including KGD reconstruction, molding, debonding, wafer-level RDL, and back-end processing such as grinding, laser marking, and package singulation [1, 4]. From the process point of view, it is very troublesome to handle the EMC wafer, which usually has a large warpage since the coefficient of thermal expansion (CTE) of the EMC is largely different from the silicon chip. In addition, the manufacturing of fine-pitch RDL on the EMC surface is also difficult as the poor coplanarity of die to mold will distort fine width and affect the RDL line continuity [5].

The process flow for eSiFO® package manufacturing is illustrated in Figure 8.2 [6]. Firstly, a silicon wafer was used to form cavities with a certain depth. A dry etch process by Bosch was used for cavity formation [7]. Secondly, thin dies with designed thickness were picked and placed into the cavities. After attachment of KGD, what is termed a "reconstructed wafer" was the result. Thirdly, the microgaps between the die and silicon carrier were filled with polymer, and the surface of the reconstructed wafer was passivated at the same time. Fourthly, the pads on die were opened by a lithography and development process. Fifthly, RDL was fabricated by standard process including seed layer deposition, photoresist (PR) formation, plating, PR strip, and seed layer etching. Then final

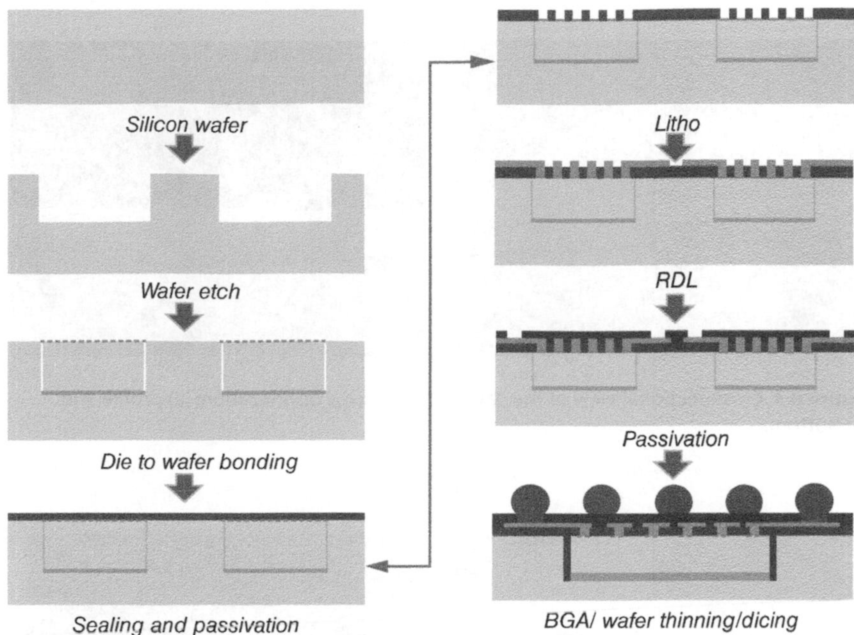

Figure 8.2 Process flow of eSiFO package manufacturing.

passivation was performed. After ball grid array (BGA) formation, wafer thinning, and dicing, finally, an eSiFO package was fabricated.

Both 8 and 12 in. wafer processes for eSiFO were developed. At first, the cavities with vertical sidewall for die embedding were formed by the Bosch process. An inductively coupled plasma (ICP) source reactor is used, and the gases used in the Bosch process are SF_6 and C_4F_8. For through-silicon via (TSV) formation, silicon etch is quite mature since the etching area is small. However, the etching of large silicon cavities with good total thickness variation (TTV) is challenging. In addition, a smooth bottom surface without any grass, the term for blades of Si at the bottom of the cavity, or bumps was required to prevent die tilt and cracks during die attach process, which were formed due to the residual passivation during Bosch process [7] and particle contamination. For device die thickness, thinning to 60, 80, and 100 μm is possible. For 100 μm die thickness, the depth of the cavities on the wafer was set as 107 μm. Figure 8.3 shows a cross-sectional view of a cavity for die embedding. The sidewall and bottom of the cavity are quite smooth. There are no "grass" or "bumps" on the bottom. The footing, the term for the angle where the sidewall meets the bottom of the cavity, variation in a single cavity is less than 5 μm without counting the region 15 μm away from the sidewall, to ensure the coplanarity of the embedded die and carrier. A flat bottom with ~1 μm roughness is achieved, which is suitable for die attachment.

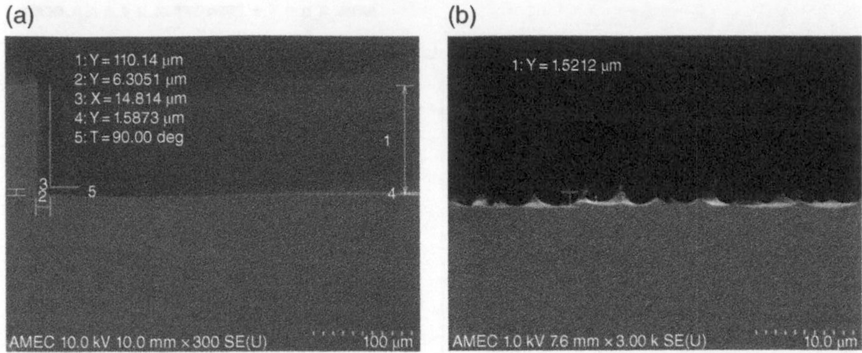

Figure 8.3 Cross-sectional view of the cavity after process optimization: (a) profile and (b) bottom.

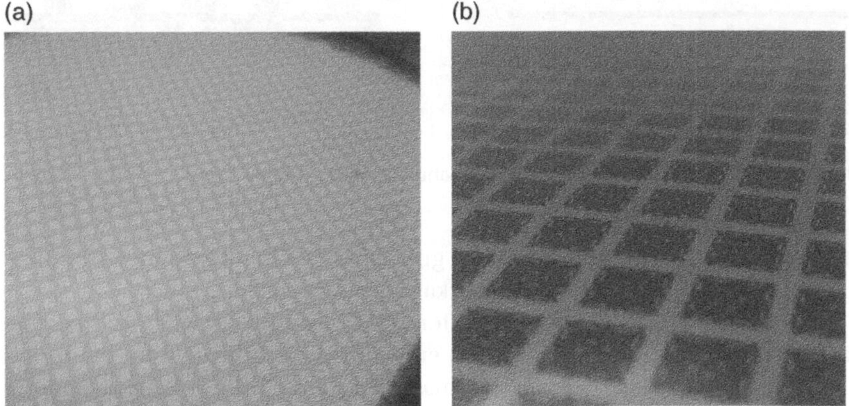

Figure 8.4 Eight inch reconstructed wafers with different die sizes: (a) 2 mm × 2 mm and (b) 7 mm × 7 mm.

A pick and place tool was used for chip to wafer attachment with an accuracy of ±4 μm. A die attach film (DAF) was applied to the backside of the die at wafer level, which is used to ensure fully bonding between the die and the silicon carrier without void. After die attachment, the wafer was hard-baked for the curing of adhesive film. As shown in Figure 8.4, two reconstructed silicon wafers were formed before polymer filling. The gaps between die and silicon carrier are not visible.

The filling of the trenches between the die and silicon carrier was a key process for eSiFO. A vacuum film lamination process was developed. The trench was filled without voids or cracks. Trenches in ~110 μm depth with 15, 20, 30,

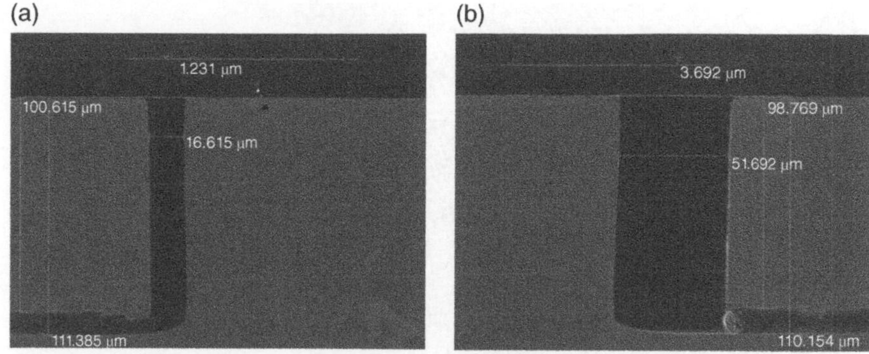

Figure 8.5 The polymer filling of the trench with different widths: (a) 15 µm and (b) 50 µm.

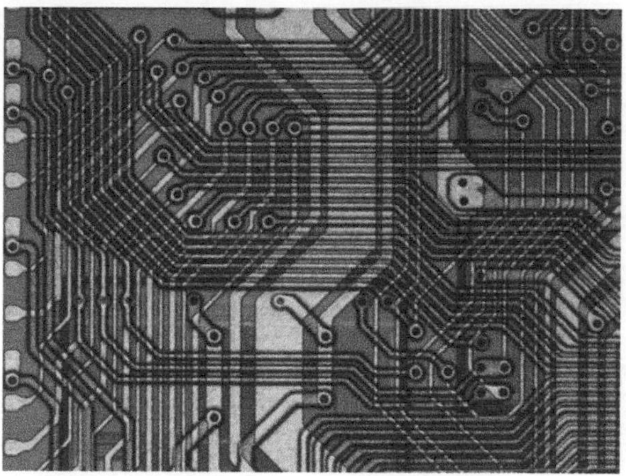

Figure 8.6 Two-layer RDL for multi-die integration.

and 50 µm width after die attach were tested for polymer filling process. As shown in Figure 8.5, void-free filling of the trenches was achieved.

Based on the current development of process, for eSiFO® package, a minimum opening size of 30 µm on first passivation, formed by the polymer filling process, can be achieved. A minimum line width of 15 µm for two RDL was developed for multi-die integration, as shown in Figure 8.6. A Ni/Au layer was plated as under-bump metallurgy (UBM) for BGA after the RDL process.

To prove the concept of eSiFO, a 3.3 mm × 3.3 mm package with 0.4 mm pitch BGA was fabricated [6]. In the package, there is one layer of Cu RDL with a thickness of 3 µm and a minimum line width of 14 µm. The die size is

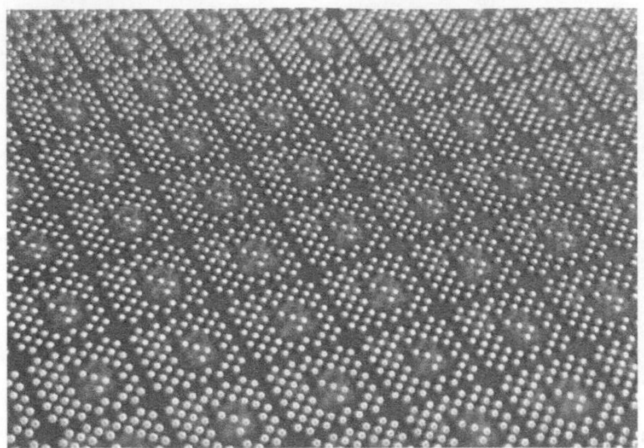

Figure 8.7 eSiFO after BGA formation.

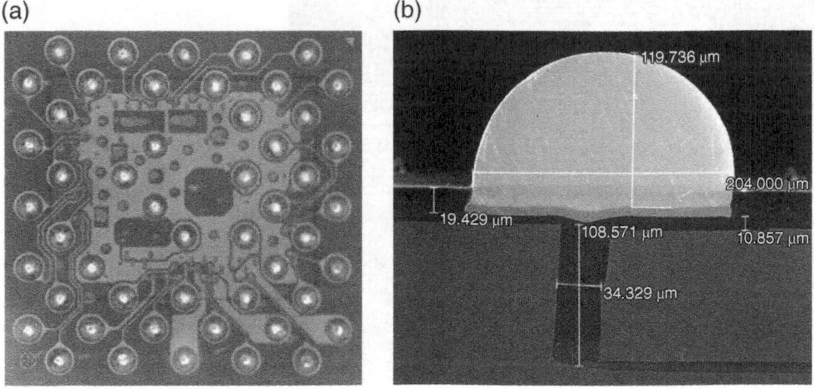

Figure 8.8 An eSiFO package: (a) the outlook and (b) the cross-sectional microstructure.

1.96 mm × 2.36 mm with a pad pitch of 90 μm. Figure 8.7 shows the completed eSiFO after BGA solder balls formation. The eSiFO package is shown in Figure 8.8a. The die is embedded in the center of the package, and the 30 μm trench is filled with polymer. The cross-sectional microstructure of the BGA, die, and trench is shown in Figure 8.8b. The filling of the trench is perfect without any voids or cracks. The BGA solder balls on the trench are not distorted due to the existence of the filling polymer. The difference in height of the die and silicon wafer surface is within 5 μm. The thickness of first passivation layer on the reconstructed wafer is about 10 μm, and the thickness of final passivation is around 20 μm.

To determine the thin package capability of eSiFO, final package thicknesses of 300, 250, 200, and 150 μm were studied and demonstrated. For wafers with

300 μm thickness, the warpage is less than 2 mm. The small warpage is due to the similar CTE between silicon wafer and embedded die.

8.4 Design Features

Design rules for the eSiFO package are listed in Table 8.1. The wafer can be thinned to 50 μm based on tool capability. But the handling of such thin die with DAF and the die attachment to silicon carrier is challenging. For silicon cavity etch, when the depth is larger than 150 μm, the depth variation will be larger than 10 μm. The opening of first passivation layer will be difficult. Therefore, a die thickness between 80 and 140 μm and a cavity depth between 90 and 150 μm were preferred. The package thickness can be thinner than extended wafer-level BGA (eWLB), a common type of FO-WLP in high volume manufacturing, since thinning is done at the final stage. The thickness of the package depends on the die thickness and warpage after thinning. For reliability, tens of microns of silicon thickness below the embedded die needs to remain. In addition, the smaller the package, the thinner die and package that can be achieved.

An RDL width of 10 μm is achieved and 7 μm RDL is under development. The processes on silicon wafer with small warpage reduced the difficulties.

The technology development roadmap for the eSiFO package is shown in Table 8.2. According to advanced technology nodes, the chip size continues to become smaller. The pick and place of smaller die with high speed and high accuracy is quite challenging. A new advanced tool needs to be developed to assemble die with size smaller than 0.3 mm × 0.3 mm for FO-WLP. For multi-layer RDL with a finer pitch than 7 μm, there are a number of difficulties. For improved adhesion between RDL and passivation, a better manufacturing environment with strict particle control is required to ensure good yield. Advanced tools for lithography, development, plating, and wet etch are required. PR materials with better resolution and chemicals for seed layer etching with less over etch are needed. The big challenge for eSiFO technology is the large die package. The board-level reliability (BLR) is a big concern since the difference of CTE between the package and printed circuit board (PCB) is large. The package thickness, size, solder ball size, underfill materials, and pad structures on PCB will affect the BLR.

8.5 System Integration Capability

FO-WLP has also become a key-enabling technology for multi-chip and 3D system integration [4]. One such example is TSMC integrated fan-out (InFO) technology, which not only provides a system scaling solution but also complements the chip scaling and helps to sustain Moore's law [8].

Table 8.1 Design rules for the eSiFO package.

Parameter	Design spec (μm)		Recommended (μm)	Development (μm)	Comments
	Min.	Max.			
Package height (without bump)	250	—	350	150	
Silicon cavity depth	90	150	108	70	
Die thickness	80	140	100	60	
Gap between die and carrier	20	50	30	15	
First passivation layer opening	30	—	—	20	
RDL line width	10	—	15	7	
BGA height	100	300	200	—	Ball drop
	100	250	130	—	Ball printer
	—	100	60	—	Electroplating

Table 8.2 Technology development roadmap for the eSiFO package.

Items	2017	2018	2019	2020
Die size (mm)	0.6 × 0.6 to 5 × 5	0.5 × 0.5 to 8 × 8	0.3 × 0.3 to 10 × 10	0.2 × 0.2 to 10 × 10
Pad pitch (μm)	≥60	≥55	≥50	≥45
Min. RDL (L/S) (μm)	10	7	5	2
RDL layers	2	3	4	4
Package size (mm)	1 × 1 to 6 × 6	0.8 × 0.8 to 10 × 10	0.8 × 0.8 to 15 × 15	0.8 × 0.8 to 15 × 15
Package thickness (μm)	250	150	150	150
I/Os	>100	>500	>1000	>1000
Min. BGA height (μm)	100	90	80	80

For eSiFO technology, multi-die integration can be realized easily in both 2D and 3D format. For 2D SiP, multiple die can be placed into one cavity or different cavities. The distance between the die can be made as small as 210 μm. So far, a SiP package with five-die integration and two-layer RDL was developed. Two kinds of chips were used, where one chip size is 1.2 mm × 1.2 mm and another chip size is 1.4 mm × 1.7 mm. A package in the size of 5.5 mm × 4.0 mm was achieved [9].

The eSiFO technology can also realize the same package structure by using TSV formed on the silicon carrier. A test package with a size of 4.14 mm × 4.14 mm was developed [9]. As shown in Figure 8.9a, on the frontside of the package, a 3 mm × 3 mm die was embedded into a silicon carrier with a 0.2 mm thickness.

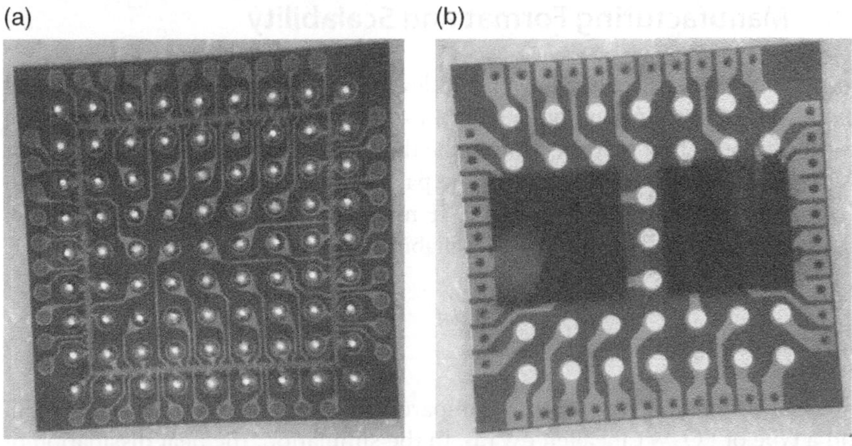

Figure 8.9 3D eSiFO package: (a) frontside and (b) backside.

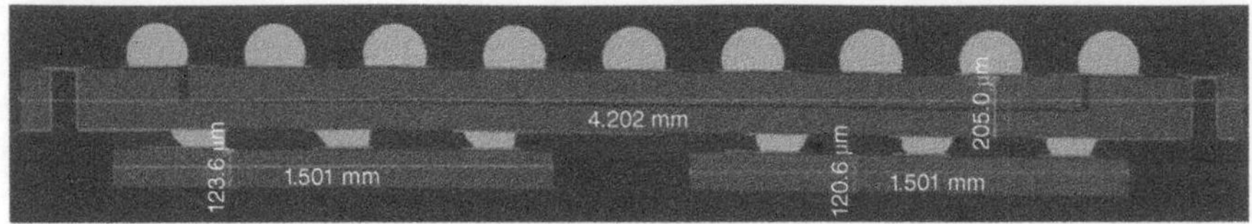

Figure 8.10 Cross-sectional view of a 3D eSiFO package.

Table 8.3 Electrical test results.

Number of units	Chains per unit	Resistance(Ω) of each chain	Results
21	18	<2.5	Passed

One-layer RDL with 25 µm line width and 81 BGA balls with 210 µm diameter and 140 µm height were formed on the top. As shown in Figure 8.9b, two die were bonded on the backside. The die size for flip chip was 1.47 mm × 1.47 mm with a thickness of 125 µm. The solder bump height was 90 µm and the UBM was 160 µm in diameter. Figure 8.10 shows a cross-sectional view of the 3D eSiFO package. The embedded die and two FC dies were connected by RDL and TSV 100 µm in diameter. There are a total of 42 TSV for the present test vehicle.

As listed in Table 8.3, electrical tests for daisy chains of 21 3D packages proved that 3D eSiFO package has good electrical connection. Future work for large package, fine-pitch RDL, and high density TSV will be studied, and reliability evaluation will also be carried out.

8.6 Manufacturing Format and Scalability

Both 8 and 12 in. processes have been developed for eSiFO technology. It will remain at 12 in. wafer manufacturing format since 12 in. silicon wafers are readily available and provide more scale than 8 in. wafers.

The limitation for eSiFO may be the package size. Package sizes larger than 15 mm × 15 mm would have a large CTE mismatch between package and PCB, which may cause failure during TC reliability testing.

8.7 Package Performance

Thermal simulation was used to compare the thermal performance of eSiFO and a type of FO-WLP called eWLB. In the simulation, the heat dissipation of the chip was assumed to be 3 W, and the package size was fixed as 6 mm × 6 mm × 0.45 mm with an embedded die 100 µm thick. Simulation results under natural convection are shown in Figure 8.11. It can be found that the thermal resistance of both eSiFO and eWLB packages would decrease as the die size increases. Furthermore, thermal resistance of eSiFO is lower than that of eWLB package by 5–40% in all die sizes. This means that the eSiFO package offers significant advantages in thermal performance.

As a wafer-level package, eSiFO enables short electrical paths from die out to package, and using high-resistance silicon as the silicon carrier can further

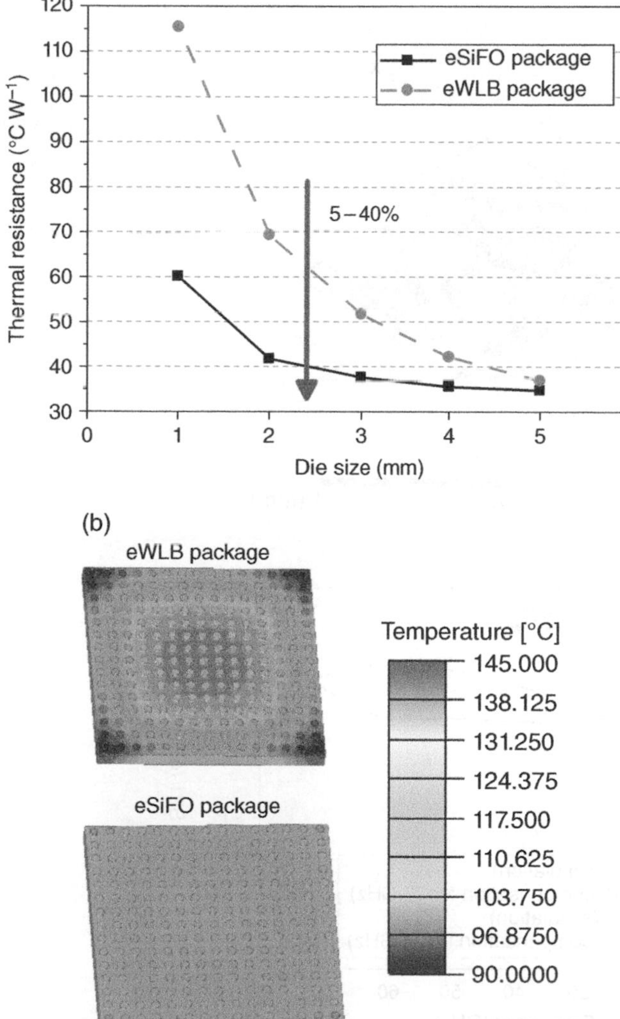

Figure 8.11 Thermal comparison between eSiFO and eWLB package: (a) thermal resistance and (b) temperature distributions for the 4 mm × 4 mm die.

improve electrical performance for low loss waveguide transmission. To prove this, a 3D full wave electromagnetic (EM) simulation of package to board transition was conducted from DC to 80 GHz. A coplanar waveguide (CPW) structure of the package side and a substrate integrated waveguide (SIW) structure of the PCB side were chosen to provide low loss and negligible radiation advantages for millimeter-wave applications [10]. Similarly, high-resistance silicon was chosen in this simulation. An EM numerical model and simulated S parameter results with/without compensation are shown in Figure 8.12. It is

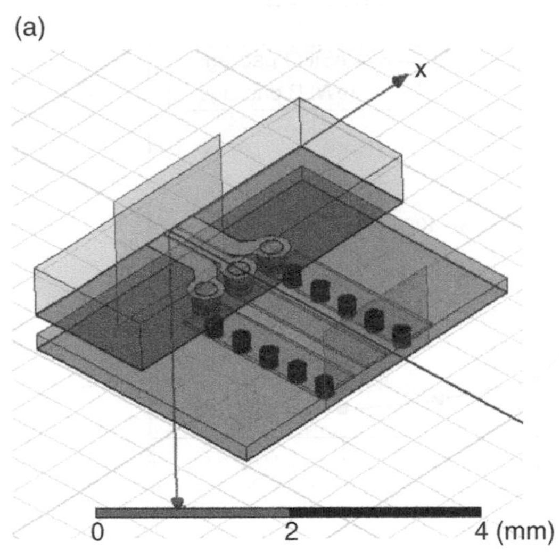

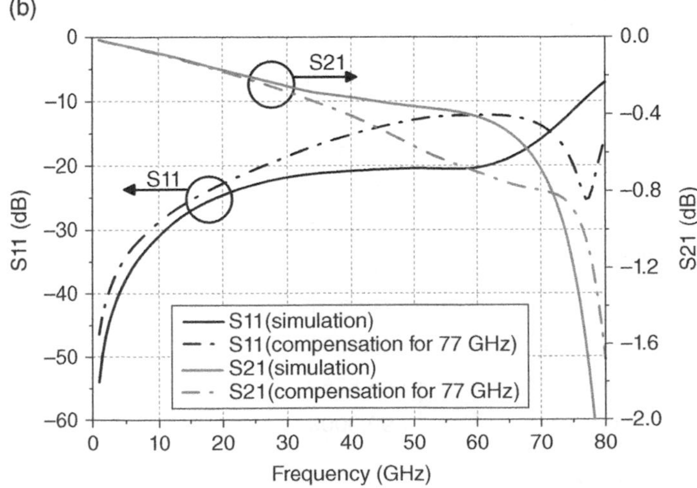

Figure 8.12 Electromagnetic simulation of eSiFO package to board transition: (a) EM numerical model and (b) S parameter results.

found that after impedance matching, eSiFO can provide ultrawide band at 20 dB return loss and 0.5 dB insertion loss from DC up to 60 GHz. In addition, through transmission line compensation at 77 GHz for automotive radar applications, an absolute bandwidth up to 6 GHz can be realized by eSiFO. The results indicate that eSiFO is qualified for packaging highly integrated millimeter-wave devices.

8.8 Robustness and Reliability Data

Both package-level and BLR tests have been performed for eSiFO. The results showed that the eSiFO package has a strong robustness.

For package-level reliability, a 3.9 mm × 3.9 mm package with a body thickness of 350 μm was used. The JEDEC reliability test specifications and conditions are shown in Table 8.4. Prior to the reliability tests, preconditioning was performed to simulate the effects of board assembly on moisturized packaging. The samples were first baked at 125 °C for 24 hours to remove the moisture inside the package and then soaked at 30 °C under 60% relative humidity (RH) for 192 hours. Last, the samples were reflowed at 260 °C for three times. Temperature cycling condition B (TC-B) test (−55 to 125 °C) up to 1000 cycles, temperature humidity storage (THS) test at 85 °C under 85% RH for 1008 hours, and highly accelerated stress test (HAST) at 110 °C under 85% RH for 264 hours were conducted for reliability evaluation. For each condition of preliminary research, 21 samples were used. According to electrical test results, all the samples passed the reliability tests, as shown in Table 8.4.

For BLR, a test vehicle of a 3 mm × 3 mm package with a single 1.4 mm × 1.4 mm embedded die was designed, manufactured, and mounted on a board [11]. Several electrical measurement structures, such as daisy chains and single bump resistance measurement, were built in the test vehicle. Four daisy chains with 24 pads were designed in one package, and there were 17 samples in one test board. A typical commercial PCB with 1 mm thickness was chosen

Table 8.4 Package-level reliability test results.

Reliability test	JEDEC specification	Test condition	Results
Preconditioning	JEDECJ-STD-020D	125 °C, 30 °C/60% RH, 3 × reflow	Passed
Temperature humidity storage (THS)	JESD22-A101	85 °C/85% RH	Passed
Temperature cycling (TC)	JESD22-A104	−55/125 °C	Passed
Unbiased highly accelerated stress test (HAST)	JESD22-A118	110 °C/85% RH	Passed

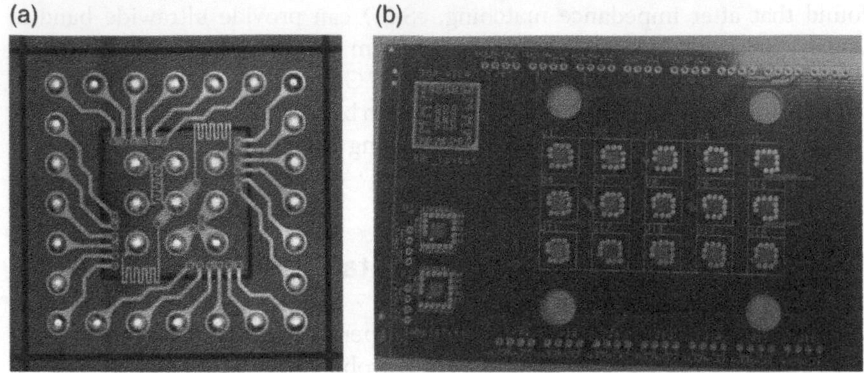

Figure 8.13 .Test vehicles for board-level reliability: (a) the designed eSiFO test vehicle and (b) the corresponding mounted board.

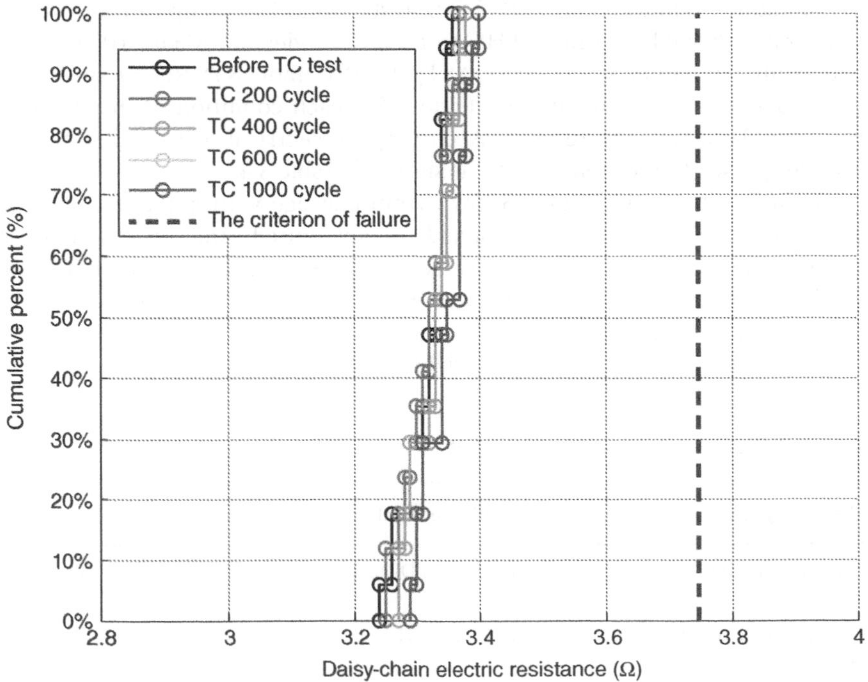

Figure 8.14 .Cumulative percentage of daisy chain electric resistances under board-level temperature cycling test.

for the test board, and two-layer copper traces with test ports were designed to connect the daisy chains to the board. The designed eSiFO test vehicle and the corresponding mounted board are shown in Figure 8.13. A cumulative distribution function is used to characterize the varied electric resistance of the eSiFO package-PCB daisy chain loop, and the criterion of failure is that the resistance value exceeds 10% of the average value of the samples. Samples with different package thickness (150, 250, 470 μm) and die thickness (70, 100 μm) were subjected to a board-level thermal cycling test (−40 to 125 °C). All 68 test samples with underfill passed 1000 cycles. The cumulative percentage of daisy chain electric resistances under TC test is shown in Figure 8.14. It is found that even after TC treatment of 1000 cycles, the average resistance change rate is less than 1%, which indicates that the reliability of the eSiFO package can meet the majority of application requirements.

8.9 Applications and Markets

The application of eSiFO is similar to eWLB, which can be used to package high performance ASIC (application-specific integrated circuit), mobile BB/APE (baseband modem/application processor engine), PMU (power management unit), RF transceiver, audio/video codec, sensors, and other multi-chip and SiP modules. Since ultrathin and ultrasmall packages and SiP integration can be achieved with high yield [9], eSiFO is particularly suitable for the markets of consumer electronics and wearable electronics. For areas with high reliability requirements, such as automotive, medical electronics, and aerospace, some proposed solutions are still in progress. Recently, RF transceiver and driver modules using eSiFO have been in volume production. ASIC, PMU, and sensors using eSiFO are also under development. In the near future, large die packages like APE and ASIC using 3D eSiFO will be also developed.

Acknowledgment

The author would like to thank Dr. Shuying Ma, Dr. Chengqian Wan, Mr. Cheng Chen, Mr. Min Xiang, Ms. Jiao Wang, Mr. Weidong Liu, and other colleagues for the paper preparation and technology development. The support by ASM Pacific Technology, AMEC, and NMC Ltd. for process development was appreciated.

References

1 Lau, J.H., Fan, N., and Ming, L. (2016). Design, material, process, and equipment of embedded fan-out wafer/panel-level packaging. *Chip Scale Review* May/June: 38–44.

2 Yu D. (2015). Embedded silicon fan-out package and the method of forming the same. Chinese Patent 201510486674.1, filed 11 August 2015.
3 Chong, S.C., Chong, S.C., Wee, D.H.S. et al. (2013). Development of package-on-package using embedded wafer-level package approach. *IEEE Transactions on Components, Packaging and Manufacturing Technology* 3 (10): 1654–1662.
4 Lin, Y., Kang, C., Chua, L. et al. (June 2016). Advanced 3D eWLB-PoP (embedded wafer level ball grid array-package on package) technology. In: *Proceedings of the 66th Electronic Components and Technology Conference (ECTC)*, 1772–1777. IEEE Press.
5 Huemoeller, R. and Zwenger, C. (2015). Silicon wafer integrated fan-out technology. *Chip Scale Review* 19 (2): 10–13.
6 Yu, D., Huang, Z., Xiao, Z. et al. (June 2017). Embedded Si fan out: a low cost wafer level packaging technology without molding and de-bonding processes. In: *Proceedings of the 67thElectronic Components and Technology Conference (ECTC)*, 28–36. IEEE Press.
7 Dixit, P. and Miao, J. (2006). Effect of SF6 flow rate on the etched surface profile and bottom grass formation in deep reactive ion etching process. *Journal of Physics: Conference Series* 34 (1): 577–582.
8 Tseng, C.-F., Liu, C.-S., Wu, C.-H., and Yu, D. (June 2016). InFO (wafer level integrated fan-out) technology. In: *Proceedings of the 66th Electronic Components and Technology Conference (ECTC)*, 1–6. IEEE Press.
9 Ma, S., Wang, J., Zhen, F. et al. (2018). Embedded silicon fan-out (eSiFO): a promising wafer level packaging technology for multi-chip and 3D system integration. *Proceedings of the 68th ElectronicComponents and Technology Conference (ECTC)*.
10 Lee, S., Jung, S., and Lee, H.-Y. (2008). Ultra-wideband CPW-to-substrate integrated waveguide transition using an elevated-CPW section. *IEEE Microwave and Wireless Components Letters* 18 (11): 746–748.
11 Cheng, C., Wang, T., Yu, D., et al. (2018). Reliability of ultra-thin embedded silicon fan-out (eSiFO) package directly assembled on PCB for mobile applications. *Proceedings of the 68th Electronic Components and Technology Conference (ECTC)*.

9

Embedding of Active and Passive Devices by Using an Embedded Interposer

The i^2 Board Technology

Thomas Gottwald, Christian Roessle, and Alexander Neumann

Schweizer Electronic AG, Schramberg, Germany

9.1 Technology Description

Currently, there are two major trends in embedding technology: embedding for system in package (SiP), which is supposed to be assembled onto commodity printed circuit boards (PCB), and system in board (SiB), where the embedding of components takes place in the main board (see Figure 9.1). The i^2 Board is an embedding technology for SiB.

The major difference between i^2 Board and other embedding technologies is that i^2 Board is a fan-out packaging technology. The i^2 Board technology was developed to enable chip embedding of fine-pitch chips with high input and output signal (I/O) count into the main boards. The i^2 Board approach uses an interposer to simplify the mechanical and electrical connection between chip and PCB. This has a significant positive influence on cost, ease of supply chain, testability, and the feasibility for chips with fine pitches. An example of an assembled interposer can be seen in Figure 9.2.

Today, several different nomenclatures are being used in chip embedding discussions: chip-first, chip-last, face-up, face-down, fan-in, and fan-out. There are several different approaches; some of them are already running in medium-scale production, e.g. ECP® of AT&S and SESUB® of TDK [1, 2].

One major problem in the embedding technology is the production yield, while another is the known good die (KGD). In a standard PCB shop, the production yield is much less than 100%, which is accepted for an inexpensive product. However, if an expensive component is embedded into a PCB, there would be high cost associated with PCB yield loss, because the embedded component would be scrapped. In addition, the PCB can be very costly, due to processing complexity and special materials. Therefore, KGD is preferred for

Advances in Embedded and Fan-Out Wafer-Level Packaging Technologies, First Edition.
Edited by Beth Keser and Steffen Kröhnert.

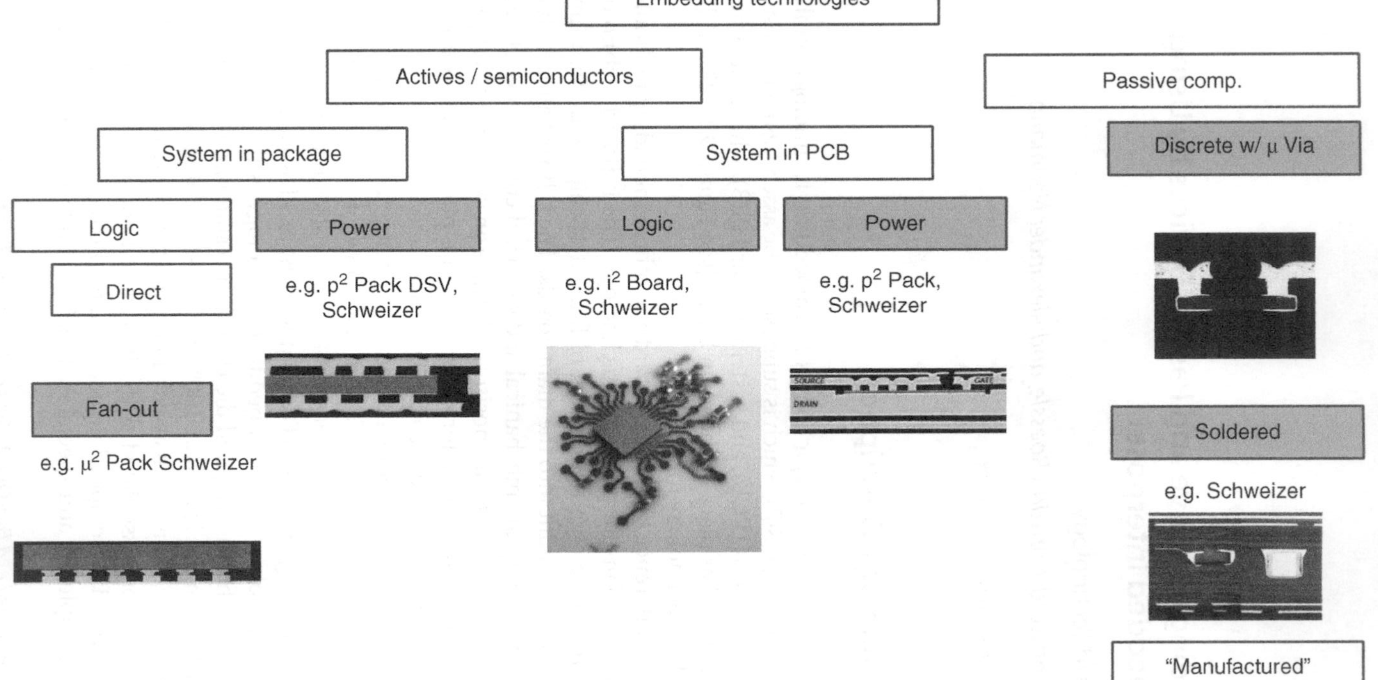

Figure 9.1 Overview of embedding technologies.

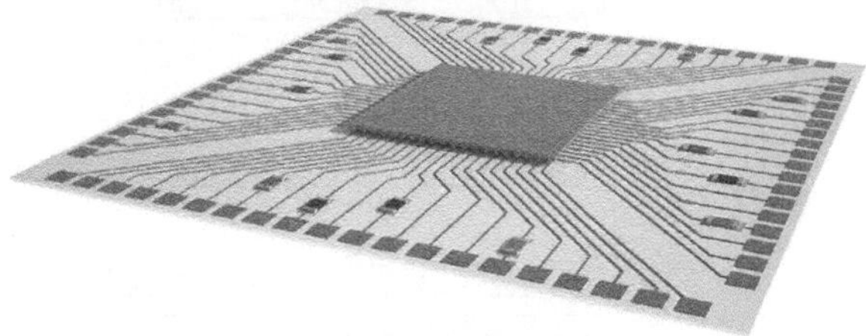

Figure 9.2 Active and passive components assembled on i^2 Board interposer before embedding.

embedding rather than an untested component, as a bad component would cause yield issues adding to scrap costs.

An embedded product typically consists of a PCB substrate, which has several buildup layers. A redistribution layer (RDL) is typically applied to the chip at wafer level. By using the area of the chip, larger contact pads can be offered for the interconnection with the PCB. The distribution of those contact pads to the chip surface leads to a fan-in design, so it is limited to the given surface area of the chip. After embedding such a chip, the escape routing of the signals to the PCB has to be carried out with several buildup layers as the design rules for PCBs are much coarser than in the semiconductor business.

In Figure 9.3, the left process sequence describes one version of a chip-last technology: First, a substrate is built using a multilayer substrate process technology and offers assembly contacts for the embedded component. Only the outer layers of the substrate are missing. In a second step, the component is assembled in a flip-chip technology to the surface of the substrate. In a third step, the assembled board is stacked with additional material layers for the embedding of the component and the later outer layers. This stack is laminated in a multilayer press to the finished board, which afterward can be treated like a multilayer in the outer layer processes. As an alternative process of the chip-last construction in Figure 9.3, a cavity for the component is formed before the assembly process.

On the right side of Figure 9.3, another variation is shown, where the substrate layers are built up around the chip. This is called chip-first, because the chip is mounted first in the process. The chip pads have to be contacted after lamination into the substrate with a process consisting of laser via drilling, through-hole plating, and outer layer etching.

The choice for the right strategy (chip-last versus chip-first) is determined by the cost of the chip component and the manufacturing cost of the substrate. An embedded construction with a small substrate of low complexity in combination with an expensive die typically would use a die last technology to maximize yield and minimize overall cost. In the case of low die cost, a chip-first

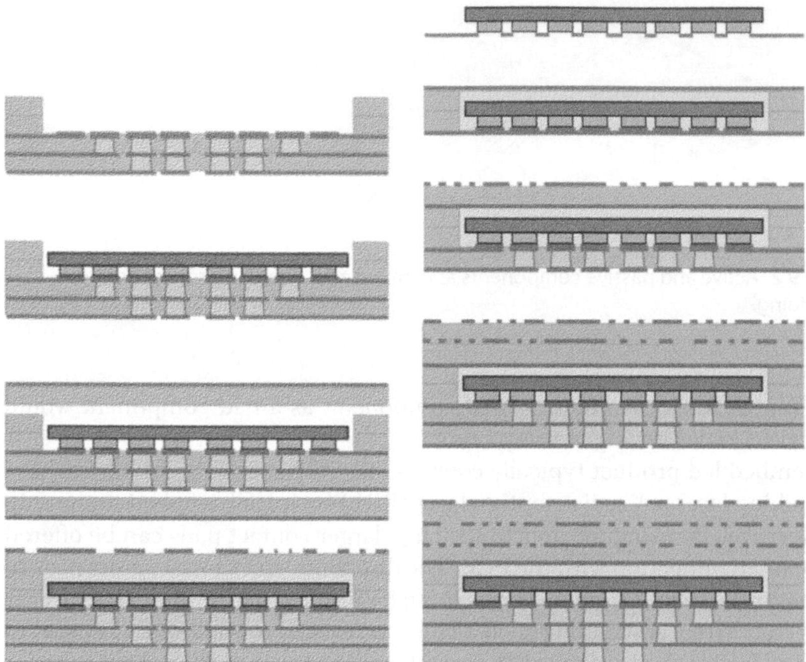

Figure 9.3 Different chip assembly methods for chip embedding: on the left is the chip-last process technology, and on the right is the chip-first process technology.

strategy with untested die might be less expensive, because bare die tests are much more costly than package tests.

In Figure 9.4, the embedding of a chip component into a PCB core is illustrated. In the face-up process (Figure 9.4 left), the backside of a component is attached to a copper foil with a conductive adhesive or by sintering [3]. In the next step, the components are covered with a resin-coated copper (RCC) foil and laminated with a multilayer process to an embedded core layer. As an alternative material, FR-4 prepregs can be used with a cutout around the components and a thin full-surface prepreg as a cover layer. After lamination, the component is contacted with laser via drilling to the contacts and electroplating. On the right side the same process is described for a face-down assembly.

Several European Union (EU)-funded projects on the topic of chip embedding have been completed, as, for example, Chip in Polymer, HIDING DIES, and SHIFT, which have been using the described technologies [4]. The approach to contact the embedded chips is quite similar. A thinned chip is fixed onto an inner layer or into a coreless substrate with high positional accuracy and then laminated or cast in a polymer matrix. The contacts, which are

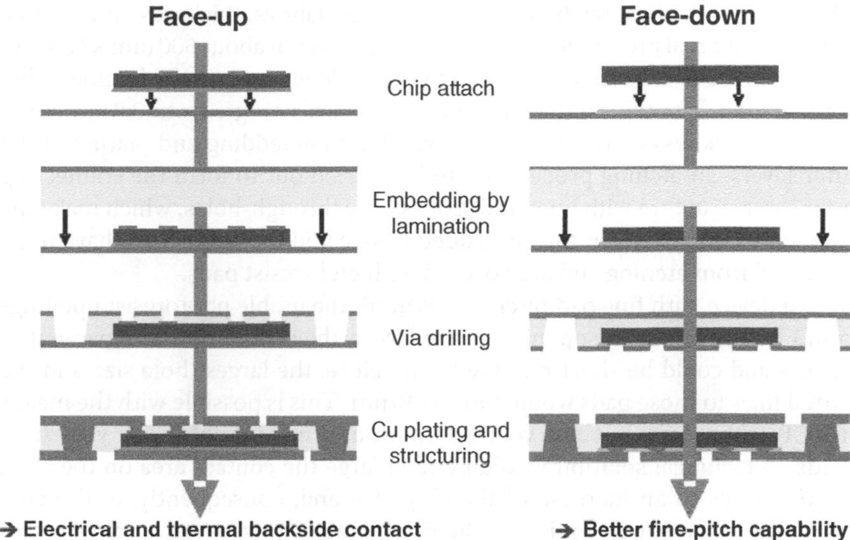

Figure 9.4 Face-up/face-down assembly. *Source:* courtesy of Fraunhofer IZM, Berlin.

oriented toward an outer layer, are opened up by laser drilling and then connected by means of through-hole plating with electroless and electrolytic copper. However, to be compatible with standard substrate and PCB processes, the chip contacts have to be plated with approximately 5 μm copper. This chip pad conversion has to be done to achieve a good adhesion in the following electroplating process, which would not work with a standard Al-surface finish of the pads. For flip-chip connections, the chip is placed face-down, reflowed, underfilled, and overmolded. All these methods have one common challenge: to match the fine contact structures of the chip with the comparatively coarse structures of the PCB. The PCB features used in these structures are as small as 20 μm.

To reach a high positional accuracy of the chip and a reliable metallization of the laser-drilled holes in subsequent processes, the following parameters are crucial:

- Accuracy of chip positioning with respect to the inner layer in x, y, theta.
- Accuracy of the thinning process of the chip.
- Control of the glue volume for a constant height of the chip with respect to the inner layer level.
- Control of the pressing conditions within the package.
- Positional accuracy of laser-drilled vias relative to the die.
- Dimensional variations of inner layer material after pressing.
- Die shift during lamination processes.

For the more critical fill direction of FR-4 glass fabrics, thickness dimensional variations of a real production batch with panel sizes of about 600 mm × 600 mm (24″ × 24″) are in the range of ±150 μm (6 mil), depending on the laminate that is being used. The typical thickness of these boards ranges from 0.8 mm up to standard thickness of 1.6 mm and above. After embedding and plating of the outer layers, an etching process has to be carried out to form the connecting outer layer structure with lines and spaces. The through-holes, which make the interconnection between the embedded device and the outer layer, have to be protected from etching and are covered with etch-resist pads.

For a device with fine pad pitches (~80 μm), the usable photoresist openings would be ~40 μm. If the openings were larger, they could not be separated in etching and could be short-circuited. Therefore, the largest hole size and the routed lines to those pads would be also 40 μm. This is possible with the means of PCB technology but not very common and not feasible with very high yields. A technical solution would be to enlarge the contact area on the chip, but this leads to an increase of the chip size and, consequently, of the chip costs. The other way around – a reduction of the pad size – is desirable to save area on the chip.

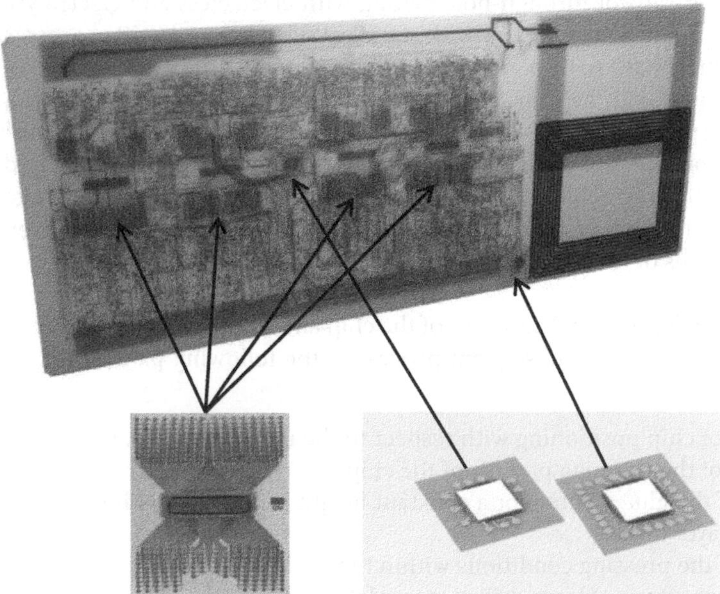

Figure 9.5 World record for embedded components: six components with 1231 I/Os with a minimum chip pitch of 50 μm.

The i^2 Board technology is not in mass production yet; however it is ready for product development. The technology itself still claims a world record for embedding in I/O count and fine-pitch devices.

In Figure 9.5, a PCB with six embedded components is shown. It was used for the control of a bi-stable e-ink display for the use in logistics in the publicly funded project "Pariflex" [5].

Embedded were the following:

- Four display drivers with 298 I/Os at pitch of 50 μm.
- RFID – front-end chip.
- Logic control chip for the display drivers.

The minimum chip pitch of 50 μm at a total I/O count of more than 1200 is still an unbroken record. The supply chain and the processes are ready to produce qualification samples.

9.2 Basic Interposer Construction

The interposer used for this technology is a thin FR-4 laminate that is etched to the necessary layout in a simple print and etch process. Therefore, a thin (e.g. 0.1 mm) FR-4 laminate with a thin (e.g. 12 μm) Cu coating is laminated with photoresist, exposed, developed, and etched. After that, a singulation is carried out to form assembly panels out of the production panel. The assembly of the interposer is carried out by surface-mount technology (SMT) and reflow soldering. Several forms of interconnect technology can be applied. Chips are typically assembled in a flip-chip process.

After assembly, an electrical test is done to assure functionality before embedding the interposer into the PCB. This prevents yield loss due to damaged components or incorrect assembly. After assembly, the tested interposer is placed onto an inner layer of a multilayer construction and fixed to assure positional accuracy. After completing the multilayer stack, the PCB is laminated in a multilayer press.

The integrated interposer (i^2 Board) can be combined with most PCB technologies. Therefore, there is no dedicated package construction, and the PCB will replace the package of the component (Figure 9.6).

9.3 Manufacturing Process Flow and BOM

The manufacturing flow is shown in Figure 9.7. For the i^2 Board, an interposer is needed, which consists of the same material as the PCB. A thin laminate, coated with copper on both sides, undergoes a photolithographic process where the interposer traces are formed. The lines and spaces needed are half of

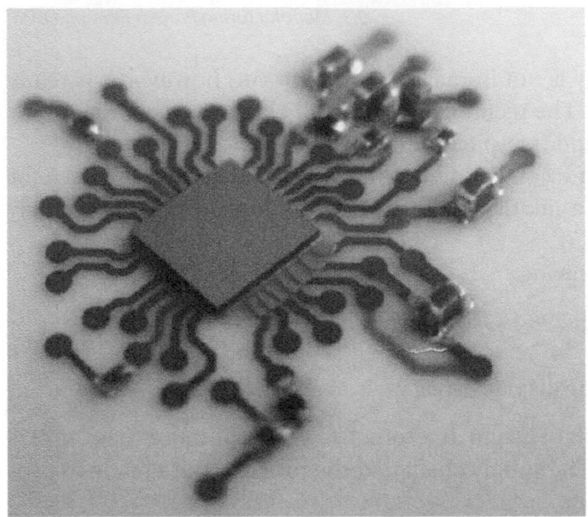

Figure 9.6 Interposer with flip-chip assembly and passive components.

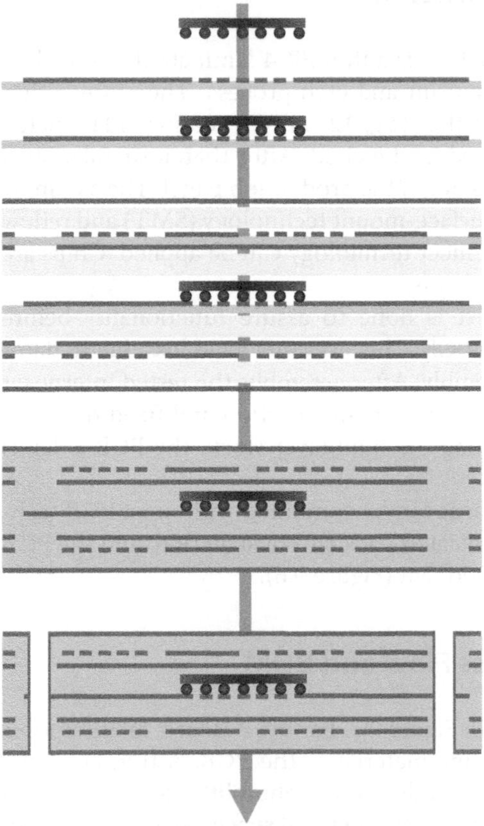

Assembly of Si-die on the routing substrate

Lay-up of multilayer

After lamination

Contacting of the chip by CNC drilling and through-hole plating of the interposer

Figure 9.7 Production process of i² Board.

the die bond pad pitch. Lines and spaces of 40 μm are used for a chip with 80 μm pitch. The lines are assembled with the component directly and lead as a fan-out design to the later interconnection pads.

After the assembly and the electrical test of the component on interposer level, the interposer itself is placed onto an inner layer of the later multilayer PCB construction. The prepregs (pre-impregnated glass fiber sheets used for bonding the inner layers) are used to embed the component, and the interposer has component cutouts to relieve stress in the laminating process from the device. After lamination the board can be treated mainly as a standard PCB. The electrical connection to the interposer is carried out with through-holes as in any conventional multilayer.

Since the i^2 Board is an SiB technology, the material of choice is derived from the target specification of the finished application. Medium (150 °C) to high (170 °C) glass transition (T_g) laminates with low CTE (12–14 $ppm\,K^{-1}$) are preferred due to higher reliability of the PCB and the need for thermal resistance due to later reflow soldering processes for the components on the outside of the i^2 Board.

The i^2 Board technology copes with plenty of different interconnect technologies like reflow soldering, bonding with conductive adhesive, isotropic or anisotropic bonding, and with plenty of different surface finishes on the chip pads like aluminum, nickel–gold, and copper finishes. It is possible to use thinned chips, but not necessary. A thinned component (e.g. 50–100 μm) can be directly placed in a single insulating layer between the other electrical layers of a multilayer construction. A thicker component could be used by cutting out not only the prepregs but also some of the inner layers. Therefore, any bare die can be embedded in combination with an interposer.

The easiest way for quick functional samples is to use gold stud bumps on the chip, because bare die can be used in this process without additional processing. The chip is then bonded to the interposer with a nonconductive adhesive (NCA). With this construction, the initial samples can be produced, and functionality tests can be carried out. If a backside connection is needed for the chip, it can be bonded with its backside to the interposer using a conductive adhesive. The chip is then simply wire-bonded to the interposer and encapsulated before the embedding process.

The assembly of the chips to the interposer is carried out on typical assembly panel formats with panel sizes of, e.g. 120 by 160 mm^2. Therefore, standard assembly equipment can be used for this process. Even fine-pitch chips can be handled with an interposer. The contacts of the chip are directly assembled on the lines of the interposer. In the case of 100 μm pitch on chip, an interposer of 60 μm lines and 40 μm spaces is used. A 50 μm pitch on chip with a 25 μm lines/spaces interposer is already being demonstrated.

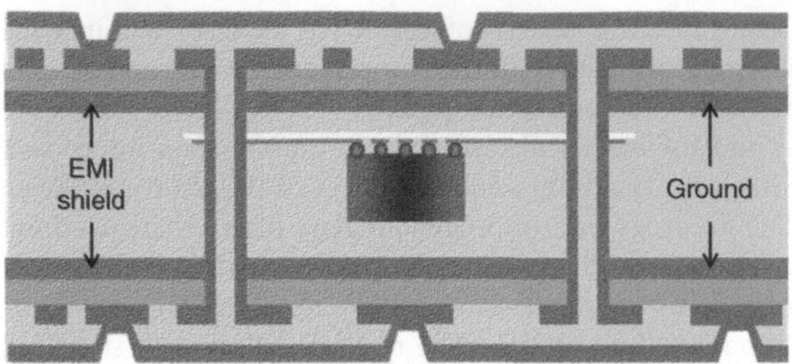

Figure 9.8 EMI shielding carried out with ground layers as shields.

9.4 Design Features

For a robust process, 80 μm bond pad pitch is reasonable but depends on the interconnection technology of choice. For solder bumps on the chip, the structures have to be kept coarser, e.g. 150 μm pitch with 100 μm ball diameter. For fine pitches down to 80 μm, Cu pillars or NCA die attach on the interposer can be chosen. With copper pillars finer structures can be addressed, because the solder volume of the solder cap is much smaller than with a massive solder ball. Therefore the likelihood of a short during soldering is lower. As the interposer can be placed between ground and power planes, an EMI shielding comes almost free with this technology (see Figure 9.8).

This technology is available for any fine-pitch chip technology, which can be assembled on an interposer. Certain limitations to power devices apply, because heavy copper design guidelines that are needed to carry power via the interposer and fine-pitch requirements are mutually excluding. Even pressure sensors can be embedded with the i^2 Board technology if special constructions for the PCB are applied.

9.5 System Integration Capability

As the technology is an SiB technology, the construction can be assembled like a conventional PCB. Therefore, additional components can be stacked on top of the embedded component. It is also possible to assemble more than one inner layer with interposers and do chip stacking inside the PCB.

9.6 Manufacturing Format and Scalability

Today a standard production panel is used for the i^2 Board technology, which is one of the advantages. The maximum format today is approximately 600 mm × 600 mm. The assembly of the interposers takes place in smaller-scale assembly panels. The manufacturing of the bare interposer is done on a large-scale panel of 600 mm × 600 mm.

9.7 Package Performance

Some components are limited in the maximum operating temperature by the package, because the thermal conductivity of the mold compound of the package is low. As most Si devices can reach a 175 °C junction temperature, the combination with thermally stable material extends the range of possible applications for a certain device. In automotive applications, 150 °C or even higher temperatures are specified as ambient temperatures. Thermal management is much easier with the i² Board. In most cases it is enough to place the component close to a ground plane with a thin isolation of 100 μm or less. The resulting thermal resistance is only a fraction of a BGA package. Whereas a BGA package can have a R_{thja} of 25–35 K W^{-1}, the thermal resistance of an embedded component can be as low as 5 K W^{-1} without any special measures.

Tests have shown that the maximum temperature of a device was 12 K lower when the device was embedded into the PCB instead of assembled as a packaged component on the board. This results in a doubling of the expected lifetime of the device.

9.8 Robustness and Reliability Data

The i² Board passed and exceeded the qualification tests of leading automotive suppliers. An internal qualification regarding the reliability of interconnections has been completed with outstanding results. Dies of 5 mm × 5 mm up to 10 mm × 10 mm with different interconnect technologies and with chip thicknesses varying from 50 to 100 μm were used in the test vehicles (Table 9.1).

A large-scale design of experiment (DOE) was performed containing parameters like CTE of laminates, glass yarn, PCB thickness, and chip standoff.

Table 9.1 Reliability test data.

Test	Number	Results
Reflow 235 °C	6×	0 defects
Reflow 260 °C	6×	0 defects
TCT −40/+125 °C	1000	0 defects
TCT −40/+140 °C	5000	0 defects
TCT −65/+150 °C	3000	0 defects

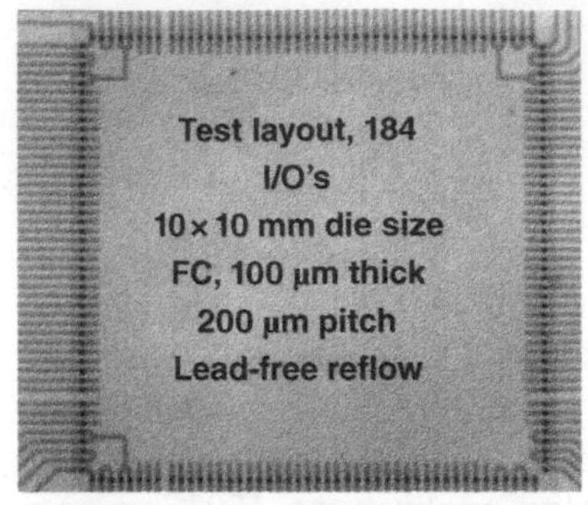

Figure 9.9 Daisy chain test principle for IR online testing during TCT.

Due to the large difference in CTE of silicon and PCB material, thermal shock is the most critical test for the interconnections between chip and interposer. Therefore, the test condition was set at 3000 thermal cycles between −65 and +150 °C at a transfer time of 10 seconds. The test layout was a daisy chain design where the signal was led from the interposer to the chip and back as shown in Figure 9.9. The interposer was routed with lines and spaces of 100/100 µm. The solder balls were 100 µm in diameter and 200 µm in pitch.

The failure criterion was defined as a rise of the electrical resistance of 0.5% (typically 10% are used in other tests) across a daisy chain of 184 interconnections on one die as shown in an X-ray image in Figure 9.10. The DOE showed very interesting interactions and influence parameters. This DOE is now a very good basis for consulting with customers regarding PCB materials, buildups, and predicted reliability (Table 9.2).

3000 thermal cycles between −65 and +150 °C and up to 5000 thermal cycles between −40 and +140 °C have already been demonstrated.

Figure 9.10 X-ray image of a 10 mm × 10 mm test vehicle. TCT: Temperature cycling test with 10 seconds transfer time.

Table 9.2 Significant terms for DOE.

R-square	0.9813
Adj *R*-square	0.91 978
RMS error	326 462
Residual df	7

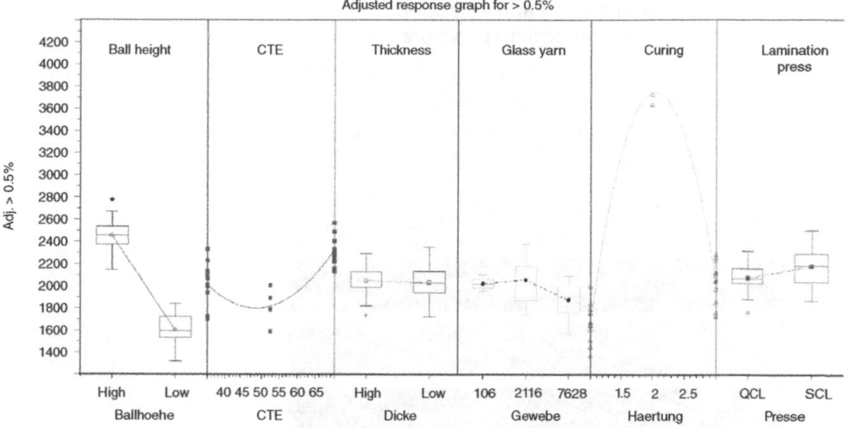

Figure 9.11 Prediction graph for resistance change <0.5% after 3000 cycles.

The result of the DOE is now a very precise predictive model for the lifetime as a function of the influencing parameters, as pointed out in Figure 9.11.

9.9 Electrical Test Considerations

9.9.1 Test Strategy

One of the biggest benefits of the interposer used in the i^2 Board technology is the ability for interim testing. The potential for testing during manufacturing offers the possibility to sort out bad components, to repair misassemblies, or not to assemble panel positions with failures in the PCB inner layers. This has a very large cost impact regarding the final yield of the product, as bad tested interposers will be sorted out and 100% good interposers will be the start of the embedded PCB production (see Figure 9.12).

9.10 Applications and Markets

There is demand for embedding in telecommunication as well as from automotive customers. In Figure 9.13 a cross section of a telecommunication PCB with embedded chip is shown. The chip thickness is 50 μm, and the total PCB thickness is 0.9 mm.

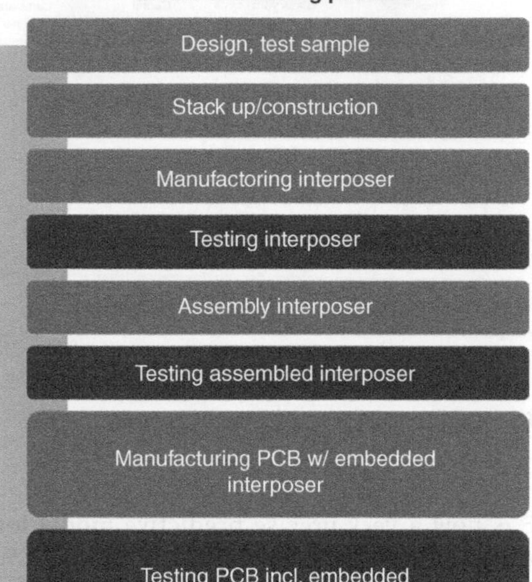

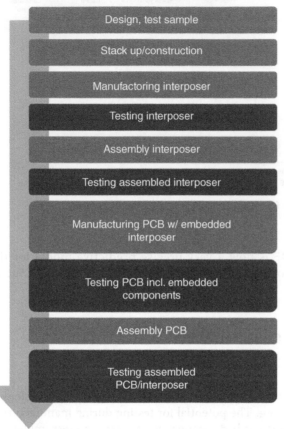

Figure 9.12 Process and test flow for i^2 Board production.

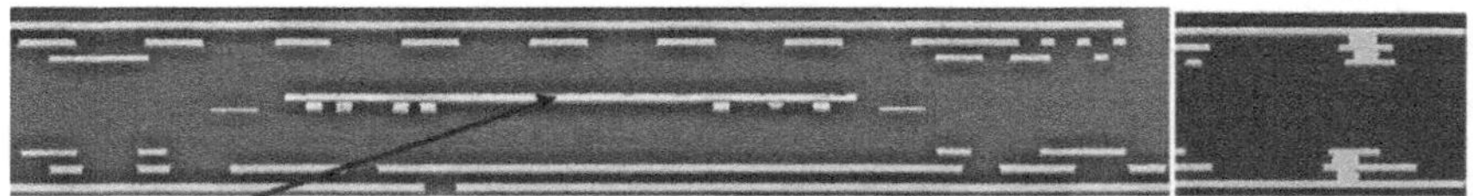

Figure 9.13 Embedded chip in an eight layer multilayer with 0.9 mm total thickness for telecommunication.

Figure 9.14 shows a demonstrator board that was developed to demonstrate the capabilities of the i^2 Board technology for RFID embedding. A RFID tag with an assembled RFID chip was embedded for the use in a fishing game for demonstration.

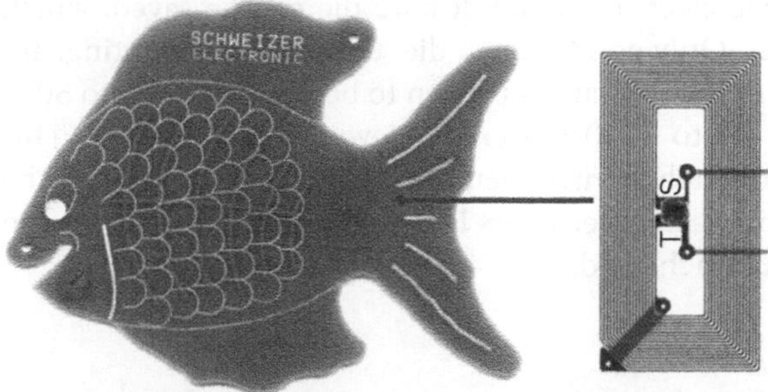

Figure 9.14 i^2 Fish. A demonstrator for embedding of RFID tags using i^2 Board technology.

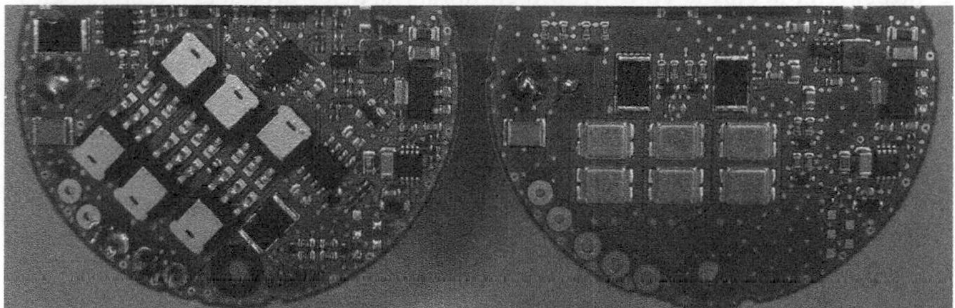

Figure 9.15 Customer samples before (left) and after (right) embedding of an IGBT driver IC.

As shown in Figure 9.15, embedding of an insulated-gate bipolar transistor (IGBT) driver into the main board in combination with a new MOSFET package type led to a theoretical free space for additional 172 passive components on the outside of the PCB.

9.11 Summary

Advantages of the i^2 Board technology include the assembly of components on an interposer, not on a full PCB panel. Therefore, high registration accuracy is limited to a small area. The embedded components have very reliable connections, as they are stabilized in a resin matrix of the PCB. Interim testing and robust processes secure high yields. All processes used for the i^2 Board technology are suitable for mass production. The technology is feasible for 80 μm chip pitch. As fine-pitch interconnection is possible, the pads on Si chip may remain small and save silicon surface area. Therefore, the total Si area stays small. Small chips mean low chip cost. The interposer can be easily electrically tested. Therefore, expensive electric testing of bare die can be saved, which lowers the overall chip cost. Only good known die, tested after mounting, are embedded. The reliability of the i^2 Board is proven to be very high. Up to 5000 temperature cycles from −40 to +140 °C were achieved with no failure. This gives a potential for use in harsh environmental conditions. Moreover, if the component is embedded between inner layers like ground and power layers, an electromagnetic shield can be achieved.

References

1 http://ats.net/products-technology/technology/ecp
2 https://en.tdk.eu/tdk-en/374108/tech-library/articles/products---technologies/products---technologies/very-small-and-extremely-flat/171672
3 Ostmann, A., Manessis, D., Boettcher, L. et al. (2009). Realisation of embedded-chip QFN packages – technological challenges and achievements. *2009 European Microelectronics and Packaging Conference (EMPC 2009)*, Rimini, Italy (16–18 June 2009).
4 Ostmann, A., Boettcher, L., and Manessis, D. (2005). HIDING DIEs: technology for embedding active dies, *Proceedings EMPC 2005*, Bruges, Belgium (12–15 June 2005).
5 BMBF-Verbundprojekt PARIFLEX – Passives RF-Identifikationssystem mit flexiblem, bistabilem Display: Beitrag: Robuste und optimale RFID-Antennensysteme für räumlich verteilte Schreib–/ Lesestationen passiver RFID-Lösungen mit optischen Komponenten (Daten- und Energieübertragung); Abschlußbericht PARIFLEX - Universität Paderborn; Berichtszeitraum: 1 October 2005–30 September 2008.

10

Embedding of Power Electronic Components

The Smart p^2 Pack Technology

Thomas Gottwald and Christian Roessle

Schweizer Electronic AG, Schramberg, Germany

10.1 Introduction

The electrification of the power train of hybrid electric vehicles (HEV) and electric vehicles (EV) increases the power demand and the power dissipation of power electronic modules. It is state of the art to use power modules based on ceramics such as direct bonded copper (DBC) or direct copper bonded (DCB) ceramic substrates because of the required combination of high ampacity, high insulation performance, and high thermal conductivity. In this conventional power module technology, bare die are assembled onto the topside copper of the DBC; the interconnection between power component and substrate and substrate to connector frame is carried out with thick aluminum (Al) wire bonds. Several wires have to be used in parallel to achieve the required ampacity of the modules. Figure 10.1 shows a typical power module for a three-phase motor that consists of three DCB substrates (DCB: direct copper bonding) – one per phase – which are soldered onto a copper base plate. On the base plate, a plastic frame with connection pins is mounted, and the wire bonds connect the die to the DCB and the DCB to the frame.

The number of process steps and the challenges for full-area soldering of DCB substrates, wire bonding, and potting with the respective cost issues were accepted due to the lack of alternatives.

Increasing power density, increasing voltage, and the demand for higher switching frequencies require a closer look at the parasitic effects of such modules, especially the inductances. Not only static losses but also switching losses are of increasing importance. By increasing the switching frequency, the volume of passive components can be minimized, and therefore the total volume of inverters decreases. Higher frequencies, however, increase the corresponding

Advances in Embedded and Fan-Out Wafer-Level Packaging Technologies, First Edition.
Edited by Beth Keser and Steffen Kröhnert.

Figure 10.1 State-of-the-art three-phase insulated-gate bipolar transistor (IGBT) power module.

switching losses. Overvoltage, which occurs after switching off the system, has a high potential of damaging the components.

One of the sources for parasitic inductances is the bond wires of conventional power modules due to the loops and their tolerances in shape, the number of bond wires, and their length. Another important issue for switching speed is the distance between gate driver and gate contact of the power component and their corresponding parasitic effects.

The bond wires are also critical with respect to reliability issues. Due to the huge difference in the coefficient of thermal expansion (CTE) between aluminum and silicon, one of the best evaluated failure modes is bond lifting from the surface of Si power devices. The higher the temperature cycle in the application, the sooner this failure will occur. The soldering joint of die to the DCB is a weak point in the design, too. This is one of the reasons why silver sintering is gaining interest for high reliability applications.

To solve the problem of the bond wires and of the soldering joints, leading manufacturers of power modules count on new technologies like silver sintering as a solder replacement for the interconnections between die and DCB, some also for the replacement of bond wires. Still, the expensive DCB substrates are part of those systems.

10.2 Technology Description p^2 Pack

The p^2 Pack technology is an embedding technology for power devices such as metal–oxide–semiconductor field-effect transistor (MOSFETs) and insulated-gate bipolar transistors (IGBTs) that produce significant power losses during operation and have a low number of electrical contacts. The source of power losses is switching losses and conduction losses of the power electronic components. The first trials of power semiconductor embedding have been made on the basis of the i^2 Board technology, but even though it worked out to be feasible, it was not a very elegant approach for a component with a vertical current flow. The die attach and the heat dissipation needed an improved technological approach.

As the p^2 Pack is very flat with a thickness of 1.2–1.7 mm, it is furthermore possible to embed this package into a PCB and realize the combination of power electronics and logic control in one single PCB without the need for additional connectors between logic and power.

As power and logic control act now in one board, this construction is called "Smart p^2 Pack." With the new p^2 Pack technology, a new architecture was developed, which is helpful for very robust, cost-efficient, and miniaturized high power inverter configurations.

10.3 Basic Package Construction

In Figure 10.2, the construction is shown in an exploded view of the layer stack. At the bottom, a base plate consisting of a copper layer is shown, which is covered by an insulating layer, made from a thermally conductive prepreg (a pre-impregnated glass fabric, resin filled with thermally conductive inorganic fillers). The second level shows a lead frame with cavities. Inside the cavities, the power electronic components are assembled. The surface finish of the components is made from copper. The gray layer above consists of another thermally conductive prepreg and a copper layer, which has been patterned. The green layer is the logic control board into which the p^2 Pack is embedded in a later step. It is a standard printed circuit board (PCB) that is used as control board for the power stage.

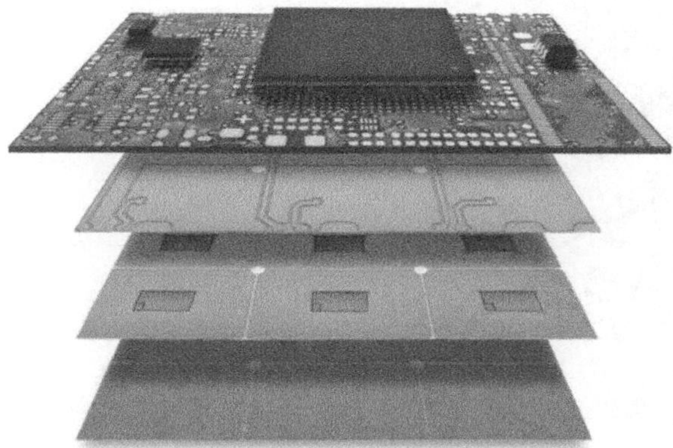

Figure 10.2 Exploded view of a Smart p^2 Pack.

10.4 The p^2 Pack Technology Process Flow

This newly developed technology uses embedding technologies and materials from the PCB industry. By using these technologies, very powerful and compact modules can be created, which lead to optimized electrical characteristics.

The starting point of this technology is lead frames with cavities that are assembled with power semiconductors, e.g. IGBTs or MOSFETs, as shown in Figure 10.3. The die attach is carried out with either diffusion soldering or transient liquid phase bonding (TLPB). Also sintering processes like silver sintering are possible. After assembly, the surface of the die and the surface of the lead frame are planar. In Figure 10.4, an assembly of six MOSFETs on a lead frame can be seen. They are already arranged in a way to form a three-phase B-6 topology for a motor drive after finishing the assembly.

Figure 10.3 Process flow: lead frame with cavities where power devices are assembled into the cavities.

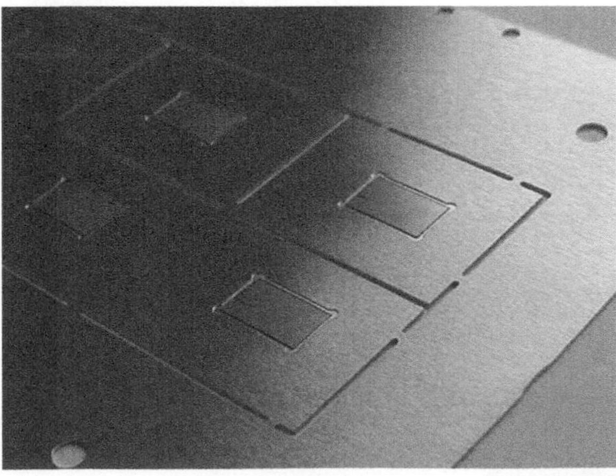

Figure 10.4 Lead frame with power MOSFETs assembled into cavities.

These assembled lead frames are laminated into a three-layer construction with an electrically insulating material on top and bottom of the lead frame (layer 2). The outer layers consist of copper sheets (layers 1 and 3) and work as a heat spreader at the bottom, while the top layer acts as a routing layer that replaces the bond wires used in ceramic power modules.

Figure 10.5 shows the top side of a B6 module, where six MOSFETs are arranged and electrically routed for a three-phase motor drive. The interconnection to the gate/source contacts is carried out with blind vias, which are drilled with laser drilling equipment through the dielectric layer and filled with copper in a galvanic Cu-plating process. The drain contact is mechanically and electrically connected with the lead frame. The lead frame can either be directly attached after routing off the cover layers to the lead frame level, or it can also be connected to the top layer with electroplated blind or through vias.

Figure 10.6 shows a cross section of a p^2 Pack. The bottom layer represents the insulated copper base plate, which can be directly attached to a heat sink. The middle layer is the lead frame with a small cavity of 70 μm with an assembled power MOSFET of the same total thickness. The thickness of the lead frame can be adapted to the needs of the application in terms of heat dissipation. The top layer is the routing layer for the interconnection of the top

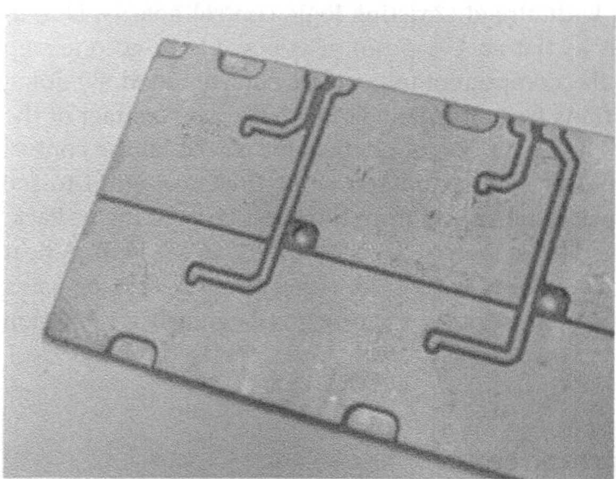

Figure 10.5 B6 bridge module. Gate/source connection with copper traces/planes.

Figure 10.6 Cross section of a p^2 Pack with copper-filled blind vias.

contact gate and source to the top layer carried out with copper-filled blind vias. The black layers are built with dielectric prepreg sheets and guarantee the insulation at those places where they are not penetrated with blind vias. The copper surface of the component ensures a highly reliable metal to metal adhesion with superior electrical and thermal conductivity.

One of the preconditions for the power devices is a surface plating of copper on the contacts of the device. This is necessary for the plating compatibility. The construction of the p^2 Pack is symmetric, which leads to a minimized warp and twist during thermal cycling and therefore minimizes pump out effects of thermal interface material (TIM), when it is directly mounted to a heat sink. The thick copper layers above and below the lead frame lead to an optimized heat spreading in each layer and in fact to a double-sided cooling of the power device. Up to one-third of the heat is transferred to the top layer and from the top layer through the complete package to a heat sink connected from the bottom side.

10.5 Smart p^2 Pack

The p^2 Pack itself can be used as a one-to-one replacement of existing DCB substrates in combination with already existing logic control boards. Due to the fact that the p^2 Pack is as flat as 1–1.4 mm, it is possible to go one step beyond and embed such a flat component into a logic control board. By doing so, very short interconnections from the gate driver to the gate contact of the power device are feasible. The driver component can be placed on the control board directly above the power semiconductor, while the connection to the gate is made with copper-plated blind vias from the outer layer to the p^2 Pack. The Smart p^2 Pack construction is open to the bottom side of the PCB. A heat sink can be easily installed to the bottom side, either using a TIM or using sintering technologies to further reduce the thermal resistance from junction to ambient.

10.6 Package Performance

10.6.1 Electrical Performance

Chip embedding technologies are used to embed thin bare die (MOSFETs or IGBTs) into a construction of lead frames and additional routing levels. The interconnections are carried out with blind vias that are filled with electrolytically plated copper that reduces the package-related resistance R_{on}. R_{on} is the resistance of the package consisting of chip resistance, bond wire, and lead frame resistance. The blind vias also reduce the parasitic inductances to a minimum. Measurements showed an improved R_{on} where the complete p^2 Pack had almost the same R_{on} as the chip itself. By using blind vias instead of bond wires, the package-related part of the R_{on} could be reduced by a factor of 100 and is difficult to be measured correctly. The difference in package resistance is pointed out in the respective data sheets for embedding components in comparison with packaged components (Figure 10.7).

Figure 10.8 shows the results of total resistance measurements of the same power electronic component. On the right, the chips were packaged in a TO 263 package with seven leads. In the middle (dotted line), the chips were mounted on a DCB substrate and wire-bonded to a connector frame. On the left, the chips were embedded into a PCB and connected with copper-filled blind vias. Each point represents the measurement result of one component. The graph is displayed as a probability plot. In this case, the total R_{dson} could be reduced from 1 to 0.5 mΩ, which is close to the theoretical chip resistance. This means there is almost no package-related resistance left.

10.6.2 Dynamic/Switching Losses

Faster switching speeds for power electronic assemblies are desirable for many reasons. The faster the switching, the smaller the component size of passive components like capacitors and inductors can be. This effect is beneficial in terms of miniaturization and cost involved. The drawback of faster switching is an increase of the dynamic losses of the system, which makes it less efficient. Several embedded dies in the p^2 Pack technology have been tested for switching speed in comparison with DCB-based power modules. Figure 10.9 shows the faster current rise and fall times that are indirect indicators for lower inductances in the p^2 Pack system.

10.6.3 Inverter Efficiency

A complete inverter for a 32 kW engine was designed and built with ETH Zurich. It uses 1200 V IGBT components and recovery diodes and consists of a classical B6 topology (see Figure 10.10). The inverter was built with two separate PCBs in this state of the development, one for the control (top) and one for the power stage (bottom) as the measurements for the characterization of the inverter were much easier to be carried out due to an easier access to measurement points on the PCBs. In a next step, this inverter will be built in one single PCB as Smart p^2 Pack. The DC-link capacitor was made from a block of foil capacitors at the bottom of the assembly. The cooling of the semiconductors was carried out by attaching a water cooling unit beneath the power stage (hose connection on the right).

The Smart p^2 Pack inverter was completely characterized and performed at a very high efficiency. Figure 10.11 shows the inverter efficiency at a switching

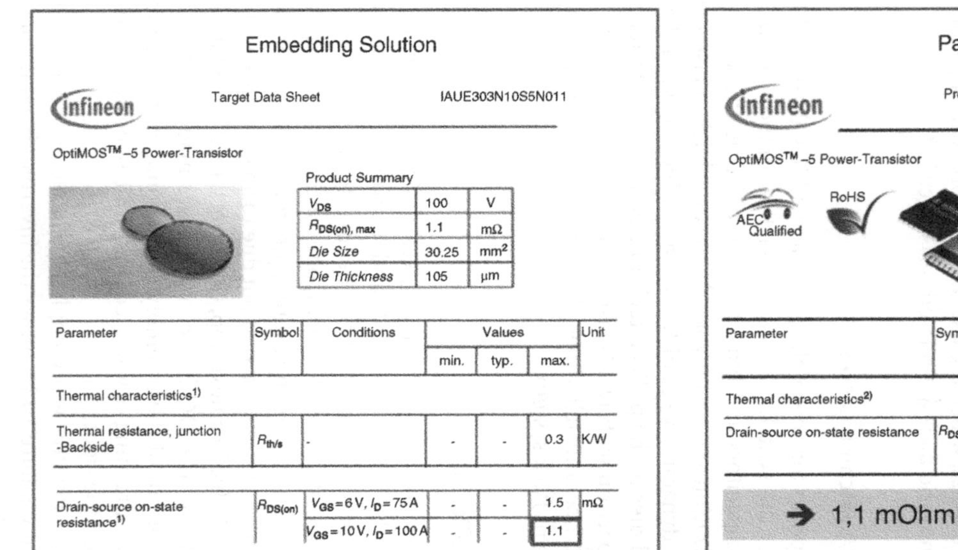

Embedding Solution

Infineon — Target Data Sheet — IAUE303N10S5N011

OptiMOS™-5 Power-Transistor

Product Summary

V_{DS}	100	V
$R_{DS(on), max}$	1.1	mΩ
Die Size	30.25	mm²
Die Thickness	105	µm

Parameter	Symbol	Conditions	Values			Unit
			min.	typ.	max.	
Thermal characteristics[1]						
Thermal resistance, junction -Backside	$R_{th/s}$	-	-	-	0.3	K/W
Drain-source on-state resistance[1]	$R_{DS(on)}$	V_{GS}=6 V, I_D=75 A	-	-	1.5	mΩ
		V_{GS}=10 V, I_D=100 A	-	-	1.1	

Package Solution

Infineon — Preliminary Data Sheet — IAUT300N10S5N015

OptiMOS™-5 Power-Transistor

Product summary

V_{DS}	100	V
$R_{DS(on)}$	1.5	mΩ
I_D	300	A

Parameter	Symbol	Conditions	Values			Unit
			min.	typ.	max.	
Thermal characteristics[2]						
Drain-source on-state resistance	$R_{DS(on)}$	V_{GS}=6 V, I_D=75 A	-	1.6	2.0	mΩ
		V_{GS}=10 V, I_D=100 A	-	1.3	1.5	

→ 1,1 mOhm * 36% → 1,5 mOhm

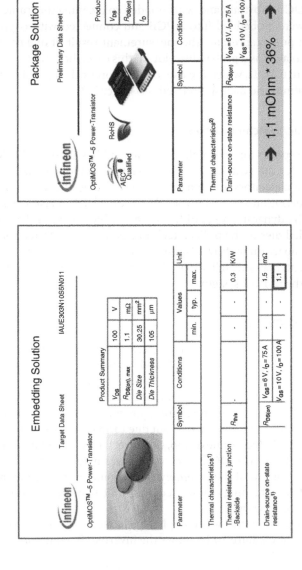

Figure 10.7 Data sheets indicating the R_{dson} (drain to source on state resistance) for embedded components (left) and packaged components (right) for the same die size and chip technology.

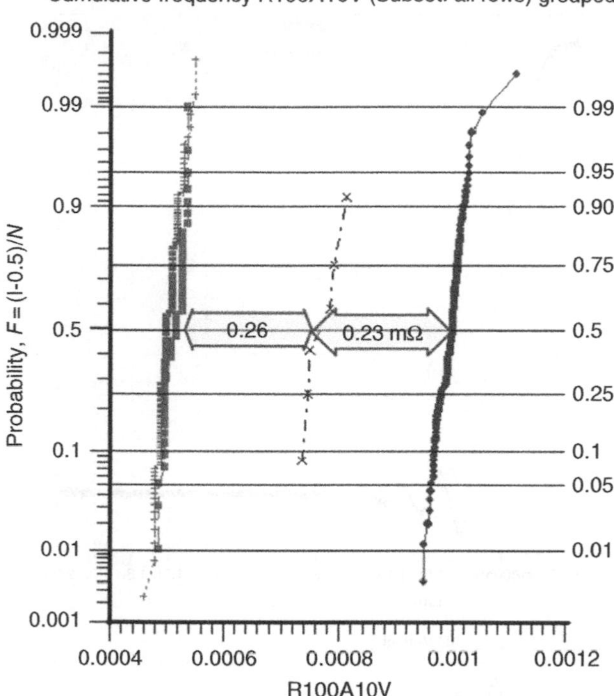

Figure 10.8 Rdson for TO 263-7 (right), die on DCB substrate (middle), and p^2 Pack (left).

speed of 10 kHz and different gate resistors. Although the inverter could only be tested up to 18 kW, it showed an efficiency of almost 98%, while the efficiency graph was still rising toward higher power; therefore higher efficiencies at full load can be expected.

10.6.4 Thermal Performance

The p^2 Pack technology was designed for superior thermal performance. The basic concept is to add a heat spreader of the same size to all power electronic components embedded into a laminate. By spreading the heat with the lead frame before transferring it through the insulating material, a very low thermal resistance (R_{th}) can be achieved.

To analyze the thermal dissipation in the p^2 Pack, a finite element (FE) simulation model was built and calculated with different variables. To reduce the calculation time for such a model, it is very common to simulate only ¼ of an object, as long as it is symmetrically built (see Figure 10.12). This was also done with the underlying simulation, whose results are shown in Figure 10.13.

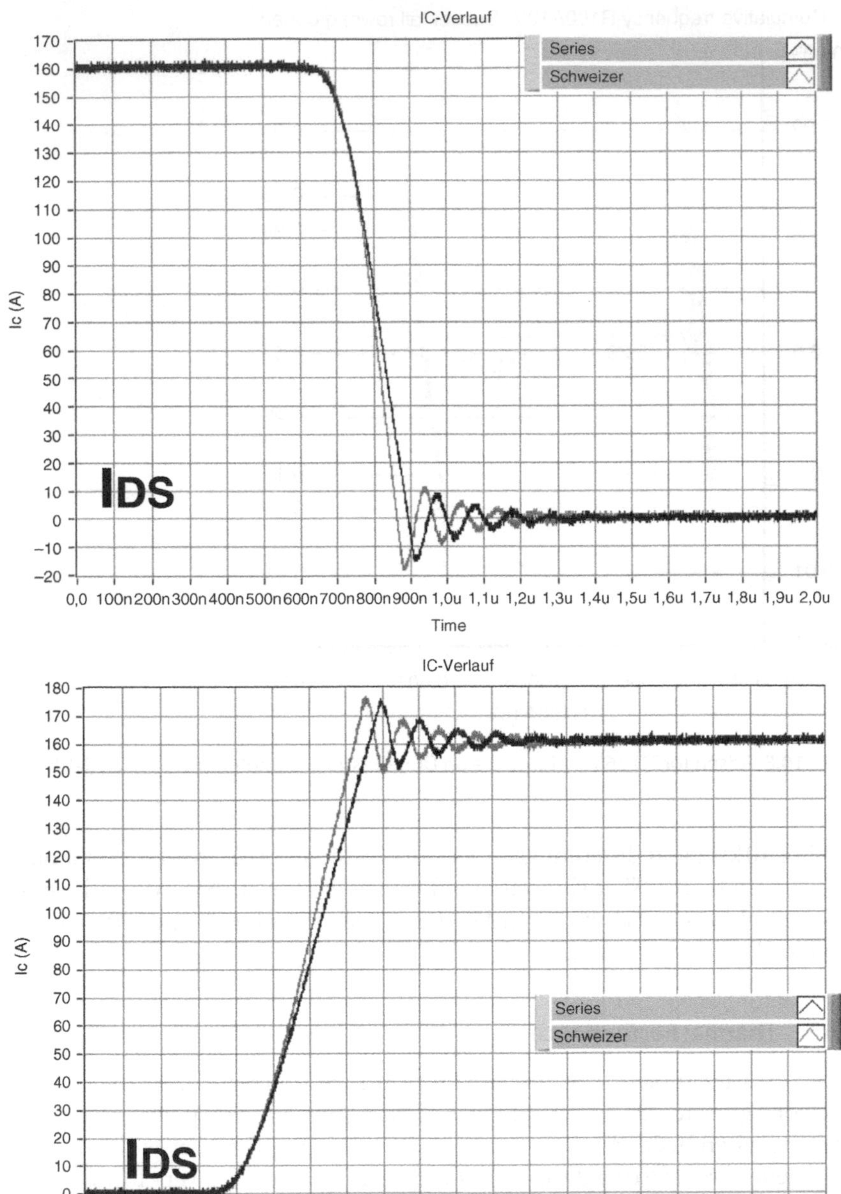

Figure 10.9 Switching characteristic of p[2] Pack (gray) vs. die on DCB (black) at 160A: 12% faster switching.

Figure 10.10 32 kW test vehicle in Smart p² Pack technology. IGBT application.

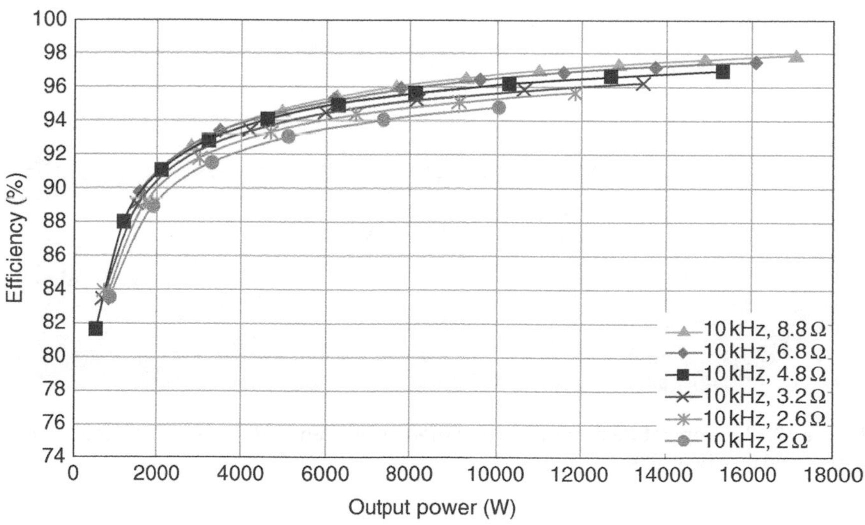

Figure 10.11 Efficiency graph of a 32 kW test vehicle in Smart p² Pack technology at partial load conditions.

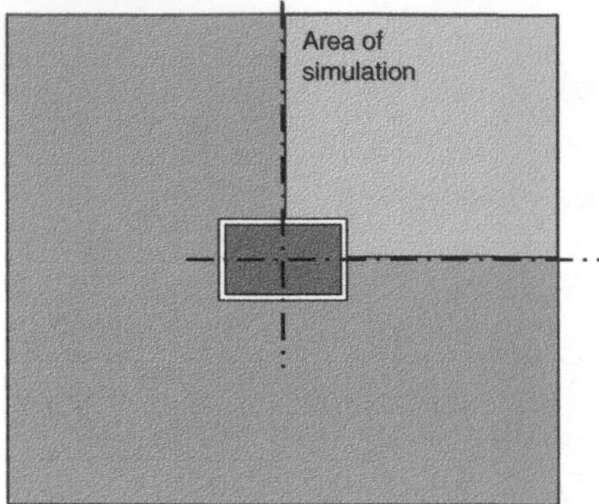

Figure 10.12 Symmetry areas for FE modeling: ¼ of a lead frame assembled with a power semiconductor in the middle.

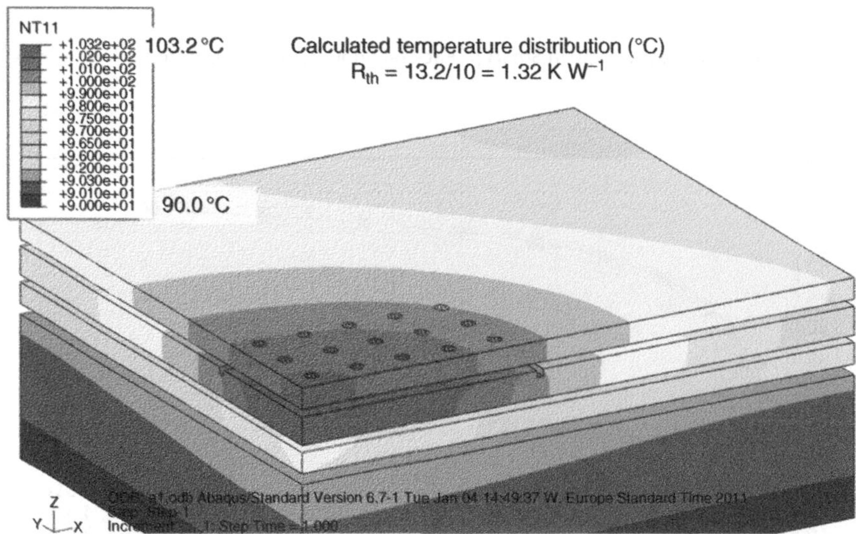

Figure 10.13 Temperature distribution and thermal resistance (R_{th}) in a p^2 Pack single cell assembled on a heat sink.

Figure 10.13 shows the calculated temperature distribution of a FE model of ¼ of a p^2 Pack in a cross section. The front left corner represents the chip center, while the edges represent the cross-sectional lines through the package. The dark layer at the bottom represents a heat sink that is installed beneath the p^2 Pack. Between the p^2 Pack and the heat sink, a TIM was designed into the model with a thermal conductivity of 1 W mK^{-1}. The TIM was the most critical material due to the highest contribution to the thermal resistance. Almost 50% of the total thermal resistance (R_{th}) came from the TIM. This gives a lot of potential for improvement of the system.

The larger the lead frame is, the better the thermal performance will be. Further simulations have shown that a factor of 4 to 1 in area for lead frame to chip area is a good starting point to balance thermal performance with cost for the lead frame and design area needed to place the embedded components.

10.6.5 Robustness and Reliability Data

The p^2 Pack went through a full component qualification as well as a full PCB qualification. Before testing, preconditioning was carried out with moisture sensitivity level 3 (MSL 3) with three reflows at a peak temperature of 260 °C [1]. The most critical tests in terms of reliability are the reflow soldering after exposure to humidity (MSL3), the thermal shock, the active load cycling, and the hot storage test. Temperature cycling was performed with two-chamber reliability equipment with 30 minutes at 150 °C and 30 minutes at −55 °C per cycle with a transfer time of <10 seconds. Active load cycling was accomplished by running high currents through the semiconductor and heating up the probe due to the component resistance R_{on}. A typical cycle was two seconds on and then two seconds off. The temperature difference between the on and off states was 80 K. Hot storage means holding the probes at a temperature of 175 °C for at least 1000 hours. Tests up to 2000 hours were also performed. All component and PCB-related qualification requirements could be achieved without failure. The reliability in active load cycling was found to be at least 10 times higher than a power module built with ceramic substrates. Up to 2 M cycles with a temperature difference of 100 K could be demonstrated, before the test was finished without damage. Typical values for ceramic-based power modules are 80k to 120k at an 80 K temperature difference.

10.7 Applications and Markets

The first generation of p^2 Pack products will be based on low-voltage MOSFETs from Infineon, which were especially designed for embedding applications. The qualification is done at automotive level as the first applications are expected to start in this market. The technology is not in production as of publication, since automotive design cycles take several years, especially for completely new technologies, but it is expected to find its first applications under the hood. At the moment Schweizer is preparing a plant for mass production. Meanwhile the next-generation p^2 Pack for wide bandgap semiconductors is in development. Very low loss components like gallium nitride

(GaN) and silicon carbide (SiC) need to have a low-inductance package like the p^2 Pack; otherwise the advantages of the new semiconductor technologies cannot be exploited.

10.8 Summary

The advantages of the p^2 Pack chip embedding technology are manifold. The technology began with the attempt to miniaturize power electronics and improve thermal dissipation; however, many additional benefits were found. It is obvious that miniaturization can be easily achieved for the *x/y*-direction as well as for the *z*-direction. The heat dissipation could be improved by 30–44% in comparison with a DCB substrate at the same size, and the thermal resistance junction to heat sink of $0.5\,\mathrm{K\,W^{-1}}$ was achieved [2].

Today's power electronic systems consist of several substrates and a collection of interconnect technologies. With the Smart p^2 Pack, the whole assembly can be manufactured by means of surface-mount technology (SMT), which makes the assembly easier to control in high volume manufacturing and more robust than conventional systems. The losses can be minimized, as the static and the dynamic losses are lower for embedded constructions compared with conventional systems. This increases the energy efficiency of the system and reduces the efforts to spend for the cooling system. The p^2 Pack and the Smart p^2 Pack can be built as electrically insulating systems. Therefore the backside of the boards can be directly assembled onto a heat sink. The choice of TIM is open to the customer and electrically conducting TIM is also available.

A very nice feature is the very low inductance of a p^2 Pack. The low inductance enables fast switching in a pulse-width modulation (PWM) mode. For some applications, faster switching offers the possibility to reduce the size of the passive components and cost involved. It also offers the opportunity to reduce the voltage class of power devices in the application. Today 650 V IGBTs are being used in drive applications although the DC-link voltage is set to 400 V. The margin of an additional 250 V blocking voltage is needed to control the voltage overshoot after switching on or off. The reason for this overvoltage is the parasitic inductance of today's systems, which is 5–10 times higher than with a p^2 Pack. If 450 V devices could be used in those systems, because overvoltage is reduced due to low inductances, the R_{on} of these components would be reduced by approximately 30%, which would increase the system's energy efficiency and reduce the cooling efforts.

One of the biggest assets of the technology is its extremely high reliability. The robustness in active load cycling could be improved by a factor of 10 compared with conventional wire-bonded power modules on DCB substrates.

Acknowledgments

The authors acknowledge the competent contribution of Infineon Technologies and ETH Zürich to this work.

References

1 IPC/JEDEC J-STD-020E (2014). Moisture/reflow sensitivity classification for nonhermetic surface mount devices (December 2014).
2 Kearney, D., Kearney, D.J., Kicin, S. et al. (2016). PCB embedded power electronics for low voltage applications. *9th International Conference on Integrated Power Electronics Systems, Proceedings* (8–10 March 2016), Nuremberg, Germany.

11

Embedded Die in Substrate (Panel-Level) Packaging Technology

Tomoko Takahashi and Akio Katsumata

J-Devices

11.1 Technology Description

With square or rectangular panels instead of circular wafers, panel-level packaging has more die per panel than an equivalent size round wafer with its edge and missing die losses, potentially providing significant cost savings. Similar to other advanced packaging techniques, a fan-out (FO) technique is required to achieve higher integration levels and a greater number of external connections. The fan-out panel-level package (FO-PLP) described in this chapter is unique, because a metal material was selected as the base plate of an FO package. The base plate was not a temporary structure to place die on, but was a part of the package as shown in Figure 11.1.

The main technology issue of panel-level assembly is how to laminate resin, similar to the substrate manufacturing process using laminate dielectric layers, on metal without delamination, warpage, or deformation during processing. To address this issue, J-Devices adds a stress release layer between the metal and the resin. This layer releases the stress caused by the coefficient of thermal expansion (CTE) mismatch among layers during the assembly process. With this approach, the warpage of the panel is well controlled and can be handled in each process step.

The FO-PLP reported in this chapter has good reliability and very low warpage. The design is extremely flexible and has several possible applications, including automotive, mobile, and antenna electronics. The initial design provides a package suitable for in-vehicle use. For mobile applications, thinner packaging is needed. For antenna use, the metal plate disturbs antenna wave irradiation. To address this issue, removing the metal plate became necessary. All three of these design targets were achieved with embedded die in substrate

Advances in Embedded and Fan-Out Wafer-Level Packaging Technologies, First Edition.
Edited by Beth Keser and Steffen Kröhnert.

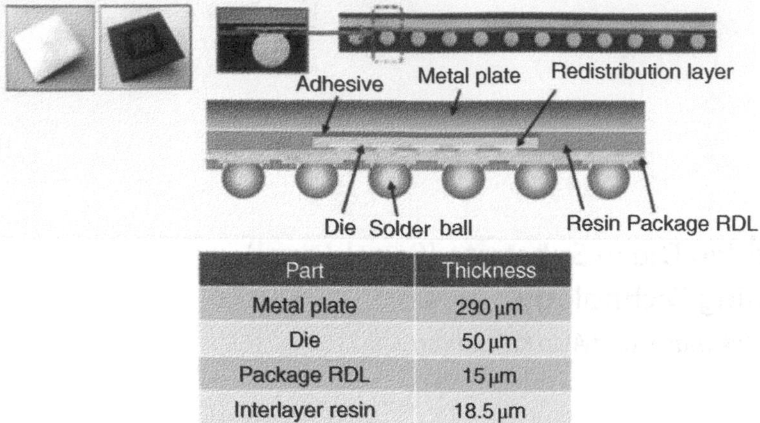

Part	Thickness
Metal plate	290 μm
Die	50 μm
Package RDL	15 μm
Interlayer resin	18.5 μm

Figure 11.1 Fan-out package using a metal base plate.

(EDS) package technology. Its detail and variations will be introduced in the following sections. Based on the design differences for different applications, there are two different EDS technology sections at the end of this chapter that explain the enhancements for smartphone modules and power devices.

11.2 Basic Package Construction

A cross-sectional view of the standard EDS package is shown in Figure 11.2. The thickness of the die can be from 35 to 100 μm, and the resin thickness can be selected from several options. The thickness of resin on the topside impacts the package warpage, so the proper thickness must be determined

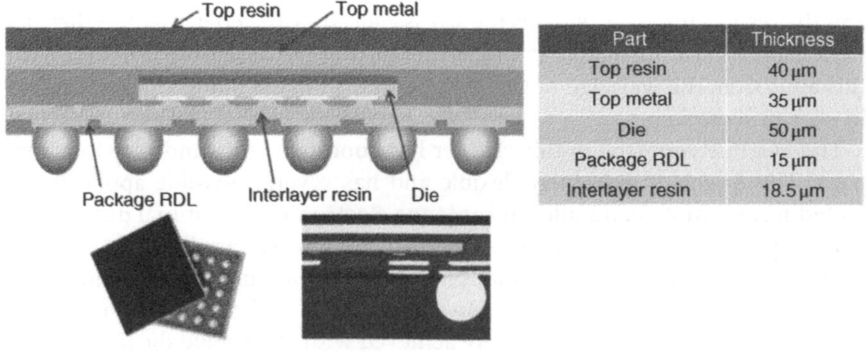

Part	Thickness
Top resin	40 μm
Top metal	35 μm
Die	50 μm
Package RDL	15 μm
Interlayer resin	18.5 μm

Figure 11.2 EDS package design characteristics.

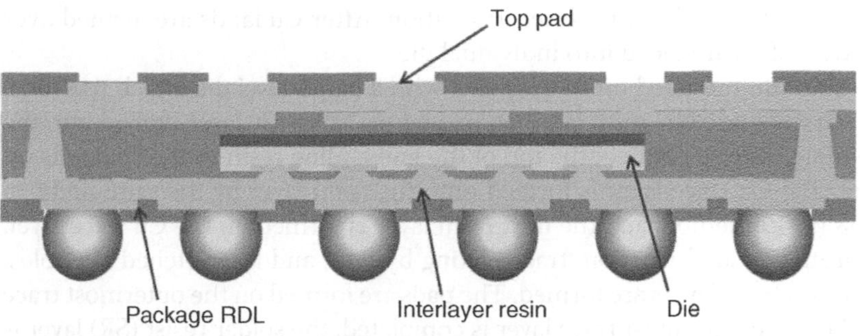

Figure 11.3 EDS with pads on both sides of the cross-sectional structure.

for each design. The thickness of each layer in Figure 11.2 is the result of extensive development efforts.

As shown in Figure 11.2, the surface of the package is covered with top resin to protect the package. It is not a base plate of the panel, so it is not the essential part of the package. If pads are needed on the topside of the package, the topside resin can be removed. The cross-sectional structure with pads on both sides is shown in Figure 11.3. The pads on both sides of the package enable a package-on-package (PoP) design.

11.3 Manufacturing Process Flow and BOM

A basic process flow for FO-PLP is shown in Figure 11.4. Wafers that have Cu lands on the die pads are prepared before assembly. Die pads should be covered by Cu metal, because a laser via is directly formed over the die pad.

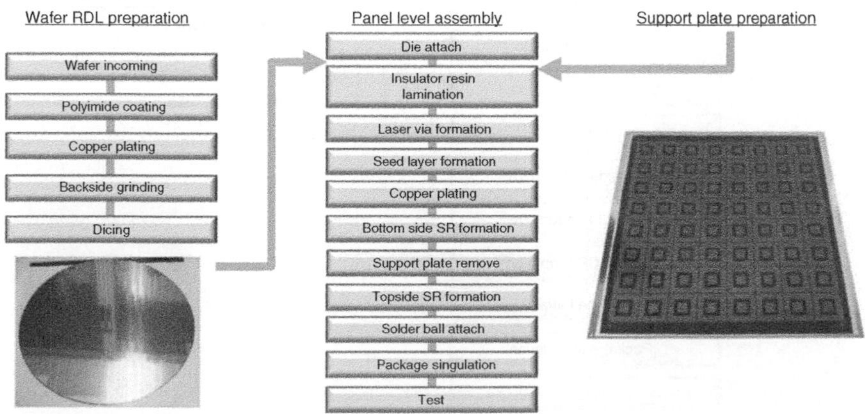

Figure 11.4 Basic FO-PLP processing steps.

Cu is the protective layer for laser irradiation. After Cu lands are formed over die pads, wafers are diced into individual die.

Next, die are mounted on to the panel with a panel-level die attach machine. They are embedded by resin lamination, and via holes are laser-etched into the embedding resin to connect the top of the embedding resin to the Cu lands on the die. The Cu trace layer is formed onto the resin layer by the semi-additive process (SAP) method, and the next resin layer is formed on the Cu trace layer. By repeating resin formation, trace plating by SAP, and laser-etched via holes, multiple Cu trace layers are formed. The pads are formed on the outermost trace layer. After the outermost trace layer is completed, the solder resist (SR) layer is formed. Following this step, solder balls are attached and packages are diced.

For EDS, the process is shown in Figure 11.5. The base plate for EDS consists of a metal layer and resin layers. EDS needs trace layers under the die to have pads on both sides of package, so the layers are assembled onto the base plate before die mount. A resin layer is formed on the base plate. Via holes are laser-etched into the resin. The Cu trace layer is formed onto the resin layer by the SAP method, and the next resin layer is formed on the Cu trace layer.

After the required layers are formed, die are mounted on the panel by a panel-level die attach machine. They are embedded by resin lamination, and via holes are laser-etched into the embedding resin to connect the top of embedding resin to the Cu lands on die. Also, deep via holes connect the top of the embedding resin to the Cu lands of traces on the bottom layers. Trace

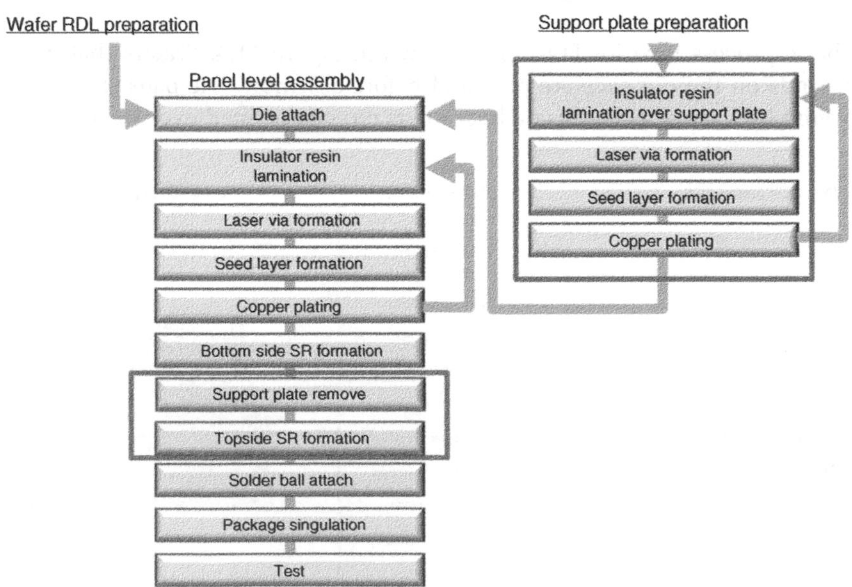

Figure 11.5 EDS processing steps.

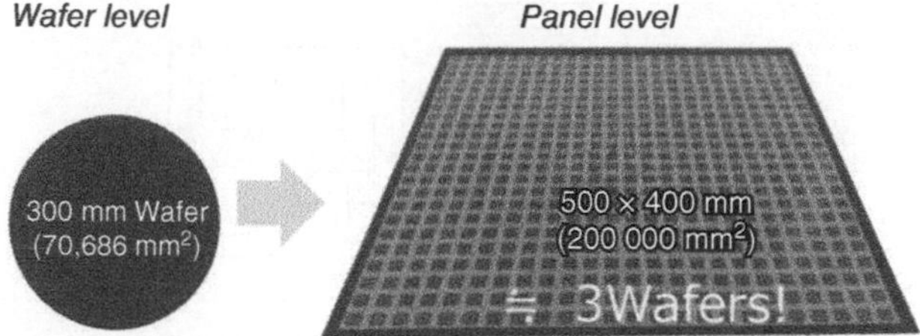

Figure 11.6 Area effect of wafer-level packages on one panel-level structure.

layers of the topside are formed by repeating resin formation, trace plating by SAP, and laser-etched via holes. A seed layer is formed by electroless plating of SAP process. The pads are formed on the outermost trace layer. After the outermost trace layer is completed, the SR layer is formed. The next step is the base plate removal. Removing the base plate, the bottomside of panel appears on the surface. To protect the package resin layer on the bottomside, an SR layer is formed. Balls are mounted in the panel level, and the panel is diced into units to complete the EDS package.

There are three key challenges in EDS process technology. The first is die mount technology. Achieving good position accuracy of die placement is very important for each company that develops FO packaging technology. For EDS technology, the required position accuracy is about 10 μm. This accuracy affects the Cu land size on the die. A small Cu land size means a small die pad pitch, and conversely, a small die pad pitch means a small die size. Consequently, accurate die mount technology is very important for shrinking the package.

The second key area is the semi-additive processing technology. SAP is a very general technology for trace formation in organic buildup substrates. The line and space (L/S) of the trace are not as small as the wafer-level photolithography technology, but substrate plating technology can achieve panel size plating. As shown in Figure 11.6, three times the number of packages on one wafer-level assembly can be achieved on one panel. This approach is very effective for high volume products.

The third key area for EDS is the method to remove the base plate. The base plate can be removed clearly and easily after almost all trace layers are assembled on the panel. At this point, the warpage of the panel is not a major problem for the remaining processes.

11.4 Design Features

The design rules of package trace layers are shown in Figure 11.7. The L/S design rule is 20 and 20 μm. This is a very standard value of a buildup-type substrate. The via land is 65 μm in diameter, so the via pitch for different signals

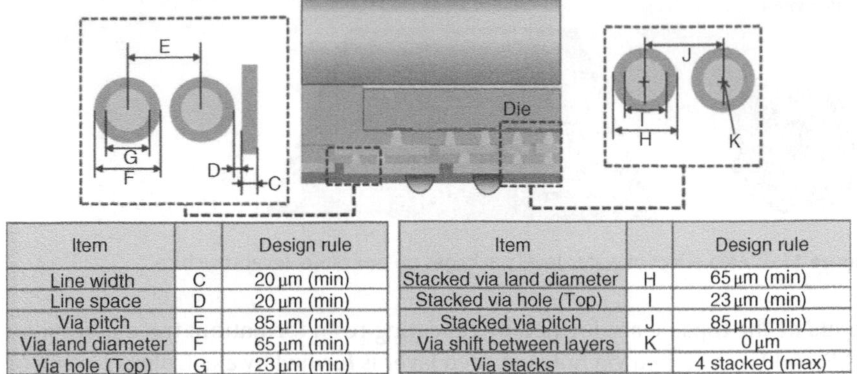

Item		Design rule
Line width	C	20 μm (min)
Line space	D	20 μm (min)
Via pitch	E	85 μm (min)
Via land diameter	F	65 μm (min)
Via hole (Top)	G	23 μm (min)

Item		Design rule
Stacked via land diameter	H	65 μm (min)
Stacked via hole (Top)	I	23 μm (min)
Stacked via pitch	J	85 μm (min)
Via shift between layers	K	0 μm
Via stacks	-	4 stacked (max)

Figure 11.7 Design rules for a typical EDS package.

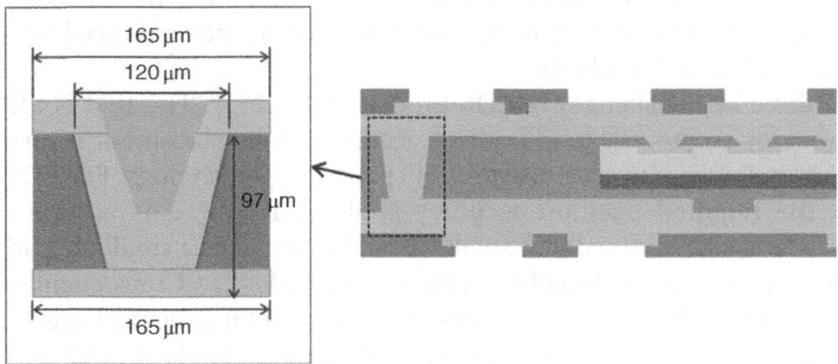

Figure 11.8 Example of a standard deep via between topside and bottomside of an EDS package.

is 85 μm. To minimize via area or for better electrical performance in via area, via stacking is the key solution. Stacked via structures have been qualified and a four-stacked via is acceptable to use. The design rule of the deep via that connects between the layer over the die and the layer under the die is determined by the die thickness and layer structure. Figure 11.8 shows a very standard example. In the case of 50 μm die thickness, via of 97 μm depth and 120 μm diameter have been qualified. The L/S are 20 μm/20 μm and via hole size is 23 μm in diameter. The via land is 65 μm for a 23 μm via hole. Considering the accuracy of die mount and reliable connection at the via bottom, the size of the via land and hole is the minimum for the process.

To determine the best package size for a panel-level assembly package, the assembly yield must be considered. In the case of a very standard design rule such as the one shown in Figure 11.8, L/S = 20 μm/20 μm and a 23 μm via hole, most of the failures are caused by foreign material in the manufacturing process, not open or shorted traces. If the package size is increased, the possibility of foreign material over one package becomes bigger.

11.5 System Integration Capability

EDS technology can have many variations of layer structures, package size, die counts, and more. Stacking die in a package, shown in Figure 11.9, has already been achieved. The EDS package can have terminals on both sides. Also, components can be mounted over the EDS package. When this is done, the EDS package becomes a module structure. The technology details of such a module are explained further in Section 11.7. Also, it is easy to form side-by-side structures, because die can be placed in 150 μm pitch and can be connected to each other by via holes and traces.

Design flexibility and the number of design possibilities are strong points of EDS technology. Design flexibility means not only layer counts but also the number of die and various die positions. Samples of various structures are shown in Figure 11.10. The top left hand is a single die, the top right hand shows two die, the lower left hand shows the interconnection of several die, and the lower right shows two die that require close proximity for improved signal communications. The capability of including and connecting several die in a package is one of the key advantages for EDS, allowing the technology to achieve a system-in-package (SiP) design. To minimize the system, connecting active die in one package is significant.

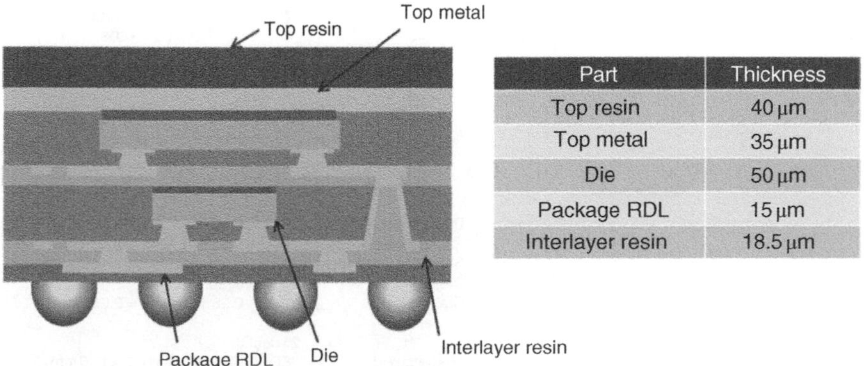

Part	Thickness
Top resin	40 μm
Top metal	35 μm
Die	50 μm
Package RDL	15 μm
Interlayer resin	18.5 μm

Figure 11.9 Stacked die EDS structure.

Figure 11.10 Possible EDS alternatives.

11.6 Package Performance

11.6.1 Thermal Performance Comparison Between EDS and FBGA

Figure 11.11 shows the results of a thermal performance comparison between an EDS and a fine-pitch ball grid array (FBGA) package. The metal plate under the EDS die provides a good heat spreader, so the thermal resistance is lower than that of an FBGA. The thermal resistance was reduced by 30% for θ_{ja} and 40% for θ_{jc}. Of course, the main heat dissipation occurs at the second-level assembly board. To reduce thermal resistance in the system, the assembly board should be designed carefully for optimum heat removal. Comparisons with FC-CSP, FO-WLP, and other embedded technologies are expected to show improvements as well, although not necessarily as great.

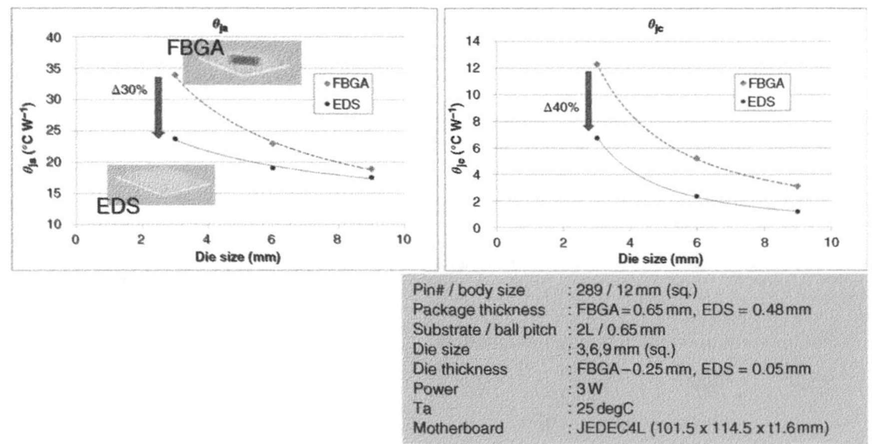

Figure 11.11 Thermal performance comparison between an EDS and an FBGA package.

11.6.2 Electrical Performance Comparison Among EDS, FC-BGA, and FBGA

EDS technology does not use a substrate in the package, like a flip-chip ball grid array (FC-BGA) or wire-bond BGA (WB-BGA). In packages using a substrate, a connecting structure between die and substrate such as a solder bump, Cu pillar, wire, or other electrical connection is required. Reflection of electrical waves happens at impedance uncontrolled points. It is difficult to match impedance completely at the connection point. The connection structure becomes a reflection point or gap for the electrical wave [1]. Consequently, the structure should be as small as possible for good signal integrity at high frequencies. In the EDS package, the via connecting the die and the signal trace is small, just 23 μm in diameter and 18 μm thick. In contrast, a WB-BGA package's connection between die and signal trace is wire, 18 μm in diameter and from 1 to 3 mm in length, and an FC-BGA's connection is solder bumps, 70 μm in diameter and around 50 μm in height. The EDS packages' connection points are smaller, when both its connectivity diameter and length are taken into account, compared with other packages' connection points.

In general, the package's substrate provides another gap point for an electrical wave. Generally, buildup substrates have core layers. The core layer is thicker than the buildup layers and includes glass cloth in the resin. Since the plating through-hole is drilled in the core layer, it is larger than a laser via hole in the buildup layers. The long length and thick plating of the drilled through-hole in the core layer becomes a large gap in the electrical wave route. In comparison EDS has no core layer and benefits from the smaller laser via hole.

The insertion losses of several packages are shown in Figure 11.12. The difference between FBGA and EDS losses is very clear at any frequency. The FC-BGA and EDS package losses are almost the same until 10 GHz, but over 10 GHz, the difference is clear. The insertion losses of the EDS package are less than others. Improvements are also expected when comparing EDS with FC-CSP, FO-WLP, and other embedded technologies although they may not be this significant. Considering this analysis, the EDS package is suitable package for millimeter wave applications. EDS packages have already been developed for 80 and 40 GHz designs.

For radio frequency (RF) applications, a high frequency package design is needed. The package also needs to integrate a variety of components. As shown in Figure 11.12, the EDS package can achieve excellent performance at over 20 GHz, and a module structure is possible for EDS technology that will be shown in Section 11.7. The combination of these two attributes makes the EDS package a suitable choice for RF applications. For now, RF applications are mainly in smartphones, but they will also be appropriate for many other Internet of things (IoT) applications as the market needs grow.

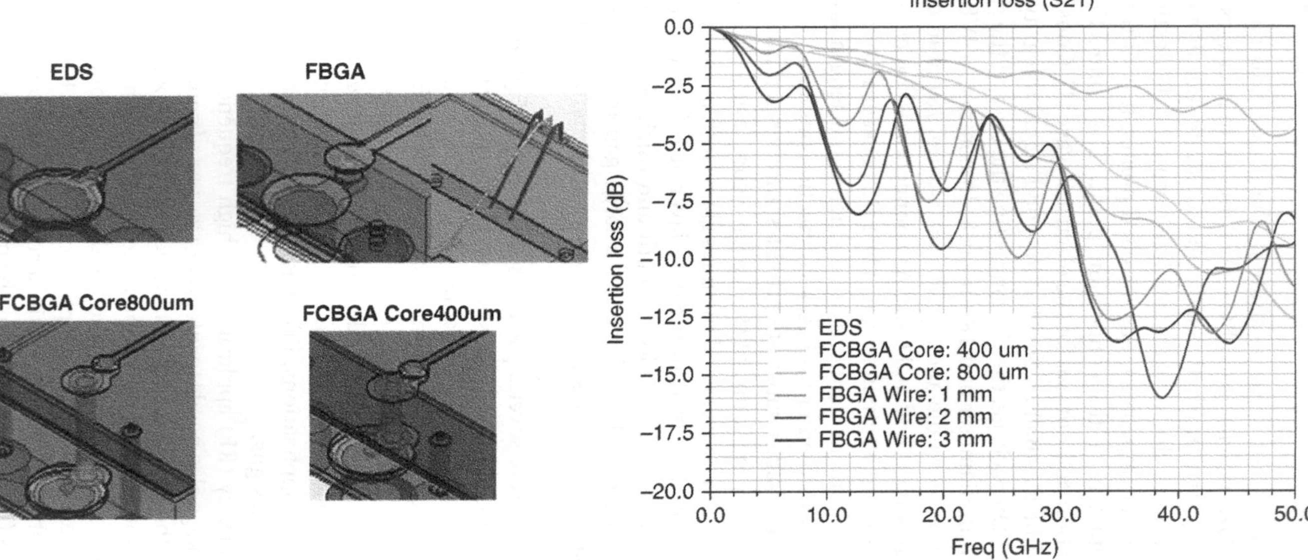

Figure 11.12 Insertion loss simulation for various packages.

11.6.3 Robustness and Reliability Data

The basic process of the FO-PLP package with metal base plate has good reliability. The data shown in Figure 11.13 is for a 12 mm × 12 mm package with 360 balls at 0.5 mm ball pitch. The test is done by electrical connection (short/open) and confocal scanning acoustic microscopy (CSAM) measurements. The conditions are based on the Automotive Electronics Council (AEC) Q-100/Grade 0. The FO-PLP with metal plate has good reliability and passed the required reliability level for automotive applications. The FO-PLP's warpage data is shown in Figure 11.14. The warpage does not change at any temperature indicating the stress in the package is very small.

The reliability data for the EDS package of Figure 11.10 is shown in Figure 11.15. In this case, the package size is 3.6 mm × 3.6 mm with 81 balls at 0.35 mm ball pitch. The testing was performed based on JEDEC standards and was checked by electrical connection and CSAM measurements. All test samples passed. More information about EDS module reliability is provided in the next section.

(a)

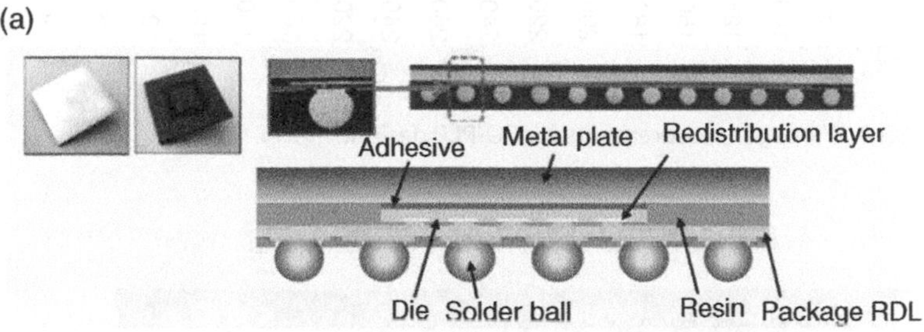

(b)

Item	Condition	S/S	500 h 500 cyc	1000 h 1000 cyc
MSL (L3) + uHAST	30 °C/60%/192 h 260 °C max 3 times + 130 °C/85%	231 pcs/3 Lot	96 h 231 OK/231 (Criteria)	168 h 231 OK/231
MSL (L3) + HTS	30 °C/60%/192 h 260 °C max 3 times + 175 °C	77 pcs /1 Lot	500 h 77 OK / 77	1000 h 77 OK / 77 (Criteria)
MSL (L3) + TCT	0 °C/60%/192 h 260 °C max 3 times + −65~175 °C	77 pcs /1 Lot	500 cyc 77 OK / 77 (Criteria)	1000 cyc 77 OK / 77
MSL (L3) + TCT	30 °C/60%/192 h 260 °C max 3 times + −50~175 °C	77 pcs /1 Lot	500 cyc 77 OK / 77 (Criteria)	1000 cyc 77 OK / 77 (Criteria)

Figure 11.13 Package cross section (a) and reliability test results (b) of FO-PLP design.

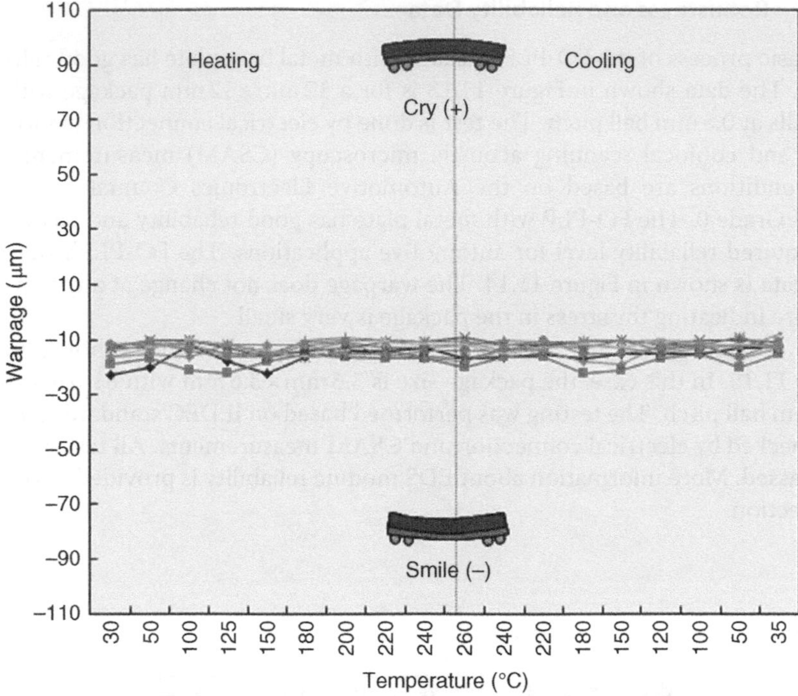

Figure 11.14 Warpage measurements for FO-PLP design.

Item	Condition	Criteria
MSL (L1)	85 °C/85%/168 h 260 °C max 3 times	–
MSL (L2)	85 °C/60%/168 h 260 °C max 3 times	–
MSL (L3) + HTS	30 °C/60%/192 h 260 °C max 3 times + 150°C	1000 h
MSL (L3) + TCT	30 °C/60%/192 h 260 °C max 3 times + –55~125°C	1000 cyc
MSL (L3) + uHAST	30 °C/60%/192 h 260 °C max 3 times + 110°C/85%	264 h

Figure 11.15 EDS reliability test matrix.

11.7 Diversity of EDS Technology: Module

11.7.1 Technology Description

For smart phone applications, design requirements include small and thin packages. To make the system smaller, the package needs to include die, components, and more, in other words, an SiP design. As shown in Figure 11.3, EDS technology can assemble a package that has outer leads on both sides. Furthermore, as introduced in a previous section, EDS technology can include more than two die in one package. With these basic technologies, a module structure using EDS technology has been developed. The details are discussed in this section.

11.7.2 Basic Package Construction

The basic structure of a module using EDS technology is shown in Figure 11.16. Components mounted on the topside of an EDS package with both side leads are protected in the overmold structure. In addition, various component heights can occur. Normally, if components are mounted in a resin layer, it is difficult to control the resin flatness over the layer where components with different heights are embedded. To avoid such issues, components were placed on the topside of EDS package. Using this design approach, various types of components can be mounted. Molding components is recommended since the mold resin protects components and their connections in the package, but molding is optional. For the molding process, compression molding with molded underfill (MUF) is recommended.

11.7.3 Manufacturing Process Flow

The process flow for the EDS package was discussed in an earlier section. After SR formation of the EDS package, the panel is cut into strip shapes for components mounting. The strip size is 74 mm × 240 mm. After components

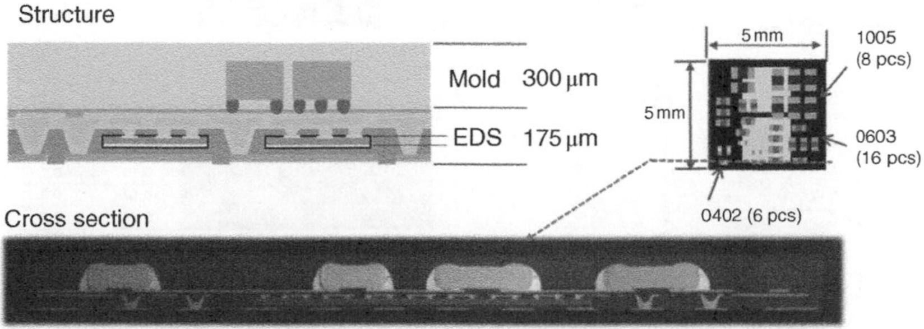

Figure 11.16 A module using EDS technology.

are mounted and reflowed, the strip is molded by compression mold with MUF, then the balls are mounted and reflowed in strip style, and the strip is diced into packages. Selecting the optimum mold resin material is one of the key aspects of this structure. Since the warpage of the panel or strip or package differs greatly between designs, the proper mold resin must be chosen for each design.

11.7.4 Design Features

For modules, the basic design rule is the same as shown in Figure 11.7. Figure 11.17 shows the design rules required for die, via, and component placement. The distance B between die was decided from mount accuracy and resin flow. Dimensions A and C were decided based on package dicing accuracy. An example of the cross-sectional analysis for each layer's thickness is shown in Figure 11.18. The mold thickness on components is normally controlled at twice the filler size in the mold resin.

11.7.5 Robustness and Reliability Data

In addition to the basic EDS testing shown in a previous section, reliability tests were conducted on basic EDS module samples with the structure shown in Figure 11.16. Figure 11.19 shows the test criteria and results. The package size was 5 mm × 5 mm with 40 land grid array (LGA) at 0.4 mm pitch. The testing was performed based on JEDEC standard and also was checked by electrical connection (continuity) and CSAM measurements. Test units passed the established criteria with no failures. This reliability testing qualifies the

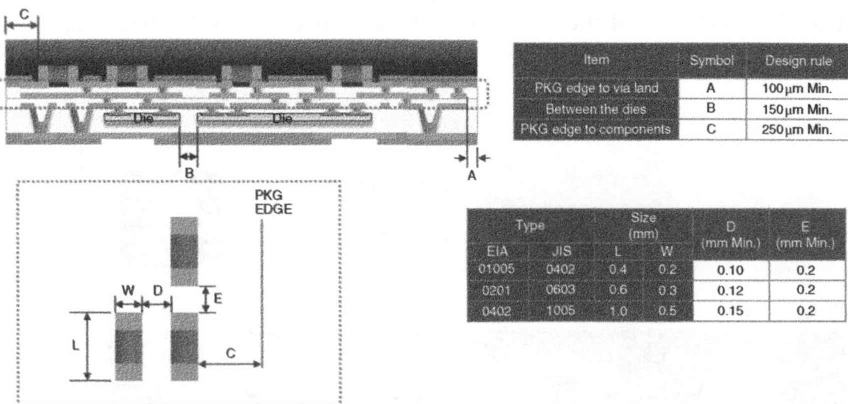

Item	Symbol	Design rule
PKG edge to via land	A	100 μm Min.
Between the dies	B	150 μm Min.
PKG edge to components	C	250 μm Min.

Type		Size (mm)		D (mm Min.)	E (mm Min.)
EIA	JIS	L	W		
01005	0402	0.4	0.2	0.10	0.2
0201	0603	0.6	0.3	0.12	0.2
0402	1005	1.0	0.5	0.15	0.2

Figure 11.17 Design rules for an EDS module.

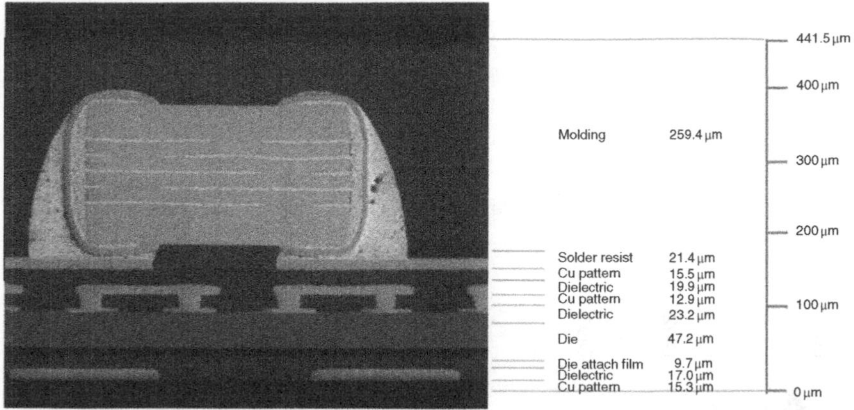

Figure 11.18 Cross-sectional analysis of an EDS module.

Item	Condition	S/S	Lap Result
MSL (L3) + TCT	30 °C/60%/192 h 260 °C max 3 times + 55~125 °C	30 pcs /1 Lot	500 cyc 30 OK/30
MSL (L3) + HTS	30 °C/60%/192 h 260 °C max 3 times + 150 °C	30 pcs /1 Lot	1000 h 30 OK/30
MSL (L3) + uHAST	30 °C/60%/192 h 260 °C max 3 times + 130 °C/85%	30 pcs /1 Lot	192 h 30 OK/30
Board-level Reliability	Temperature Cycle -40°C~125°C (10 minute soak)	30 pcs /1 Lot	1500 cyc 30 OK/30
Drop	JEDEC Standard (JESD22B111)	30 pcs /1 Lot	60 Drop 30 OK/30

Figure 11.19 Module reliability test matrix and results.

basic EDS module structure. However, since the module design is customized for each system, reliability should be verified for each system.

11.7.6 System Integration Capability

Another EDS module example is shown in Figure 11.20. This system includes two die, wafer-level chip-scale packaging (WLCSP), and various passive components. The die are stacked in the package and not mounted side by side. WLCSP and components are mounted on the package.

In this SiP design, the system's size was reduced by 35% in area from the original system that contained several packages. EDS technology provides considerable design flexibility, so a variety of SiP designs can be easily achieved.

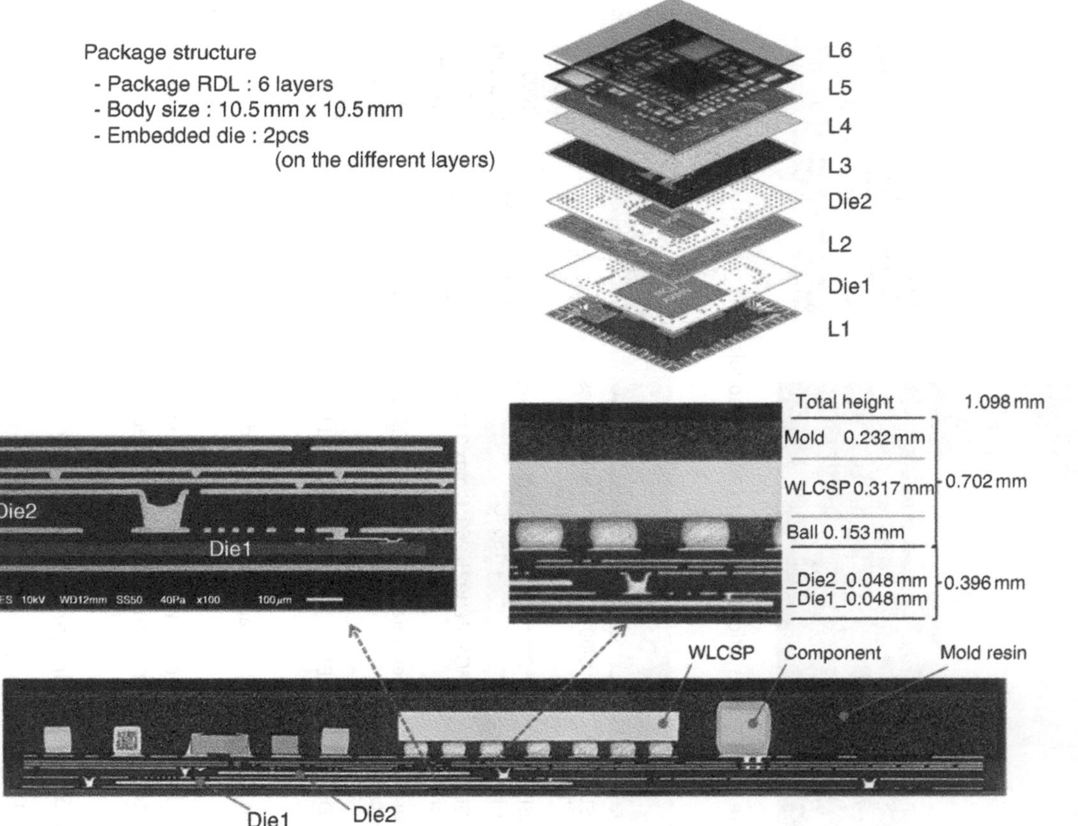

Figure 11.20 Example of an EDS module with WLCSP.

11.8 Diversity of EDS Technology: Power Devices

11.8.1 Technology Description

Power electronics is very important in many industrial fields [2]. Power semiconductors, including MOSFETs, insulated-gate bipolar transistors (IGBTs), and other devices, have been used for many years as essential parts of power electronics for controlling or switching the current from the power source. Recently, in automotive applications, the demand for electric and hybrid vehicles has increased because of their fuel efficiency and concerns about environmental pollution problems. The power devices in the power electronic circuits in those automobiles require high density assembly to obtain intelligent, high powered, and reliable control [3]. Additional requirements for each power device dictate that its packaging has small size, high thermal performance and is fabricated with low cost manufacturing [4]. However, conventional power device packages have miniaturization and cost limits due to the use of a wire-bonding process where each bond is fabricated one at a time. Moreover, the thermal performance is also restricted because the thermal flow inside the package only has a single path through the narrow metal wires to dissipate outward. Consequently, a new package with both small size and high thermal performance is desirable.

With the above background, a new and expandable advanced package using EDS technology for power devices that matches the desired requirements is discussed in this section. The package uses a type of multilayer laminated construction and can be fabricated using a printed circuit board (PCB) process. Power device chips are embedded in the laminated resin layer. The package size is very small and very thin with precise trace patterning. The traces are formed by process integration of insulator lamination, via formation, metallization, and subtract etching. At insulator lamination, the thickness of insulator on the embedded die should be well controlled because it strongly affects the quality of via size and metal filling in the following process. And at the metallization process, precise control of the thickness of the metal is very important for fine pattering at the following subtract etching. In addition, the package has good thermal performance with low thermal resistivity and high power dissipation as discussed on later in this section. Moreover, the package design has very low drain to source on-resistance with low resistivity connections using many via holes as shown in Figure 11.22. The package uses a redistribution layer (RDL) of a trace and multilayer structure, so the pattern layout is easy to customize. With this design methodology, there can be many types of devices including discrete or power module systems with integrated control driver chips.

11.8.2 Basic Package Construction

Figure 11.21 shows the outline of a basic EDS package for discrete power devices. The discrete MOSFET power device has only three pads: source, drain, and gate. Most chips for power devices are a vertical-type design that has a source pad and gate pad on one side (usually the top) and a drain pad on the other (usually the bottom).

The cross-sectional view in Figure 11.22 is a facedown design. The chip is embedded in an insulator resin, and the gate or source pads are connected on the bottomside of the package through shallow vias. The drain pad is connected on top through a bottom via and connected to the bottom package side through the package's deep via. Three types of contact vias are designed for each of the pads: on the top of and near the die through the insulator resin and under the die through the die attach film (DAF).

11.8.3 Manufacturing Process Flow

The process flow of the EDS power package is shown in Figure 11.23. Starting with the formation of layer 1 (L1, die attach mark) to layer 2 (L2, top trace layer), the process is same as a typical EDS process. The only difference is the

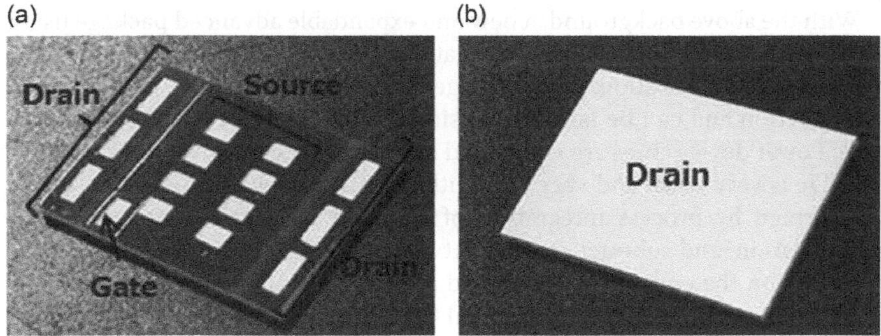

Figure 11.21 Outline of a basic EDS package for power devices: (a) bottom and (b) top.

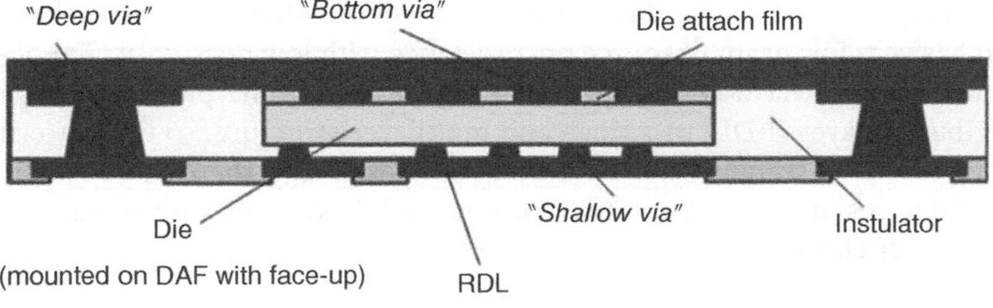

Figure 11.22 Cross-sectional view of a facedown EDS power package.

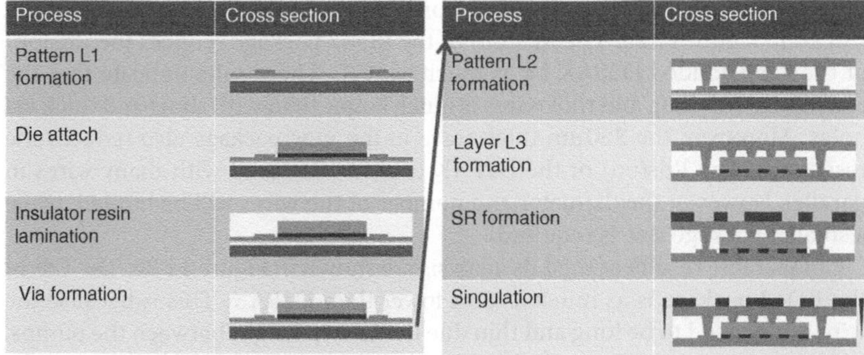

Process
Cross section
Pattern L1 formation
Die attach
Insulator resin lamination
Via formation
Process
Cross section
Pattern L2 formation
Layer L3 formation
SR formation
Singulation

Figure 11.23 Fabrication process flow.

formation of layer 3 (L3): bottom vias and bottom trace layer. In the L3 formation, after peeling off the substrate using a carrier sheet, vias are formed in the DAF from the bottomside of the die with copper plating used to fill the bottom die and to form the top of the package. In the SR formation step, the SR layer is laminated and patterned using an exposure and development method to open the windows for pads. Singulation at the end of the process results in each package being cut off the panel. This is an LGA package.

11.8.4 Electrical Characteristics

Figure 11.24 shows four package models compared in the electrical characteristics design analysis. The RDL packages, RDL1 and RDL2, have trace thicknesses of 50 and 250 μm, respectively. The loss-free package (LFPAK) represents one of packages that has a copper clip as the outer trace [5]. A D2PAK (or TO263) is used to compare the new package with a wire-bond-type package [6]. The drain to source on-resistance (Rds[on]) and drain to source inductance (Lds) are key parameters for comparing MOSFET losses [7]. Rds(on) and Lds were calculated that are, respectively, defined to determine the DC drain-source resistance on DC and AC inductance on AC while the FET is on. Rds and Lds indicate power loss on direct current and frequency, respectively [6].

RDL1
(Cu 50 μm)
RDL2
(Cu 250 μm)
D2PAK
(LF)
D2PAK
(Wire)
model

Figure 11.24 Package models compared in the electrical characteristics analysis.

The results of the electrical simulation comparing the packages' Rds(on) are shown in Figure 11.25. The Rds(on) of the RDL2 package reduces the Rds(on) of the wire-bonded D2PAK by as much as 97%. The results indicate that the resistance with long and thin wires is much larger than with short and thick via holes. Moreover, the 250 μm thick trace in the new package also reduces the resistance. The Rds(on) of the D2PAK could be reduced with many wires in parallel. However, the diameter and number of the wires will be limited, while a smaller package size is required.

Comparable results of the Lds analysis are shown in Figure 11.26. The Lds of the RDL2 package is as much as 80% lower than D2PAK. This is because the wire traces need to be long and thin due to the step height between the bumps, which also increases the switching noise. In contrast, the RDL trace can be

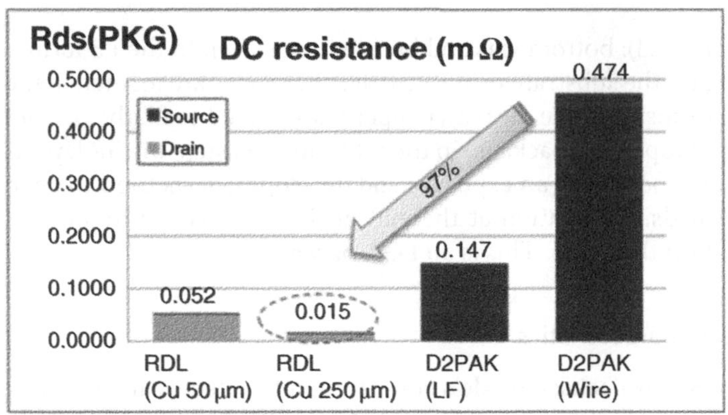

Figure 11.25 Results of Rds comparison.

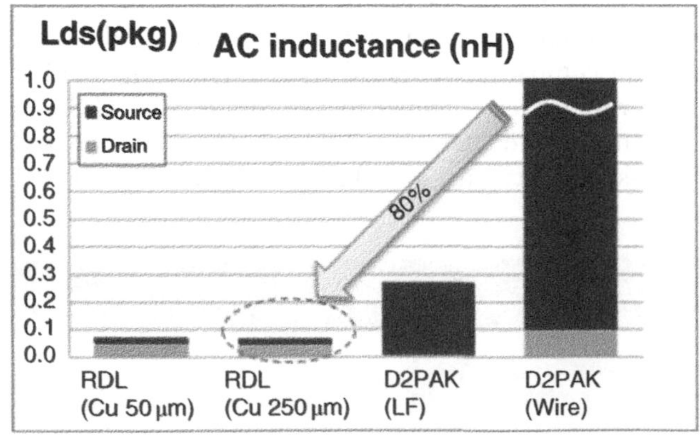

Figure 11.26 Results of Lds comparison.

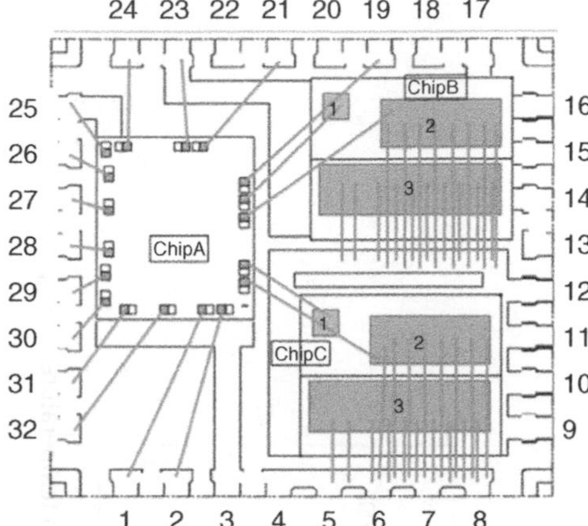

Figure 11.27 Three-in-one power module packages. (a) PQFN type and (b) embedded multilayer type.

short and thick with a layer-by-layer structure, so the noise through the trace is much lower. Similar to the case of the resistance, the wire's curved form is also limited in a smaller package. As a result, the RDL structure has advantages for electrical characteristics required in a small package.

11.8.5 Size Miniaturization and Thermal Characteristics on Power Module Package

As an example of size reduction, two packages are shown in Figure 11.27a and b. The package structure used is a 3-in-1 module package, which has two MOSFET chips and one control driver chip. (a) is power quad flat no-lead (PQFN) type using wire bonding, and (b) is an embedded-type comprising multilayer I and II embedding the chips and RDL. The package size of the embedded type can be reduced by 28% of PQFN type as shown in Figure 11.30. The chips can be embedded in different layers, I or II, and interconnected using RDL, so the package size can be smaller than a PQFN.

Thermal simulations were conducted on the two types of module packages. Figure 11.28 shows the results of the thermal simulation of the thermal resistance of heat flowing through the junction in the die to the ambient, θ_{ja}, and the maximum junction temperature, T_{jmax}. Both parameters vary with the velocity of air flow. The embedded power package has better heat dissipation than the conventional package by 20 or 30%.

The temperature distribution of the two packages is shown in Figure 11.29. The circles indicate the maximum temperature points in each model. These exist around the wires in PQFN model and around the via holes in the

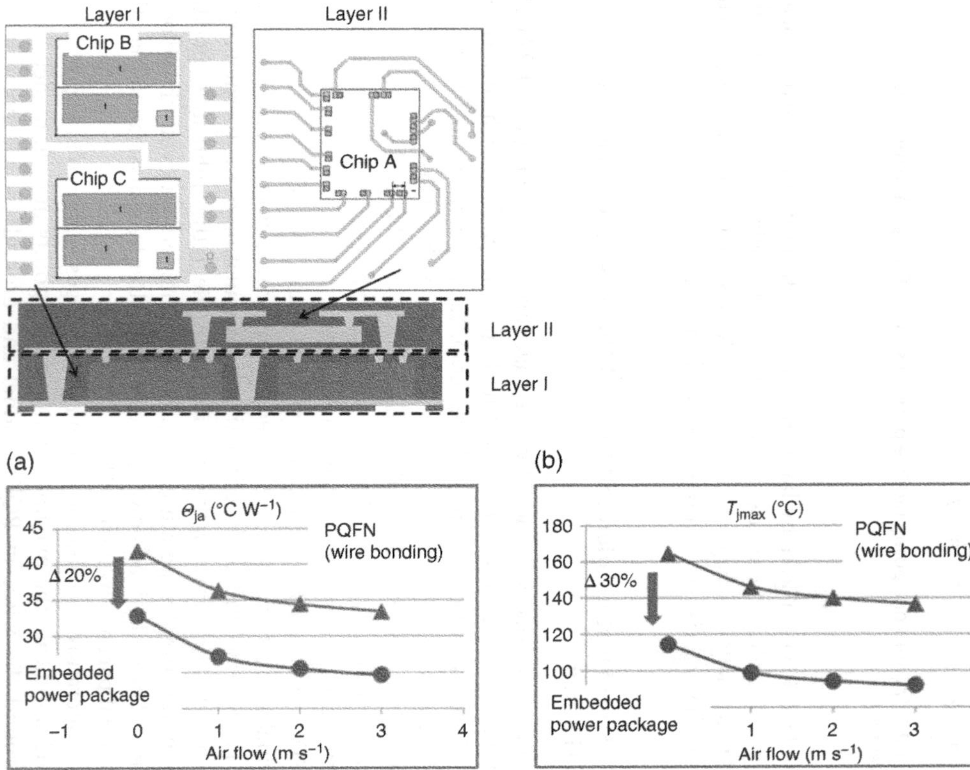

Figure 11.28 The results of the thermal simulation. (a) Thermal resistance and (b) maximum junction temperature.

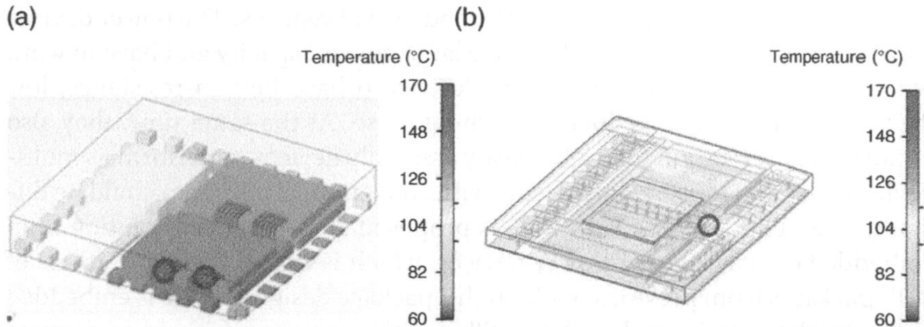

Figure 11.29 The temperature distribution of the two packages: (a) PQFN as reference and (b) embedded power package. Circles indicate the area with the maximum temperature.

	PKG size (mm)	θ_{ja} (°C W^{-1})	T_{jmax} (°C)
PQFN	5.0×5.0	41.8	164.5
Embedded	4.5×4.4 (▼28%)	32.8 (▼30%)	114.3 (▼20%)

Figure 11.30 Size miniaturization and thermal resistance reduction on power module package.

embedded power package model. The many via holes and wide plain trace in the power package has more effective thermal dissipation than the thick and long wires in PQFN package. These results are summarized as Figure 11.30.

11.9 Applications and Markets

EDS technology provides significant advantages to FO-PLP. In addition to basic EDS technology with a metal base plate that has been qualified for and implemented in RF applications, mainly in smartphones, enhancements provide additional applications for smartphone modules and for power devices. Modifying EDS technology to have outer leads on both sides and including more than two die in one package provides an EDS module for an SiP design. This approach has demonstrated significant area reduction. Compression molding with MUF is recommended to provide protection to the added components. Since EDS technology provides considerable design flexibility for modules, a variety of SiP designs can be easily achieved.

Power devices are used in advanced products in several market segments. In automotive applications, the revenue is increasing rapidly, especially with the introduction of more and more electric and hybrid vehicles. The power devices used in those applications need to have large power capacity and have to work at high frequency. This requires the devices to have low on-resistance, low thermal resistance, and reduced switching noise. At the same time, they also require miniaturization for high density assembly devices to perform as multi-devices. Meeting those requirements with conventional packages could be difficult, so an EDS package with RDL is proposed. The proposed solution is an embedded die package for power devices, which is one of the applications of RDL packages from previous work. In this package design, the die is embedded in the insulator resin, and via holes filled with copper are formed to connect the die pads with the RDL. This package has three different types of vias to connect both top- and bottomside of the die with the RDL and between the RDLs. Specifically, the vias for the bottomside of the die are formed through the DAF. Simulations have been performed to confirm the electrical and thermal performance of this package. Better characteristics were obtained compared with other packages, proving that the vias are effective if reliable fabrication occurs.

References

1 Charles, A. (2005). *Harper, Electronic Packaging and Interconnection Handbook 4/E*. McGraw Hill.

2 Rashid, M.H. (2007). *Power Electronics Handbook, Devices, Circuits, and Applications*. Academic Press.

3 Uesugi, T. (2000). Power devices for automotive applications – reviews of technologies for low power dissipation and high ruggedness. *R&D Review of Toyota CRDL* 35 (2).

4 Maliniak, D. (1996). Power Device Packaging Beats The Heat. *Electronic Design* September.

5 LFPAK/LFPAK-iPower MOSFET Exciting loss-free package portfolio, Renesas Electronics Europe (April 2010). www.grupelektronik.com.tr/images/PDF/HAT2270.PDF

6 D2PAK or DDPAK – Double Decawatt Package, EESemi (2006). http://eesemi.com/d2pak.htm

7 Raab, F.H. and Sokal, N.O. (1978). Transistor power losses in the class E tuned power amplifier. *IEEE Journal of Solid-State Circuits* 13 (6).

12

Blade

A Chip-First Embedded Technology for Power Packaging

Boris Plikat and Thorsten Scharf

Infineon Technologies AG, Regensburg, Germany

12.1 Technology Description

The name "Blade" for Infineon's laminate chip embedding technology was chosen as the outward appearance of the very thin package bears a resemblance to the blade of a knife or a similar tool. The discrete metal–oxide–semiconductor field-effect transistor (MOSFET) package Blade3x3 and the package of the integrated buck-converter product DrBlade2, where Dr stands for Driver, even share the shiny metallic surface at their topsides with a typical metallic cutting blade. A buck converter is a DC-to-DC power converter that steps down voltage from its supply (input) to its load (output), where DC stands for direct current.

The main application field of Blade products is buck converters, e.g. for use in servers. The integrated buck converter DrBlade1 (see Figure 12.1) was branded with its high peak efficiency, the small package size (only $5 \times 5\ mm^2$) and low profile (0.5 mm), and the compact and simplified board layout that it enables. In addition, DrBlade2 comes with temperature sensing and thermal warning as well as integrated high precision load current sensing. Together with the multiphase pulse width modulation (PWM) controller from Infineon, these allow optimizing the operation of several parallel DrBlade2 in a multiphase configuration as well as protection features.

The first Blade product, announced by Infineon in 2013, was DrBlade1 [1]. The brand name DrBlade is a fusion of the name for the Blade technology and the well-known DRMOS standard, a combination of a MOSFET half bridge and driver. The development of the Blade package technology was triggered by the MOSFET chip shrink when it started in the middle of the previous decade. Originally, it targeted the packaging of vertically conducting MOSFETs with a few mm^2 area or even less than $1\ mm^2$ area and about $60\ \mu m$ thickness oriented

Advances in Embedded and Fan-Out Wafer-Level Packaging Technologies, First Edition.
Edited by Beth Keser and Steffen Kröhnert.

Figure 12.1 DrBlade1 with small package size and low profile of $5 \times 5 \times 0.5\,\text{mm}^3$.

facedown to the application board to achieve low parasitics. The new package technology overcomes the high contribution of wire-bond interconnects to the product's on-resistance and inductivity and the limitation of clips to bigger chip sizes. Additionally, the exposed die pad on the topside allows efficient topside cooling. With changed market demands in the low voltage area, the focus changed from discrete MOSFETs to integrated DC-to-DC converters. For the integration of a driver chip and the two MOSFETs, the technology's strength in terms of complex and compact redistribution with low inductance came into play. DrBlade1 with its 5 mm × 5 mm footprint hosts more silicon area than its molded wire-bond and clip-bond predecessor with its 6 mm × 6 mm package area.

12.2 Development and Implementation

As of 2017, Infineon has brought three Blade packages to the market. In addition to the two different integrated DC-to-DC converter packages of DrBlade1 and DrBlade2, a discrete MOSFET package with a 3.0 mm × 3.4 mm package size was launched. Targeting different markets, the three packages were produced in different volumes. The highest contribution to the overall volume of Blade packages stems from the youngest product DrBlade2.

The Blade package technology merges conventional packaging technology and printed circuit board (PCB) technology. Furthermore, it contains dedicated chip embedding process steps. The use of a copper lead frame and the diffusion solder or glue die attach are also part of conventional power package technologies, and the package separation by mechanical dicing is similar to the process used for molded array packages assembled in strip format. Although the lead frame already has certain functionality in terms of redistribution, the Blade package technology can be classified as chip-first embedding technology, because the actual embedding and dominant part of the redistribution are realized with PCB processes after die attach.

12.3 Basic Package Construction

The basic package construction is described in this section. Figure 12.2 shows a schematic cross section of the DrBlade1 package mainly indicating the different materials, but not to scale. The actual dimensions can be inferred from the SEM cross section given in Figure 12.3.

Besides the laminate layers, the copper lead frame makes up the dominant volume of the package. With 250 μm thickness at DrBlade1, it accounts for about half of the package thickness. Its shape reveals that it is an etched lead frame. Laterally, the die pads and leads are electrically insulated to each other by laminate epoxy resin and its filler particles. Resin and fillers were pressed out of the surrounding prepreg, layers of uncured glass fiber reinforced epoxy, above and below the lead frame during the first lamination step to fill these cavities. Hence, there are no glass fibers in these areas except for those slightly bent into the lead frame's etched openings. Areas without glass fibers are also the cutout regions of the prepreg around the dies and the etched areas of the redistribution layers (RDLs) (see Figure 12.3). All lamination layers used in this package contain glass fibers.

The lead frame, the last metal of the die pads, the vias, the RDLs, and footprint of the package are all made of copper (Cu) with Cu-to-Cu interconnects. Hence, there is no driving force for any diffusion or phase formation at the corresponding interfaces, contributing to the high reliability of the Blade package technology. Only the diffusion solder die attach, final finish, and bumping – if applicable – establish interfaces of the package to other metals than copper.

The Blade packages contain two via sizes. The small about 70 μm thick via type connects the chip pads, two adjacent RDLs, and the lead frame to the footprint layer. The ~120 μm thick via type connects the first RDL over the die

Figure 12.2 Schematic cross section of DrBlade1; the color codes indicate the materials used.

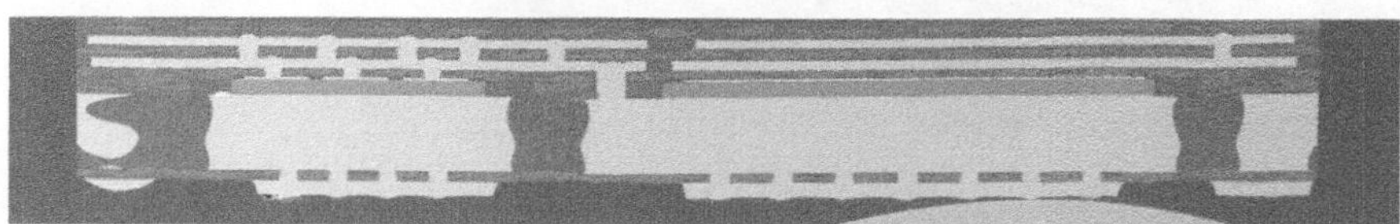

Figure 12.3 SEM image of a mechanical cross section through a DrBlade1 package.

to the closest lead frame side. These large vias are needed because of the vertical current flow through the MOSFETs. In DrBlade1 only, they also serve as interconnect elements from the package topside to the leads.

The die attach layers with low bond line thickness of a few μm for the diffusion solder are almost invisible in the cross section in Figure 12.3. The low bond line thickness supports the low package R_{on} and R_{th} and enables thin chip die attach. The not shown glue die attach layer of the driver chip is slightly thicker, which has to be counterbalanced by a thinner chip. The well-defined bond line thickness and correspondingly low die tilt as well as die thickness are essential for the chip embedding process, as the surrounding geometry of layers and micro-vias limits the degrees of freedom.

All segments of the lead frame of DrBlade1 and DrBlade2 have an exposed part to the package surface at the dicing street. Originally, they form one metal body to keep all parts in their position before the laminate material connects them after the first lamination process. Only at package dicing is the lead frame cut through, and hence has exposed surfaces. Although these are at different electrical potentials, for the addressed products operating at low voltage, this is not an issue. Nevertheless, the process flow can be adapted to create insulated sidewalls of the packages as well. This would be needed for high voltage products to ensure sufficient creepage distances.

The cross sections of DrBlade1 (Figure 12.2) and also the micro-computer tomography (μ-CT) picture in Figure 12.4 illustrate that every electrical potential to be connected from the topside of the package to the footprint requires an individual lead (or pad) of the lead frame. This adds complexity to the package construction and consumes package area.

The DrBlade2 package (see Figure 12.5) trades the symmetry of a redistribution on both sides to improve this drawback, as the lead frame is not located

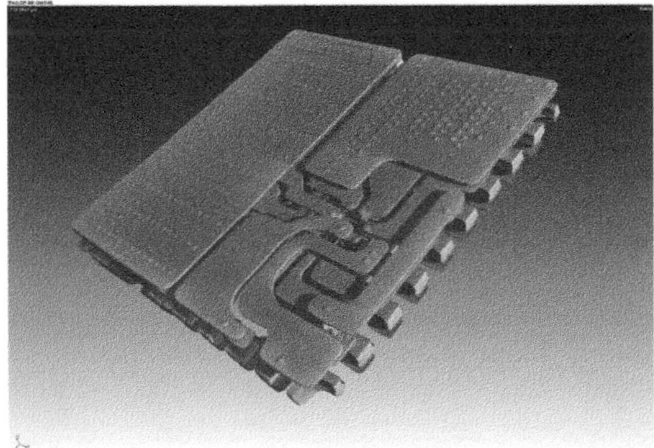

Figure 12.4 μ-CT image of DrBlade1, topside view.

Figure 12.5 Schematic cross section of the DrBlade2 package.

between the die and the footprint toward the application board, but above the die. The larger vias are only needed to carry the vertical current flow of the MOSFETs.

Obviously, the lead frame makes the Blade package a one-sided package with limitations for the integration of components on the package top or package-on-package (PoP) solutions. Although terminals can easily be created on both package sides, the electrical path through the lead frame layer needs big vias, dedicated pads on the lead frame, and therefore package area. This usually limits the topside to low pin count connections, like passives.

Chip embedding laminate packages without lead frames are technically feasible and already in the market [2]. Not only to carry electrical current but also for thermal reasons, vias can be connected to the backsides of the die. Care must be taken with applications depending on a low package R_{on} and a high thermal impedance or heat spreading capability within the package, if the lead frame is omitted.

However, without the lead frame, the laminate chip embedding package becomes a true double-sided package. With a copper via interconnect on both sides of the die, it can contain die with different orientation ("face-up/face-down"). A laminate chip embedding package without lead frame is basically symmetric with regard to interconnects and redistribution on package top and bottom side. This is the base for a much higher flexibility in terms of its 3D capability and enables more stacked configurations (PoP, passives on top) compared with the Blade package technology. Basically the two laminate chip embedding technologies, with and without lead frame, both address different specific points for their respective application. While the technology without lead frame has advantages in terms of 3D capability, the Blade technology with lead frame has superior high current and high thermal performance.

With typical minimum line/space widths of roughly 60 μm/60 μm at typically about 35 μm thickness of the RDLs, the Blade package technology is made to realize DC-to-DC converters with load currents of 60 A and more. Compared with eWLB-like embedding technologies, fields of application and design rules are complementary. Pitches, cross sections of traces, and cooling options are much smaller in these technologies.

12.4 Manufacturing Process Flow and BOM

12.4.1 Manufacturing Equipment

The production line for the Infineon Blade technology consists of a mixture of a few semiconductor packaging machines and several PCB production machines (see Figure 12.6). The die bond process follows classical approaches for solder and glue die bonding, modified for the diffusion soldering. Also, the dicing and testing is done with standard machines, although large production formats require some special consideration. Standard PCB machines, on the other hand, are horizontal wet benches for cleaning, roughening, and develop, etch, strip (DES). Also, resist lamination, laser drilling, and FR-4, a standard laminate material with flame retardant grade 4, and precut prepreg lamination are done in typical PCB line equipment, with moderate adaptions to allow the high requirements for quality and traceability. The precutting of the prepreg is done with a laser cutter or mechanical driller. For the Cu plating, a vertical plating line with more than 10 different steps is applied, which can be divided into desmear, activation, and via filling. The requirements for the process are higher than for typical PCB processes, especially as two different via geometries are filled at the same time. This requires a much higher effort in control of the bath performance, making an automatic analysis and dosing system advantageous.

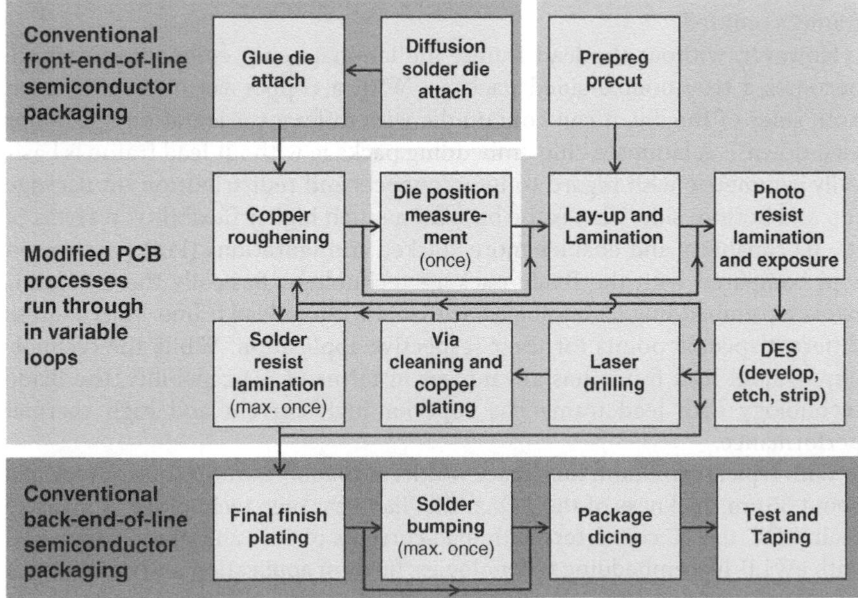

Figure 12.6 Process flow overview.

In general, the lithographic exposure can be done with standard masks or foils, but a mask-less exposure has significant advantages for the assembly flow.

12.4.2 Basic BOM

The material inside the Infineon Blade package is a mix of classical packaging material and PCB material. The main component material is copper. A lead frame of typical thickness for small power packages of ~250 μm and all RDLs, pads, and micro-vias consist of Cu. Cu on the top or bottom surface is covered with NiP/Au as solderable finish. In between the Cu, high-glass transition temperature (T_g) FR-4 material is used. Also special for power packages is the Au-containing diffusion solder die lead frame interconnect, which makes the whole device lead-free and green according to the Restriction of Hazardous Substances (RoHS) (EU directive).

12.4.3 Wafer/Die/Assembly Preparation

The Infineon Blade technology was developed as a thin chip package. While test vehicles were built with more than 200 μm thick chips, the main focus definitely is on the thickness range below 100 μm total chip thickness. This implies that one major preparation step is the thinning of the chip. A new challenge for the thinning process is the tight tolerance. While for usual packages their wires are bonded on the die surface, allowing a wide chip height tolerance, a chip in the Blade technology has to fit in a tight space [3]. This limits the process window for the chip thickness and requires unusual awareness for the grinding processes.

Another central point is the chip metallization. For technologies that embed in laminate, Cu pads are more or less unavoidable. The high heat capacity and high reflectivity at IR wavelengths combined with a sufficiently high melting temperature make it a performant laser stop layer for via drilling. On the other hand, it provides chemical robustness and avoids unnecessary material interfaces with the micro-vias, which are Cu anyhow. Many tests were done with differing metallization. They always faced problems with insufficient robustness to the intense use of different aggressive chemicals or poor adhesion of the laminate or even the micro-vias.

Open Cu areas are attacked by the adhesion-promoting step, a Cu roughening. This etch has to be considered in the planning of the pad thickness. To avoid damage during laser drilling, the minimal Cu thickness at the micro-via positions has to be planned thoroughly, including process variations, reductions by etching, and testing needle imprints. In summary, there is never too much Cu [3].

For the rear side of the chip, diffusion solder was chosen to achieve a very thin, stable bond line thickness with extreme electrical and thermal performance. The respective metallization has to be applied.

Special care has to be taken for the die separation, as very thin chips with thick metals tend to chip, crack, and break. Therefore, dicing blades and feed speed have to be selected carefully. For some versions, laser dicing was also applied.

12.4.4 Die Attach and Adhesion Promotion

Attaching die to a lead frame with solder is typical in the semiconductor assembly industry. Very thin die and diffusion solder are advanced, yet not unknown. Where the die have to be insulated from the lead frame potential, a standard glue die bonding with thin insulating glue was used.

After the die bonding, the adhesion-promoting step follows (see Figure 12.6). For this, a Cu roughening is done as it is known from the PCB industry. Unfortunately, the lead frame material, which is rolled Cu, has a different, slower roughening behavior than the chip metallization. Therefore, the roughening step has to compromise between sufficient adhesion on the lead frame and not too deep etching on the chip.

A major challenge of chip embedding in laminate is the design rules of the chip pad sizes. While for wire bonding bond pad openings significantly below 100 μm are well known, this is currently not possible with the Blade technology, where 180 μm is a typical value. An enormous contributor to the needed area is the tolerance chain of the micro-via positioning as well as the mere size of the via of 70 μm. The first can be strongly reduced by measuring the die positions before lamination and later correction of the micro-via positions. By this, the tolerance chain is reduced to the single measurement error and the distortion of the panel during lamination. The position data is stored for later use.

12.4.5 PCB Processes

The next step is the lamination of the chip and lead frame into the FR-4 material. For this, the prepreg layers, which are at chip height, are cut out at the die positions by laser cutting or mechanical drilling. A stencil is put between the lead frames and the first stack is built up including several prepreg layers and Cu foils. This is then pressed as in standard laminate substrate and PCB manufacturing.

The chips are contacted by forming vertical electrical interconnects. For this, different PCB via formation technologies can be selected like dual beam laser drilling, direct laser drilling (including a black oxide process), or lithographic definition of the micro-via positions in the Cu layer and CO_2 drilling through the laminate afterward. It was mentioned before that the adaptation of the via positions to the die positions is highly advantageous for the narrowing of design rules. Therefore, lithography should be done with mask-less tools and laser drilling with individualized drilling positions for each chip. Having a fixed mask would not align to individual die positions and would hence impact yield, cause much worse design rules, or in the worst case of not properly adapted design rules cause fails in the field. No matter which process is used, the software landscape as well as the machine capability to do a lot to lot correction for the individual die positions is complex.

In all cases, the final step of drilling through the laminate toward the chip is done with a CO_2 laser. The advantage of this wavelength is the almost full reflection at the Cu surfaces, which allows for a safe process without chip damage [3].

Another complex process step is the galvanic Cu deposition in the laser-formed vias mentioned above, which can be divided into desmear, activation, and filling. The desmear process, a cleaning of the laser drilled holes to remove debris, can be done chemically or with plasma, the activation either by a palladium (Pd) activation or a direct plating approach. Also for the filling, different suppliers offer solutions that differ more by the chemical approach than by the result. The most significant difference between a standard PCB and the Blade process is the filling of two different via geometries at the same time (parallel process). This is only solvable by careful process selection. A perfect filling of vias would result in a completely flat surface above. Higher filling results in small elevations above the vias, lower filling in so-called dimples. At this point it has to be carefully and individually considered, which via shape is necessary and at which point neither quality nor function is harmed. The whole range of shapes was thoroughly tested for robustness. Therefore, certain compromises in the dimple shape could be released.

After via plating, the RDL already reaches its final thickness. Now it can be lithographically structured to form pads and conductive traces. For this, standard PCB exposure equipment and mask-less exposure are possible. The high value per area supports investment in machines, which can deliver a high yield. The development of the lithographic structures is done in a standard DES horizontal line.

Depending on the complexity of the product, these steps can be done on only one side or both sides of the panel and can also be repeated to form additional RDLs.

For the finish of the Blade technology, several approaches are followed, depending on the needs of the specific product. The simplest one just applies a final finish, typically e-less NiP/Au. It could be shown that also other finishes work well. In some configurations, where the pads can be electrically connected during plating and later be separated, e.g. at package dicing, even galvanic plating like Ni/Sn can be applied.

The more complicated approach adds an additional half RDL by applying a solder resist. This would result in a negative standoff of the pads, which is challenging for the soldering behavior on board. Therefore, an additional solder

Figure 12.7 Top and bottom views of DrBlade2 (left) and DrBlade1 (right).

depot is applied by solder printing and reflow at DrBlade2 (see Figure 12.7) and Blade3x3. DrBlade1 is without solder resist and bumps but has Ni/Au plated copper terminals.

After this the individual packages are separated by mechanical dicing. Several process flows were successfully introduced for different products. Dicing on standard dicing tape on wafer frames allows parallel probe card testing afterward (test of components still on wafer frame), while tapeless dicing requires single device handling into the tester. Also, the laser marking can be done in frame or on the single device. Finally, the products are packed into tape and reel.

12.4.6 Inspection and Process Controls

To ensure a high product quality and extremely low failure rate, a very high rate of process control is used. This includes an automated optical inspection (AOI) after every lithography step. The final inspection is done from all six sides and measures the specified package outline. For the other processes, controls on a statistical basis are applied. The line was set up to allow for a full backward traceability from the final product back to the position on the panel and back to the chip on the front-end wafer. For high process stability, special care on the galvanic baths is necessary, realized by an automated analysis and dose system.

12.5 Design Features

12.5.1 Mature Design Rules and Roadmap

Qualified products are produced in a range from about 3 mm × 3 mm to 4.5 mm × 6 mm. On the test vehicle level, devices as small as 2 mm × 2 mm and as large as 11 mm × 11 mm were tested. For special purposes, even 30 mm × 35 mm was demonstrated [4, 5]. The thickness of a Blade package is typically in the range of 500 μm and above, depending on the number of layers, the chip thickness, and, very importantly, the application of solder depots. The naming therefore sometimes results in USON or UIQFN (ultrathin: >500 and

≤650 μm) or WIQFN (very very thin: >650 and ≤800 μm) [6]. The package families of the Blade products are for DrBlade1 "Laminate Green, Ultra Integrated Quad Flat Nonleaded Package (LG-UIQFN)," for Blade33 "Laminate Green Ultra-thin Small Non Outline Non-Leaded Package (LG-USON)," and for DrBlade2 "Laminate Green, very very thin Integrated Quad Flat Nonleaded Package (LG-WIQFN)."

Pin counts reach from 3 to ~40. So far no ball grid arrays are used as they are not typical for power packages. Pad sizes are more limited by the typical second-level design rules.

In general there is no technological limitation for line/space design rules to be different from those of high density integration (HDI) PCB production; however, the main application for the Infineon Blade technology is for power packaging. Therefore, the focus so far was on low ohmic connections more than high density. This results in RDL thicknesses of >30 μm and line/space of 60 μm/60 μm for high volume production. Mature design rules include two RDLs on each side plus a half layer realized by solder resist. Also, asymmetric assemblies are qualified.

The minimum allowed pad size, which is contacted by micro-vias, is 180 μm. The minimum pad pitch results from the line/space and the overlay accuracy between the layers.

12.6 System Integration Capability

12.6.1 2D and Side-by-Side Packaging

The first announced product of the Blade technology, DrBlade1, was a multichip package, including power chips and a logic chip side-by-side. Complex routing can be achieved over several layers, including layers on top and bottom, which is absolutely necessary for vertical power dies, as both sides of the die have to be contacted.

12.6.2 3D and Package on Package

A basic design element of the Infineon Blade technology is vertical interconnects, realized by micro-vias in the FR-4 laminate toward the die and lead frame. Due to the low thickness of the packages, a PoP approach is obvious to fill the gained space. At the same time it is easy to realize if only the existing lines in the RDL have to be exposed at the required positions and need a solderable surface. 3D chip stacking is not part of the Blade package technology design features and would imply a much higher complexity in laminate chip embedding technology than a PoP approach. The piggyback package on package as well as passives on top of packages was demonstrated. A typical application also uses a large inductor bridging over the package, not electrically connecting the topside of the same. This saves almost the complete board space of the package itself, and at the same time the inductor aids as a heat sink for the power die.

12.7 Manufacturing Format and Scalability

Blade products were produced in different formats with different lead frame strip sizes and numbers of strips per panel arranged into a rectangular panel. Also, different panel sizes were used. A typical HDI board has a size of 24″ × 21″ (~600 mm × 500 mm). An early production concept worked with quarter panels hosting about 10 lead frame strips. The use of small strips was originally triggered by the availability of diffusion solder die bonders with sufficient placement accuracy. Within later production concepts, the lead frame area was increased by a factor of 2–3, and the panel size moved toward the size of a typical HDI PCB production panel.

The applied lead frame strip sizes and panel formats have a huge impact on the panel occupation with active devices. A cost-optimized production concept has to consider this carefully and monitor the yield as well as the higher engineering effort with larger formats. Larger formats will increase the challenges of the alignment concepts required to maintain attractive design rules, for example, in terms of chip pad sizes. Chip position measurement routines applied before lamination do not take into account panel distortions during lamination and further processing.

Nevertheless, even a lead frame strip size close to the panel size itself will never lead to an occupation close to 100% as long as the outer rim of the lamination panel has to be discarded as an inherent limitation of PCB technology.

The Blade package technology has a very high potential concerning the scalability of the packages themselves. Packages can be as small as the Blade3x3 (3 mm × 3.4 mm) or even smaller. In terms of the production technology, a complete lead frame strip can also make up an individual package. Test vehicles with several centimeters of edge length were easily built on a Blade package production line [4].

12.8 Package Performance

12.8.1 Electrical Performance

The Infineon Blade technology was developed as a power technology. Therefore, electrical performance is an obligatory basic feature. The microvia connection to the chip has a resistance of less than 1 mΩ per via, at the

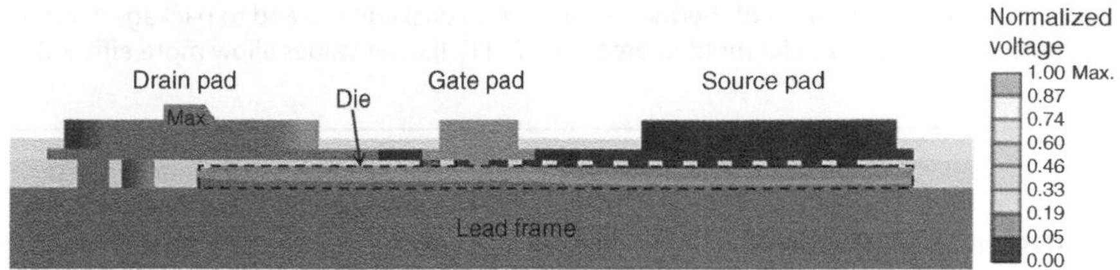

Figure 12.8 Cross section through simulated voltage (normalized values) within an open gate of the MOSFET (die).

same time allowing high parallelization and a good coverage of the pad area, reducing sheet resistances. The typical RDL thickness of 35 μm and wide lines allow low resistances as well. Good designs typically avoid horizontal routing of the high current lines and take advantage of the vertical capabilities. The die are best placed directly facing their according potential on the board. Only minimal horizontal routing distances over the RDL and usage of the high cross section of the lead frame allow superior electrical performance.

As an example, the simulated voltage over the Blade3x3 shows no significant voltage drop within the package metal. Almost the whole resistance lies in the silicon, which is a 1 mOhm chip after all (cf. Figure 12.8).

So far, half bridges with integrated drivers with current ratings up to 80 A on 20 mm^2 were built.

12.8.2 Thermal Performance

The thermal performance is absolutely crucial for power packages. The preferred setup, with the most heat-producing chip facing toward the PCB, allows an extremely low R_{th} (thermal resistance) from junction to board. Additionally, the included lead frame and the wide RDLs deliver a relatively high thermal mass, short wired over Cu vias or diffusion solder, which is advantageous for the Z_{th} (thermal impedance, time-dependent thermal resistance). The thin package, either with exposed lead frame or with a thin insulation layer, allows double-sided cooling. The Blade products have a significantly lower thermal resistance junction to package topside than corresponding molded products (see Table 12.1). Hence, forced air convection can significantly improve topside cooling for Blade packages (see Figure 12.9).

Simulations have shown that forced air convection can significantly improve the cooling of the device over the topside. An additional heat sink on the top is much more efficient than for thick standard molded packages.

Table 12.1 Data sheet values of thermal resistances to package top and to package bottom for Blade products and similar molded products [7–11]. (Lower values allow more efficient cooling.)

Product	Package type	Remark	R_{th} junction case top (KW^{-1})	R_{th} junction case bottom (soldering point) (KW^{-1})
Blade3x3	Blade, exposed lead frame, die face-down	Large die products	1.0	1.6
Blade3x3		Small die products	1.0	3.2
DrBlade1	Blade, lead frame embedded, die face-up	Large MOSFET (low side)	2	1
		Small MOSFET (high side)	7	2
DrBlade2	Blade, lead frame embedded, die face-down	Typical	3.7	3.5
DrMOS	Molded package, die face-up with soldered clip(s)	Typical	20	5
Discrete MOSFET in TDSON8		Typical	20	1.3

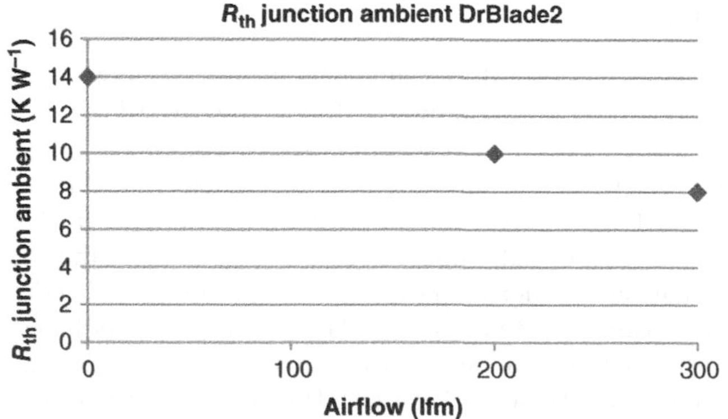

Figure 12.9 Thermal resistance to ambient; Ploss = 4.5 W, TA = 70 °C, eight-layer server board with 2 oz. copper per layer (lfm: linear feet per minute). Source: data taken from data sheet DrBlade2 [9].

12.8.3 Thermomechanical/CTE, Moisture, and Warpage Issues

The Blade package uses laminate materials and copper. Therefore, the CTE closely matches that of the PCB to which the package is attached, thereby reducing some standard problems such as thermo-mechanical package-to-PCB mismatch. Although it is very thin and somewhat bendable, dedicated bending stress tests on test boards showed resistance to mechanical stress to almost the same level as heavier molded packages. This is probably because the thin chips bend under stress instead of breaking.

However, FR-4 materials are known to absorb much more moisture than mold compound. This requires special care for corrosion-resistant chips and materials. If some material combinations like chlorine-containing laminates and non-noble metals are avoided, the moisture resistance is the same or less as for the PCB and therefore fully sufficient.

An ultrathin package can show some warpage due to CTE mismatch between the materials used, but the included relatively thick Cu lead frame stiffens the system significantly. At least for the small packages, warpage is therefore within the typical package specifications of less than 80 µm [8]. Warpage is more an issue in the panel during manufacturing in the line than in the final singulated component, which requires a high level of knowledge in materials and process engineering to overcome.

12.9 Robustness and Reliability Data

12.9.1 First Level/Component Level (CLR)

Blade products are basically qualified according to Infineon's industry standard, similar to AEC Q100. Only autoclave stress tests are not used, since the Blade package falls under the category of laminate-/PCB-based packages and a uHAST (130 °C, 85% RH) is applied instead [12, 13].

The Blade3x3 was additionally exposed to combined and extended stress tests to check for potential weaknesses concerning component reliability. These tests covered active and passive tests.

The IPC (IPC – Association Connecting Electronics Industries) has published the industry standard IPC-TM-650 2.6.25. It defines stress test conditions for the investigation of conductive anodic filament (CAF) at PCB materials. Among others, a 50-hour storage at a temperature of 85 °C and at 85% relative humidity with an applied voltage of 100 V is described. This stress test was applied to Blade3x3 products for 500 h instead of the above defined 50 h. Nevertheless, within this test exceeding the stress test duration as defined in the standard by a factor of 10, no CAF could be observed. Also at 175 °C, 20 V for more than 1000 h, extended H3TRB (high temperature reverse bias with T_A = 85 °C/85% relative humidity with device reverse biased at 80% of rated breakdown voltage), corresponding gate stress, and HAST (biased highly accelerated stress test) tests, neither CAF nor other package failures could not be provoked. This high reliability is related to the high quality lamination material in the BOM of the Blade packages.

12.9.2 Second-Level/Board-Level Reliability (BLR)

Blade products deliver excellent board-level reliability (BLR) test results. For temperature cycling on board (TCoB), tests devices were soldered to a four-layer PCB 1.6 mm thick with Sn final finish and green SnAgCu solder paste. TCoB was performed between −40 and 125 °C with a duration of one hour per cycle according to IPC-9701 [14]. All Blade products withstood far more than 1000 TCoB without electrical fails in the online monitoring.

In a study with DrBlade2, none of the devices showed fails in the electrical online monitoring up to the test end at 6000 cycles. Cross sections were made after 1000 cycles showing no degradation of the solder joint (see Figure 12.10).

As Table 12.2 shows, all Blade products withstood far more than 1000 TCoB cycles without fails in online readouts.

Compression tests were performed on the DrBlade2 package LG-WIQFN-38 (4.5 mm × 6.6 mm) containing two MOSFET chips and a driver chip and the Blade3x3 package LG-USON-6-1 (3 mm × 3.4 mm), which is a single MOSFET

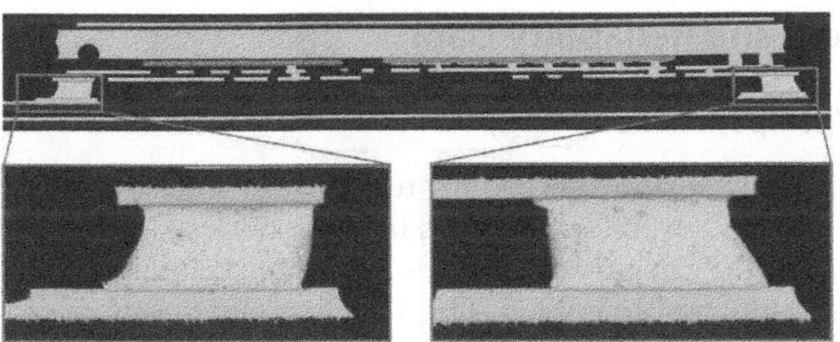

Figure 12.10 Mechanical cross section through a DrBlade2 package at outer row along the long package edge after 1000 TCoB on a four-layer 1.6 mm thick PCB.

Table 12.2 Blade solder joint robustness in TCoB test.

Product	Package	Cu layers of board	First fails in online readout
DrBlade1	LG-UIQFN-32-2	4	2000 cycles
DrBlade2	LG-WIQFN-38-1	4	3000 to >6000 cycles depending on board assembly
Blade3x3	LG-USON-5-1	4, 8, 10	No fails at 5000 cycles

package. For the tests, the devices were soldered to a high T_g FR-4 PCB with 1.6 mm thickness and four copper layers with chemical Sn final finish with a SAC305 solder (96.5% tin, 3% silver, and 0.5% copper). Online readout equipment was connected to the test PCB. Then a force ramp was applied perpendicular to the PCB and package plane until an electrical fail was detected or a maximum force was reached. For both packages no fails could be induced with the maximum applied forces of 2000 N (Blade33) and 2500 N (DrBlade2). This behavior is similar to molded power packages. Applied to the plastic encapsulated package with a copper lead frame, a single MOSFET die, and a soldered clip as source interconnect PG-TDSON-8 (Plastic Green Thin Dual Small Outline Non-leaded Package; size: 5 mm × 6 mm), no damage could be provoked up the maximum applied force of 2500 N in the same test setup. Also bending tests gave similar results as for molded packages.

Electromigration stress tests were performed on a dedicated Blade test vehicle in a modified Blade3x3 package to test single via connections and a newly developed nonstandard test setup, where a single micro-via of ~60 μm diameter was subjected to a current flow of 30 A at 165 °C at each device under test (DuT). The DuTs were soldered to a board with 105 μm thick outer copper traces with Ni/Au finish with a SnAgCu solder. No defects could be provoked at the vias toward the chip, copper traces in the package, or other package internal Cu-to-Cu interconnects. With end-of-life tests by far exceeding the components mission profile after more than 3000 h test duration at 165 °C, the second-level interconnect was finally destroyed, and the board-level solder dissolved part of the copper of the package pads and copper vias (see Figure 12.11).

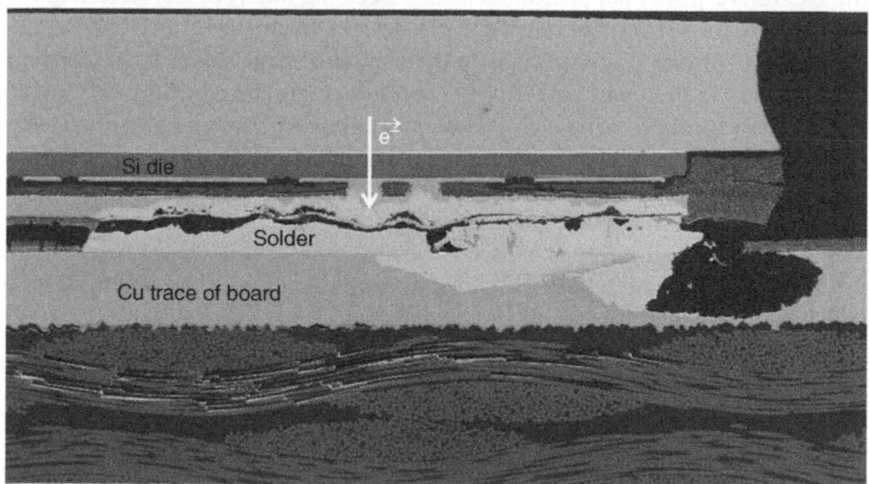

Figure 12.11 Cross section through electromigration test vehicle after more than 3000 hours at 165 °C with 30 A current flow showing diffusion of solder into the Cu trace of the test board as well as into the Cu pad and vias of the package.

No drop tests were performed for the Blade package technology, as the targeted application range so far does not include mobile applications and these tests are obviously irrelevant for servers.

Overall, the Blade package technology has the potential to be qualified for conditions well exceeding standard industrial qualification conditions.

12.10 Electrical Test Considerations

All Blade and DrBlade products are 100% electrically tested. This can only be done after package dicing, because the lead frame shorts the devices. There are many different options for dicing as well as for testing. Dicing on dicing frames on foils allows frame testing, which is efficient for devices with a high logic content. Cheaper foil-less dicing requires single device handling and testing later, which makes high current tests simpler, as they do not require test needle cards.

Fortunately for testing, the Blade package is a fully robust package, which can be handled and contacted without special risk of damage. Therefore, the test can be done on the device pads. E-less Ni/Au surfaces are very easy to contact. On the other hand, the solder depots of the Blade3x3 and DrBlade2 with their uneven surface challenge the test and needle design.

12.11 Applications and Markets

Following the general trend of miniaturization and the reduction of system cost, the DC-to-DC power supplies for central processing units (CPUs) started in the early 2010 to move from discrete MOSFET devices toward integrated half bridges including the driver ICs. Especially in the market of large server farms, current consumption is the main cost driver. Therefore, high efficiency for DC-to-DC conversion, especially in the high current parts like CPU and graphics processing unit (GPU), is a strong market requirement. Due to the high switching frequencies above 100 kHz, low parasitics are significantly improving the efficiency.

Additional advantage of the Blade technology can be taken for applications, which require small outlines, either as board space or in package height. The product in Blade3x3 package has a size of only 10 mm^2 with an R_{on} of 1.2 mΩ, which is unbeaten at this technology generation.

Acknowledgments

This work was partly supported by the research and development project ProPower (Grant Number 13N11879), funded by the German Federal Ministry of Education and Research (BMBF), and in part by the project eRamp (Grant Agreement Number 621270), co-funded by grants from Austria, Germany, Slovakia, and the ENIAC Joint Undertaking.

We want to thank colleagues at Infineon Technologies AG for their support by fruitful discussions, review of the manuscript, and provision of pictures: T. Both, R. Fischer, A. Gruber, G. Haubner, J. Höglauer, A. Keßler, G. Lohmann, X. Schlögel, F. Treutinger, S. Weiß, and R. Wombacher.

References

1 Infineon Technologies AG (2013). Infineon Introduces DrBlade, the New Generation DrMOS in Innovative Chip-Embedded Packaging Technology, Neubiberg, Germany and Long Beach, CA, USA (2013). www.infineon.com (accessed 18 March 2013).

2 Yole Development (2017). *Embedded Die Packaging: Technology and Market Trends*. Yole Development.

3 Munding, A., Kessler, A., Scharf, T. et al. (2017). Laminate chip embedding technology – impact of material choice and processing for very thin die packaging. *67th Electronic Components and Technology Conference (ECTC).*

4 Munding, A., Gruber, M., Both, T. et al. (2014). Laminate based LED Module with embedded MOSFET Chips, *Electronics System-Integration Technology Conference (ESTC).*

5 Munding, A., Gruber, M., Both, T. et al. (2015). Thermal Improvement of Plastic Laminate Based LED Modules with Embedded Chips. *Proceedings of the European Microelectronics Packaging Conference (EMPC).*

6 JEDEC Solid State Technology Association (2008). JESD30E: Descriptive Designation System for Semiconductor-Device Packages, Arlington, VA, USA, 2008.

7 Infineon Technologies AG (2013). Blade 3x3, Pushing the Boundaries for Discrete 25V/30V Power MOSFETs, Product Brief, Villach, Austria, 2013.

8 Infineon Technologies AG (2013). High Performance DrBLADE, TDA21310, Data Sheet, Revision 2.1, Munich, Germany.

9 Infineon Technologies AG (2015). High-Performance DrBLADE, TDA21320, Data Sheet, Revision 2.4, Munich, Germany.

10 Infineon Technologies AG (2013). High-Performance DrMOS, TDA21220, Data Sheet, Revision 2.5, Munich, Germany.

11 Infineon Technologies AG (2013). OptiMOS(TM) Power-MOSFET, BSC010NE2LS, Rev. 2.2, Munich, Germany.

12 JEDEC Solid State Technology Association (2005). JESD22-A102E: Accelerated Moisture Resistance – Unbiased Autoclave, Arlington, VA, USA, 2005.

13 JEDEC Solid State Technology Association (2015). JESD22-A118B: Accelerated Moisture Resistance – Unbiased HAST," Arlington, VA, USA, 2015.

14 IPC – Association Connecting Electronics Industries (2006). Performance Test Methods and Qualification Requirements for Surface Mount Solder Attachments, IPC 9701A, Bannockburn, Illinois, USA, 2006.

13

The Role of Liquid Molding Compounds in the Success of Fan-Out Wafer-Level Packaging Technology

Katsushi Kan[1], Michiyasu Sugahara[2], and Markus Cichon[3]

[1]*Nagase ChemteX Corporation*
[2]*Nagase & Co., LTD.*
[3]*Nagase (Europa) GmbH*

13.1 Introduction

Semiconductor packaging technologies are state of the art for various industries and applications including automotive, machinery, entertainment, communication, sensor, security, authentication, medical and environment systems, and many more. Today, many variations of electronics are available, and indeed the entire world and environment could not function without these in the manner that people have become used to. Recently, fan-out wafer-level packaging (FO-WLP) has attracted attention as a high performance and very cost-efficient packaging solution [1]. This packaging is highly suitable for wireless devices, most common in mobile and automotive sensor applications [2]. Nagase, as a diversified epoxy material supplier, began active development for these applications in the mid-2000s, and LMC was successfully implemented into FO-WLP at the end of 2007.

Epoxy resin has two specific features: excellent adhesion to various surfaces and superb electrical insulation. Therefore, epoxy compounds became a very popular encapsulant for semiconductor packaging some time ago. During its development history, epoxy compound encapsulant has been used in many different forms of packaging in various applications with very specific requirements. Generally, epoxy molding compound (EMC) is a solid-type resin, which consists of epoxy and phenol chemicals and was originally developed for transfer molding systems. In contrast, liquid-type epoxy molding compound was developed for several specific applications that inevitably require a liquid state such as unique flowability and/or a dust-free requirement.

Advances in Embedded and Fan-Out Wafer-Level Packaging Technologies, First Edition.
Edited by Beth Keser and Steffen Kröhnert.

There are various liquid epoxy compound products including LMC for compression molding, capillary underfill (CUF), pre-applied underfill called NCP (nonconductive paste), and a glob top material that facilitates the dam and fill process for encapsulation and environmental protection of various electronic modules, devices, and other components. Liquid-type epoxy molding compound (LMC) was developed for the latest applications and process adoption of FO-WLP where it has become indispensable.

13.2 The Necessity of Liquid Molding Compound for FO-WLP

The original chip-first FO-WLP manufacturing process developed by Infineon Technologies AG using LMC is called extended wafer ball grid array (eWLB), and will be described as shown in Figure 13.1 [1].

1) Temporary adhesive tape is laminated on a metal carrier.
2) Die are mounted with active side face-down on the temporary adhesive tape.
3) LMC is dispensed over the die on the carrier.
4) LMC is molded using a compression molding system, creating a die embedded resin wafer.
5) The molded die embedded resin wafer is then post mold cured.
6) The molded die embedded resin wafer is separated from the temporary adhesive tape using thermal treatment process.
7) A redistribution layer (RDL) process is created, providing insulation between the copper (Cu) metal routing layers.
8) Solder balls are then mounted on die embedded resin wafer.
9) Finally, package singulation or dicing of die embedded resin wafer into final FO-WLP is completed.

There are three major advantages of using LMC in this process. Firstly, FO-WLP manufacturing takes place in a class 1000 clean room environment, which is much stricter than traditional assembly, and since LMC is in a liquid state, it is free of any dust. This kind of dust-free material can be used in the entire semiconductor manufacturing processes including sensitive clean room environments. In comparison, solid-state EMC and granule type resin used for traditional flip-chip ball grid array (fc-BGA) and wire-bond ball grid array (wb-BGA) packages are not dust-free and thus cannot be used in such clean dedicated areas. Recently, EMC suppliers have been controlling fine particles to minimize the dust in the EMC manufacturing process itself, but it cannot be guaranteed that fine particle dust powder is completely eliminated because during transportation and handling an unavoidable collision of epoxy parts generates fine powder. To minimize this, molding equipment manufacturers developed specific dust preventing measures for their equipment, but it

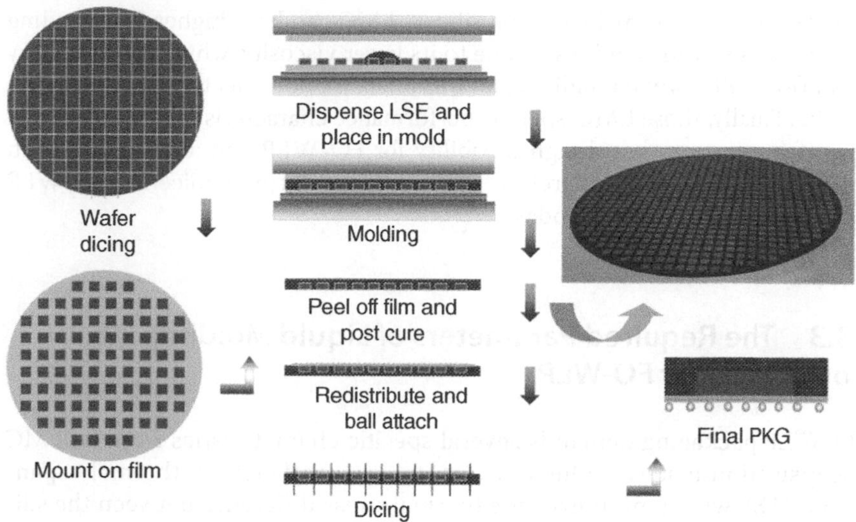

Figure 13.1 Procedure of a chip-first FO-WLP process also known as embedded wafer-level BGA (eWLB) developed by Infineon Technologies AG.

remains a major concern in large-scale mass production. Once a machine stops or requires adjustment, a technician has to fix it. Thus, it is generally impossible to prevent contamination by fine particles of dust while under repair. This pollution would generate major problems in further processes, contaminating the treated wafers and their devices, thus reducing the production yield dramatically. So, dust and contamination control and dust and contamination-free materials are indispensable in the semiconductor assembly manufacturing world today.

Secondly, LMC has, due to its specific properties, the ability to cure at lower temperatures in the range of 110–125 °C in comparison with transfer molding systems and solid-state EMC that typically require a curing temperature of 175 °C. The lower temperature is in general a major advantage for many process-related issues. It minimizes the thermal stress between the curing temperature and room temperature, which results in lower warpage and better warpage controllability. Another advantage of LMC is the prevention of so-called flying die. The temporary adhesive tapes' bonding declines at increased temperatures, and especially smaller die are easily detached from the adhesive, becoming so-called "flying die." By keeping the temperature moderate, the die do not move, enhancing the processability for smaller die, which is one of the main merits of FO-WLP and fan-out process efficiency.

Thirdly, LMC has the excellent characteristic of high inorganic filler loading in the final compound. High filler loading is one of the key characteristics of LMC performance and very important for warpage control due to minimization of thermal strain on silicon. LMC can have higher filler loading in comparison with solid resins due to its lower viscosity, which provides very good flowability while molding, and this phenomenon effectively prevents flying die. Finally, those LMC-specific performance characteristics provide larger manufacturing process design flexibility for FO-WLP and also for the resin formulation capability. Therefore LMC is the preferred choice for FO-WLP worldwide manufacturing today.

13.3 The Required Parameters of Liquid Molding Compound for FO-WLP

FO-WLP packaging demands several specific characteristics from an LMC encapsulation material. One of these is a low coefficient of thermal expansion (CTE), which minimizes the thermal stress difference between the silicon die and the LMC, resulting in less thermomechanical strain on the die, and leads to less warpage. The current target value of CTE below T_g (glass transition temperature) is less than 10 ppm °C^{-1}. A second requirement is to have a well-balanced viscosity level, which enables excellent handling and flowability during the molding process. LMC has significantly lower viscosity than solid epoxies in the molding process. However, LMC has a high viscosity value for dispensing due to typical characteristics and especially the filler loading (filler content). This has led to the development of a new dispenser technology that enables fluids with high viscosity to be integrated in a fully automatic compression molding machine where the material should not exceed the viscosity value of 1000 Pa s. Filler content is one of the key factors for the two parameters shown in Figure 13.2. On the one hand CTE decreases when the filler content is increased, while on the other hand viscosity increases. Therefore a well-balanced material in the molding process is a must to enable dispensing with viscosities below the critical value of 1000 Pa s.

Another very important factor is LMC's excellent flowability during the molding process. The aim to be ever smaller, thinner, and lighter in the semiconductor industry has always been a standard requirement. In some cases, when the thickness of LMC over the die becomes very thin, failure can occur. This failure, known as flow marks, is shown in Figure 13.3. Flow marks are caused in LMC by filler clogging during compression molding as shown in Figure 13.4. In a standard compression molding system using LMC, the LMC will be dispensed in the center of the wafer, and during compression it flows radially toward the wafer edge. When there is a narrow space for LMC between the die and the machine tool, normally one can observe at the edge of the wafer larger filler particles clogging, and a flow mark can be observed.

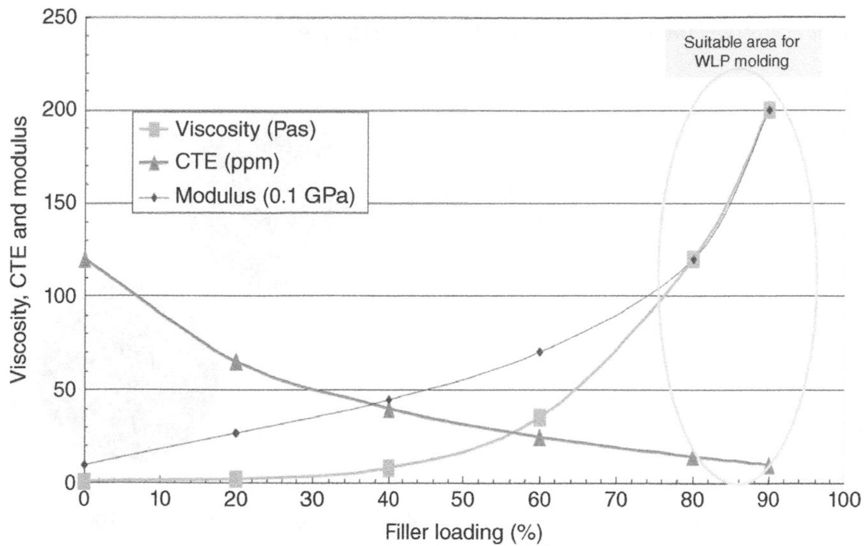

Figure 13.2 Influence of filler loading (content) on LMC properties.

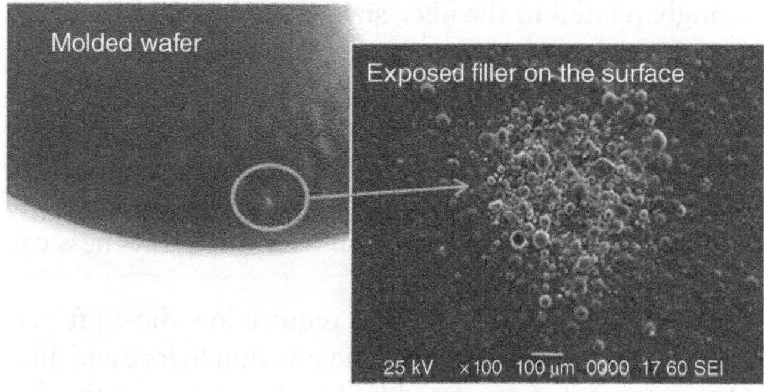

Figure 13.3 Flow mark appearance.

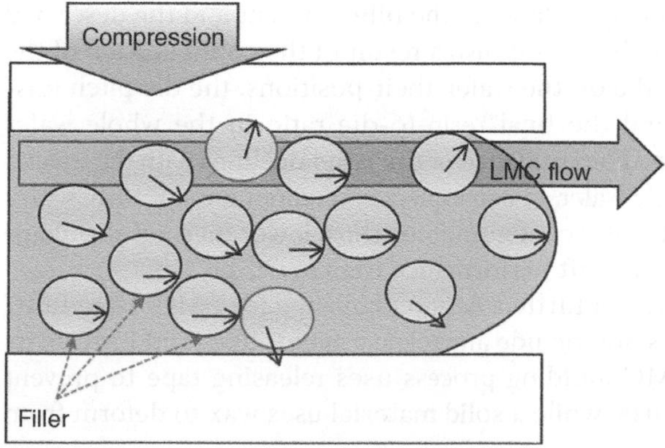

Figure 13.4 Cross-sectional image of flow mark.

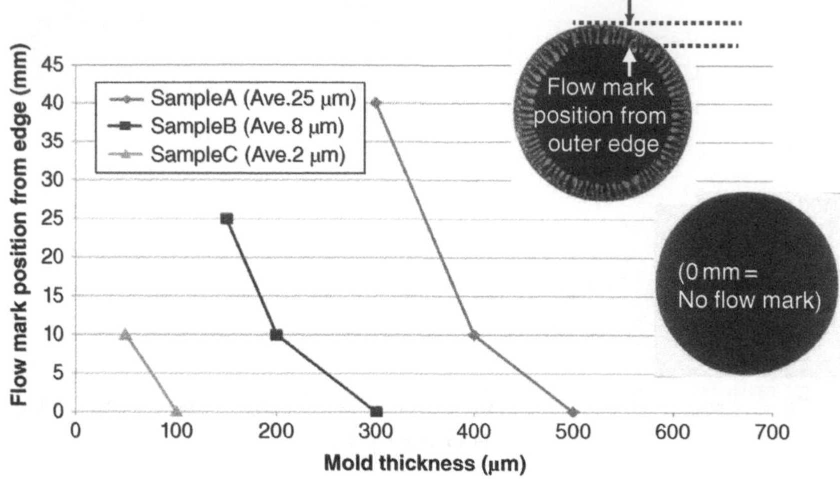

Figure 13.5 The influence of mold thickness and filler size on flow mark appearance.

This flow mark is strongly related to the filler size of the LMC as shown in Figure 13.5. Three kinds of LMCs with different filler size averages have been evaluated regarding their flowability performance to verify the flow mark dependency of the filler size. This exercise shows that 25 μm average filler size can be used if the gap between the die top and mold machine tool, the so-called mold thickness, is at least 500 μm. If this dimension declines, the flow mark will appear. In the case of a smaller filler average size, a smaller mold thickness can be achieved.

In addition, redistribution layer (RDL) processes require low die-shift performance of LMC. Die shift is the difference of the die position before and after molding and curing. One main factor generated by LMC material for the die-shift phenomenon is physical (thermal) and chemical shrinkage shown in Figure 13.6. Generally, the FO-WLP process preferably uses LMCs with lower thermal and chemical shrinkage related to the filler content and the described CTE parameters. However, die shift is also a result of the configuration of the whole arrangement of the die on the wafer, their positions, the die pitch (distance between the die), and the final resin-to-die ratio in the whole wafer (occupation) after molding. The influence of the resin declines with the smaller resin-to-die ratio in the whole wafer. In consequence, requirements of LMCs vary depending on the different wafer configurations; thus, lower thermal shrinkage is more advantageous for die-shift performance than low CTE.

RDL compatibility presents a further design challenge for LMC. Here LMC is preferred because it does not include any release agents like solid EMC in its formulation. Since the LMC molding process uses releasing tape to prevent sticking to the machine parts, while a solid material uses wax to deform from

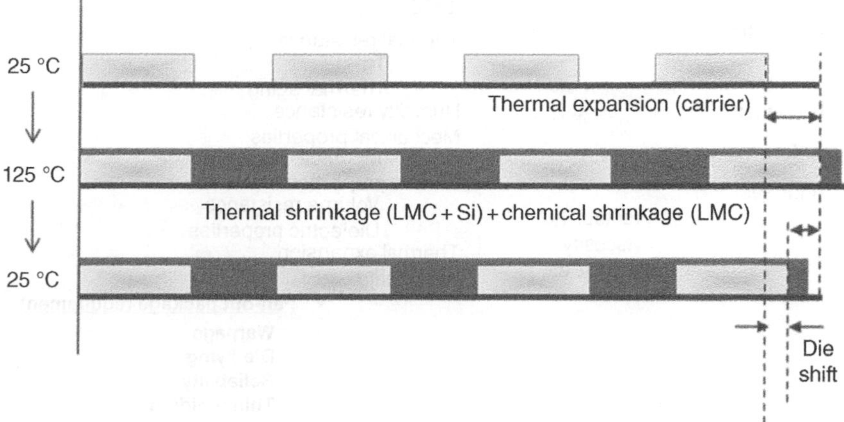

Figure 13.6 Die-shift phenomenon in molding process.

the molding machine die (from the cavities), LMC does not require a release agent. Should the surface quality of the molding material be poor due to the internal release agents, the RDL chemicals during the coating process will shed, and the quality of the process decline drastically.

13.4 Design of LMC Resin Formulation

Generally, a liquid molding compound consists of epoxy resin, hardener, filler, and some additives such as coloring agent, accelerator, adhesion promoter, and further special chemicals as required. All these components and chemicals have their own functions and influence on the performance of the final LMC. FO-WLP requirements demand the best possible balance of materials and substances so the final LMC fulfills all the requirements shown in Figure 13.7. Acid anhydride derivatives that are suitable as hardeners for LMCs provide low viscosity and low temperature fast curing as well as maintaining thermal performance such as T_g. Recently, some derivatives of acid anhydride have already been listed by the European Chemicals Agency (ECHA) as substances of very high concern (SVHC). As a result of this situation, an LMC was proactively developed with non-acid anhydride to comply with European regulations. Meanwhile, the Anhydride Joint Industry Taskforce (AJIT) was established by acid anhydride producers, importers, formulators, and end users to re-enhance the safety precautions concerning the usage of acid anhydride [3]. The FO-WLP market continues to observe AIJT activities and the future possibility to continue using acid anhydride in the European Union.

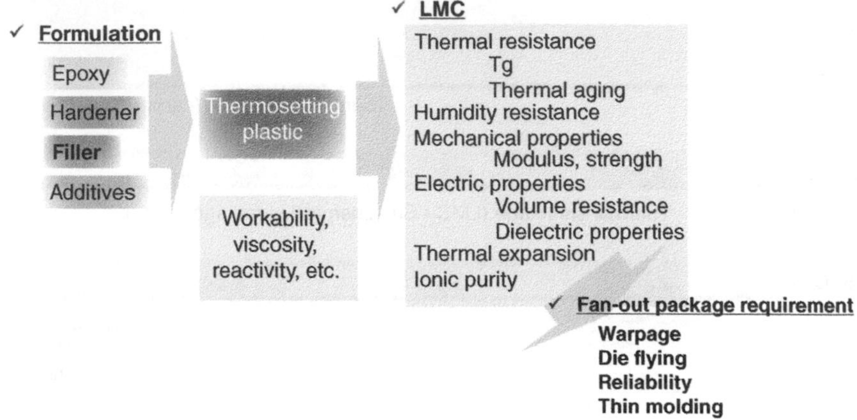

Figure 13.7 Formulation image for requirement.

Silica is suitable as a filler for LMCs. It influences the CTE and thermal shrinkage properties. As explained above, high filler loading leads to lower CTE but can also generate higher viscosity of LMCs, and a balance of those is very important for FO-WLP application. Therefore selection of fillers is a key factor to help provide high filler loading with appropriate viscosity values. The main mechanism of viscosity changes is the friction between filler and the liquid component of the LMC [4]. High filler loading has a large friction area and generates higher viscosity. The method to minimize this influence is to use low friction fillers. The suitable filler has a spherical shape, smooth surface, specific size distribution, and surface compatibility to liquid components. Furthermore, the control of filler size and the filler loading in LMC is very important for FO-WLP technology. The fine filler technology in LMC is suitable for thinner molding, fine-pitch die-to-die distance where LMC is required to fill in-between die, and better accuracy of through-mold via (TMV). TMV is a laser-drilled channel in the molding compound to enable further connection in multidimensional scale, e.g. 3D packaging. In contrast, fine filler loading causes higher viscosity because it is increasing surface area in LMC. Therefore, next LMC generation has to overcome this obstacle between filler size and filler loading for future FO-WLP.

Recently, functional fillers such as thermal conductors and electromagnetic shielding are coming into focus as well. Beside these components, LMC includes more additives such as carbon black that is important for laser marking on the final package, adhesion promoters to enhance adhesion, stress release agents to improve warpage behavior, and others depending on the requested functionality modification.

13.5 Development of LMC in Connection with Latest Requirements

The current packaging trends are mainly large molding areas, fine-pitch lines and space (L/S), and complex packages like system in package (SiP). To be able to follow the technology, these trends require very challenging LMC development. There is fine filler loading LMC development to fulfill the thinner molding and fine TMV performance to fill into narrow spaces between the die, but at the same time, it has the obstacle of increased viscosity due to the friction increase caused by the decreased filler size, resulting in a dramatic surface area increase. Therefore, developing new fillers that have special spherical shapes, smooth surfaces, and a specific size distribution to minimize the roll resistance and provide dedicated surface compatibility to liquid components is imperative [4].

13.6 Current LMC Representative Proprieties

Three types of standard LMC's with different filler cuts (maximal filler size) are shown in Table 13.1. These are produced by Nagase ChemteX Corporation for FO-WLP. All these materials have high filler loading above 85% of volume and maintain the dispensability requirement of a viscosity below 1000 Pa s. Nagase ChemteX has developed, launched, and released several kinds of LMCs into the semiconductor market for more than 10 years where the main application is the FO-WLP. Using this experience and knowledge, Nagase ChemteX

Table 13.1 Standard LMC properties.

Item	Unit	R4212-2C	R4202-N1	R4511
Application			FO-WLP	
Filler content	%	89	88	87
Filler top cut	Mm	75	55	25
Specific gravity		2.02	2.00	1.98
Viscosity	Pa ·s	600	200	250
Flexural modulus (25 °C)	GPa	22	22	18
Tg (DMA)	°C	165	165	170
CTE1	ppm K^{-1}	7	8	9

continually works on further development and new LMC materials to support the semiconductor market with its fast-moving trends and technologies and challenging environmental requirements.

13.7 Conclusions

FO-WLP requires LMC and has been using it now for a decade. At the beginning there were many different issues that required time to be solved [1, 2, 4]. Today, the LMC is well adjusted for current standard FO-WLP, but new applications and functions arise, and the requirements for the LMC are changing with these new developments. For further progress in the semiconductor electronic world, there is a need for design and development cooperation between LMC formulators, equipment, and packaging designers.

Acknowledgment

Authors would like to thank Nagase ChemteX Semiconductor Packaging team, Nagase High Performance Material Section, and our partners for their excellent support.

References

1 Brunnbauer, M., Fürgut, E., Beer, G., and Meyer, T. (2006). Embedded wafer level ball grid array (eWLB). *Proceedings of 8th Electronic Packaging Technology Conference* (2006).
2 Obori, T., Kan, K., Nishikawa, Y. (2009). Development of liquid molding compound for wafer level package. *Microelectronics Symposium*, Japan (2009).
3 Anhydride Joint Industry Taskforce (AJIT). http://anhydrides.eu (accessed 6 August 2018).
4 Kan, K. (2016). The novel liquid molding compound for FAN-OUT wafer level package. *IWLPC* (2016).

14

Advanced Dielectric Materials (Polyimides and Polybenzoxazoles) for Fan-Out Wafer-Level Packaging (FO-WLP)

T. Enomoto, J.I. Matthews, and T. Motobe

HDMicroSystems

14.1 Introduction

This chapter provides an overview of advanced dielectric materials developed by HD MicroSystems (HDM) that are based on polyimide (PI) and polybenzoxazole (PBO) technologies and that are targeted for use as redistribution layers (RDL) in fan-out wafer-level package (FO-WLP) applications where lithographic and reliability performances are important requirements. The PI/PBO dielectric materials are supplied in liquid form and, as such, are typically applied onto reconstituted wafers using standard spin-coating processes that include spin coating, soft baking, exposure, development, and final cure.

14.2 Brief History of PI/PBO-Based Materials in Semiconductor Applications

The first PI products for semiconductor applications were introduced in the early 1970s for use as stress buffers or passivation layers on integrated circuits as well as interlayer dielectrics in high density interconnects on multi-chip modules. These products, later termed non-photo-definable PI (non-PDPI), were based on polyamic acids (PI precursors) synthesized by reacting dianhydrides with diamines dissolved in a suitable solvent such as *N*-methyl pyrrolidone (NMP) and where the dianhydride and diamine were chosen for end-use performance [1]. These materials were typically processed on a silicon wafer by spin coating, soft baking, and patterning using a conventional photoresist or equivalent process and then cured at temperatures >350 °C to produce the PI polymer (see Figure 14.1).

Advances in Embedded and Fan-Out Wafer-Level Packaging Technologies, First Edition.
Edited by Beth Keser and Steffen Kröhnert.

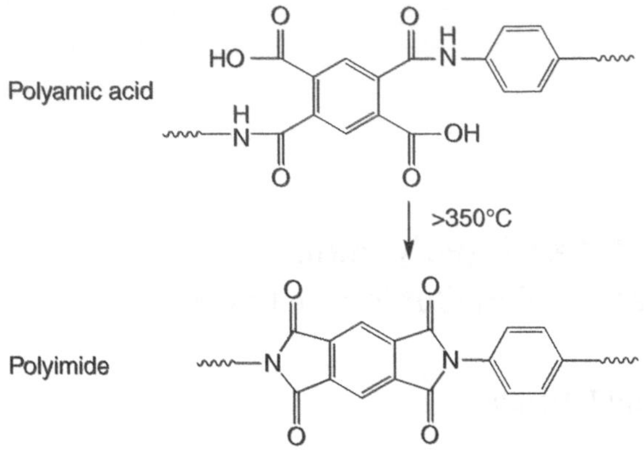

Polyamic acid

>350°C

Polyimide

Figure 14.1 Conversion of polyamic acid to polyimide.

It should be noted that some commercial non-PDPI materials are not self-priming (no adhesion promoter present in the formulation), which would mean that an additional step is first required to prime the silicon wafer with an adhesion promoter (typically aminosilane-based materials diluted in a suitable solvent) to achieve optimum adhesion of the PI to the silicon wafer.

In order to simplify the process and subsequently reduce costs, photo-definable polyimides (PDPI) were developed in the late 1980s to reduce the number of process steps by eliminating the need to use photoresist. It should also be noted that PDPIs were generally self-priming, which further reduced the number of process steps. The first PDPIs to be commercialized were negative-acting, solvent-developable PI (PI-Gen1) that, during the exposure step (i-line or broadband), utilized chain polymerization using radical-generating photoinitiators to cross-link acrylate groups present on the polymer backbone (in the form of an ester or an ionic salt) as well as monomers in the formulation [2]. This resulted in the exposed area being relatively insoluble to the developer solution so that only the unexposed areas were removed during development (termed negative acting). The remaining pattern was then cured at temperatures >350 °C to produce the PI film.

In the late 1990s, positive-acting, aqueous-developable materials were introduced for additional cost and environmental benefits as well as improved resolution [3–5]. These materials were based on either PI or PBO precursor polymers containing phenolic or acid moieties that provided solubility to alkaline developers such as 2.38% tetramethylammonium hydroxide (TMAH). Patterning was obtained by using diazonaphthoquinone (DNQ) photoacid generators to provide contrast during the exposure and development steps. In the unexposed form, the DNQ interacts with phenolic moieties on the backbone precursor to reduce the dissolution rate [6]. However, on exposure to UV light energy (i-line or broadband), the DNQ undergoes a Wolff rearrangement to form indene carboxylic acid (ICA) that increases both the dissolution rate and contrast during development (termed positive acting) [7]. The remaining unexposed pattern is then cured at temperatures >300 °C to produce either the PBO or PI film (see Figure 14.2).

It should be noted that positive-acting, aqueous-developable products based on PBO precursors (PBO-Gen1) tend to be more widely used as compared with those based on PI precursors. There are no overriding performance differences between the two polymer types although it should be noted that hydroxyl groups on the PBO precursor that allow solubility during development are removed after cure, which can result in a lower moisture uptake (see Figure 14.2). With PI precursors, the hydroxyl or acid moieties that provide solubility during development and that remain on the polymer backbone after development can be reduced by reaction with a suitable thermal cross-linker during the curing process to improve end-use properties. Formulation optimization is required to avoid reduction in mechanical properties such as % elongation due to excessive cross-linking.

As a general comparison, positive-acting systems have the advantage of resolving smaller via diameters, while negative-acting systems have more flexibility with respect to processing higher film thicknesses. This is due to the fact that positive-acting systems using DNQ photoacid generators typically have very high UV light absorption (optical densities >4) at i-line or broadband wavelengths and are bleached when the DNQ is converted to the ICA, allowing the light to penetrate down into the film during the exposure process. While this can provide more control on resolution, it can be a challenge for thicker films that would require higher exposure energies, resulting in more sloping

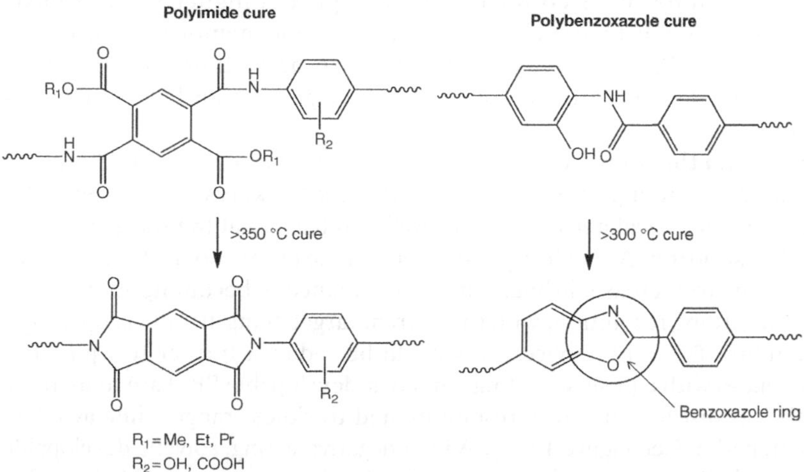

Figure 14.2 Positive-acting PI and PBO.

sidewalls. With negative-acting systems, the optical density is lower (typically 0.4–0.5) and can be readjusted for thicker films by optimizing the concentration of the photoinitiators to maximize resolution while maintaining an acceptable balance between delamination (in exposed areas) and residue formation (in unexposed areas) [8].

14.3 Dielectric Challenges in FO-WLP Applications

It is well known that dielectrics based on PI and PBO technologies, due to their proven end-use performance when cured at temperatures >300 °C, are widely used as RDL in fan-in wafer-level packaging (WLP), flip-chip chip-scale packaging (FCCSP), and other applications to relocate I/O connections and reduce stress as well as allowing die stacking. However, for FO-WLP applications, as the reconstitution process requires the use of a molding compound that has a T_g around 150–170 °C, new PI and PBO materials are needed that can be cured at lower temperatures while still meeting process, end-use, and reliability requirements in FO-WLP applications [9, 10]. The challenges for dielectric materials in FO-WLP applications based on PI and PBO polymer technologies are listed as follows:

1) **Low temperature cure.** As indicated above, for FO-WLP applications, the molding compound used in the reconstitution process typically has a T_g around 150–170 °C, which subsequently limits the cure temperature used for processing dielectrics. While the current cure temperature range for PI or PBO materials in FO-WLP applications is 200–230 °C, <200 °C will be required for lower warpage and higher yield in next-generation FO-WLP devices designed to include memory chips, and, in this respect, PI and PBO materials need to be further redesigned for lower cure temperatures (<200 °C) for next-generation FO-WLP applications [11–16].
2) **Cured film thickness.** Cured film thicknesses varying from 4 to 20 μm are required depending on the copper (Cu) track thickness, which can influence both electrical performance as well as reliability at the solder joint.
3) **High resolution.** As devices get smaller and the number of I/O connections continues to increase, lithographic performance is becoming more stringent regarding resolution, with the current target being 10 μm via openings. For future FO-WLP devices, this could be reduced further to 5 μm. The challenges with a positive-acting, aqueous-developable PBO are to increase both contrast for improved resolution and thickness range while avoiding delamination (see Figure 14.3a). With a negative-acting, solvent-developable PI, the challenge is to reduce residue (footing) due to scattered light at fine resolutions (see Figure 14.3b).

4) **Sidewall shape.** Regarding sidewall shape, there is no specific target for the slope angle, although too steep an angle (like a cliff) should be avoided to prevent concentration of stress from the solder bump on the top corner edge of the sidewall and also to produce a more even distribution of Cu during the plating process. In this respect, a smooth profile with no crowning at the top edge of the sidewall and a sidewall angle of 70–80° to the substrate surface is preferred.
5) **Deep gap formability.** For some applications, deep gap formability is required to develop pre-baked films up to 25 μm to open scribe lines when processing multiple RDL (see Figure 14.4).
6) **Elongation.** A high % elongation is required to enhance crack resistance (see Figure 14.5) between Cu and the dielectric interface and also between dielectric layers during temperature cycle test (TCT) and pressure cooker testing (PCT). Typical test conditions for TCT and PCT are given in Table 14.1.
7) **Chemical resistance.** High chemical resistance to selected chemicals (resist strippers, etchants, fluxes) used in both the RDL and bumping process that can result in delamination (see Figure 14.6) provides a wider process window for the PI and PBO dielectrics and subsequently avoids performance issues during reliability testing.

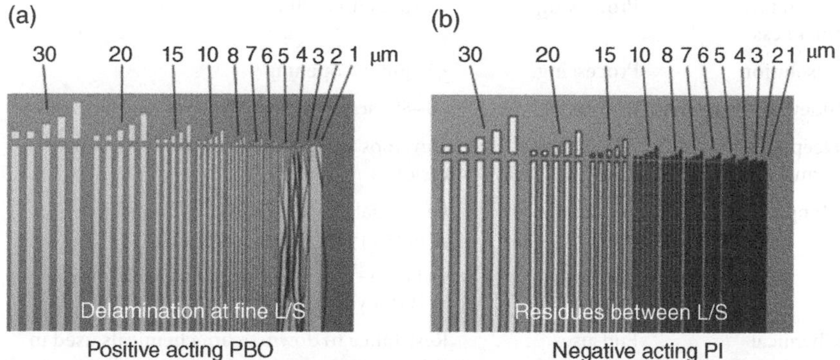

Figure 14.3 Defects in fine patterning. (a) Positive acting PBO. (b) Negative acting PI.

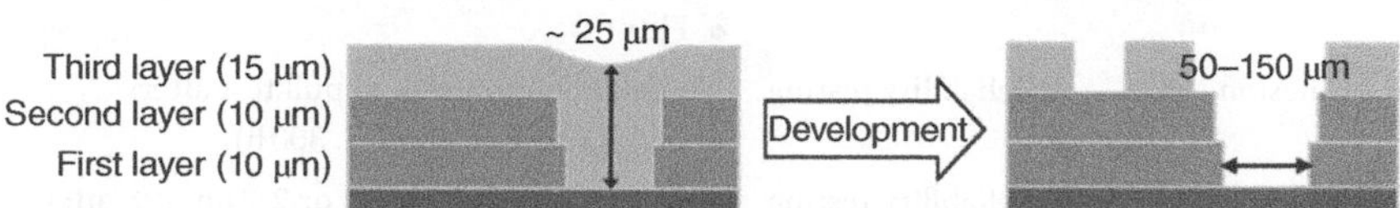

Figure 14.4 Deep gap formability for multilayer structures.

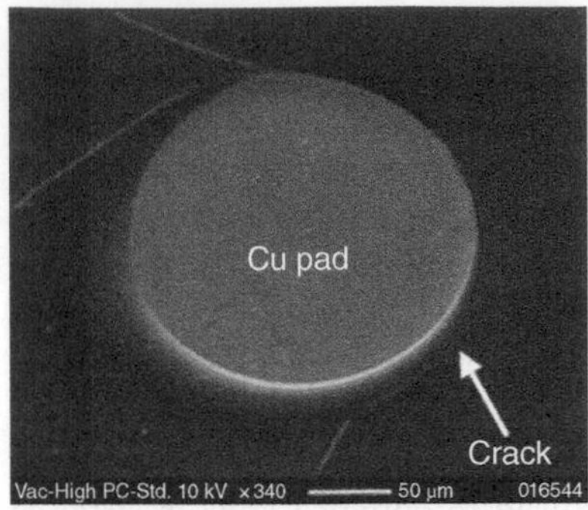

Figure 14.5 Cracking at Cu pad area after TCT.

8) **Adhesion to Cu.** The adhesion between Cu and the dielectric needs to be strong enough to avoid any delamination occurring at the Cu and dielectric interface (see Figure 14.7) that can act as the origin of crack propagation during PCT and TCT (see Table 14.1 for typical test conditions). The adhesion strength is typically measured either using the stud pull test or through visual examination under a microscope for delamination or cracking.

Table 14.1 PI and PBO dielectric challenges for FO-WLP applications.

Challenge	Application	Target specification
1. Cure temperature	Processing	<200 °C
2. Cured film thickness	Processing	>15 μm cured film
3. Resolution	Processing	<10 μm via opening
4. Sidewall shape	Processing	70–80° acute angle to the substrate surface
5. Deep gap formability	Processing	Develops ~25 μm pre-bake films with no residues
6. Elongation	End-use properties	Pass reliability testing (>40% at ambient) • PCT (121 °C/100% RH, 168 h) • TCT (−65 °C/15 min ↔ 150 °C/15 min 1000 cycles)
7. Chemical resistance	End-use properties	Resistance to downstream chemicals used in Fab • Rework solvent • Resist stripper • Resist cleaner • Flux
8. Adhesion to Cu	Reliability testing	No delamination at stud pull test after: • PCT (121 °C/100% RH, 300 h)
9. Insulation	Reliability testing	No short circuit with 5/5 or 2/2 μm L/S after: • bHAST (130 °C/85% RH/3.3 V, 200 h)

9) **Insulation.** In order to maintain insulation between Cu tracks with finer line and space (L/S) patterns down to 5/5 μm or even 2/2 μm L/S, no short circuiting as observed by changes in electrical resistance should occur during biased highly accelerated stress testing (bHAST). In addition, any dendrite formation, corrosion, or delamination as checked for through visual examination (see Figure 14.8) under a microscope should be avoided. The test conditions for the bHAST are given in Table 14.1.

An overview of the main challenges is given in Table 14.1.

In order to meet these challenges for next-generation FO-WLP applications, a new generation of positive-acting, aqueous-developable and negative-acting, solvent-developable materials has been developed that will be outlined in the next section.

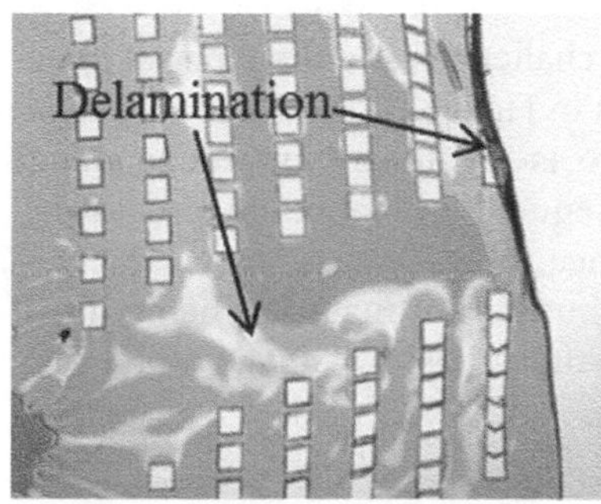

Figure 14.6 Delamination after immersion in resist stripper.

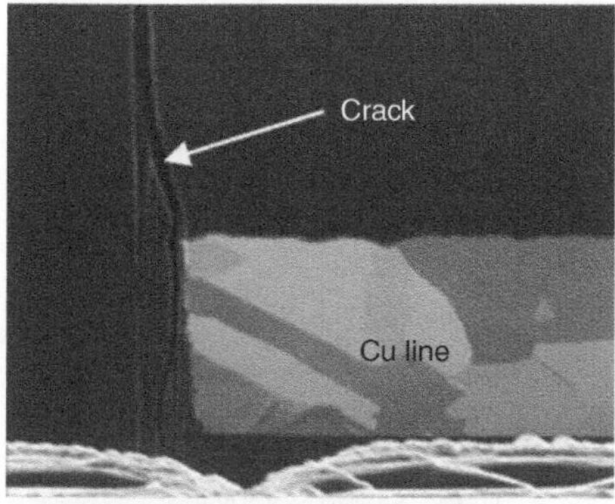

Figure 14.7 Crack propagation from delamination at the Cu surface.

14.4 HDM Material Sets for FO-WLP

As indicated previously, PI-Gen1 and PBO-Gen1 were among the first photo-definable PI and PBO materials, respectively, to be introduced as dielectrics for semiconductor applications that are still widely accepted for use as stress buffers, RDL for WLP, FCCSP, and others where both materials are typically cured at >300 °C for optimum end-use and reliability properties. It should be noted that the high cure temperatures used in the processing of PI-Gen1 and PBO-Gen1, in addition to producing complete ring closure of the PI/PBO precursors, also resulted in the emission of the photo-package and other additives to varying levels to allow the back-bone polymer to achieve end-use properties similar to non-PDPI materials.

In order to address the need to reduce cure temperatures for

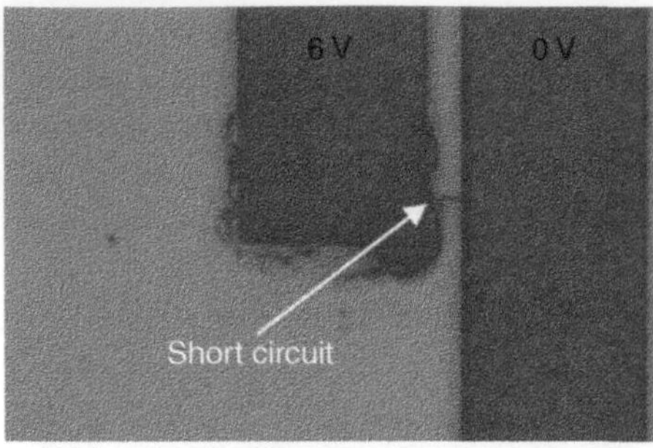

Figure 14.8 Short circuit after bHAST (20 μm space pattern).

FO-WLP and other semiconductor packaging applications, the development concept as outlined in Figure 14.9 was to redesign the backbone polymers for lower temperature cure as well as introducing novel photoinitiators, cross-linkers, and additives to assist in addressing the challenges described in the previous section.

The correlation between the performance challenges and the types of materials used to meet those challenges are given in Figure 14.10. From a formulation perspective, it can be seen that there is significant interdependency between raw material choice and the subsequent influence on processing, end-use properties, and reliability performance.

This development program resulted in a number of materials being introduced where the cure temperature was reduced below 300 °C and where

PBO Gen1 and PI Gen1 (high temperature cure)

- ✓ 100% imidization/cyclization
- ✓ No ingredients other than the polymer remain in the cured film.
- ✓ Polymer backbone and interaction between polymer chains are important to achieve desired cured film properties.

PBO Gen3 and PI Gen2 (low temperature cure)

- ✓ Polymer is designed to cyclize easily at low temperature cure.
- ✓ New photo-package increases lithographic performance.
- ✓ Cross-linker reinforces the entanglement of polymers.
- ✓ Cross-linker has an unique structure that interacts with Cu.
- ✓ New additive also helps to enhance the adhesion to Cu.

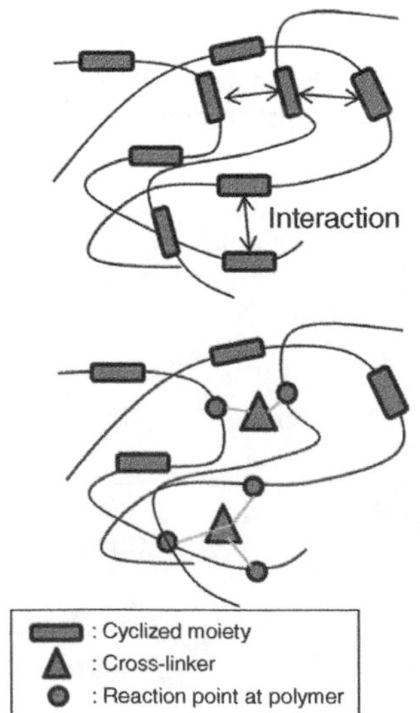

Figure 14.9 Design concept of a low temperature cure material.

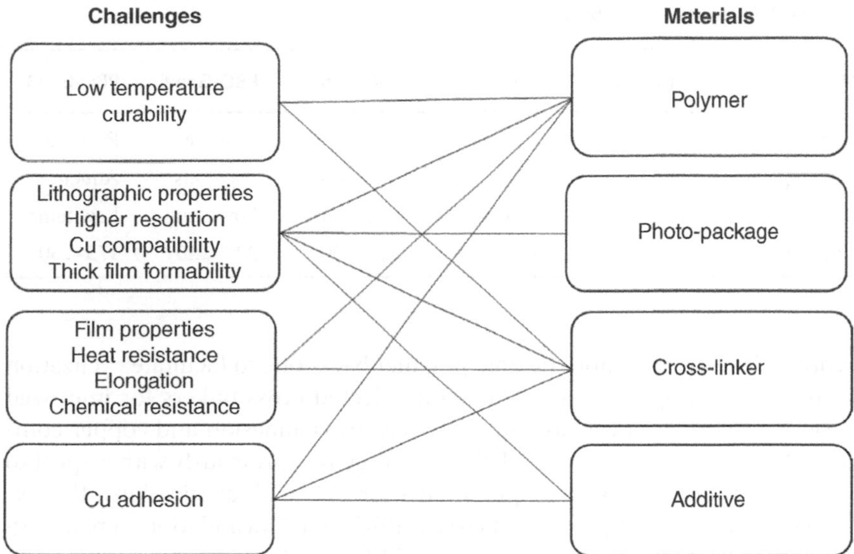

Figure 14.10 Correlation between challenges and materials.

PBO-Gen2 was the first product introduced by HDM that could be cured at 200 °C (see Table 14.2).

In order to reduce further the cure temperature to <200 °C to address the current challenges for next-generation FO-WLP applications described above, further development work has resulted in two new low temperature cure materials:

- **PBO-Gen3 (positive-acting, aqueous-developable PBO)**
- **PI-Gen2 (negative-acting, solvent-developable PI)**

Both materials have also been developed in solvents other than NMP. Process and end-use reliability data for both PBO-Gen3 (with PBO-Gen2 used as a reference) and PI-Gen2 are outlined in detail in the following sections.

14.5 PBO-Gen3 (Positive-Acting, Aqueous-Developable Material)

This section details the process flow, lithography, material properties, and reliability performance that have been generated for PBO-Gen3 to date and where PBO-Gen2 is used as a reference. As indicated in a previous section, PBO-Gen2 was the first product introduced by HDM that could be cured at 200 °C and where the design concepts used in the development of PBO-Gen2

Table 14.2 Overview of PI/PBO material sets.

Material	PI-Gen1	PI-Gen2	PBO-Gen1	PBO-Gen2	PBO-Gen3
Tone	Negative	Negative	Positive	Positive	Positive
Development	Solvent	Solvent	Aqueous	Aqueous	Aqueous
Cure	High temp	Low temp	High temp	Low temp	Low temp
Cure (°C)	350–390	175–250	320–350	200–250	175–250

involved developing a more flexible polymer backbone to facilitate cyclization at lower cure temperatures, introducing selected cross-linkers for improved thermal properties as well as additives to address adhesion and copper compatibility. However, to meet the challenges discussed previously with respect to both process and reliability requirements for FO-WLP applications, the following performance improvements to the PBO-Gen2 formulation were undertaken that resulted in the development of PBO-Gen3:

- An alternative photo-package was selected to improve lithographic performance.
- A novel cross-linker was introduced to enhance chemical resistance, moisture resistance, and adhesion to Cu.
- A new additive was added to improve lithographic performance on Cu as well as adhesion to Cu at low cure temperatures.

14.6 PBO-Gen3 Process Flow

The process flow for PBO-Gen3 is outlined in Table 14.3 where spin speeds, pre-bake conditions, exposure energies, development times, and curing conditions are given for 5, 7, and 10 μm cured films together with thicknesses after pre-bake, development, and cure.

Setting up the PBO-Gen3 process requires balancing coating thickness, pre-bake, exposure, development, and cure conditions to provide the desired final film resolution, cured film thickness, and film properties. It should be noted that the process conditions outlined in Table 14.3 can vary depending on equipment used and clean room conditions. Please note some general points regarding the spin-coating process:

- The spin-coating speed generally controls the range of cured coating thicknesses although the final coating thickness can also be influenced to a lesser extent through surface topography, processing equipment, and clean room environment. With current Cu track thicknesses in FO-WLP applications

Table 14.3 PBO-Gen3 process flow.

Process		Unit	Condition			Remarks
			5 μm	7 μm	10 μm	
Coating	Pre	rpm/s		1000/5		
	Main		3000/30	2000/30	1200/30	
Pre-bake		°C/s	110/180			Hot plate
Thickness after pre-bake		μm	8.1	11.2	16.4	
Exposure dose (i-line)		mJ/cm^2	160	220	300	
Development		s	30 × 2	40 × 2	55 × 2	2.38% TMAH
Film thickness after dev.		μm	6.2	8.7	12.1	
Cure profile		—	Step 1: ramp up to 100 °C (10 °C/min)			Under N_2
			Step 2: hold for 30 min			
			Step 3: ramp up to 175–250 °C (10 °C/min)			
			Step 4: hold for 120 min			
Film thickness after cure		μm	5.1	7.1	10.3	

typically being around 5 μm, acceptable coverage of the Cu track can be achieved with a 7 μm cured film thickness.

- The pre-bake step removes the bulk of the solvent, resulting in a film that is "dry to the touch," which can be handled in subsequent processes.
- Both i-line (365 nm) and broadband steppers can be used for the exposure process.
- The development process (typically puddle) removes all of the exposed film as well as some unexposed film on the wafer. The ratio of the remaining unexposed film thickness to the initial pre-baked film thickness is known as the percentage film retention (% FR) after development.
- Curing is typically performed as a batch process using programmable ovens or furnaces under a nitrogen atmosphere (oxygen concentration < 100 ppm).
- In setting up the process for PBO-Gen3, based on the conditions outlined in Table 14.3, the pre-bake and development conditions are first adjusted to a predetermined % FR after development (see Table 14.4). The exposure energy is then optimized (avoiding any residue formation due to underexposure or film lifting due to overexposure) for resolution and pattern profile.

Regarding the processing of PBO-Gen3 and PBO-Gen 2, both materials, as indicated in Table 14.4, produce similar film thicknesses after pre-bake,

Table 14.4 Film retention (FR) comparison between PBO-Gen3 and PBO-Gen2.

Item	Unit	PBO-Gen2	PBO-Gen3	
		200 °C/1 h	175 °C/2 h	200 °C/2 h
Thickness after pre-bake	μm	15.8	15.8	16.4
Thickness after development	μm	12.3	11.9	12.1
Thickness after cure	μm	10.3	10.0	10.3
% FR after development	%	78	75	74
% FR after cure	%	84	84	85
Overall % FR	%	65	63	63

development, and cure that are further reflected in the % FR after development (74–78%), % FR after cure (84–85%), and overall % FR (63–65%).

14.7 PBO-Gen3 Lithography

14.7.1 PBO-Gen3 Resolution

Using the process conditions given in Table 14.3, high resolution down to 2 μm L/S and 2 μm via openings are obtained with PBO-Gen3 on both Si and Cu with a 7 μm cured film thickness (see Figure 14.11) that also resulted in a smooth sidewall with no crowning of the dielectric being observed at the interface of the top edge and top surface of the sidewall (see Figure 14.12). Regarding the sidewall slope, too steep an angle is unsuitable due to stress concentration at the edge of the pattern, which can also reduce copper plating distribution, while too shallow an angle is not preferable for fine patterning. In this respect, the sidewall angle of 70–80° that was obtained is an acceptable compromise.

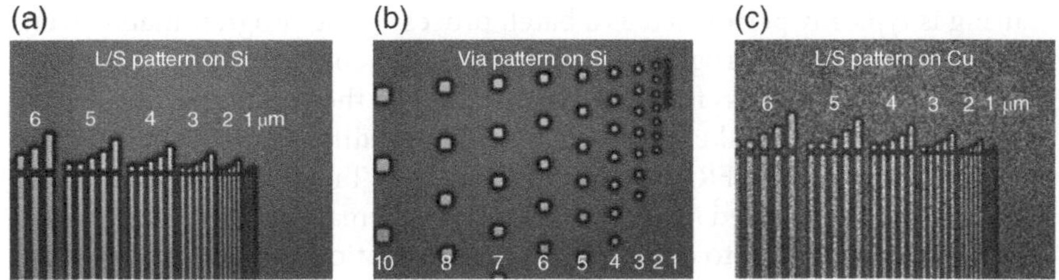

Figure 14.11 PBO-Gen3 resolution on Si and Cu (7 μm thickness after cure). (a) and (c) show L/S patterns on Si and Cu, respectively. (b) shows a via pattern on Si.

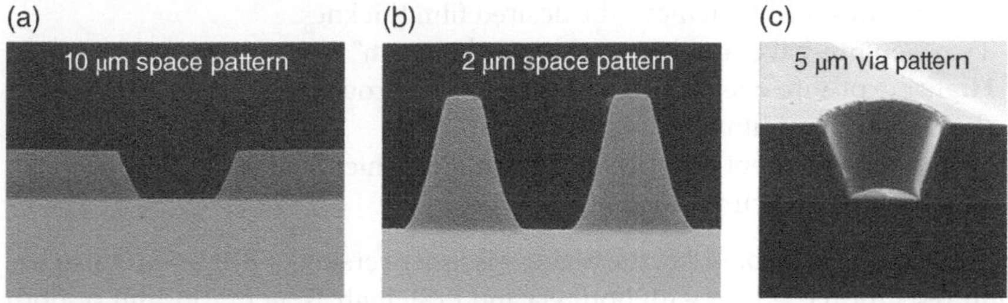

Figure 14.12 PBO-Gen3 lithographic performance (cross-sectional 7 μm thickness after cure).

14.7.2 PBO-Gen3 Thick Film Formability

A higher viscosity version of PBO-Gen3 was developed for thick film formability where >15 μm cured film thicknesses are required. The process flow for cured film thicknesses ranging from 15 to 20 μm is outlined in Table 14.5.

As compared with the standard PBO-Gen3 process outlined in Table 14.3, the process for the higher viscosity version requires:

Table 14.5 PBO-Gen3 (high viscosity version) process flow for thick film formability.

Process		Unit	Condition			Remarks
			15 μm	17.5 μm	20 μm	
Coating	Pre	rpm/s	1000/5			
	Main		1800/30	1500/30	1200/30	
Pre-bake		°C/s	80/180 + 110/270			Hot plate
Thickness after pre-bake		μm	25.8	29.5	34.2	
Exposure dose (i-line)		mJ/cm^2	560	640	820	
Development		s	43 × 4	47 × 4	55 × 4	2.38% TMAH
Film thickness after dev.		μm	18.6	21.4	24.4	
Cure profile		—	Step 1: ramp up to 100 °C (10 °C/min)			Under N_2
			Step 2: hold for 30 min			
			Step 3: ramp up to 175–250 °C (10 °C/min)			
			Step 4: hold for 120 min			
Film thickness after cure		μm	15.4	17.7	20.2	

- Lower spin speed to achieve the desired film thickness.
- Two-step pre-bake to achieve a "dry to the touch" film.
- Higher exposure energies to allow exposure through the whole thickness of the thicker dried film.
- Longer development times to develop out the thicker film.
- No change in the cure process.

As indicated in Table 14.6, the higher viscosity version of PBO-Gen3 also has high resolution (2–3 μm with both via and L/S), high aspect ratio, and smooth sidewalls with no crowning with cured film thicknesses up to 15 μm (a 10 μm via opening with a 20 μm cured film thickness is shown in Figure 14.13). This is in comparison with PBO-Gen2 that has lower resolution at higher film thicknesses and still requires higher exposure energies.

14.7.3 PBO-Gen3 Deep Gap Formability

PBO-Gen3 can be also used for deep gap formability applications to open scribe lines after processing multiple RDL (see Figure 14.4) and where no residues were observed after exposure and development with a 25 μm pre-baked thickness film as compared with PBO-Gen2 where residues were found after development (see Figure 14.14). It should be noted that, as outlined in Table 14.5, higher exposure energies and development times are required during processing of pre-bake thicknesses up to 25 μm.

14.8 PBO-Gen3 Material Properties

Mechanical and thermal end-use properties as well as residual stress are measured to provide performance indicators for dielectric materials in downstream processes used in the manufacture of FO-WLP devices as well over the lifetime of the semiconductor device. Mechanical properties (modulus, tensile strength, and % elongation) are measured as freestanding films on a tensile tester using the following test procedure:

1) The test material is processed on 6″ Si wafers as follows:
   - Spin coat and pre-bake to target a 10 μm cured film thickness.
   - Expose (1000 mJ cm^{-2}) using a BB stepper through a glass mask designed with 10 mm wide rectangular strip patterns.
   - Develop (puddle) in 2.38% TMAH to target 75% FR.
2) Cure in a furnace under a nitrogen atmosphere at the desired temperature for two hours. The wafers with the cured test material are immersed into a 4.9% HF aqueous solution to release the films from the Si wafer and then washed with distilled water and dried in air.

Table 14.6 PBO-Gen3 and PBO-Gen2 thick film patterning capability.

Sample	Thickness after PB (μm)	Exposure dose (mJ/cm^2)	Thickness after dev. (μm)	% FR after dev.	Thickness after cure (μm)	Resolution via and L/S (μm)	Aspect ratio[a]
PBO-Gen2	11.2	230	8.5	75.9	7.3	2	>3.5
	19.8	1040	15.2	76.8	12.2	10–15	1
PBO-Gen3	11.4	180	8.5	74.6	7.2	2	>3.5
	20.3	500	15.6	76.8	13.1	3	>4
	27.4	640	19.9	72.6	15.6	3	>5

[a] Aspect ratio = thickness after cure/resolution.

Figure 14.13 PBO-Gen3 thick film formability.

Material	PBO-Gen2	PBO-Gen3
Overall % FR after development of third layer	75	75
Appearance		
Scanning profile at deep gap	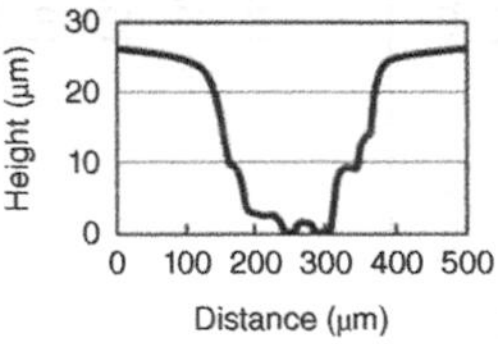 Height (μm) 0–30; Distance (μm) 0–500	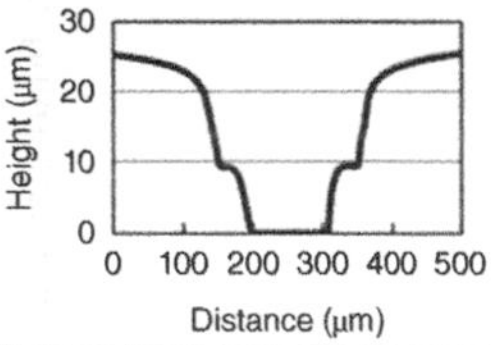 Height (μm) 0–30; Distance (μm) 0–500

Figure 14.14 Deep gap formability performance.

3) The cured freestanding films (dimensions: thickness 10 μm, width 10 mm, length 70–100 mm) are then mounted onto a tensile tester (chuck distance 20 mm, pull speed 5 mm min^{-1}) at ambient temperature and tested for mechanical properties.

The mechanical and thermal end-use properties as well as residual stress for PBO-Gen3 and PBO-Gen2 are given in Table 14.7. From the data, it can be seen that, due to the combination of appropriate cross-linkers with the base polymer that produces a robust and ductile 3D cross-linking structure, both PBO-Gen3

Table 14.7 PBO-Gen3 and PBO-Gen2 cured film properties.

Item	Unit	PBO-Gen2				PBO-Gen3			
Cure temp.	°C	175	200	225	250	175	200	225	250
Tensile strength	MPa	170	170	170	170	160	150	120	120
% Elongation (ave.)	%	85	80	80	80	65	70	70	55
Young's modulus	GPa	1.9	1.8	1.8	1.7	2.1	2.0	1.8	1.7
T_g	°C	240	240	245	245	240	245	255	270
CTE	$\times 10^{-6}$/°C	80	80	80	80	75	75	80	84
Weight loss temp. (5%)	°C	280	310	345	360	300	320	340	365
Residual stress	MPa	25				27			

and PBO-Gen2 achieve acceptable end-use properties, resulting in a wide cure temperature margin over the cure temperature range of 175–250 °C.

One important point to note is that mechanical properties and in particular % elongation can be linked to performance in TCT, which is conducted to determine the effect of extreme changes in temperature on the overall performance of a device as well as the materials used to manufacture the device. In this respect, the % elongation of both the PBO-Gen3 and PBO-Gen2 when measured at ambient temperatures after TCT testing (JESD22-A104 Condition C) is not significantly changed after 1000 cycles (see Table 14.8).

However, any brittleness that occurs with organic polymers at low temperatures can result in a drop in % elongation when measured at that low temperature, which can subsequently produce cracks in the dielectric and damage to the copper tracks. This means that high % elongation when measured at low temperature is an important property in ensuring maximum reliability performance.

Table 14.8 Mechanical properties after TCT (−65 °C/15 min to 150 °C/15 min).

Sample	Cure temp. (°C/2 h)	TCT (cycles)	Tensile strength (MPa)	Elongation (%)	Modulus (GPa)
PBO-Gen2	200	0	125	86	1.8
		1000	141	82	1.8
PBO-Gen3	200	0	148	55	2.0
		1000	111	58	2.1

Table 14.9 Mechanical properties (ambient versus –50 °C).

Sample	Cure temp. (°C/2 h)	Measurement temp. (°C)	Tensile strength (MPa)	Elongation (%)	Modulus (GPa)
PBO-Gen2	200	25	134	54	2.6
		–50	120	16	2.6
PBO-Gen3	200	25	147	65	2.7
		–50	164	44	2.6

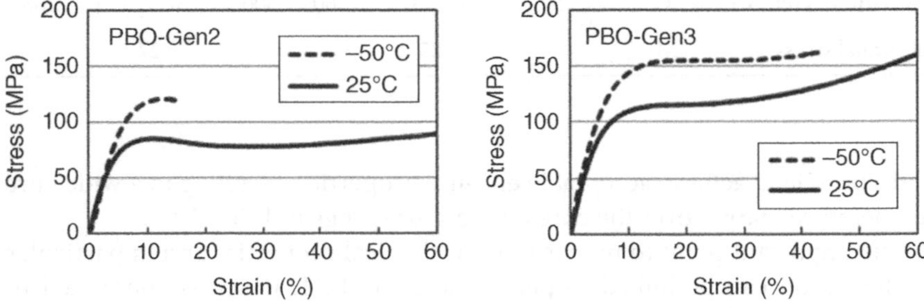

Figure 14.15 Mechanical properties. Examples of stress–strain curves (25 vs. –50 °C).

The effect of measuring % elongation at different temperatures is subsequently shown in Table 14.9 where the % elongation of PBO-Gen3 and PBO-Gen2 decreases when measured at –50 °C (limit of the equipment used) as compared with 25 °C. However, the drop in % elongation with PBO-Gen3 (65% at ambient dropping to 44% at –50 °C) is not as prominent as compared with PBO-Gen2 (54% at ambient temperature dropping to 16% at –50 °C), which would suggest an improved performance with PBO-Gen3 as compared with PBO-Gen2 (see Figure 14.15). From these tests, the main conclusion is that the % elongation of 40–45% obtained with PBO-Gen3 when measured at –50 °C is ductile enough to provide crack resistance and so avoid performance issues during component and board reliability testing involving TCT (see section on PBO-Gen3 package reliability performance). These results further indicate that the temperature at which the films are being measured for mechanicals is a critical factor in determining reliability performance.

In addition, after PCT, while a slight drop in % elongation (65–60%) and modulus (2.7–2.3 GPa) was observed with PBO-Gen3 after 168 hours PCT (see Table 14.10), no brittleness was observed due to any interaction with moisture. This is in comparison with PBO-Gen2 where, after PCT, the cured film was very brittle with a hazy appearance such that mechanical testing could not be conducted on the cured films.

Table 14.10 Mechanical properties after PCT (121 °C/100% RH, 168 h).

Sample	Cure temp. (°C/2 h)	PCT (hours)	Tensile strength (MPa)	Elongation (%)	Modulus (GPa)
PBO-Gen2	200	0	156	73	1.9
		168	No data		
PBO-Gen3	200	0	147	65	2.7
		168	149	60	2.3

One important point to note is that the % elongation obtained is not an absolute measurement and can vary from one test location to another depending on processing conditions of the cured film to be tested, test equipment, and test procedure. However, when measured side-by-side, % elongation data can be used as a guide to performance during reliability testing and where it can be seen that the % elongation of the PBO-Gen3 is superior to PBO-Gen2 when measured at −50 °C (see Table 14.9 and Figure 14.15) and after PCT (see Table 14.10).

14.9 PBO-Gen3 Dielectric Reliability Testing

Reliability testing is undertaken to determine that the dielectric is not degraded in any way during downstream processes that can result in yield losses and also that the performance of the dielectric does not change over the lifetime of the semiconductor device. In this respect, the following tests were conducted on PBO-Gen3 with PBO-Gen2 as a reference:

- Adhesion strength to Cu and PBO/PBO after PCT.
- Resistance to chemicals used in downstream processing.
- bHAST.

14.9.1 PBO-Gen3 Adhesion After PCT

In the FO-WLP device, delamination at the interface between the dielectric and either the Cu track or another layer of dielectric can induce cracking of both the dielectric material and the Cu track in the RDL layers. In addition, delamination can also induce electrochemical copper migration between Cu tracks in the presence of moisture as Cu ions can easily be generated from the surface of the anode, subsequently move freely through the moisture, and be immediately converted to Cu at the cathode. This is a typical propagation mechanism for Cu dendrite formation where moisture can act as a medium for accelerating Cu migration. Therefore, good adhesion of the dielectric to Cu as well as between two dielectric layers is important for package reliability.

The PCT was originally adopted to shorten the time to failure of a semiconductor chip in a molded package due to the erosion of Al lines. Once delamination between the molding compound and the chip occurs due to degradation of the molding compound under high humidity conditions, moisture that penetrates into the molded package collects at the delamination site and accelerates the erosion of the Al lines. In this respect, PCT treatment was found to be a good method to evaluate the adhesion performance of molding compounds and has subsequently been used to accelerate the adhesion performance of organic dielectric materials in electronic applications.

The adhesion strength of both PBO-Gen3 and PBO-Gen2 to Cu and PBO/PBO was measured using a stud pull test before and after PCT (121 °C/100% RH, 300 hours, JESD22-A102). The samples used to test for PBO/Cu adhesion were prepared as follows:

- The stud pin with an epoxy adhesive is fixed to a cured test film processed on a silicon wafer electroplated with Cu. The epoxy adhesive is then cured at 120 °C for one hour.
- The test specimen is placed on the stud pull tester, and the stud pin is gradually pulled down vertically until breakage occurs. If breakage of the epoxy adhesive occurs, the adhesion strength of the dielectric to Cu is higher than the epoxy cohesion strength, and this breakage mode indicates good adhesion performance to Cu. In contrast, if delamination at the interface between Cu and the cured dielectric film occurs, the adhesion strength to Cu is lower than the epoxy cohesion strength, and this breakage mode is unacceptable.
- The same test procedure is then conducted with the cured film after PCT treatment in order to evaluate the adhesion performance before and after PCT treatment.

For PBO/PBO adhesion testing, the test dielectric is first processed to cure (200 °C/2 h) on a silicon wafer electroplated with Cu and an additional layer of the test dielectric processed to cure (200 °C/2 h) over the first cured film. The stud pull test is then conducted using the procedure described above. It should be noted that the limitation of using the stud pull test is that the actual adhesion strength above the epoxy cohesion strength cannot be measured. In addition, adhesion can only be measured at ambient temp. and cannot be measured at high temperatures that simulate reflow conditions.

It can be seen from Figures 14.16 and 14.17 that PBO-Gen3 cured at 175 and 200 °C passed the stud pull test to both Cu and to itself after 300 hours PCT. This is in comparison with PBO-Gen2 cured at 175 and 200 °C, which failed on Cu after 200–300 hours PCT and PBO/PBO after 300 hours PCT.

The adhesion strength of PBO-Gen3 and PBO-Gen2 to Cu L/S after PCT was also tested using the following procedure:

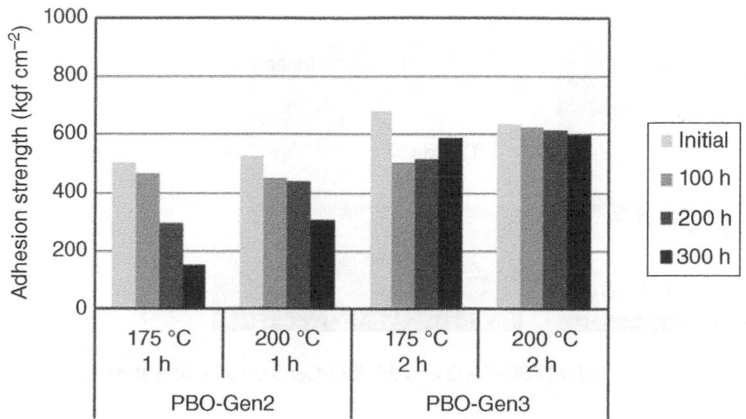

Figure 14.16 Adhesion stud pull test results on Cu after PCT.

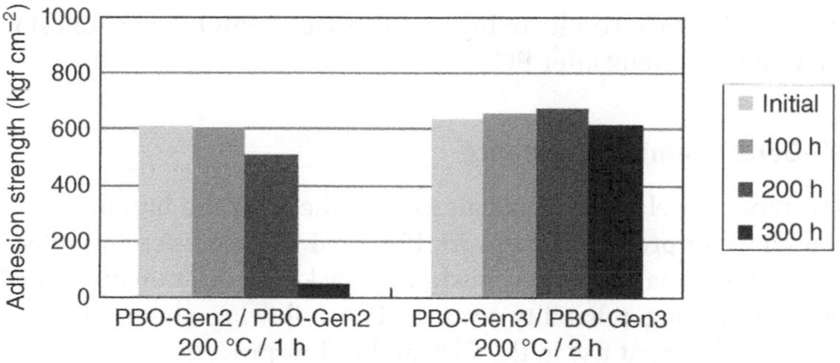

Figure 14.17 Adhesion stud pull test results on PBO/PBO after PCT.

- The test dielectric is spin-coated over the 10 μm Cu L/S, pre-baked, and cured (200 °C/2 h).
- The coated test sample is then placed in a PCT chamber set to 121 °C/100% RH for 300 hours.
- Cross sections before and after PCT are inspected for delamination using scanning electron microscopy (SEM).

The adhesion results for PBO-Gen3 and PBO-Gen2 on 10 μm Cu L/S are shown in Figure 14.18. Although simulation analysis of stress distribution in dielectric layers indicated that the tensile stress is concentrated at the sidewall of the Cu line, which can induce delamination at that point, it should be noted that no delamination was observed with PBO-Gen3 in contrast to PBO-Gen2.

The improved adhesion performance of PBO-Gen3 over PBO-Gen2 on Cu is due to PBO-Gen3 having improved Cu compatibility and higher interaction

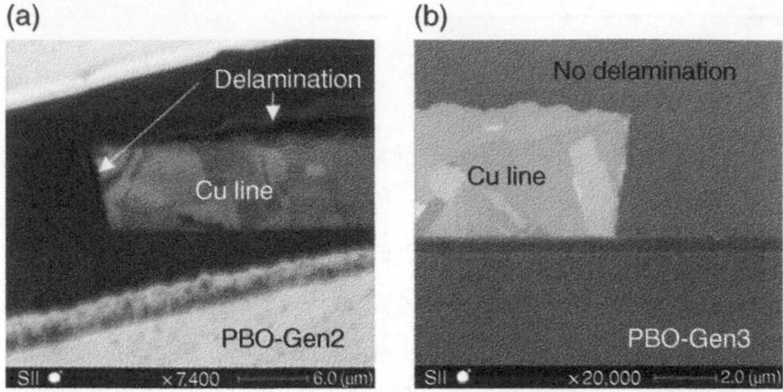

Figure 14.18 Cross sections of (a) PBO-Gen2 and (b) PBO-Gen3 on a Cu line after PCT.

with the Cu surface due to the choice and optimization of both cross-linker as well as additives used in the PBO-Gen3 formulation to improve adhesion. In addition, as discussed previously, PBO-Gen3 is more ductile and does not degrade during PCT, which results in higher PBO-Gen3/PBO-Gen3 adhesion as compared with PBO-Gen2 after PCT.

14.9.2 PBO-Gen3 Chemical Resistance

Chemical resistance to selected chemicals used in the RDL and bumping processes provides a wider process window for PBO and PI dielectrics and subsequently avoids performance issues such as cracking, delamination, and thickness changes during reliability testing. The following chemicals were chosen due to their frequent use in the RDL and reflow process:

1) Rework solvent (NMP) used to remove pre-baked PBO films as well as a cleaner for removing residues.
2) Photoresist stripper (Dynastrip 7700, Dynaloy) used to remove photoresist after the Cu plating process.
3) Resist cleaner (OK-73, Tokyo Ohka Kogyo) used to remove any residues remaining after the stripping process.
4) Flux (WS-600, Cookson Electronics) used to remove any oxide on the surface of the solder ball before reflow (WS-600 is a strong acid that is applied directly to the surface of the dielectric material before reflow).

The test procedure for chemical resistance is as follows:

- Process 7 μm cured films with a 100 μm square pattern on Si wafers.
- Dip the cured film into the selected chemical under the test conditions outlined in Table 14.11.
- Remove the cured film, wash with distilled water, and dry in air at ambient temp.

Table 14.11 Chemical resistance of PBO-Gen2 and PBO-Gen3.

Chemicals	Treatment conditions		Check item[a,b]	PBO-Gen2	PBO-Gen3	
	Temp. (°C)	Time (min)		200 °C 1 h	175 °C 2 h	200 °C 2 h
Rework solvent (NMP)	25	30	Appearance	Cracking	No change	No change
			Film thickness change	OK	OK	OK
Resist stripper (Dynastrip 7700)	70	30	Appearance	Delamination	No change	No change
			Film thickness change	OK	OK	OK
Resist cleaner (OK-73)	25	60	Appearance	Cracking	No change	No change
			Film thickness change	OK	OK	OK
Flux (WS-600)	245	1	Appearance	25% swelling	No change	No change
			Film thickness change	OK	OK	OK

[a] Appearance No change = no cracking, no delamination, and no hazing.
[b] Film thickness change % change within ±10% is acceptable.

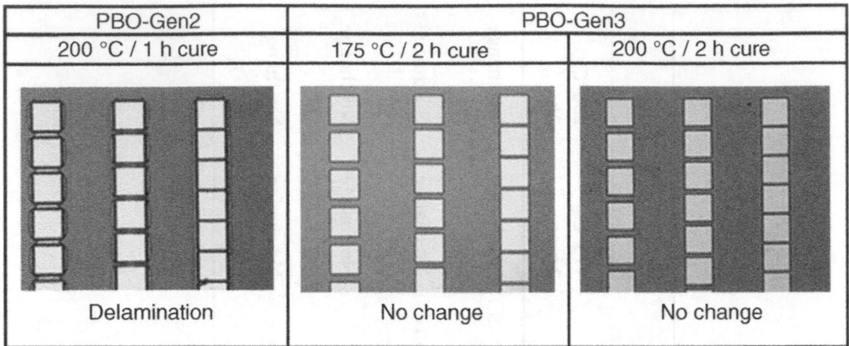

PBO-Gen2
PBO-Gen3
200 °C / 1 h cure
175 °C / 2 h cure
200 °C / 2 h cure
Delamination
No change
No change

Figure 14.19 PBO-Gen2 and PBO-Gen3 appearance (100 μm sq. pattern) after immersion in resist stripper.

- Measure thickness of the cured film by a stylus type thickness measurement tool.
- The film thickness change is then estimated by comparing with the initial thickness.
- Conduct a visual check using an optical microscope (100× magnification).

The chemical resistance results are given in Table 14.11 where no changes were observed in either appearance or film thickness of the PBO-Gen3 even when cured at 175 °C after immersion in the chemicals described above for a given temperature and time. This is in contrast to PBO-Gen2 that shows delamination and cracking at 200 °C with NMP, Dynastrip 7700, and OK-73 (see Table 14.11 and Figure 14.19). It should also be noted that no thickness change occurred with PBO-Gen3 after treatment with WS-600 flux, while PBO-Gen2 swelled by 25%.

14.9.3 PBO-Gen3 bHAST

bHAST uses temperature, humidity, and bias to accelerate the penetration of moisture through a dielectric, which can cause performance loss or electrical breakdown (JESD22-A110). For FO-WLP applications, bHAST is used to determine the insulation reliability of the dielectric between Cu tracks and is becoming more important with the recent trend toward finer L/S designs [17–19].

The test vehicle used for the test was processed in cooperation with the Hitachi Chemical Packaging Solution Center and consists of 5/5 and 2/2 μm Cu L/S processed on PI-Gen1 (see Figure 14.20). The test procedure is as follows:

- The test dielectric is spin-coated, pre-baked, and cured at 200 °C/2 h so as to cover the 5/5 or 2/2 μm L/S Cu comb pattern.

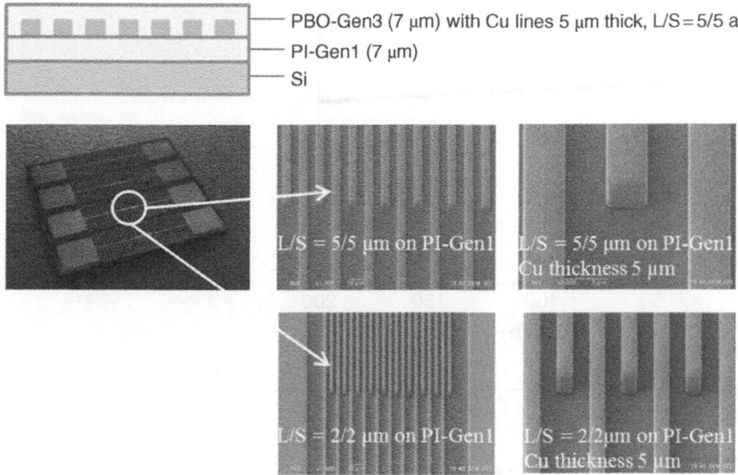

Figure 14.20 Test vehicle for bHAST.

- The anode and cathode of the test sample are connected, respectively, by wires to a migration tester equipped with HAST chamber and monitoring unit.
- The HAST chamber is set to 130 °C and 85% RH.
- DC voltage at 3.3 V is applied constantly, and *in situ* insulation resistance is monitored during testing.

bHAST results for PBO-Gen3 indicated that no short circuits, dendrite formation, or delamination was observed with both 5/5 and 2/2 µm Cu L/S after 200 hours (longer than the required standard test time of 168 hours) as well as no change in the insulation resistance (see Figure 14.21). This is in comparison with PBO-Gen2 that failed bHAST after 200 hours.

14.10 PBO-Gen3 Package Reliability Performance (TCT Testing at Component and Board Level)

The main purpose in conducting dielectric reliability testing at the component level is to define or determine the influence of any degradation of the dielectric materials under severe environmental conditions (moisture, temperature, and bias voltage) on device performance, which has already been discussed in part in a previous section. On the other hand, conducting dielectric reliability tests at the board level is focused more on the mechanical stress around solder joints as well as the durability to the stress induced by CTE mismatches between the device and board.

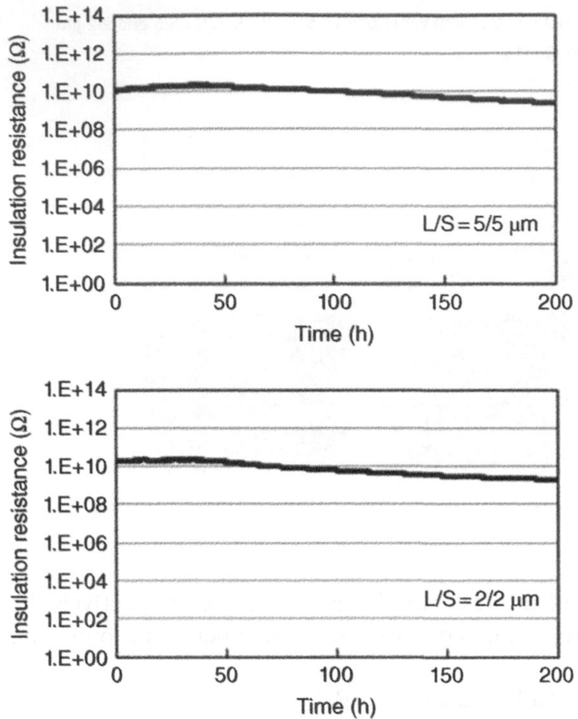

Figure 14.21 PBO-Gen3 insulation resistance during bHAST (130 °C/85% RH/3.3 V, L/S = 5/5 and 2/2 μm).

There are a number of reliability tests that can be conducted at the component and board level that include moisture sensitivity levels (MSL), TCT, high temperature storage (HTS) test, unbiased highly accelerated stress test (uHAST), and the drop test. However, for initial component and board-level testing, TCT was regarded as the most useful test to provide performance on actual stress produced due to temperature variation [20–22].

The test vehicle used for the TCT was assembled in cooperation with the Hitachi Chemical Packaging Solution Center. The processing of the FO-WLP test vehicle is shown in Figure 14.22 together with a cross section of the test vehicle in Figure 14.23. Details of the vehicle structure are outlined in Table 14.12.

For the component-level evaluation, the test vehicle was preconditioned at 85 °C/85% RH for 168 hours followed by 10× N_2 reflow at 260 °C and then subjected up to 1000 TCT cycles (−65 °C/15 min ↔ 150 °C/15 min). The test vehicle was visually inspected using an optical microscope for any defects at

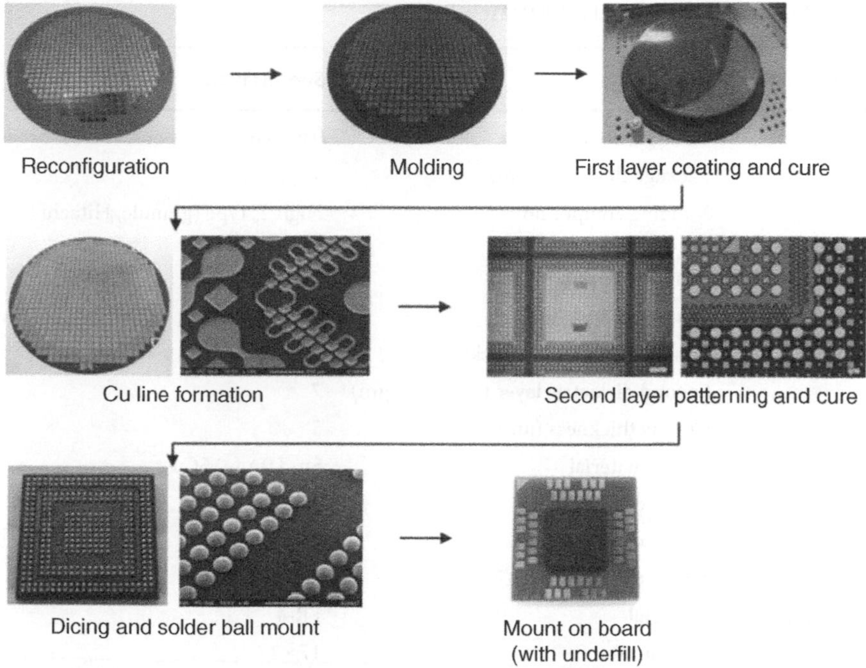

Figure 14.22 Assembly process of FO-WLP test vehicle used for reliability testing (12 in. process).

Figure 14.23 Cross section of FO-WLP test vehicle.

200 cycle intervals and no cracking or delamination was observed after 1000 TCT cycles (see Table 14.13).

For the board-level evaluation, a one-time reflow treatment (260 °C max) without moisture soaking was used to mount the test vehicle onto the board and then subjected to 1000 TCT cycles (−65 °C/15 min ↔ 150 °C/15 min). As the test vehicle was designed with a daisy chain pattern to measure the electrical resistance across the dielectric between Cu lines, any electrical failure due to solder cracking or cracks to the Cu lines will be detected. In this test, the electrical resistance was measured up to 1000 cycles TCT at intervals of 200 cycles, and no changes in electrical resistance were observed with PBO-Gen3 as the dielectric (see Table 14.13).

Table 14.12 Specification of FO-WLP test vehicle.

Classification		Specification
FO-WLP	Package size (mm)	9.6 × 9.6
	Package thickness (μm)	450
	Molding compound	High T_g type (granule, Hitachi Chemical)
	Chip size (mm)	7.3 × 7.3
	Chip thickness (μm)	400
	First dielectric layer thickness (μm)	7
	Second dielectric layer thickness (μm)	7
	Cu line thickness (μm)	5
	Bump material	Sn-3.0Ag-0.5Cu
	Bump diameter (μm)	250
	Bump number	336
	Bump pitch (μm)	300
Substrate	Material	FR-4
	Size (mm)	17 × 17
	Thickness (mm)	0.8
	Surface finish	E-less Ni/Au plating
Underfill		High T_g type (Hitachi Chemical)

Table 14.13 PBO-Gen3 (cured 200 °C/2 h) reliability test results after TCT at component and board level.

	Reliability test result (component level)	Reliability test result (board level)
TCT cycles	Appearance after TCT	Cumulative electrical failure rate (%)
0	No crack or delamination	0
200	No crack or delamination	0
400	No crack or delamination	0
600	No crack or delamination	0
800	No crack or delamination	0
1000	No crack or delamination	0

14.11 Performance Comparison Between PBO-Gen3 and PBO-Gen2

An overview of the test data outlined above indicates that PBO-Gen3 has improved lithographic and reliability performance as compared with PBO-Gen2 with respect to meeting the challenges and requirements for FO-WLP applications.

Regarding lithographic performance, PBO-Gen3 has a higher resolution over a wider thickness range where a L/S resolution of 3 μm with a 15 μm cured film can be obtained as compared with a resolution 10–15 μm with PBO-Gen2. In addition, PBO-Gen3 has an improved deep gap formability in that a 25 μm post-bake thickness can be developed with no remaining residues as compared with PBO-Gen2 where residues remain after development.

Regarding reliability performance (see Table 14.14), PBO-Gen3 as compared with PBO-Gen2 has:

- Higher % elongation at −50 °C and after PCT.
- Improved adhesion to both Cu and PBO/PBO after 300 hours PCT (PBO-Gen2 failed after 200 hours).
- Higher chemical resistance, in particular, to strong solvents and resist strippers.
- Improved insulation resistance in bHAST.

In addition, PBO-Gen3 passed TCT component (visual) and board (electrical) testing, while no tests were conducted on PBO-Gen2.

14.12 PI-Gen2 (Negative-Acting, Solvent-Developable Material)

This section details the process flow, lithography, material properties, and reliability performance that have been generated to date for PI-Gen2. The design concept used in the development of PI-Gen2 was to introduce:

- An alternative photo-package to improve lithographic performance.
- A novel cross-linker to enhance both chemical and moisture resistance.
- A new additive to improve adhesion to Cu at low temperature cure.

14.13 PI-Gen2 Process Flow

The process flow for PI-Gen2 is outlined in Table 14.15 where spin speeds, pre-bake conditions, exposure energies, development times, and curing conditions are given for 5, 7, and 10 μm cured films together with thicknesses after pre-bake, development, and cure.

Table 14.14 Reliability test results comparing PBO-Gen3 with PBO-Gen2.

Reliability		Test conditions		PBO-Gen2	PBO-Gen3
Elongation	JESD22-A104	TCT	0 cycles	86%	55%
			1000 cycles	82%	58%
		PBO film temp.	Ambient	54%	65%
			−50 °C	16%	44%
		PCT	0 hour	73%	65%
			168 hours	Not measured	60%
Adhesion	JESD22-A102	PBO on Cu after PCT	300 hours	Fail	Pass
Stud pull		PBO on PBO after PCT	300 hours	Fail	Pass
Chemical resistance	Visual	Rework solvent (NMP)	25 °C/30 min	Fail	Pass
		Resist stripper (Dynastrip 7700)	70 °C/30 min	Fail	Pass
		Resist cleaner (OK-73)	25 °C/60 min	Fail	Pass
		Flux (WS-600)	245 °C/1 min	Fail	Pass

bHAST	JESD22-A110	130 °C/85% RH/	5/5 μm L/S	Fail	Pass
		3.3 V/200 h	2/2 μm L/S	Fail	Pass
Component	Visual	Precondition/TCT	1000 cycles	No data	Pass
Board	Electrical	TCT	1000 cycles	No data	Pass

TCT −65 °C/15 min ↔ 150 °C/15 min. PCT 121 °C/100% RH.

Table 14.15 PI-Gen2 process flow.

Process		Unit	Condition			Remarks
			5 μm	7 μm	10 μm	
Coating	Pre-coat	rpm/s	1000/10			
	Main		3600/60	2800/30	2000/30	
Pre-bake		°C/s	105/120 + 115/120			Hot plate
Thickness after pre-bake		μm	8.2	9.7	12.9	
Exposure dose (i-line)		mJ/cm^2	400–600			
Development		s	10 × 2			
Film thickness after dev.		μm	6.1	7.9	11.1	
Cure profile		–	Step 1: ramp up to 175–250 °C (5 °C/min)			Under N_2
			Step 2: hold for 120 min			
			Step 3: cool down to 100 °C (5 °C/min)			
Film thickness after cure		μm	5.1	7.1	10.3	

Setting up the PI-Gen2 process requires balancing coating thickness, pre-bake, exposure, development, and cure conditions to provide the desired resolution, cured film thickness, and final film properties. It should be noted that the process conditions outlined in Table 14.15 can vary depending on equipment used and clean room conditions. One main point to note is that PI-Gen2 shows a higher overall FR of 75–80% after the cure step (see Table 14.16), in particular when compared to high temperature, negative-acting, solvent-developing materials that have a FR around 50%, which is due to the lower cure temperature that retains a portion of the cross-linker. The higher film retention after cure is also advantageous for improving:

- Resolution due to a lower pre-bake thickness for a target cured film thickness.
- Planarity after curing, which can be important in multilayer RDL structures. This is illustrated in Figure 14.24 where a 10 μm thick cured PI layer (second layer) was processed over a 100 μm space pattern of an already 10 μm thick cured PI layer (first layer) and where it can be seen that PI-Gen2 shows noticeably higher planarity as compared with a high temperature cured PI (PI-Gen1).

Table 14.16 PI-Gen1 and PI-Gen2 film thickness and % FR variation during processing.

Item	Unit	PI-Gen1	PI-Gen2			
		375 °C/1 h	175 °C/2 h	200 °C/2 h	225 °C/2 h	250 °C/2 h
Thickness after pre-bake	μm	22.0	12.7	12.9	13.3	13.5
Thickness after development	μm	20.7	10.8	11.1	11.4	11.7
Thickness after cure	μm	10.0	10.0	10.0	10.0	10.0
% FR after development	%	94	85	86	86	87
% FR after cure	%	48	93	90	88	85
Overall % FR	%	45	79	78	75	74

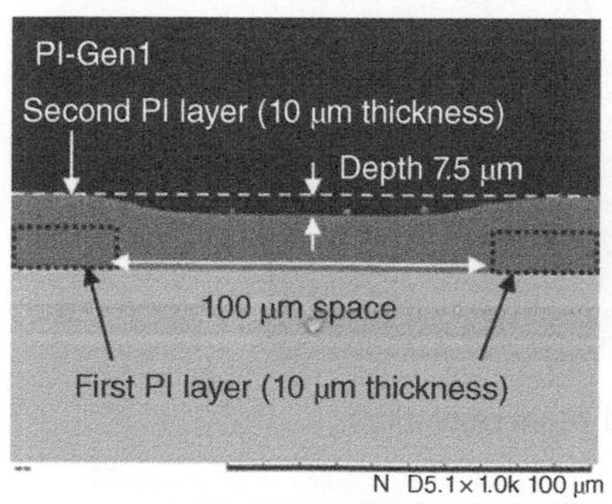

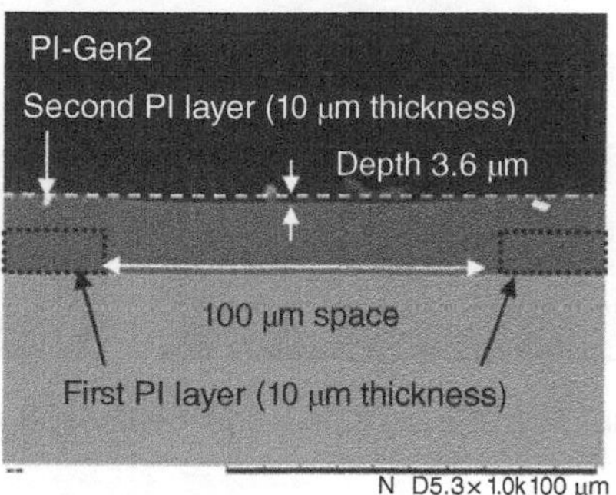

Figure 14.24 PI-Gen1 and PI-Gen2 topography comparison (after cure).

14.14 PI-Gen2 Lithography

Regarding lithographic performance, a high resolution of 5–6 μm over an exposure energy range of 300–500 mJ cm^{-2} was obtained with a 5–10 μm cured thickness (see Figures 14.25 and 14.26). In addition, a smooth profile (70–80° acute angle to the substrate surface) with no crowning can be obtained:

- Over a range of cure temperatures 175–250 °C (see Figure 14.27).
- On Si and Cu surfaces (see Figure 14.28).

14.15 PI-Gen2 Material Properties

Mechanical and thermal end-use properties as well as residual stress are measured to provide performance indicators for dielectric materials in downstream processes used in the manufacture of FO-WLP devices as well

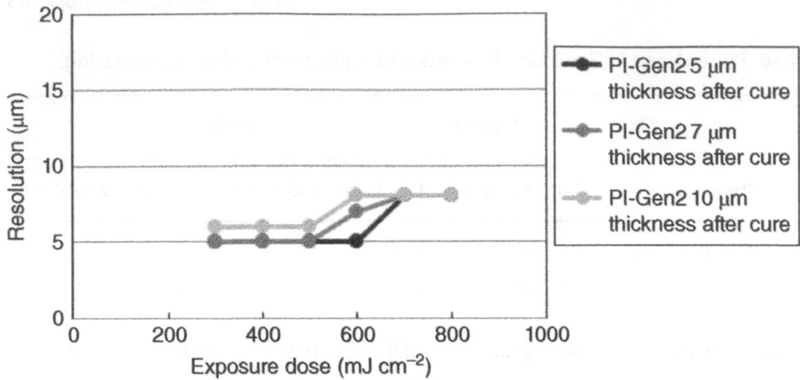

Figure 14.25 PI-Gen2 resolution at varying thicknesses after cure.

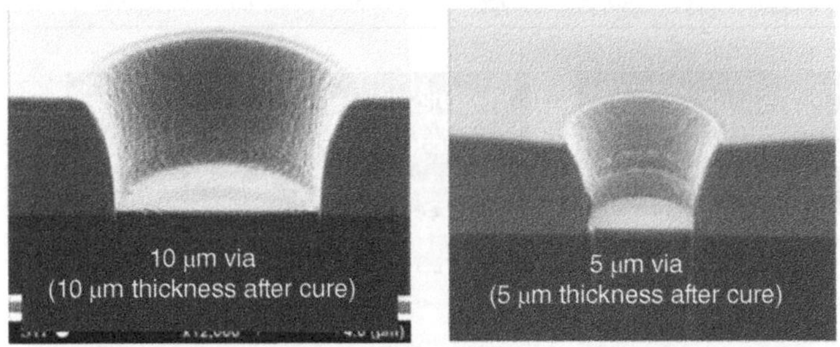

Figure 14.26 Cross sections of PI-Gen2 vias (5 and 10 μm resolution).

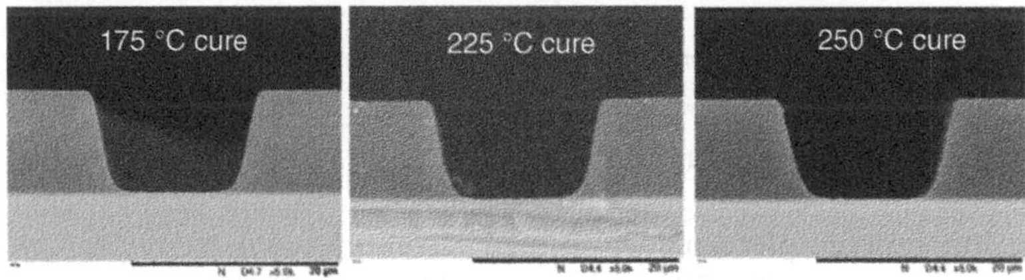

Figure 14.27 PI-Gen2 cross sections at varying cure temp. (10 μm space, 10 μm cured thickness).

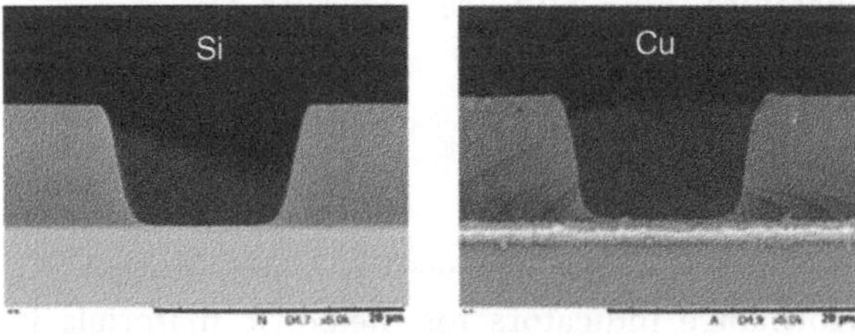

Figure 14.28 PI-Gen2 cross sections on various substrates (10 μm space, 10 μm cured thickness).

Table 14.17 PI-Gen2 cured film properties.

Item	Unit	PI-Gen2			
Cure temp.	°C	175	200	225	250
Tensile strength	MPa	190	177	175	174
Elongation (ave.)	%	39	41	40	42
Young's modulus	GPa	3.4	3.3	2.9	2.8
Tg (TMA)	°C	225	233	233	242
CTE	$\times 10^{-6}$/°C	63	61	60	59
Weight loss temp. (5%)	°C	317	336	347	347
Residual stress	MPa	25			

Table 14.18 PI-Gen2 mechanical properties measured at low temperature (−50 °C).

Sample	Cure temp. (°C/2 h)	Measurement temp. (°C)	Tensile strength (MPa)	Elongation (%)	Modulus (GPa)
PI-Gen2	200	25	141	24	3.7
		−50	178	20	3.7

as over the lifetime of the semiconductor device. The mechanical and thermal end-use properties as well as residual stress for PI-Gen2 are given in Table 14.17 where it can be seen that, due to the formation of a robust and ductile 3D cross-linking structure, PI-Gen2 achieves acceptable end-use properties, resulting in a wide cure temperature margin over the cure temperature range of 175–250 °C.

PI-Gen2 showed no significant change in % elongation when measured at ambient temperature and −50 °C (see Table 14.18 and Figure 14.29) on the same mechanical tool, indicating that the % elongation of PI-Gen2 did not deteriorate at temperatures down to −50 °C. It should be noted that the mechanical tool used was different to that used in generating the data given in Table 14.17 and is the reason why differences in % elongation and modulus were obtained.

In addition, while a drop in the modulus and tensile strength was obtained with PI-Gen2 after 100 hours PCT (121 °C/100% RH, 2 atm.), no change in the % elongation was observed, indicating that PI-Gen2 has a high moisture resistance that suppresses degradation of mechanical film properties (see Table 14.19).

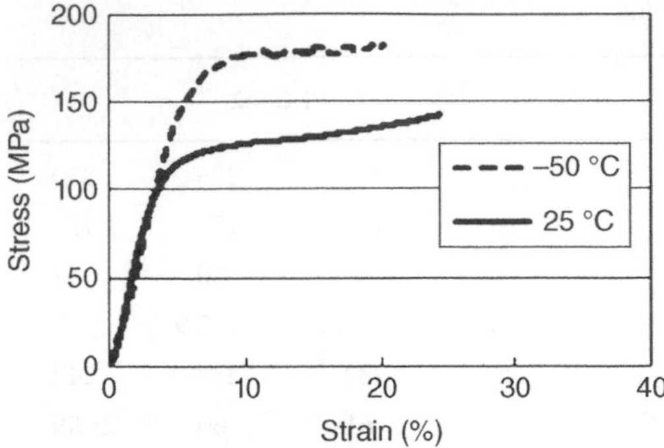

Figure 14.29 PI-Gen2 mechanical properties. Example of stress–strain curve at 25 °C versus −50 °C.

Table 14.19 PI-Gen2 mechanical properties after PCT.

Sample	Cure temp. (°C/2 h)	PCT (h)	Tensile strength (MPa)	Elongation (%)	Modulus (GPa)
PI-Gen2	175	0	190	39	3.4
		100	140	37	3.1
	200	0	177	41	3.3
		100	157	42	3.1
	225	0	158	40	2.9
		100	146	36	2.9

From the above tests, the main conclusion is that a similar % elongation obtained with PI-Gen2 at low temperature (−50 °C) as compared to ambient indicates that the cured film is ductile enough to provide crack resistance and so avoid performance issues during component and board reliability testing involving TCT (see section on PI-Gen2 package reliability performance). In addition, no noticeable change in % elongation after PCT indicates that high moisture resistance is also obtained with PI-Gen2.

14.16 PI-Gen2 Dielectric Reliability Data

Reliability testing is undertaken to determine that the dielectric is not degraded in any way during downstream processes that can result in yield losses and also that the performance of the dielectric does not change over the lifetime of the semiconductor device. In this respect, the following tests were conducted on PI-Gen2:

- Adhesion strength to Cu and PI/PI after PCT.
- Resistance to chemicals used in downstream processing.
- bHAST.

14.16.1 PI-Gen2 Adhesion After PCT

As discussed previously, delamination at the interface between the dielectric and either the Cu track or another layer of dielectric can induce cracking of both the dielectric material and the Cu track in the RDL layers. In addition, delamination between two dielectric layers can also produce electrochemical migration of the Cu between Cu tracks in the presence of moisture. Therefore, good adhesion of the dielectric to Cu as well as between two dielectric layers is important for package reliability.

The adhesion strength of PI-Gen2 to Cu and to itself both before and after PCT was measured using the stud test procedure outlined previously. From the test results, it can be seen from Figures 14.30 and 14.31 that PI-Gen2 cured at various temperatures passed the stud pull test to both Cu and to itself after 300 hours and 200 hours PCT, respectively. In addition, the adhesion results for PI-Gen2 on 10 μm Cu L/S after 100 hours PCT indicate no delamination when inspected by SEM (see Figure 14.32).

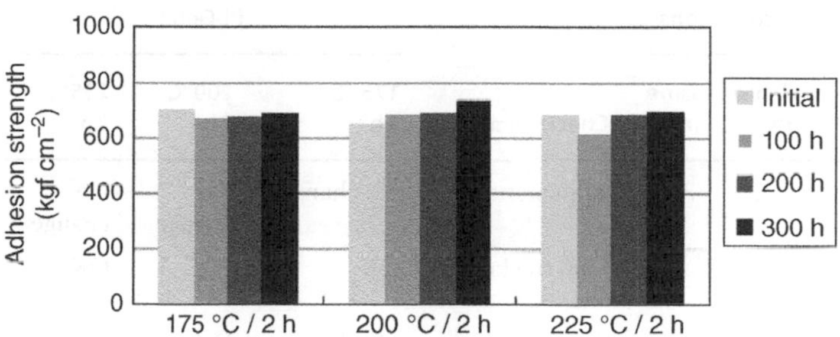

Figure 14.30 PI-Gen2 adhesion stud pull test results to Cu after PCT.

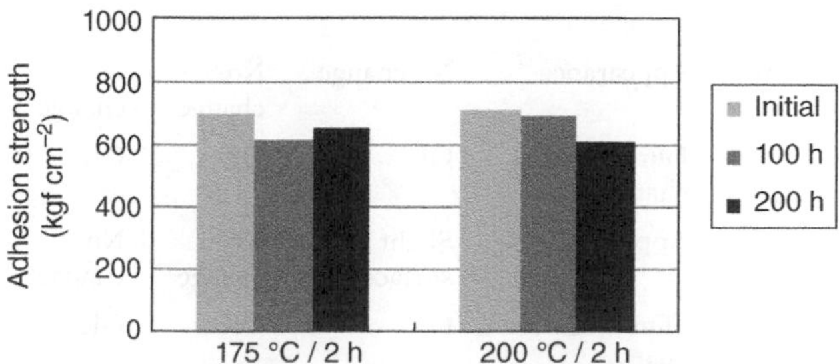

Figure 14.31 Adhesion stud pull test results to PI-Gen2/PI-Gen2 after PCT.

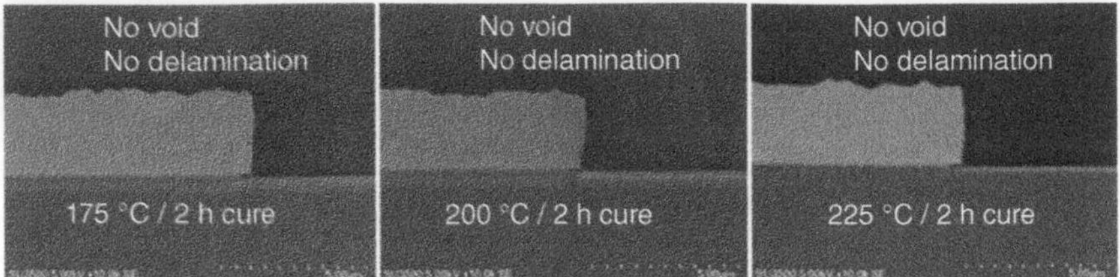

Figure 14.32 PI-Gen2 cross sections of Cu line after 100 hours PCT at varying cure temperatures.

14.16.2 PI-Gen2 Chemical Resistance

PI-Gen2 chemical resistance was tested with the same selection of chemicals as described previously. The results (see Table 14.20 and Figure 14.33) indicated that no changes in either appearance or film thickness were observed with PI-Gen2 films cured at 175, 200, and 225 °C after immersion in the chemicals tested at a given temperature and time.

Table 14.20 PI-Gen2 chemical resistance.

	Treatment conditions			PI-Gen2		
Chemicals	**Temp (°C)**	**Time (min)**	**Check item[a,b]**	**175 °C 2 h**	**200 °C 2 h**	**225 °C 2 h**
Rework solvent (NMP)	25	30	Appearance	No change	No change	No change
			Film thickness change	OK	OK	OK
Resist stripper (Dynastrip 7700)	70	30	Appearance	No change	No change	No change
			Film thickness change	OK	OK	OK
Resist cleaner (OK-73)	25	60	Appearance	No change	No change	No change
			Film thickness change	OK	OK	OK
Flux (WS-600)	245	1	Appearance	Slight rough surface	No change	No change
			Film thickness change	OK	OK	OK

[a] Appearance No change = no cracking, no delamination, and no hazing.
[b] Film thickness change % change within ±10% is acceptable.

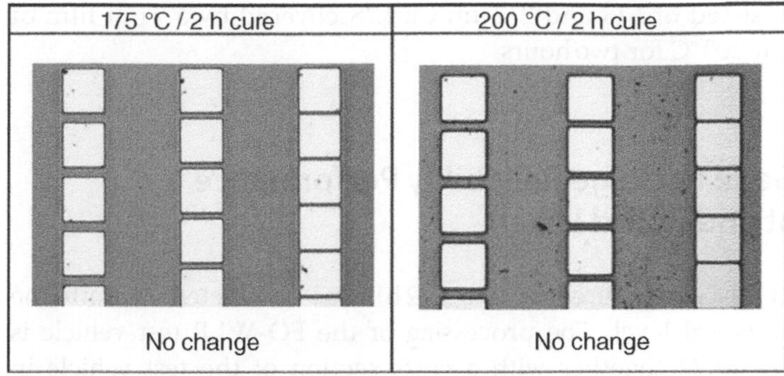

Figure 14.33 PI-Gen2 appearance (100 μm square pattern) after immersion in resist stripper.

14.16.3 PI-Gen2 bHAST

bHAST conducted on PI-Gen2 showed that, with 5/5 and 2/2 μm Cu L/S, no short circuiting, dendrite formation, corrosion, or delamination was observed for up to 200 hours as indicated by little to no change in the insulation resistance (see Figure 14.34). The test procedure has been described previously, and the

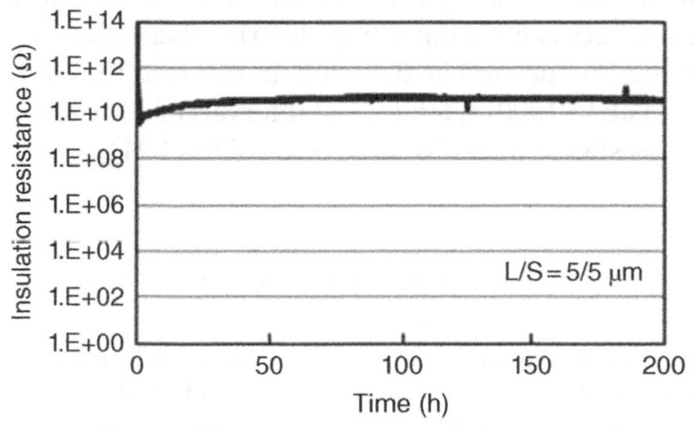

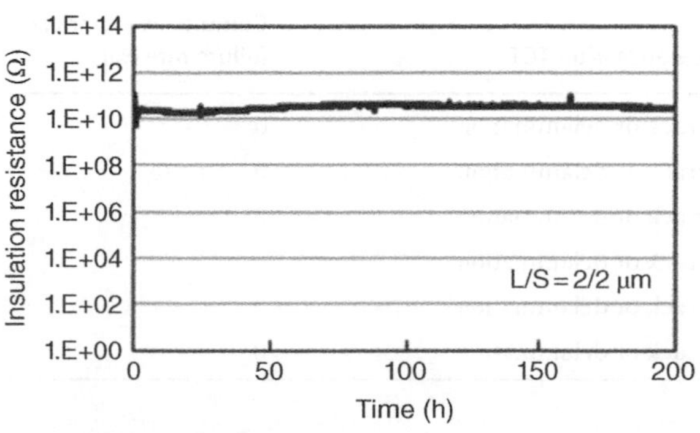

Figure 14.34 PI-Gen2 insulation resistance during bHAST (130 °C/85% RH/3.3 V, L/S = 5/5 and 2/2 μm).

test vehicle consisted of 5/5 and 2/2 μm Cu L/S covered by a 7 μm film of PI-Gen2 cured at 200 °C for two hours.

14.17 PI-Gen2 Package Reliability Performance (Component and Board Level)

TCT testing of PI-Gen2 (cured at 200 °C/2 h) was conducted at both the component and board level. The processing of the FO-WLP test vehicle is shown in Figure 14.22 together with a cross section of the test vehicle in Figure 14.23. Details of the test vehicle structure are outlined in Table 14.12.

For the component-level evaluation, the test vehicle was preconditioned at 85 °C/85% RH for 168 hours followed by 10× N_2 reflow (260 °C max) and then subjected up to 1000 TCT cycles (−65 °C/15 min ↔ 150 °C/15 min). The test vehicle was visually inspected using an optical microscope for any defects at 200 cycle intervals and where no cracking or delamination was observed after 1000 TCT cycles (see Table 14.21).

For the board-level evaluation, a one-time reflow treatment (260 °C max) without moisture soaking was used to mount the vehicle onto the board and then subjected up to 1000 TCT cycles (−65 °C/15 min ↔ 150 °C/15 min). As the test vehicle was designed with a daisy chain pattern to measure the electrical resistance across the dielectric between Cu lines, any electrical failure due to solder cracking or cracks to the Cu lines will be detected. In this test, the electrical resistance was measured up to 1000 cycles TCT at intervals of 200 cycles, and no changes in electrical resistance were observed (see Table 14.21).

Table 14.21 PI-Gen2 reliability test results after TCT at component and board level.

	Reliability test result (component level)	Reliability test result (board level)
TCT cycles	Appearance after TCT	Cumulative electrical failure rate (%)
0	No crack or delamination	0
200	No crack or delamination	0
400	No crack or delamination	0
600	No crack or delamination	0
800	No crack or delamination	0
1000	No crack or delamination	0

14.18 Comparison Between PBO-Gen3 and PI-Gen2

Both PBO-Gen3 and PI-Gen2 meet the challenges previously described at the beginning of the chapter (Table 14.1) for dielectric materials in FO-WLP applications. The challenges are described as follows, and a more detailed comparison between PBO-Gen3 and PI-Gen2 is given in Table 14.22:

1) **Cure Temperature:** PBO-Gen3 and PI-Gen2 meet the target cure temperature of <200 °C. In addition, both materials can also be cured over a wide temperature margin ranging from 175 to 250 °C with acceptable end-use properties.
2) **Thick Film After Cure:** Both materials meet the target maximum cured film thickness of >15 μm, and, while the maximum cured thickness for PI-Gen2 is currently 15 μm, PBO-Gen3 can be cured up to 20 μm. This is a noticeable improvement as, historically, PBO-based positive-acting, aqueous-developing systems are difficult to process above cured film thicknesses of 10–15 μm and require noticeably higher exposure energies and longer

Table 14.22 Comparison between PBO-Gen3 and PI-Gen2.

Challenges	Target specification	PBO-Gen3	PI-Gen2
1. Cure temp.	<200 °C	175–250 °C	175–250 °C
2. Thick film after cure	>15 μm cured film thickness	>15 μm	15 μm
3. Resolution	10 μm via opening (7 μm cured film)	2 μm via opening	10 μm via opening
4. Sidewall shape	70–80° acute angle to substrate	75	74
5. Deep gap formability	Develop 25 μm pre-bake thickness	Max. 25 μm	25 μm possible
6. Mechanical properties	High elongation at low temp. and after PCT/TCT	>50% at ambient 30–40% at −50 °C and no change after PCT/TCT	>40% at ambient 20–30% at −50 °C and no change after PCT
7. Chemical resistance	Resistance to chemicals used in FO-WLP process	Pass	Pass
8. Adhesion to Cu	No delamination at Cu and PI/PBO interface after PCT	Pass after PCT on Cu and PBO/PBO	Pass after PCT on Cu and PI/PI
9. Insulation (bHAST)	No Cu migration during reliability testing	Pass	Pass

development times as compared with PBO-Gen3. In fact, negative-acting, solvent-developing materials, due to the lithography technology used (described previously), are typically easier to design for thicker films by optimizing the photo-package, and, in this respect, development work is in progress to develop a thicker film version of PI-Gen2 by adjusting % solids and photo-package levels as well as optimizing pre-bake conditions.

3) **Sidewall Shape:** After processing, PBO-Gen3 and PI-Gen2 both have a smooth profile, no crowning at the top edge of the sidewall, and a sidewall angle of 70–80° to the substrate surface. In this respect, both materials meet the target for sidewall shape with good process control.
4) **Resolution:** Both materials meet the target resolution of <10 μm via opening. PBO-Gen3 can resolve 2 μm via opening with a 7 μm cured thickness, and PI-Gen2 can resolve 10 μm via openings with a 10 μm cured thickness as well as 5 μm via openings with a 5 μm cured thickness. As indicated previously, due to the photoinitiator system utilized, positive-acting systems generally have the advantage of resolving smaller via diameters as compared with negative-acting systems.
5) **Deep Gap Formability:** PBO-Gen3 can meet the target of developing a 25 μm pre-bake thickness after exposure with no residues for applications where it is necessary to open scribe lines after processing multiple RDL. While PI-Gen2 has not been fully tested to date, being a negative-acting material, it should be possible to develop a 25 μm pre-bake thickness with no exposure and, in particular, where high resolution is not a requirement.
6) **Mechanical Properties:** PBO-Gen3 and PI-Gen2 have a % elongation >40% when measured at ambient and 20–40% when measured at −50 °C, indicating that both materials are ductile enough to provide crack resistance and so avoid performance issues during component and board reliability testing involving TCT. In addition, no significant change in % elongation was observed after PCT with either PBO-Gen3 or PI-Gen2, indicating high moisture resistance with both materials.
7) **Chemical Resistance:** PBO-Gen3 and PI-Gen2, when cured between 175 and 225 °C, are resistant to a selection of chemicals used in RDL and bumping processes, which subsequently avoids performance issues such as cracking, delamination, and thickness changes during reliability testing as well as providing a wider process window in downstream processes.
8) **Adhesion to Cu:** PBO-Gen3 and PI-Gen2 passed the stud pull test for adhesion after PCT at the Cu/dielectric and the dielectric/dielectric interfaces. In addition, no visual voids or delamination was observed on 10 μm Cu L/S after PCT.
9) **Insulation (bHAST):** No short circuits, dendrite formation, delamination, or Cu migration was observed with either PBO-Gen3 or PI-Gen2 after bHAST on test vehicles designed with either 5/5 or 2/2 μm Cu L/S.

14.19 Summary

To meet the demands of current FO-WLP applications, it has been necessary to redesign established high temperature cure negative-acting, solvent-developing PI and positive-acting, aqueous-developing PBO materials for low temperature cure (<200 °C) while, at the same time, addressing challenges in lithography, end-use performance, and reliability at both the component and board level. As described in this chapter, new generation dielectric materials based on both PI (PI-Gen2) and PBO (PBO-Gen3) technologies have been developed that can meet the current requirements of this emerging technology. However, as devices continue to shrink in size while requiring even more I/O connections, there will be an ongoing need to continue development of both PI and PBO materials to meet future performance requirements.

While this chapter has been focused on liquid PI/PBO dielectric materials for FO-WLP applications, there is also potential for these dielectric materials to be used in embedded die packaging applications where dielectric thin films are typically used. Currently, one of the challenges in using dielectric thin films to process additional prefabrication layers (multilayers) is in laser drilling via holes <20 μm, and while there is a current need for photosensitive dielectric thin films to improve resolution, liquid-based PI/PBO dielectric materials that have been adjusted for panel processing could meet lithographic and reliability requirements for this application.

References

1 Coburn, J.C. and Pottiger, M.T. (1996). Thermal curing in polyimide films and coatings. In: *Polyimides Fundamentals and Applications*, Chapter 8 (ed. M.K. Ghosh and K.L. Mittal), 207–247. Dekker.

2 Rubner, R. (2004). Innovation via photosensitive polyimide and poly(benzoxazole) percursors – a review by inventor. *Journal of Photopolymer Science and Technology* 17 (5): 685–691.

3 Makabe, H., Banba, T., Hirano, T. et al. (1997). A novel positive working photosensitive polymer for semiconductor surface coating. *Journal of Photopolymer Science and Technology* 10: 307–312.

4 Tomikawa, M., Suwa, M., Yoshida, S. et al. (2000). Novel positive-type photosensitive polyimide coatings "PW-1000". *Journal of Photopolymer Science and Technology* 13: 357–360.

5 Nunomura, M., Sasaki, M., Ohe, M. et al. (2000). *Hitachi Chemical Technical Report* 34: 25–28.

6 Hanabata, M., Oi, F., and Furuta, A. (1992). Novolak design concept for high performance positive photoresists. *Polymer Engineering & Science* 32 (20): 1494–1499.

7 Süs, O. (1944). *Liebigs Annalen der Chemie* 556: 65.

8 Monroe, B. and Weed, G. (1993). Photoinitiators for Free-Radical-Initiated Photoimaging Systems. *Chemical Reviews* 93: 435–448.

9 Brunnbauer, M., Fürgut, E., Beer, G., et al. (2006). An embedded device technology based on a molded reconfigured wafer. *56th Electronic Components and Technology Conference* (2006), pp. 547–551.

10 Tseng, C. F., Liu, C. S., Wu, C. H., and Yu, D. (2016). Info (wafer level integrated fan-out) technology. *66th Electronic Components and Technology Conference* (2016), pp. 1–6.

11 Töpper, M., Fischer, T., Bader, V. et al. (2011). Ultra low temperature PBO polymer for wafer level packaging application. *International Conference on Electronics Packaging* (2011), pp. 452–455.

12 Windrich, F., Malanin, M., Eichhorn, K.J. et al. (2014). Low-temperature photo-sensitive polyimide processing for use in 3D integration technologies. *Materials Research Society Symposium* 1692: 1–6.

13 Shoji, Y., Masuda, Y., Hashimoto, K. et al. (2016). Development of novel low-temperature curable positive-tone photosensitive dielectric materials with high elongation. *66th Electronic Components and Technology Conference* (2016), pp. 1707–1712.

14 Sasaki, T. (2016). Low temperature curable polymide for advanced package. *Journal of Photopolymer Science and Technology* 29 (3): 379–382.

15 Windrich, F., Kappert, E.J., Malanin, M. et al. (2016). In-situ imidization analysis in microscale thin films of an ester type photosensitive polyimide for microelectronic packaging applications. *European Polymer Journal* 84: 279–291.

16 Enomoto, T., Abe, S., Matsukawa, D. et al. (2017). Recent progress in low temperature curable photosensitive dielectrics. *International Conference on Electronics Packaging* (2017), pp. 498–501.

17 Tomikawa, M., Matsumura, K., Shoji, Y. et al. (2017). Development of photosensitive polyimide B-stage sheet having high Cu migration resistance. *Journal of Photopolymer Science and Technology* 30 (2): 181–185.

18 Mitsukura, K., Abe, S., Toba, M. et al. (2016). Highly reliable Cu wiring layer of 1/1 m line/space using newly designed insulation barrier film. *49th International Symposium on Microelectronics (IMAPS2016)*, pp. 165–170.

19 Toba, M., Mitsukura, K., Ejiri, Y. et al. (2017). Ultra-fine Cu wiring surrounded by electroless-plated Ni: effective structure for high insulation reliable wiring applicable to panel level fabrication. *50th International Symposium on Microelectronics (IMAPS2017)*, pp. 742–746

20 Anzai, N., Fujita, M., and Fujii, A. (2014). Drop test and TCT reliability of buffer coating material for WLCSP. *64th Electronic Components and Technology Conference* (2014), pp. 829–835.

21 Fujita, M., Fujii, A., Shimoda, S., and Kariya, Y. (2015). TCT reliability of organic passivation layer for WLCSP. *48th International Symposium on Microelectronics (IMAPS2015)*, pp. 505–509.
22 Chen, Y.C., Wan, K., Chang, C.A., and Lee, R. (2017). Low temperature curable polyimide film properties and WLP reliability performance with various curing conditions. *67th Electronic Components and Technology Conference 2017*, pp. 2040–2046.

15

Enabling Low Temperature Cure Dielectrics for Advanced Wafer-Level Packaging

Stefan Vanclooster and Dimitri Janssen

Fujifilm Electronic Materials, NV

15.1 Description of Technology

General packaging trends require wafer-level chip-scale packages (WLCSP) and fan-out wafer-level packaging (FO-WLP), often referred to as embedded wafer-level BGA (eWLB) packages, to cope with increasing functionality (e.g. higher number of inputs and outputs [I/Os]), larger and thinner package sizes, thermal and electrical performance, and lower cost of ownership combined with more demanding reliability requirements. Enhanced photosensitive dielectric materials are key building blocks to address the improved functionality and reliability expectations of these advanced packages. This chapter will focus on the development and the integration flow of low temperature photosensitive materials, providing FO-WLP packages the enhanced functional requirements.

A wide range of chemical polymer platforms like epoxies, acrylates, phenolic-based resins, benzocyclobutenes (BCB), silicones, fluorinated polymers, polyimides (PI), and polybenzoxazoles (PBO) are available for integration as dielectric materials in wafer-level packages. Table 15.1 provides a limited overview of the chemical platforms and their key material properties applicable to WLCSP.

Generally PI and PBO materials outperform the other chemical platforms in various material properties, especially in terms of mechanical and thermal properties. However these properties are typically obtained after high temperature cure conditions in the range of 350–380 °C. Among the various FO-WLP technologies, the "chip-first" approach constitutes a prominent group. Chip-first refers to the process where the chip is placed with the device side

Advances in Embedded and Fan-Out Wafer-Level Packaging Technologies, First Edition.
Edited by Beth Keser and Steffen Kröhnert.

Table 15.1 Overview of the key material properties of typical photosensitive chemical polymer platforms (DI, dielectric; Cu, copper).

Polymer	Epoxy	BCB	PI	PBO
Chemical structure	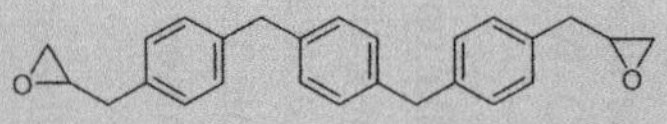	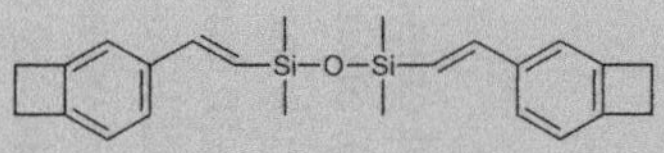 Si–O–Si	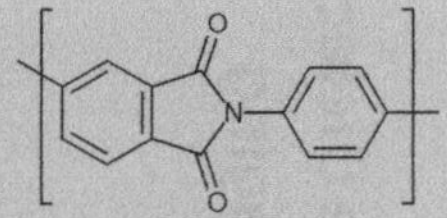 O, N, O	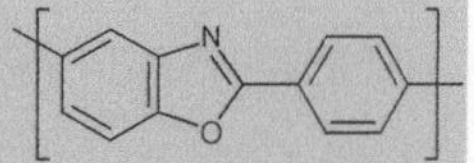 N, O
Type	Thermoset	Thermoset	Thermoplastic	Thermoplastic
Imaging tone	Negative tone	Negative tone	Negative tone	Positive tone
Mechanical properties	+	+	+++	+++
Electrical properties	++	+++	++	++
Chemical compatibility	+	+	+++	++
Adhesion strength (DI/DI or DI/Cu)	++	+	+++	++

facing down on a temporary substrate or carrier prior to building the package around it. In this chip-first FO-WLP process flow, the chip is embedded in an epoxy mold compound (EMC), which limits the thermal budget to cure temperatures below 250 °C. The EMC has a typical glass transition temperature (T_g) below 200 °C, and the change in thermal expansion coefficient results in an unpredictable wafer warpage when heated above 200 °C. To a lesser extent, the EMC may also start to degrade at higher temperatures, even under inert atmosphere (e.g. N_2).

The first generation of eWLB packages typically integrated epoxy as dielectric material because these materials are cured at low temperatures in the range of 200 °C [1]. To enable higher reliability performance of eWLB packages, dielectric materials with improved mechanical properties, higher decomposition temperatures, and improved chemical resistance are required while maintaining the low temperature bake step. In addition these new dielectric products need to comply with the ever more stringent environmental and safety legislations like RoHS and REACH directives and with the extensive lists of banned components from IDMs.

From the various chemical platforms, PI or PBO materials are in the best position to meet these enabling requirements if their cure temperature can be reduced below 220 °C and if the outstanding material properties are preserved when cured at these low temperatures. Whether the integration scheme is a chip-first or a chip-last (more commonly referred to as redistribution layer first [RDL-first]) approach has no impact on the material challenges for these dielectrics, although the latter scheme does not strictly require a low temperature cure dielectric. In the RDL-first approach, like the SWIFT® package from Amkor, the high density RDL is built up on a carrier platform using conventional WLP technology, followed by die bonding and encapsulation [2].

The nonstandardized carrier sizes and dimensions also introduce a variety in coating technologies for the advanced dielectric material like spin coating, slot die/slit coating, or even dry film lamination. The next section will discuss a possible product development pathway to obtain low temperature cure PI materials with enabling material properties.

15.2 Material Challenges for FO-WLP

The dielectric material is a major component of FO-WLP packages and has therefore a large influence on the final package reliability performance. Current FO-WLP packages that are manufactured in high volume typically use two dielectric layers and one copper redistribution layer (Cu RDL) with a cured film thickness between 5 and 12 μm. Advanced FO-WLP packages may apply up to four dielectric layers or even dual side redistribution for three-dimensional (3D) package-on-package (PoP) stacking.

The key material requirements for these enabling photosensitive dielectric materials are:

- Outstanding mechanical properties: Young's modulus >2 GPa, ultimate tensile strength >150 MPa, and elongation at break >30%.
- Thermal material properties:
  - Glass transition temperature at approximately or above the solder ball reflow temperature.
  - Coefficient of thermal expansion (CTE) below <65 ppm °C^{-1} to minimize warpage.
  - Decomposition temperature >400 °C.
- High chemical resistance against solvents, acids, bases, and solder fluxes.
- Good adhesion to copper and no copper migration during reliability tests.
- Low film shrink upon cure to reduce wafer warpage.
- Final cure temperature below 220 °C and target cure temperature below 200 °C.
- Limited outgassing above 300 °C.
- Lithographic performance:
  - High via resolution, high cured aspect ratio, high photospeed, robust process window, and sloped profile.
- Good electrical properties: low dielectric constant and dissipation factor over frequency range from several kHz up to hundreds of GHz.
- Compliance with current environmental, health, and safety (EHS) legislations (RoHS, REACH, etc.).

These material challenges are similar for both chip-first and chip-last approaches though the former additionally requires a higher planarization capability to cope with the typically higher topography during copper wiring and the associated chip-to-epoxy height offset. Chip-to-epoxy height offset is the typical protrusion of the chip above the EMC, introducing additional topography as shown in the schematic in Table 15.2. Chip-to-epoxy height offset is inherent to chip-first technologies because the die are placed active side down onto double-sided adhesive tape and then molded.

Despite other rather obvious but nevertheless challenging constraints such as cost, purity, and stability, the chemical platform is preferably adjustable to various coating technologies used in the FO-WLP process flows as shown in Table 15.2.

Table 15.2 Material requirements comparison between "chip-first" and "chip-last" (RDL-first) integration schemes.

	FO-WLP process	
	Chip-first approach	RDL-first Chip-last approach
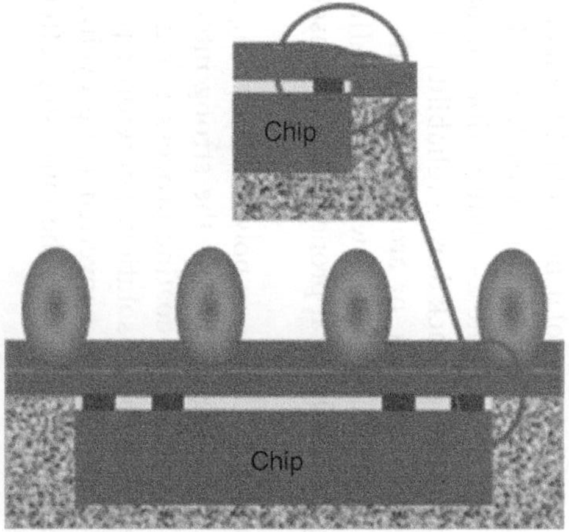 Chip-to-epoxy height offset Dielectric needs to smooth out the topography	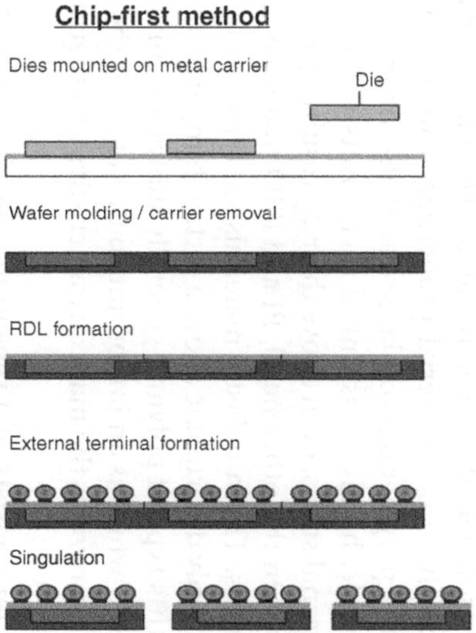 Chip-first method Dies mounted on metal carrier Die Wafer molding / carrier removal RDL formation External terminal formation Singulation	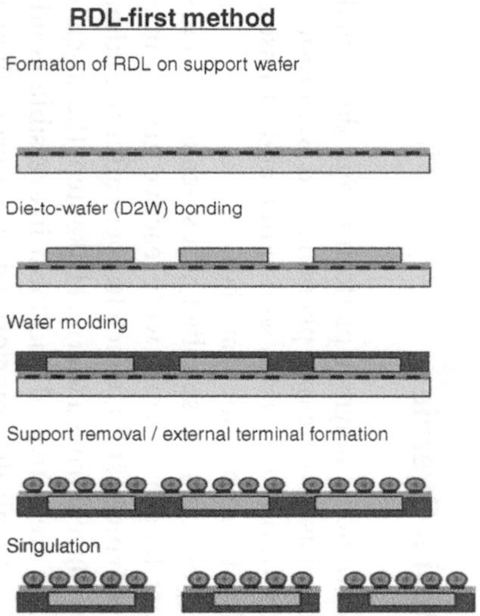 RDL-first method Formaton of RDL on support wafer Die-to-wafer (D2W) bonding Wafer molding Support removal / external terminal formation Singulation
Degree of planarization required	High	Moderate
Coating technology	Spin coating Slot die/slit coating	Slot die/slit coating Spin coating Dry film lamination

15.3 Material Overview

Several semiconductor polymer platforms were developed for RDL applications. Materials based on epoxy, phenolic, and BCB resins can commonly be classified as thermoset materials, while PBO- and PI-based materials are usually classified as thermoplastic materials.

Thermoset polymers are supplied as a so-called precursor consisting of a relatively short length polymer with reactive groups in an organic casting solvent. The polymerization, which happens after the lithographic patterning process, takes place during a thermal bake step, the so-called cure, where the reactive groups are activated and form irreversible bonds. In this step the prepolymer is chemically transformed into a cross-linked 3D polymer network, which fixes the polymer chains in a certain position ("setting") and provides the isotropic dielectric behavior. The glass transition temperature (and many of the other material properties) will not only depend on the structure of the polymer but to a large extent also on the degree of cross-linking. For example, highly cross-linked thermosets will be difficult to deform under an external stress, even above their T_g, resulting in a rather low elongation at break. On the other hand, PI and PBO materials are mostly thermoplastic materials. Thermoplastic materials are typically long-chain polymers that are reversibly deformable above a certain temperature. Not restrained by cross-links, this type of polymer generally shows a significantly higher elongation at break, which is an important contributor for the package reliability [3]. Table 15.3 provides the main characteristics of a thermoset versus a thermoplastic polymer.

A PI polymer consists of a repeating [R–CO–N–CO–R] unit group typically with a molecular weight (MW) in the range of 20–100 K (Da). The MW and polydispersity index can be measured by gel permeation chromatography [4]. The presence of strong polar bonds and the fact that the lone electron pair on the nitrogen is conjugated with the carbonyl group (C=O) enable strong interchain dipolar and π–π interactions, making PI resistant to chemical agents and moisture attack during reliability stressing. The type of hydrocarbon units (R as aliphatic or aromatic) and the presence of other functional groups (Cl, F, NO_2, OCH_3) will further influence the material properties including its mechanical properties and consequently their final application.

In the case of PI and PBO, the long polymer chain length will have a very low solubility due to the high MW and the strong interchain interactions (dipolar and π–π). However, for coating processes (e.g. spin coating), the polymer needs to be available as a solution. Therefore a precursor polymer is typically used that is chemically converted (ring cyclization) to the final PI or PBO polymer structure during the cure step after the coating and lithography steps.

Table 15.3 Overview of the main characteristics of thermoplastic and thermoset as thin film dielectric.

	Thermoplastic polymer 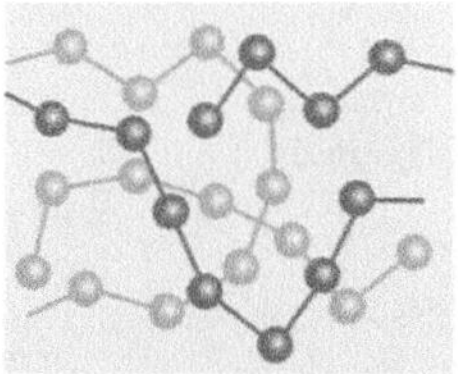	Thermoset polymer 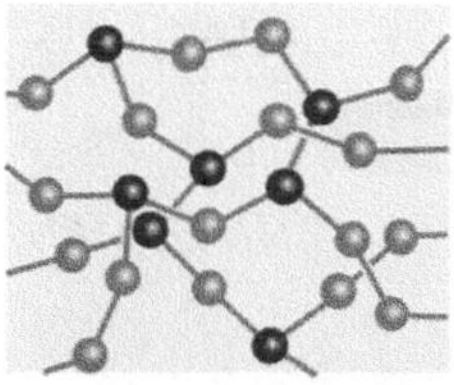
Under mechanical stress	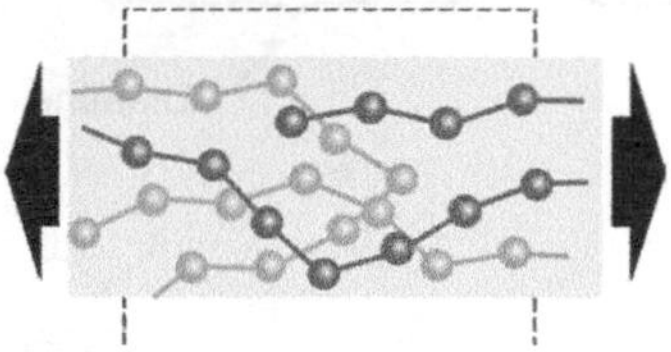	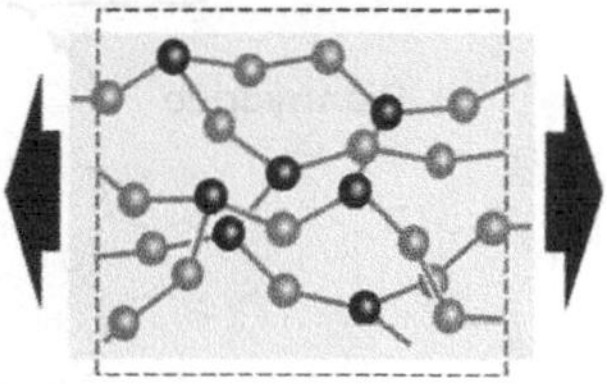
Advantages	High elongation at break Retains properties when heated above T_g	Low cure temperature High T_g
Disadvantages	Higher cure temperature	Low elongation at break Mechanical properties degrade when heated above T_g (less suitable for multiple heating cycles)

In order to render the precursor polymer photosensitive, further modifications are required to the polymer structure by attaching reactive groups that allow a change in polymer solubility (by cross-linking or a polarity change) upon reaction with an activated photoinitiator.

In the following sections, we will mainly focus on the PI chemical platform for low temperature cure dielectrics for advanced FO-WLP packages. A photosensitive PI precursor is generally prepared by the chemical reaction of an aromatic/aliphatic diamine with an aromatic/aliphatic dianhydride or multicarboxylic acid/ester in a dipolar aprotic solvent (see Figure 15.1). The polyamic acid or ester precursor is formed by the nucleophilic attack of the amino group on the carbonyl carbon of the anhydride unit.

A subtle variance in the selection of the diamine and dianhydride monomer structure will have a large effect on the mechanical and other material properties of the final PI. The selected monomers' rigidity, overview given in Figure 15.2, will determine the final polymer morphology, either like a rigid rod or a flexible spaghetti as schematically presented in Figure 15.3. A rigid

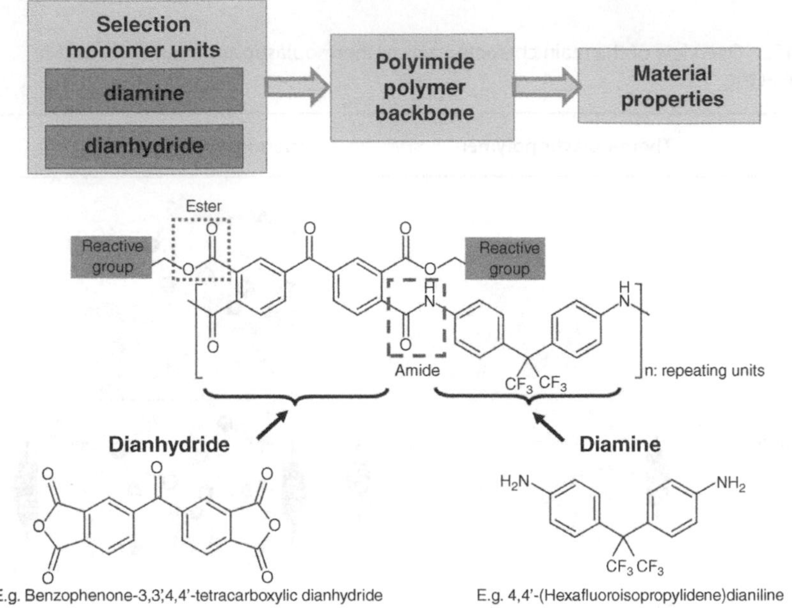

Figure 15.1 Schematic presentation of building blocks for a photosensitive polyimide precursor.

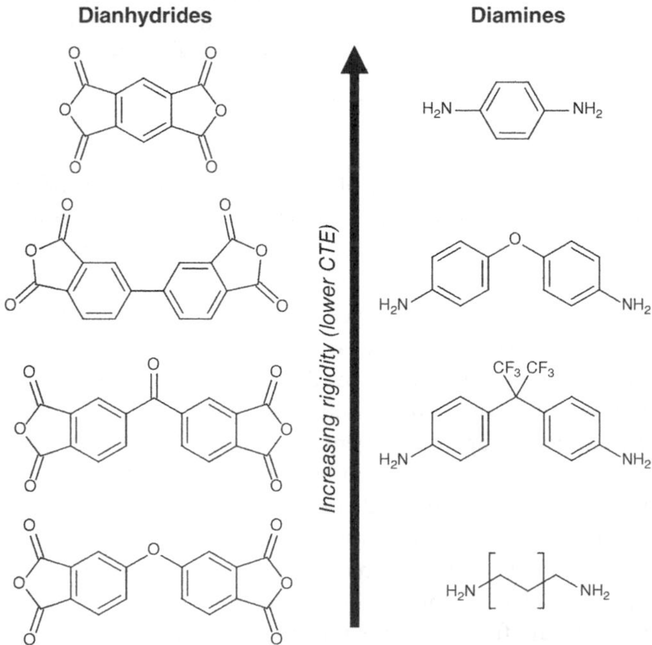

Figure 15.2 Typical monomers used for the synthesis of polyimide precursors.

Figure 15.3 Schematic presentation of a rigid stiff and flexible polymer backbone.

Type	Cure temp	Mechanical properties	Glass transition (T_g)	Warpage (CTE)
Rigid	High	High	High	Low
Flexible	Low	Low	Low	High
Semirigid	Medium	Medium	Medium	Medium

Figure 15.4 Correlation table between the cure temperature (high >300 °C, low <300 °C), mechanical (high/low E, UTS, and E_b), and material properties (high/low CTE, T_g) as a function of the polymer backbone.

polymer backbone requires a significantly higher cure temperature to obtain the same level of imidization as a flexible polymer backbone because sterically it becomes increasingly difficult for the functional groups to react and cyclize. On the other hand, a rigid polymer backbone will usually exhibit outstanding mechanical properties compared with the flexible polymers. Enabling dielectric materials therefore requires a delicate balance between rigidity and flexibility and thus between lower cure temperature and improved physical properties as shown in Figure 15.4. A semirigid polymer backbone is therefore beneficial and is commonly used in low temperature cure PI materials.

After lithographic patterning of the PI precursor, the PI film will be thermally heated or baked. High thermal activation energy is required for the nucleophilic ring closure. Depending on the rigidity of the polymer backbone, a cure temperature of 350 °C or even higher will be required to complete ring cyclization or "imidization."

To achieve a low temperature cure PI, ways need to be found to force the ring closure at a much lower temperature. Several possibilities have been proposed in the literature and applied in industry. First of all, different curing techniques have been suggested to reduce the cure temperature such as microwave irradiation [5]. This approach however requires new, specialized equipment while most assembly lines prefer to utilize their existing infrastructure of thermal cure ovens. Another development path is to incorporate pre-imidized PI units into the polymer backbone to avoid the need for an imidization reaction [6]. The strong interchain interactions and close stacking between the polymer chains are loosened by introducing bulky units in or on the backbone to keep the polymer soluble.

An obvious and important way that was mentioned earlier is by optimizing the flexibility of the backbone to facilitate imidization at lower cure temperature.

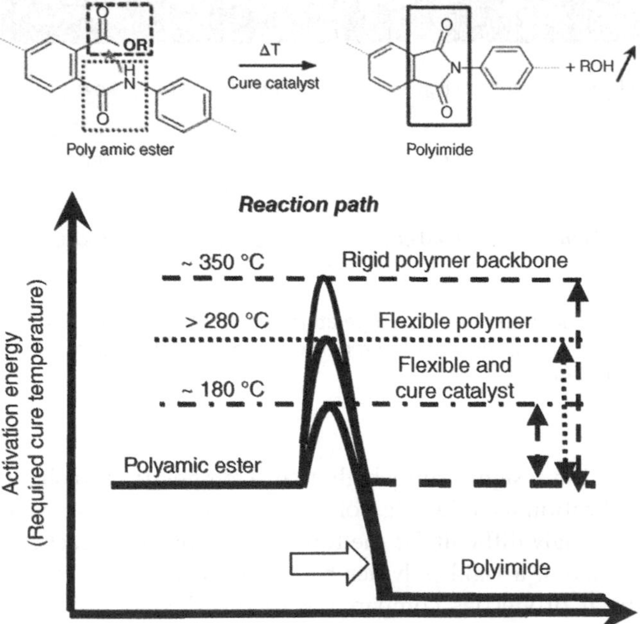

Figure 15.5 Chemical reaction path and corresponding activation energy for imide ring closure.

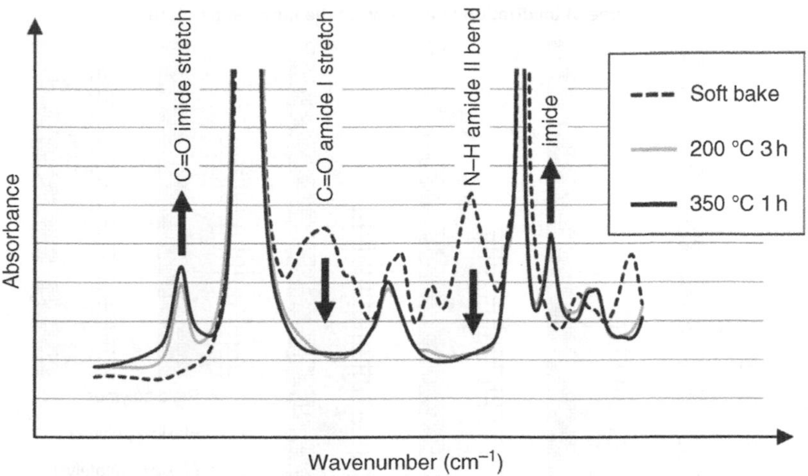

Figure 15.6 FTIR analysis indicates that the polyimide precursor is fully imidized after a cure cycle of 200 °C.

Chemical catalysis is another customer-friendly approach to lower the activation energy enhancing the imide ring closure. A chemical catalyst is typically an acid or a base that is thermally activated at the beginning of the cure cycle and significantly lowers the final cure temperature. The required activation energy needed to transform the polyamic ester into the cyclic imide as a function of polymer backbone rigidity and chemical catalysis is schematically presented in Figure 15.5.

Fourier transform infrared spectroscopy (FTIR) is a practical tool to determine and track the chemical composition and changes in a material. In FTIR, the sample is exposed to infrared (IR) light, and the transmitted or reflected light is measured. If the IR wavelength matches the energy of a chemical bond (and the bond has a net dipole), that bond will absorb the IR light. Chemical changes can thus be tracked by following the absorption changes of the characteristic bonds under study. In this case, for example, the degree of imidization as a function of the cure temperature can be determined by investigation of the imide peaks of the spectra at different cure temperatures. Figure 15.6 illustrates the changes in absorption peaks for a dedicated design of polymer backbone and cure catalyst, resulting in a fully imidized PI at 200 °C cure temperature. The degree of imidization of the enabled PI as a function of the cure temperature is given in Figure 15.7. Consequently, enabled PI platforms are suitable for low or extreme temperature cure applications similar to epoxy, acrylate, or BCB materials.

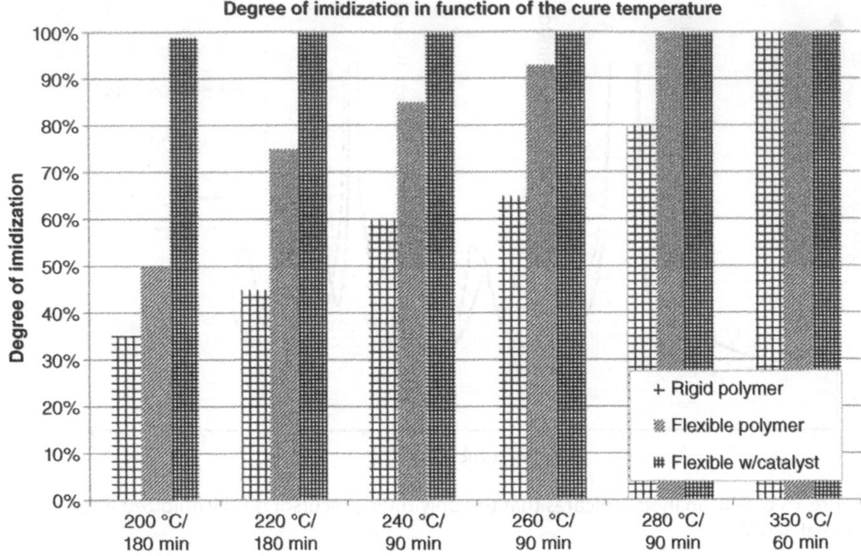

Figure 15.7 Degree of imidization as a function of cure temperature for a rigid polymer, a flexible polymer, and a specially designed polymer formulation.

15.4 Process Flow

The integration of low temperature cure PI in an FO-WLP production process flow with either chip-first approach or RDL-first approach requires some process optimization in the coating, exposure, development, and cure process steps. Figure 15.8 provides the generic process steps. For photosensitive materials, the imaging type (positive/negative) is a consequence of the chemical system and an important aspect in a lithographic process: a positive tone material will become soluble in developer upon light exposure, while a negative tone material will become insoluble in developer upon light exposure (thus generative of the

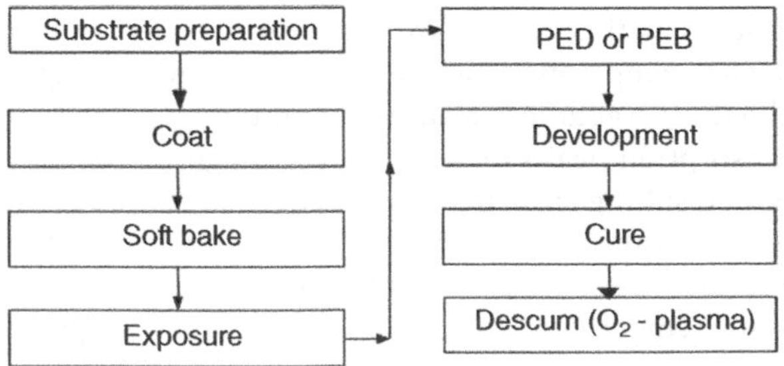

Figure 15.8 Typical process flow for a negative tone low temperature cure polyimide.

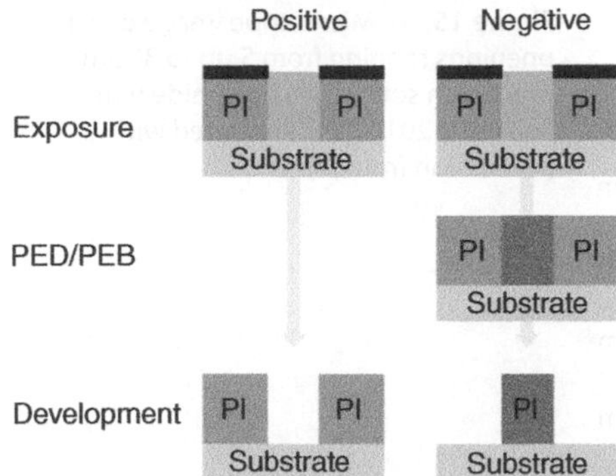

Figure 15.9 Comparison between positive tone and negative tone photosensitive low temperature cure dielectrics.

opposite image of the mask). Low temperature cure photosensitive polyimide (PS-PI) materials are typically negative tone PI, while low temperature cure materials like PBO, epoxy, or BCB can be either positive or negative tone (see Figure 15.9). The negative tone PI materials generally use an organic solvent system for development, while the positive tone materials use an aqueous solution of tetramethylammonium hydroxide (TMAH) as developer.

The lithographic process of a low cure PS-PI starts with substrate preparation, which is important to remove organic contamination from the surfaces and bring the substrate surface into the preferred chemical state to enhance the chemical interaction (i.e. adhesion) between substrate and PI material. The material (viscous liquid) is then applied onto the substrate by a spin-coating process followed by a soft bake process to partially remove the solvent. After determination of the film thickness as a function of coating and bake conditions, the optimum spin-speed and spin-time conditions are selected to achieve good topography coverage with high wafer uniformity over a large soft bake temperature range (±10 °C) to compensate for possible reconstituted substrate warpage. These coated substrates then head for exposure. The lithographic recipes to define the minimum resolution capability are defined by setting up a standard focus exposure matrix (FEM), followed by optical inspection. Good resolution of round vias can be obtained with dimensions between 5 and 25 μm in a 12 μm thick soft baked film (see Figure 15.10) [1]. The exposure process conditions indicate large process latitude as demonstrated by the Bossung plot in Figure 15.11 with the (220 mJ cm^{-2}) line as the best isofocal dose [1]. A linearity check is performed under the best process conditions to determine the mask bias between dimension on mask (DOM) and dimension on substrate (DOS).

After exposure and post-exposure delay or post-exposure bake, the exposed substrates are transferred to a developer module where an organic solvent system is used to develop the low temperature cure PI. An atomized spray or a

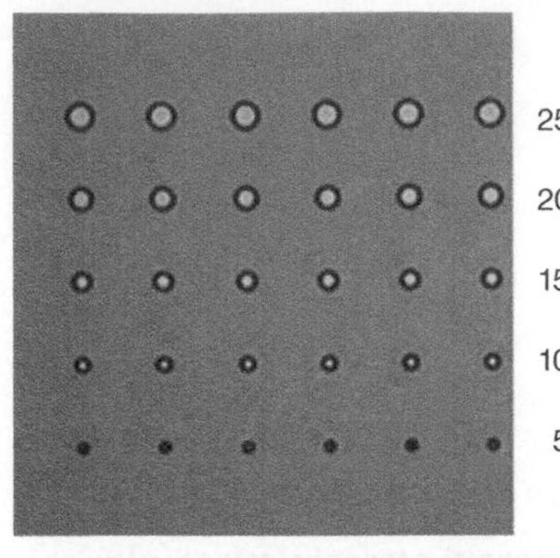

Figure 15.10 Microscope image of via openings ranging from 5 up to 30 μm in a 12 μm soft baked polyimide film. *Source:* © 2014 IEEE. Reprinted with permission from [1].

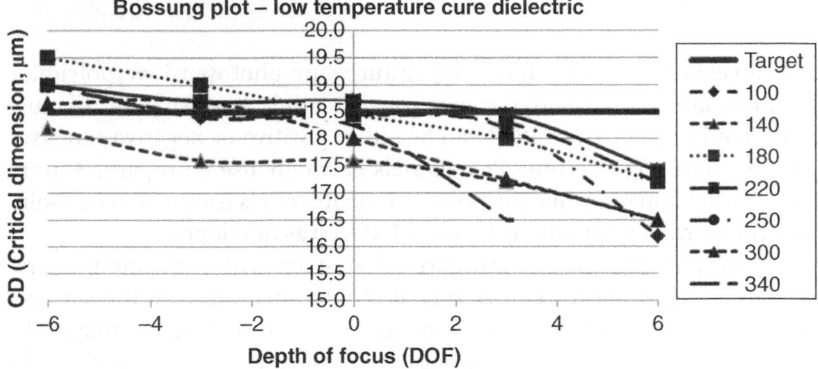

Figure 15.11 Bossung plot – low temperature cure dielectric. *Source:* © 2014 IEEE. Reprinted with permission from [1].

multiple spray and puddle development process is applicable although the former process provides the highest resolution with minimum footing. The development time is not only a function of the soft baked film thickness but also of the soft bake temperature and time influence in a second-order bulk PI dissolution speed. Cyclopentanone and propylene glycol monomethyl ether acetate (PGMEA) are typically the developer and rinse system for low temperature cure PI systems. After development a post-development bake at 100 °C is optional to remove the excess of developer from the film.

The final cure process is performed in a furnace or oven during which the polymer will convert to the final PI. The imidization reaction is a polycondensation reaction with the formation of light alcohol components. These by-products are removed together with any residual casting solvent by gentle nitrogen (N_2) purge in the oven. The oxygen (O_2) level in the oven and the N_2 carrier gas need to be as low as possible and preferably below 100 ppm. Higher oxygen levels in the furnace can oxidize the dielectric material, potentially

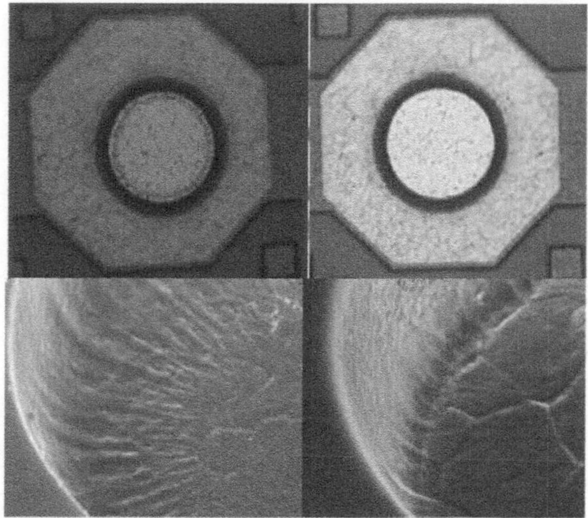

Figure 15.12 Optical microscope inspection and scanning electron microscopy of a via opening in a cured polyimide film on an aluminum bond pad before and after RIE plasma descum. *Source:* © 2014 IEEE. Reprinted with permission from [1].

resulting in reduced mechanical properties, and a darker PI film will be observed. During this final cure, the strong adhesion reaction of the PI film to the underlying substrates like EMC, copper RDL lines, cured PI film, or aluminum bond pads is established. The final shrinkage from an exposed soft baked film to cured film for a cure temperature <250 °C is less than 25%, generating a low residual stress film.

A descum process is finally performed to remove some residual PI ("footing") in the small feature sizes. These PI residues are caused by UV light reflection during exposure on reflective areas of the substrate (e.g. aluminum bond pads). This footing extends 1–3 μm into the opening and reduces the available area for electrical contact. A reactive ion etching (RIE) reactor is used for an O_2-anisotropic descum process, which removes the residues and increases via opening diameter and profiles. Then, the plasma removes any residual adhesion promoter of the PI formulation and cleans the metal pads' surface, enabling a low contact resistance with the metal RDL (Figure 15.12).

15.5 Material Properties

Low temperature cure PIs have been successfully integrated in FO-WLP high volume production process flows, mainly in the chip-first approach, but also in fewer cases in the RDL-first approach. The general process flow for the

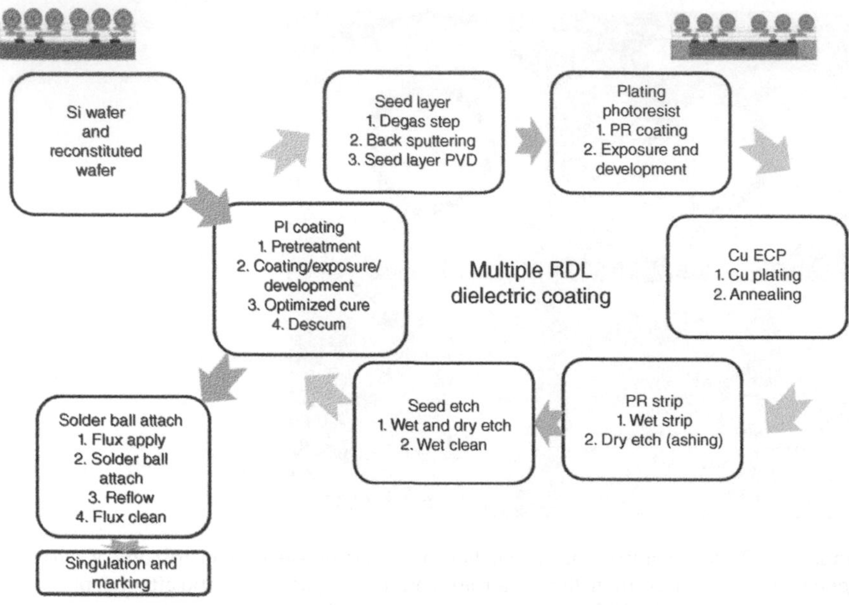

Figure 15.13 Schematic integration flow of photosensitive polyimide material in FO-WLP process flow.

integration of photosensitive low temperature cure PI in the FO-WLP process is given in Figure 15.13. Typically, two dielectric layers are used for FO-WLP packages, but advanced applications require up to four RDL dielectric layers or even redistribution layers on both sides of the package.

The high level FO-WLP process flow is very similar for the different dielectric material classes. An advantage of the low temperature cure PI is the limited outgassing level after final cure. This is important for the copper plating seed layer deposition process: during physical vapor deposition (PVD), the substrates can heat up significantly, and extensive outgassing could contaminate the substrate surfaces, potentially leading to reduced adhesion and delamination. Thermogravimetric analyses (TGA) show a 2 and 5 wt% loss temperature, respectively, above 300 and 340 °C for a cured PI film cured at 230 °C and even a decomposition temperature above 500 °C (Figure 15.14). This reduced outgassing of the dielectric material during PVD will improve the seed and dielectric layer adhesion strength and reduce the risk of delamination. The high 2 and 5 wt% loss temperatures also favor shortening the outgassing time during the PVD seed layer sputter process, which allows for faster cycle times.

Regarding film deposition techniques, low temperature cure PI chemical platforms require relatively little adjustment to cover both liquid spin-coating processes and slot die/slit coating processes. The viscosity value usually needs to be tuned with additional casting solvents and wetting additives for improved

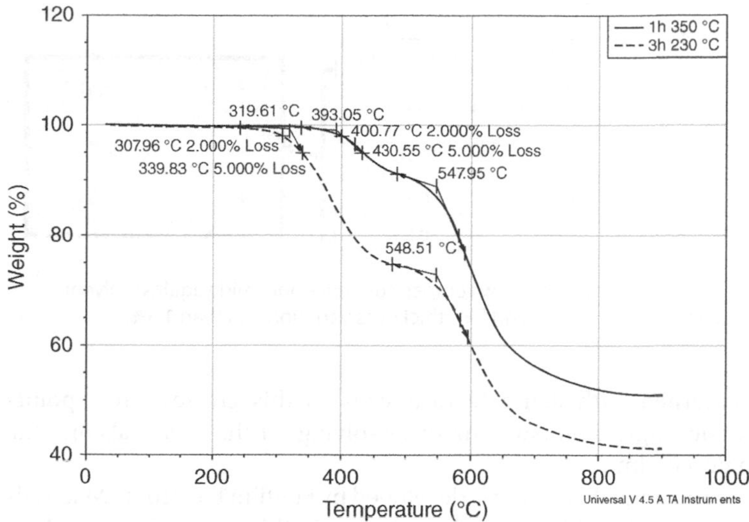

Figure 15.14 TGA analyses of polyimide film cured at 230 °C (dotted line) and 350 °C (solid line).

flowability. Also the soft bake temperature profile normally requires some optimization for gentle outgassing of the abundant casting solvent to avoid possible voids or defects in the soft baked film.

The commonly used WLP exposure tools (e.g. Ultratech AP300, Canon FPA-5500 series, Rudolph Instruments JetStep series, SÜSS MicroTec MA300) are also suitable for the patterning of the low temperature cure PI. Broadband (BB) exposure is used for the larger feature sizes with high wafer throughput, while i-line exposure is typically used for the finer via structures in the first RDL dielectric layer. Besides stepper or mask aligner exposure, these dielectric materials are also suitable for laser direct imaging processing. Laser light (from a YAG solid-state laser with a wavelength of 355 or 405 nm) can expose and pattern the soft baked film. A variety of development equipment and furnaces have been successfully applied for the low temperature cure PI.

Easy integration of a dielectric material in the FO-WLP process flows requires outstanding material compatibility with acids, bases, and solvents that are used in various process steps (e.g. plating, seed etch, etc.). In order to evaluate chemical compatibility, a common method is to prepare coupons of, e.g. a Si wafer coated with the material of interest and place these coupons in a wide range of chemicals (solvents, strippers, acids, bases) for a certain time and at a certain temperature (often elevated). The coupons are then inspected for changes like discoloration, delamination, cracks, etc., and the film thickness (change) is determined. To make the test even more realistic, the dielectric film

Film thickness loss		MS 3001 (amine/DMSO)	
		55 °C/30 min	75 °C/15 min
E07	230 °C/180 m	−0.1%	0.7%
E67	200 °C/180 m	0.6%	1.2%
	200 °C/120 m	0.7%	1.2%
	230 °C/120 m	0.1%	0.3%
	350 °C/60 m	0.1%	0.2%

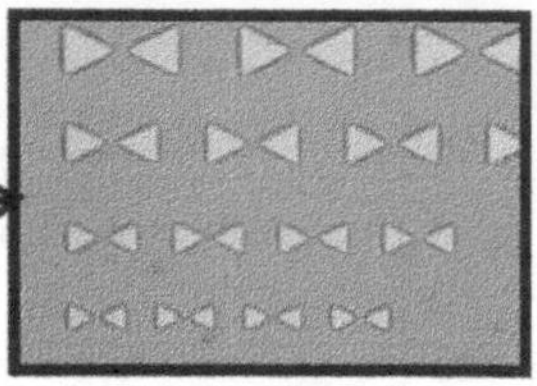

Figure 15.15 Chemical resistances of low temperature cure polyimide against solvent strong polyimide stripper: no cracks and film thickness variation less than 1.5%.

is usually patterned with dense feature sizes as this creates stress points and allows easier ingression (swelling or dissolving) of the chemicals into the dielectric material film.

Low temperature cure PI materials developed by Fujifilm Electronic Materials named LTC 9300 E07 and E67 have been tested in this way and do not show any discoloration, delamination, or cracks around the features. Also the film thickness change stays less than 1.5%. Figure 15.15 shows these results for a rather aggressive stripper based on a mixture of amines and dimethyl sulfoxide (DMSO). The outstanding chemical resistance is a consequence of the strong PI interchain interactions (dipolar and $\pi-\pi$), which lead to a close stacking and packing of the PI polymer backbones, thereby hindering penetration of solvent molecules into the dielectric.

A key set of material properties for a dielectric material are the mechanical properties. Young's modulus (E), ultimate tensile strength (UTS), and elongation at break are determined by tensile testing. In a tensile test, a freestanding film specimen of the material under study is clamped and pulled while measuring the extension and applied force from which the parameters mentioned above can be determined. The freestanding film specimen is typically prepared by coating and curing a dielectric film on a wafer with a sacrificial layer like SiO_2, TiW, or copper. After cure, the wafer is placed in an etchant to remove the sacrificial layer, thereby releasing the dielectric film from the substrate. The film can then be cut in test strips of the desired size. Alternatively for a photosensitive dielectric, strips (straight or dog bone shaped) can be patterned lithographically in the film, which usually leads to better test results as the edges are smoother compared with a cut film (microscopic tears are stress points where the specimen can break prematurely, leading to low repeatable results). These sample preparation methods allow easy evaluation of various process parameters of the cure such as ramp-up rate, final cure temperature and time, and the cool-down rate.

Tensile tests are typically performed at room temperature (25 °C) but can also be done at reduced temperature (−55 °C) with special accessories to simulate

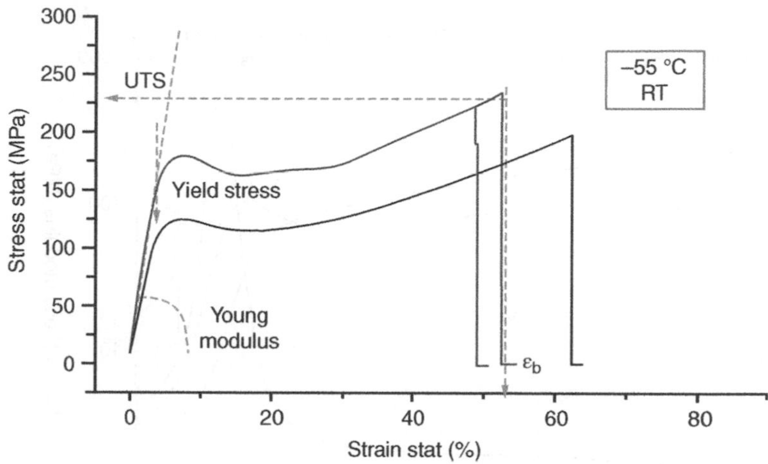

Figure 15.16 Stress–strain curve of a 230 °C cured polyimide film.

the material behavior during thermal cycling test (TCT). Figure 15.16 shows the stress–strain cure of a low temperature cure PI material (LTC 9300 series cured at 230 °C) at room temperature and at −55 °C.

Other important material properties like glass transition temperature (T_g) and CTE are measured by dynamic mechanical analysis (DMA) and thermomechanical analysis (TMA) (see Figure 15.17). The glass transition temperature is the temperature (range) at which the material (reversibly) transitions from a glassy (hard) to a rubbery (soft) state (related to motion of segments of the polymer backbone), while the CTE is quite simply the rate of change in dimensions as temperature changes.

In TMA the elongation of a film specimen is measured while a certain force is applied, and the temperature is gradually changed. At T_g, the elongation will rapidly increase as the material changes to a rubbery state, so from this measurement, both the CTE and T_g can be determined.

DMA is quite similar in setup, but instead of applying a stationary force, the external force is oscillated, and the material response is recorded while temperature changes. When the material under study changes to a rubbery state, the response will change in amplitude and frequency from which the modulus (loss and storage) and the loss tangent δ can be calculated. Again the temperature at which this change occurs is indicative of the glass transition.

As the (low temperature cure) dielectric material serves a specific function in the final FO-WLP package where it is subjected to electrical fields, the electrical material properties are also of key importance. These electrical properties need to be characterized over a wide frequency range since device applications range from DC to EHF. Several electrical measurement techniques are needed to cover the frequency range of interest from several kHz to hundreds of GHz.

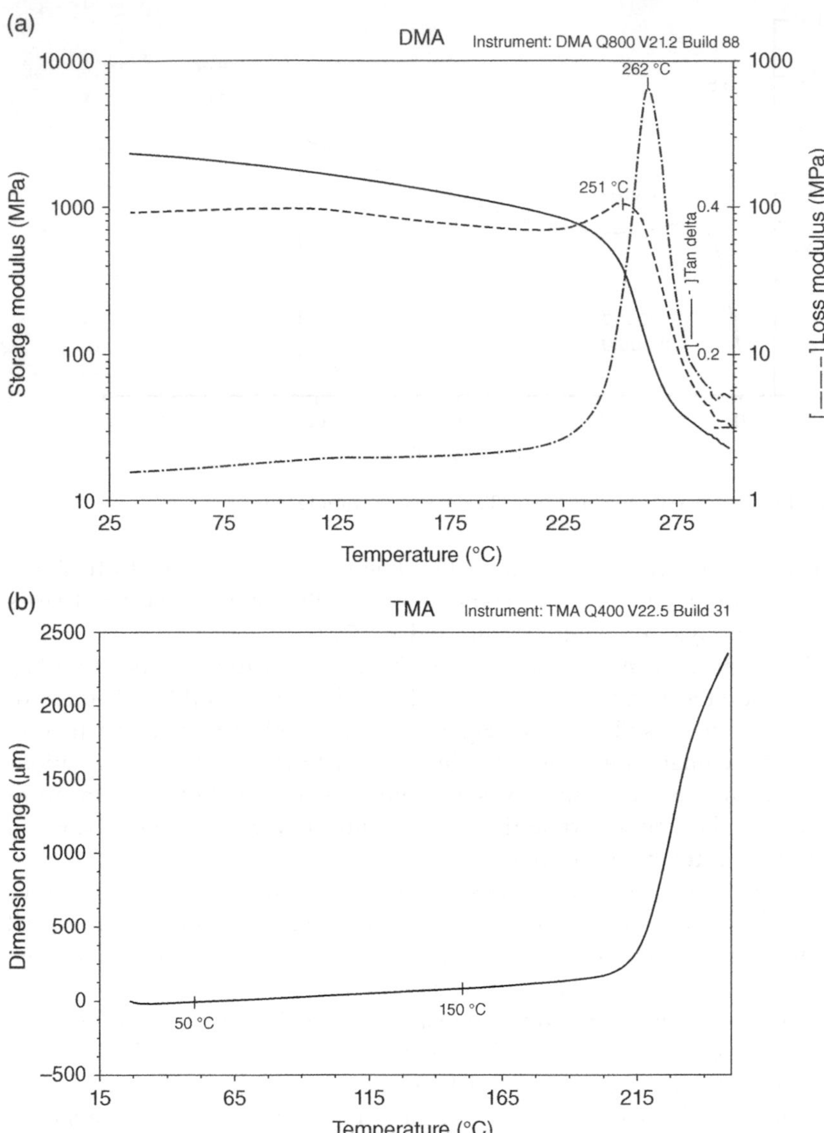

Figure 15.17 (a) DMA and (b) TMA curves of a 230 °C cured polyimide film.

Mercury probe analysis is commonly used at low frequency from 3 kHz up to 3 MHz. Transmission line models in combination with a vector network analyzer (VNA) are used for medium frequency ranges from several MHz up to 3 GHz, while ring resonator models allow measurement of high frequencies ranging from 3 up to 110 GHz [7].

These techniques have been applied to Fujifilm's low temperature cure dielectric material, and Figure 15.18 summarizes the results.

The dielectric constant of low temperature cure PI varies from 3.4 to 2.95 over the full frequency range, and the dissipation factor is below 0.01. The decreasing dielectric constant with increasing frequency is attributed to reduced polarization of polarizable groups and dipoles (permanent and non-permanent) on the PI backbone as these can no longer follow the rapidly changing electric field at these high frequencies.

If the electrical measurement setup is mounted inside an environmental chamber, the influence of temperature and humidity can be investigated. The electrical characterization as a function of temperature (−40 to 125 °C) reveals that the temperature has little impact on the electrical parameters for low temperature cure PI. The hygroscopic nature of PI increases the dielectric constant by 20% under high (80% relative humidity [RH]) humidity conditions [8].

As a summary of the material parameters described above, Table 15.4 provides an overview of low temperature cure PI in comparison with other commercially available low temperature cure materials.

Last, but certainly not least, is the adhesion of dielectric materials: considering the complexity and the multitude of materials and layers in WLP packages, strong adhesion of the dielectric material is critical for achieving outstanding device reliability.

PIs in general have good intrinsic adhesion properties due to the high amount of dipoles on their backbone, which can interact with other materials or substrates. An oxide- or hydroxyl-terminated surface will therefore enhance the adhesion of PI. Other substrates (e.g. metals) might require a treatment (dehydration bake, plasma treatment, primer) to achieve the required adhesion level. Most low temperature cure PI formulations are also optimized with dedicated adhesion promoters to improve the chemical interaction between organic and metallic substrates.

To evaluate adhesion, many test methods are available as many of these have been developed by the paint industry. A very common method is the tape test where an adhesive tape is applied to the coated substrate and removed. If the film stays on the substrate, the adhesive force is larger than that of the tape. To improve the sensitivity of this method, often a crosscut is made in the film, or in the case of photosensitive materials, a checkerboard pattern is printed as delamination is often initiated at the edges. Counting the number of lifted fields from the checkerboard also provides a way to semiquantitatively judge adhesion quality.

Another test method is the scratch method whereby a stylus is moved over a coated surface, and the force is gradually increased. The force at which a scratch or some sort of damage occurs is indicative of the adhesion toward the substrate.

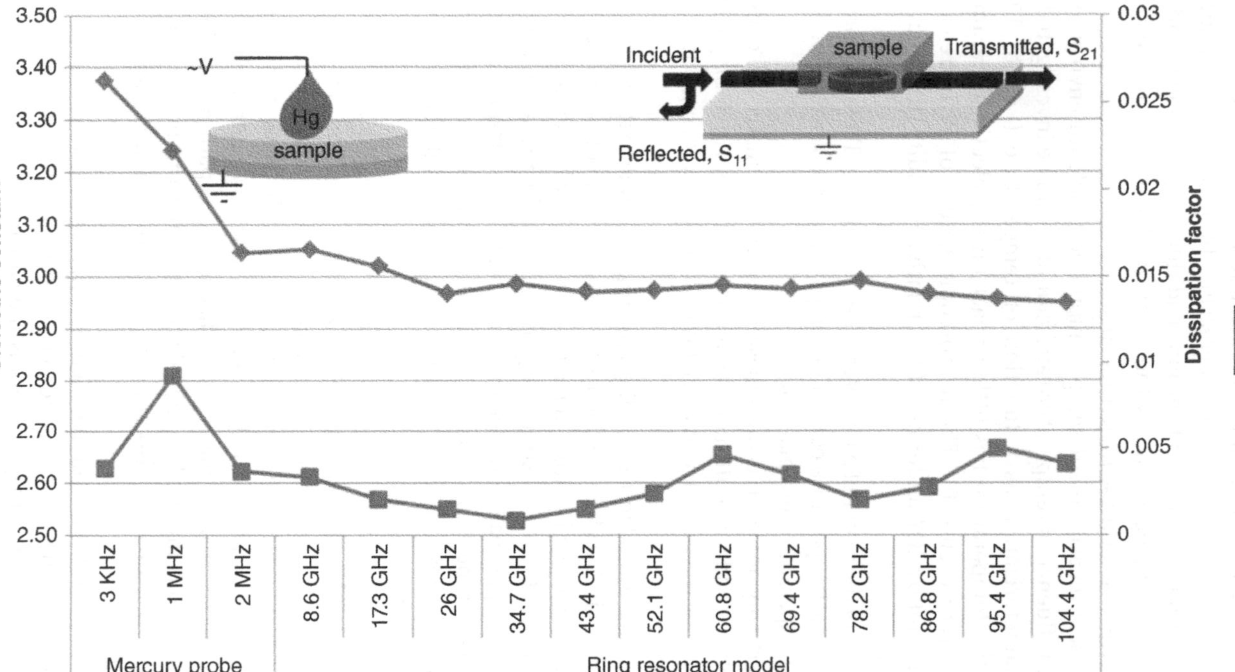

Figure 15.18 Dielectric constant and dissipation factor over the full frequency range for the low temperature cure polyimide material LTC 9300 series.

Table 15.4 Material properties overview of various dielectric materials for FO-WLP in the market.

Supplier		DIM	TARGET	DOW	FFEM	FFEM	HD	JSR	JSR
Product name				Cyclotene 4000	LTC 9320 E07	LTC 9320 E67	HD 8930	WPR 1201	WPR 5100
Photosensitivity				Negative	Negative	Negative	Positive	Negative	Positive
Developer				Organic	Organic	Organic	Aqueous	Aqueous	Aqueous
Base chemistry				Benzocyclobutene	Polyimide	Polyimide	Polybenzoxazole	Nanofilled phenol resin, CA, contains Cl	Nanofilled phenol resin, NQD based
Cure temperature		°C	Low	210–250 °C	200–350 °C	180–350 °C	175–225 °C	190 °C	190 °C
Electrical properties	Dielectric constant		Low	2.7	3,3	3.3	3.1	3.6	3.5
	Loss factor		Low	< 0.01	0.03	0.03		0.03	0.02
	Dielectric strength	$V\,\mu m^{-1}$	High	–	450	450	–	–	–
Mechanical properties	Young's modulus	GPa	High	2.9	>2.5	> 2.5	1.8	2.2	2.5
	Tensile strength	MPa	High	87	>150	> 150	170	90	80
	Elongation at break	%	High	8	>50	> 50	80	6.4	6.5
	Residual stress	Mpa	Low	28	25	25	25	–	–
Material properties	Glass transition temperature/ decomposition temperature	°C	High	> 350 °C	230 °C/548 °C	255 °C/548 °C	240 °C	210 °C	210 °C
	CTE	$ppm\,°C^{-1}$	Low	45	50	55	80	56	54
Water uptake		%	Low	< 0.2	1.2–1.5	1.2–1.5		1.5	1.5

Source: Overview table adapted after © 2010 IEEE. Reprinted with permission from [3].

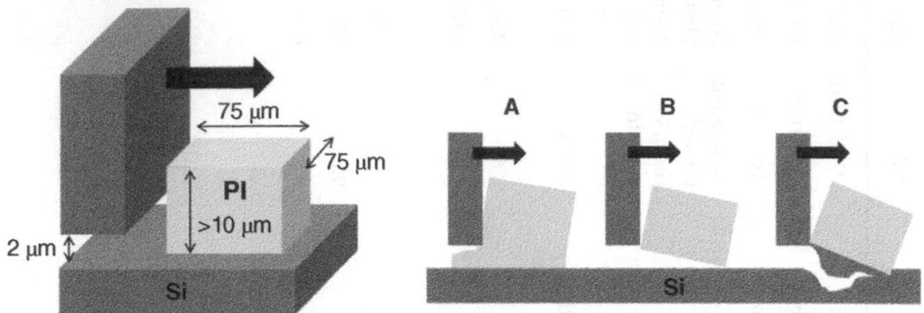

Figure 15.19 Shear test of PI stud on different substrate types. Study of the failure modes to improve the adhesion strength of the low temperature cure polyimide.

Pull or peel off tests are also not uncommon: a loose piece of the coating or a metal stud that is glued to the coating is pulled perpendicular to the substrate, and the force at which the film/stud is detached is indicative of the adhesion toward the substrate.

Finally, a shear tester, such as those by Nordson DAGE and XYZTec, can be used to measure the force needed to shear a patterned feature of the dielectric film from a substrate. This test is schematically depicted in Figure 15.19.

Using this method, it is possible to study and optimize the adhesion strength as a function of various parameters such as substrate preparation, adhesion promoter type, and concentration.

15.6 Design Rules

The design guidelines for the target dielectric film thickness in FO-WLP packages are typically in the range of 7–12 µm film thickness. The viscosity of the formulation determines the film thickness range and is primarily set by the polymer concentration level. The standard version is a high viscosity version and covers the cured film thickness range from 6 to 35 µm. Thinner film thickness ranges are possible with a lower viscosity version, which is a dilution of the high viscosity version with casting solvent and which provides identical material properties. The scribe lines or sawing streets between the FO-WLP packages require special attention during the RDL dielectric design phase. As the low temperature cure PI dielectric has a film shrinkage around 25% in all directions, some consideration regarding RDL dielectric stacking layout is required to avoid negative profiles at the sawing streets as shown in Figure 15.20. The first drawing, as shown in Figure 15.20a, results in negative dielectric profiles because this layout favors a larger shrinkage in the vertical direction than in the horizontal direction. A negative profile is easily overcome by increasing the horizontal overlap of dielectric layer (N) to dielectric layer ($N-1$) film

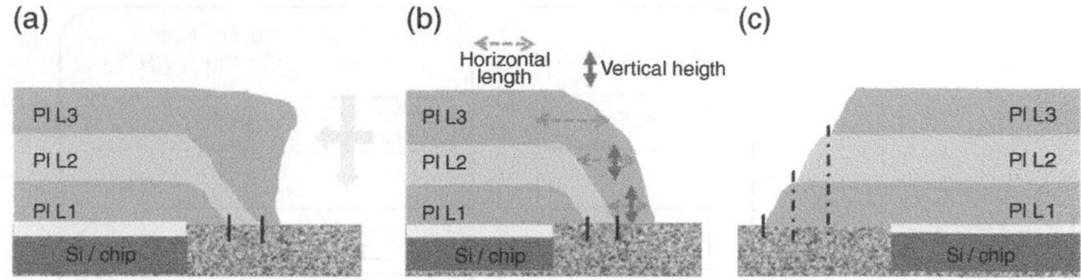

Figure 15.20 Design guideline at the sawing street to avoid negative dielectric profiles. (a) Negative dielectric profile due to larger shrinkage in the vertical direction. (b) Pull-in design. (c) Pull-out design.

compared with the (vertical) film thickness as shown in Figure 15.20b. The preferred ratio of horizontal length to vertical height should be preferably larger than a factor of 1.2. Another option is to avoid any dielectric overlap of dielectric (N) to dielectric ($N-1$) as shown in the third drawing of Figure 15.20c.

15.7 Reliability

The dielectric material plays a key role in the final reliability performance of the advanced FO-WLP packages. The standard package reliability test consists of a preconditioning test followed by either a TCT, (un)biased highly accelerated stress test (uHAST), high temperature storage test (HTS), or temperature humidity bias test (THB). Secondly, board-level tests are performed including temperature cycle on board (TCoB) and drop test. These tests, described in various JEDEC standards (EIA/J-STD-020C, JEDEC JESD22-A103, JEDEC JESD22-A104, JEDEC JESD-22-A104/IPC-9701, JEDEC JESD-22-B111 daisy chain), are however very time-consuming, labor-intensive, and therefore expensive analyses.

In order to avoid the above lengthy and expensive reliability test, a prescreening on dielectric level is proposed. The behavior of material and mechanical properties under reliability stress test can be evaluated on blanket dielectric films. The test consists of first determining the material and mechanical properties at time zero and then after a defined time in an autoclave (i.e. pressure cooker test [PCT] – JEDEC JESD22-A102) ranging from 0, 250, and 500 hours up to 1000 hours. PCT conditions are 121 °C with 100% RH and 2 atm pressure. A schematic presentation of the test on blanket cured PI film is described in Figure 15.21.

The Young's modulus, UTS, and elongation-at-break data as a function of the PCT test time has been determined for LTC 9320 cured at 230 °C for three hours and is shown in Figure 15.22. This data clearly shows no significant change in material properties over PCT time, and this is indicative that potential chemical changes such as hydrolysis of the imide units and scission of the polymer do not take place during the extended exposure to humidity and temperature.

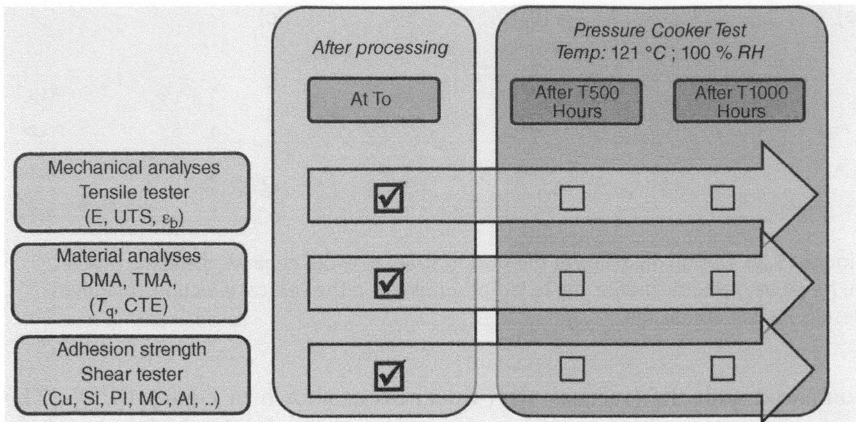

Figure 15.21 Schematic presentation for the short loop reliability test matrix for dielectric materials.

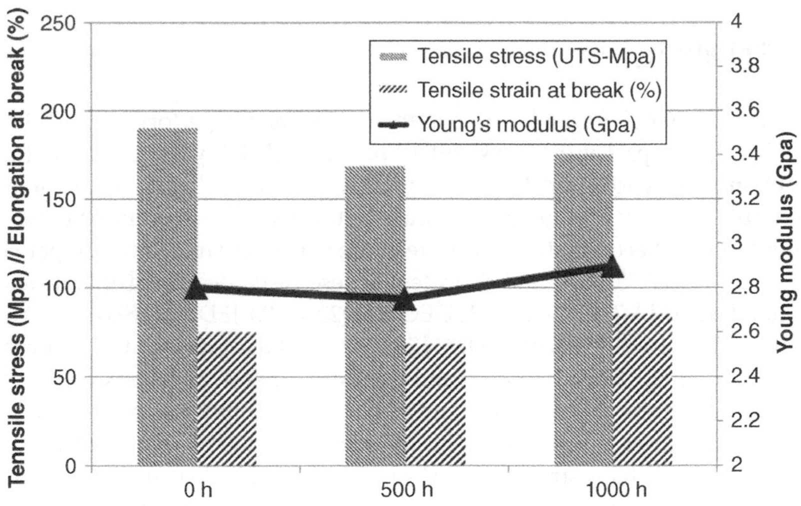

Figure 15.22 Mechanical properties of the LTC 9320 as a function of the PCT time.

Analysis of the FTIR spectra also proved that there is no degradation of the PI under PCT stress conditions.

As low temperature cure PI materials maintain their mechanical properties during accelerated aging test, this predicts that the material will be able to provide the required mechanical support during the package reliability tests. Therefore, the low temperature cure PI was reliability tested on FO-WLP packages according the JEDEC standards. Table 15.5 summarizes the results of the test vehicles without UBM layer and the applied test conditions for package and board reliability level.

Table 15.5 Details of both test vehicles TV A and TV B used for reliability characterization, package-level, and board-level reliability test conditions.

Test vehicle	Package type	Package size	Die size	Bump/ pad pitch
TV A	BGA	9.25 × 8.8 × 0.8 mm	5.6 mm × 5.3 mm	0.5 mm
TV B	BGA	7.5 × 7.5 × 0.8 mm	5.0 mm × 4.96 mm	0.4 mm
Stress (standard)		**Condition**	**Criteria**	**Status**
PRECON (JESD22-A113/ J-STD-020)		MSL 1	Level 1(T_{peak}: 260 °C)	Pass
PRECON + TC (JESD22-A104)		Condition B −55 °C < => +125 °C2 cycles h^{-1}	0 fail 1000/1500 cycles	Pass
PRECON + uHAST (JESD22-A118)		Condition A:130 °C/85% RH	0 fail 96/188 h	Pass
HTS (JESD22-A103)		Condition B:150 °C	0 fail 1000 h	Pass
THB (JESD22-A101)		85 °C/85% RH; V_{cc}: 5 V	0 fail 1000 h	Pass
Stress (standard)		**Condition**	**Criteria**	**Status**
TCOB (IPC-97-01)		Condition C −40 °C < => +125 °C1 cycle h^{-1}	FF > 500/850 cycles	Pass
Drop test (JESD22-B11)		B1500 GS; 0.5 ms duration; half-sine pulse	<10% fails at 20 drops	Pass

Source: © 2014 IEEE. Reprinted with permission from [1].

Unlike the standard low cure dielectrics under thermal cycling, it was found that the low temperature cure PI has the capability to slow down the propagation of cracks, which originate in the intermetallic that is formed between the solder and copper pad at the outer circumference around the solder ball. The new low temperature cure PI material has the capability to stop the crack propagation at the bottom dielectric layer such that the crack will not propagate in the die active area for both test vehicles [1]. Therefore, the electrical functionality of the unit is preserved at least until 1000 cycles, and the cross sections carried out during characterization indicate that this limit can be even exceeded [1].

TCoB for TV A was carried out with continuous *in situ* electrical monitoring to estimate the parameters of the Weibull distribution. The Weibull cumulative distribution function (cdf) model predicts 0% failures until 500 cycles and 1% for 766 cycles, and the mean time to failure (MTTF) is 1225 cycles. Failure analysis of the solder balls revealed solder ball fatigue damage with cracks developing from the corner of the intermetallic formed between solder ball and copper pad of the PCB board. The failure mode is more likely due to the surface-mount technology (SMT) process (Figure 15.23) [1].

A drop test with *in situ* electrical monitoring during stress test was performed for TV A. The pass criteria of less than 5% fails at 30 drops based on the Weibull cdf require 140 drops, and the MTTF is 1059 drops. Physical failure characterization shows that the only failure mechanism observed was solder joint fracture at the component side, specifically in the intermetallic layer, which became the more fragile interlayer connection (Figure 15.24).

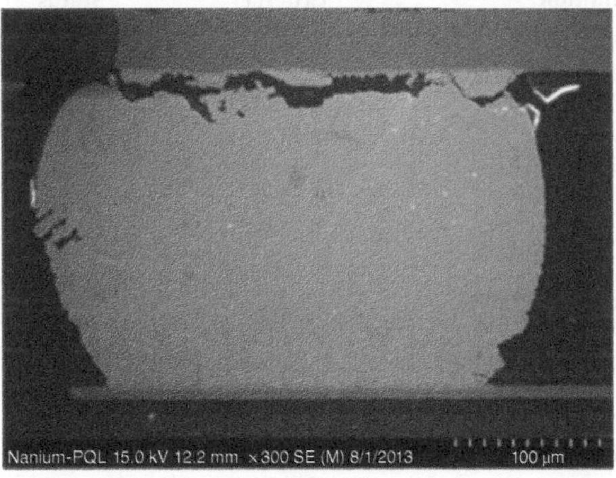

Figure 15.23 Solder ball bulk crack at PCB side, which is the top of the picture for TCOB. FO-CSP package on the bottom side of the SEM picture for TV A. *Source:* © 2014 IEEE. Reprinted with permission from [1].

Figure 15.24 Intermetallic crack at the component side for TV A for drop test. *Source:* © 2014 IEEE. Reprinted with permission from [1].

15.8 Next Steps

The implementation of low temperature cure PI in FO-WLP packages resulted in a significant improvement of the reliability performance of these packages. Nevertheless the general FO-WLP package roadmap with multiple RDL dielectric layers will demand further improvements to the low temperature cure PI chemical platforms as shown in Figure 15.25.

A first key challenge is to reduce the cure temperature from >200 °C toward a cure temperature in the range of 170 °C while maintaining the outstanding mechanical properties. The target temperature of the low temperature dielectric should be similar to the mold compound cure temperature to reduce wafer warpage. Lower wafer warpage reduces wafer handling issues and improves the overall yield.

Secondly, improved resolution or higher cured aspect ratio will be required to meet scaling roadmaps. This seems likely to be achievable by optimization of the photo-package system of the formulation.

Copper migration under electrical bias will become critical with reducing line/space dimensions below 5/5 μm downward. Copper diffusion as a function of the cure temperature of the low temperature PI will need to be modeled. The influence of copper lines/spaces, substrate preparation, environmental conditions, and applied voltage during stress test on copper diffusion needs to be fundamentally understood for future product optimization. Additionally a reduction in the moisture uptake of PI may retard copper corrosion processes, e.g. copper discoloration.

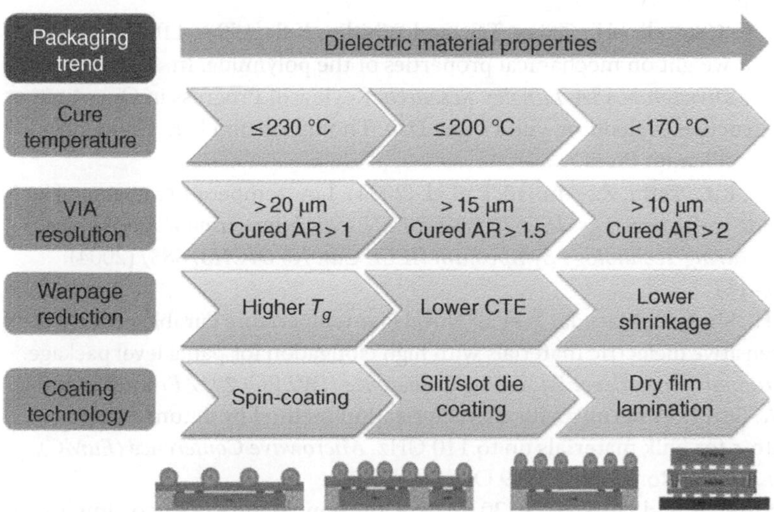

Figure 15.25 Key dielectric trends for advanced packages.

The manufacturing processes of low temperature cure PI will be challenged to cope with the continuous trend of reduced trace metals and particles in the formulation. The batch-to-batch variability needs to be limited to allow the assembly suppliers to reduce the in-line statistical process control (SPC) check and reduce the overall manufacturing cost.

Finally, the pathways for these new enabling dielectric materials need to be in compliance with the ever stringent EHS legislation set by regulatory bodies such as the EU's European Chemicals Agency (ECHA) or the United States' Environmental Protection Agency (EPA). This puts a high burden on the research and development divisions of the material companies to propose technical solutions without significant cost adders in a short time period.

References

1 Almeida, R., Barros, I., Campos, J. et al. (2014). Enabling of fan-out WLP for more demanding applications by introduction of enhanced Dielectric material for higher reliability. *2014 IEEE 64th Electronic Components and technology conference (ECTC)*, Orlando, FL (2014), pp. 935–939.

2 Huemoeller, R., Zwenger C. [Amkor Technology, Inc.] (2015). Silicon wafer integrated fan-out technology. Reprinted from March April, 2015. http://ChipScaleReview.com (accessed 6 August 2018).

3 Toepper, M., Fischer, T., Baumgartner, T., and Reichl, H. (2010). A comparison of thin-film polymers for wafer level packaging. *2010 Proceedings 60th Electronic Components and Technology Conference (ECTC)*, Las Vegas, NV (2010), pp. 769–776.

4 Yost, W.T., Cantrell, J.H., Gates, T.S., and Whitley, K.S. (1998). Effects of molecular weight on mechanical properties of the polyimide. In: *SJTM Materials Division NASA-Langley Research*, Review of Progress in Quantitative Nondestructive Evaluation, vol. 17 (ed. D.O. Thompson and D.E. Chimenti). New York: Plenum Press.

5 Hubbard, R.L., Fathi, Z., Ahmad, I. et al. (2004). Low temperature curing of polyimide wafer coatings. *IEEE/CPMT/SEMI 29th International Electronics Manufacturing Technology Symposium (IEEE Cat. No.04CH37585)* (2004), pp. 149–151.

6 Araki, H., Shoji, Y., Masuda, Y. et al. Novel low-temperature curable positive tone photosensitive dielectric materials with high elongation for panel level package. *International Wafer-Level Packaging Conference (IWLPC) 2017 Proceedings.*

7 Talai, A. (2014). A permittivity characterization method by detuned ring resonators for bulk materials up to 110 GHz. *Microwave Conference (EuMC), 44th European*, Rome, Italy (6–9 October 2014).

8 Vanclooster, S. and Janssen, D. (2016). Are low temperature cure polyimides suitable for high frequencies? *Fujifilm Advanced Lithography Workshop*, Dresden, Germany (September 2016).

16

The Role of Pick and Place in Fan-Out Wafer-Level Packaging

Hugo Pristauz, Alastair Attard, and Harald Meixner

BESI, Austria

16.1 Introduction

Pick and place (P&P) is a crucial subprocess for fan-out and embedded wafer- and panel-level packaging, as it is an essential contribution to the package cost of ownership. It also has a significant influence on the yield as P&P is challenged by high placement accuracy requirements over a large working area and by the absence of local alignment marks in several fan-out technologies. Stable die P&P accuracy plays a key role for product feasibility [1]. To introduce some important P&P capabilities, a set of well-documented fan-out and embedded packaging processes can be consulted, like embedded wafer-level ball grid array (eWLB) [2, 3], redistributed chip package (RCP) [4], wafer-level integrated fan-out (InFO) [5], M-Series (a fully molded fan-out packaging technology) [6], silicon wafer integrated fan-out technology (SWIFT) [7], silicon-less integrated module (SLIM) [8], and embedded multi-die interconnect bridge (EMIB) [9]. This set of processes has no claim on completeness (some more can be found in [10]), but is representative to highlight important P&P requirements for fan-out and embedded wafer- and panel-level packaging processes.

Early deployed fan-out packaging processes like eWLB and RCP require die attach with the active side down (face-down) onto a sticky carrier [2–4]. Since the die is placed before processing of the redistribution layer (RDL), this approach is classified as chip-first. A considerable challenge for P&P equipment is that eWLB and RCP carriers do not support local alignment marks, which leads to a global accuracy alignment requirement, a concept that will be explained in detail in the next section. With the absence of wafer bumping, these "chip-first" or "face-down" fan-out processes are claiming low packaging costs, improved electrical and thermal performance, reduced

Advances in Embedded and Fan-Out Wafer-Level Packaging Technologies, First Edition.
Edited by Beth Keser and Steffen Kröhnert.

package thickness, and ultra-low-k device compatibility [2–4]. However, these packages have lower board-level reliability than traditional wire-bond chip-scale package (WB-CSP) or flip-chip chip-scale package (FC-CSP) due to higher BGA solder ball stress caused by mismatch of coefficients of thermal expansion (CTE) between chip and printed circuit board (PCB) in combination with the lack of a sufficient thick buffer zone between silicon die and solder balls [2].

To overcome the drawbacks of limited board-level reliability, one of the approaches is to introduce bumped die for fan-out packaging, with bumps and epoxy material between bumps serving as a stress relief buffer layer between die and RDL. To deal with bumped die, some alternative fan-out packaging technologies have been developed like InFO [5] and M-Series [6]. While these technologies still require chip-first placement, the die has to be attached with the active side up (face-up). EMIB, a silicon bridge embedding process that supports cost-effective high density 2.5D interconnection [9], similarly requires the silicon bridges to be placed faceup.

A drawback of chip-first packaging processes is that the die has to be committed to the package at a stage where the RDL still does not exist. Yield issues in the subsequent RDL buildup process would consequently lead to a waste of expensive functional die. To avoid this problem, an alternative approach has been proposed by placing die onto a known good RDL. Such approach is classified as chip-last and is utilized in SWIFT and SLIM packaging processes, where the die has copper pillar bumps with solder caps and die orientation is naturally face-down [7, 8]. Since RDL processing is already completed before die attachment, a suitable interconnection method is required between die copper pillar bumps and RDL: this is achieved by mass solder reflow for SWIFT and SLIM [7, 8] with flux dipping capabilities during P&P. Due to the presence of RDL patterns, a local alignment is possible during P&P.

With a focus on P&P for fan-out and embedded packaging, it makes sense to introduce four process types labeled by the face-up or face-down orientation and additionally by the chip-first or chip-last assembly process. Such a categorization provided with key advantages and P&P requirements is summarized (without claim of completeness) in Table 16.1.

Initially, it seemed that standard flip-chip bonders fulfill all P&P requirements for fan-out wafer-level and panel-level packaging (FO-WLP and PLP) [4]. However, these standard P&P tools needed to be customized and refined with new capabilities over time so that currently, about one decade after the first deployment of FO-WLP and PLP, there are dedicated P&P machines that are called fan-out bonders [11].

The next sections of this chapter will cover three key topics related to fan-out bonders, starting with a description of equipment requirements for fan-out bonders, which are best explained in terms of the core capabilities of advanced die attach equipment [11]. The following two sections are of practical

Table 16.1 List of different fan-out and embedded wafer- and panel-level packaging process types with important representatives, highlighting key advantages and P&P requirements.

Process type	Process examples	Key advantages	P&P requirements
Chip-first/ face-down	eWLB [2, 3], RCP [4]	Very cost-effective package due to absence of bumps	Face-down P&P, global accuracy
Chip-first/ face-up	InFO [5], M-Series [6]	Good board-level reliability, full encapsulation, utilizing copper pillar bumps without solder caps	Face-up P&P, global accuracy, optional: constant bond heat
Chip-last/ face-down	SWIFT [7], SLIM [8]	High yield by utilization of known good RDL	Face-down P&P including flux dipping, local accuracy
Chip-last/ face-up	EMIB (bridges) [9]	Cost-effective high performance, localized high density interconnect	Face-up P&P, global accuracy

importance and cover aspects of how to avoid fan-out bonding pitfalls and introduce field-proven procedures for fan-out bonder qualification.

16.2 Equipment Requirements for Fan-Out Bonders

16.2.1 Core Capabilities of an Advanced Die Attach Equipment

In order to get some basic understanding about fan-out bonders, it should be realized that a fan-out bonder belongs to the family of advanced die attach equipment. Going through the eight core capabilities of advanced die attach equipment [11], one can get a better understanding of what comprises a fan-out bonder (Figure 16.1).

16.2.1.1 Die Feeding

Die feeding consists of the subprocess of presenting die and releasing die from a presentation medium (film frame carrier, waffle pack, tape and reel [T&R]) to a die handler, which might be a flip unit or a bond head. The various die feeding methods can be applied to any FO-WLP and PLP process type. High volume die feeding includes picking directly from diced wafers (mounted on film frame carriers), which are typically 300 mm (in some exceptions 200 mm) in diameter. Suitable thin die handling capability can be important for a fan-out

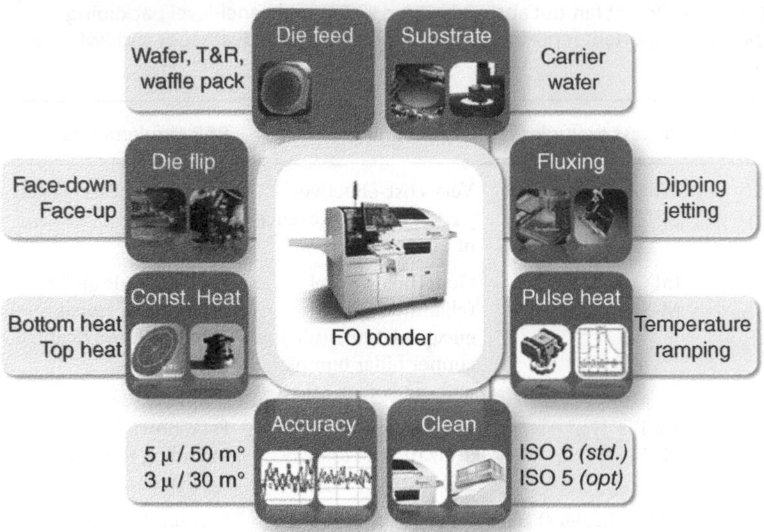

Figure 16.1 Core capabilities of advanced die attach equipment.

bonder, since FO-WLP and PLP offer very low package thickness possibilities and have become even more interesting for 3D fan-out packages, which require very thin die. A die thickness capability down to 50 μm is the current standard.

Picking from waffle pack is an additional option that was a standard requirement for the first eWLB P&P applications. This is due to the use of two glass die for the subsequent RDL mask alignment, which are picked from a waffle pack, even in mass production. Using a waffle pack for glass die is now a legacy option, but in times of large panel-level packaging, this approach is again considered.

The third required option, feeding from T&R, has recently gained more interest. One of the main reasons is that memory stacks are shipped in T&R format, and thus fan-out bonders need to be able to handle this supply format too. There are some ideas to place passives with a fan-out bonder; however, this is unlikely in mass production because of the high cost-of-ownership contribution of the fan-out bonder compared with a chip shooter.

16.2.1.2 Substrate Handling

The second core capability of an advanced die attach machine is related to the type of substrate used and its handling. In the context of fan-out bonders, the term substrate refers to round or rectangular carriers for fan-out packaging [2–8] and to PCB in panel format for embedded packaging [9]. The exact requirements will depend on the type of process used to create each package type.

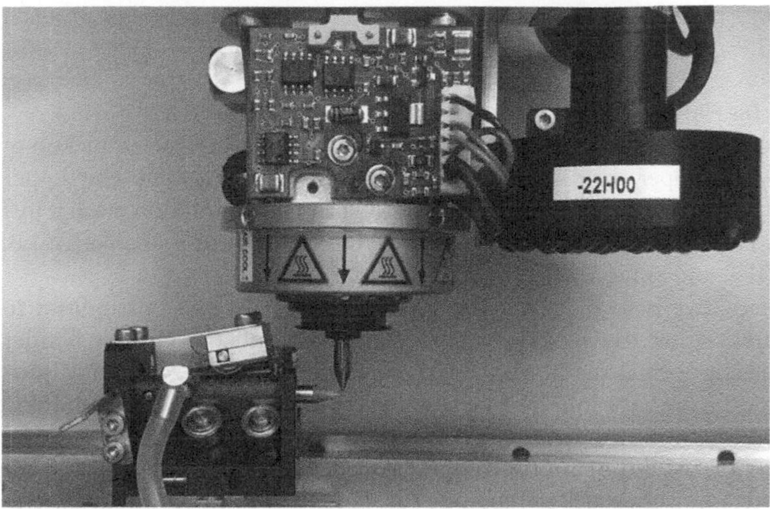

Figure 16.2 90° flipping of a component.

For a FO-WLP process like eWLB, a circular steel carrier plate is in use, which has flats on four sides and is covered with double-sided adhesive tape [1]. The flats are provided to allow a proper transport of the carriers on belt conveyors and to be loaded and unloaded to and from standard magazine handlers (Figure 16.2). This approach provides significant cost saving compared with the higher costs associated with the use of wafer handling robots.

An FO-WLP process can also run on panels, which are based on square or rectangular carrier plates [4, 6]. If the edge lengths of the panel do not exceed the diameter of the carrier, the same fan-out bonder can be used for the FO-PLP process, which also means that the complete magazine-to-magazine-based conveyor handling system can remain exactly the same. The same applies for the EMIB process where the substrate is a rectangular PCB.

Another alternative is to populate the die onto a round (wafer-like) carrier as happens with InFO [5], SWIFT [7], and SLIM [8]. Round carriers cannot be handled directly on belt conveyers; thus the trick is to transport the wafer in a rectangular transport carrier from magazine to magazine via belt conveyor. This requires, however, extra handling of the fan-out carriers into the transport carriers, and vice versa. Robot handling with carriers presented by front opening universal pods (FOUP) is thus preferred in this case, although FOUP and robot handlers add more costs compared with magazine-to-magazine belt conveyor systems.

Large panel handling in the size of 600 mm × 600 mm and bigger is still an area of research. To control warpage there are carrier thickness specifications up to 10 mm in request, which is significantly higher than what is typically used in FO-WLP applications.

16.2.1.3 Die Flip

A face-down fan-out bonder is basically a flip-chip bonder. During pickup of the die by the flipper, the die is oriented faceup on the dicing tape. The flipper turns the die around by 180°, so the bond head can take it over and place it in a face-down orientation. This is the case for all face-down processes, like eWLB, RCP, SWIFT, and SLIM.

The alternative is to place the die face-up. There are two possibilities to achieve this: either the flipping step is eliminated and the fan-out bonder behaves like a normal die bonder, or there is an additional re-flipping step, which flips the die back into face-up orientation (Figure 16.3) [11]. Both InFO [5] and M-Series [6] processes are face-up processes. Re-flipping has better conditions for fast throughput since the die ejection happens in parallel to the bonding step. As a drawback re-flipping adds more complexity to the equipment, with the need of more nozzles and flipping hardware.

Regarding flipping, one might not expect that there is a third option: 90° flipping (see Figure 16.2). Such an approach is utilized in mass production for microelectromechanical systems (MEMS), like magnetometers. Even though 90° flipping is not currently in mass production for FO-WLP and PLP applications, there is increasing interest from the packaging industry in this option.

16.2.1.4 Fluxing

While the application of flux (fluxing) is not a requirement for chip-first processes, fluxing is a crucial process step for chip-last processes, such as SWIFT and SLIM, which are based on mass reflow solder interconnect [8]. Flux is

Figure 16.3 Parallel pick handling, implemented in terms of a flipper and a re-flipper for face-up bonding, in combination with a constant heat bond head. The die is transferred from the pick flipper to the re-flipper (left), from where it is fetched with the heated bond head (right).

required for two key reasons: first as a temporary adhesion agent for holding die in position after placement and second for oxide reduction during the reflow process in order to get reliable solder joints.

A proven method of flux application is via a dipping process, where a fluxing unit prepares a flux film with consistent thickness in a cavity into which the die is dipped before placement. In terms of die attach process, SWIFT and SLIM are nothing but wafer-level mass reflow flip-chip attach processes with an intermediate dipping step for flux application.

Being an additional step in the P&P sequence, fluxing slows down machine throughput and therefore adds cost of ownership. However, for high-end devices with expensive die, this extra cost is compensated by the yield cost savings leveraged from the chip-last approach.

16.2.1.5 Constant Bond Heat

For face-up FO-WLP and PLP processes, new temporary adhesive materials are under investigation, which require the use of heat to improve die adhesion. With these materials, face-up FO-WLP and PLP require die placement under hot conditions (50–200 °C). In these cases, only constant temperature is required, so it is sufficient to provide constant temperature at the top by the P&P tool and at the bottom side by the stage. There are no further requirements to temperature ramping or cooling during the bonding sequence.

Currently, constant bond heat is a standard request for fan-out bonders and is a field-upgradable option. As the material development for temporary adhesives is still underway, the manufacturing industry requires the availability of such capability or at least the possibility to upgrade the equipment to constant bond heat in case of high volume readiness of new attractive temporary bonding materials. The targeted temperature range for top and bottom heat is both up to 200 °C.

16.2.1.6 Pulse Heat

Pulse heat refers to heating where the temperature has to follow a dynamic temperature ramp during the die attach process. Such capability is needed in thermo-compression processes for local reflow of a solder joint and/or curing of pre-applied underfill materials. Pulse heat is currently not used in mass production for FO-WLP and PLP. In general, pulse heat slows down P&P throughput due to the time required for the transient phases during temperature ramping. Since the FO-WLP and PLP cost target is to beat the cost of wire-bonded lead frame packages, pulse heat, which slows down throughput and adds additional equipment cost, is typically not considered.

However, pulse heat may be justified. The first reason is that the temperature conditions for direct die picking from a dicing foil or blister tape would not allow process temperatures at 200 °C. If die is picked directly with the bond head from a dicing foil or blister tape, then the bond head temperature would have to be lowered for die picking and ramped up again during travel to the bond location. An alternative work-around in such a scenario would be parallel pick handling, as it is the case with a flipper, a double flipper (see Figure 16.3), or an additional pick handling system that picks the die from dicing foil or blister tape and puts it onto an intermediate stage, from where it is fetched by the heated bond head. In all these cases picking is performed in a cold condition using a separate head, which eliminates the need for temperature ramping on the bond head.

A further application area for pulse heat is in chip-last processes. Although in this case mass reflow flip-chip attach is a cost-effective method, thermo-compression-based chip-last attach, which requires pulse heat, would have the advantage of better yield and warpage control, since the bond and solder interconnection is performed *in situ* in the fan-out bonder.

16.2.1.7 Accuracy

P&P equipment for fan-out packaging must support a very accurate and stable P&P process on a large working area with high throughput. One of the reasons for high accuracy is to support high routing density of the RDL with minimized landing areas for die interconnection pads, while another important reason is to have some tolerance buffer for absorption of die shift in the subsequent molding process [1] in chip-first processes.

The large working area of FO-WLP started with 200 mm carriers and is now running in high volume on 300 and 340 mm carriers, and in the future FO-PLP is targeted to run on areas of approximately 600 mm × 600 mm. Large panels are being developed as a means to increase the number of components present per panel and thus bring down the per-device fraction of the batch costs of the fan-out packaging process, which are in majority caused by RDL processing. High throughput is also necessary to bring down the cost of ownership of the sequential P&P process, which comprises another major cost factor of the fan-out package.

In the case of chip-first processes, due to the absence of local alignment marks on the carrier, fan-out bonders can only utilize global marks to perform the alignment (e.g. in the eWLB case, two holes that are mechanically drilled into the steel carrier are used). Such an approach is known as a global accurate placement process. A fan-out bonder has to place a very consistent matrix of die onto a carrier with an adhesive layer, where the consistency of the die matrix is needed to support downstream mask processes for building the RDL. A global accurate placement process follows a machine sequence that is displayed in Figure 16.4b.

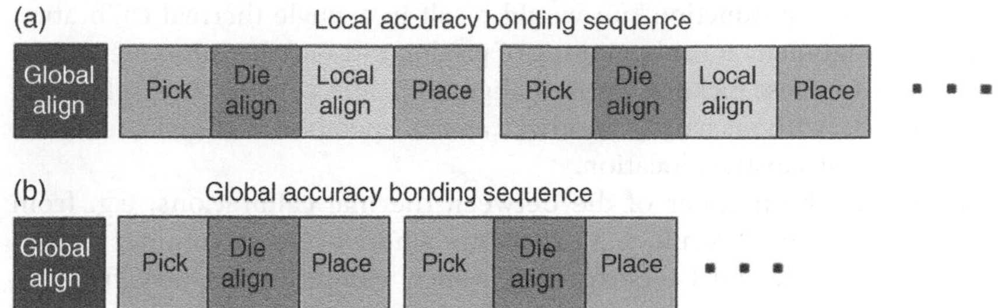

Figure 16.4 Machine sequences for (a) local and (b) global accurate placement process.

To dive into some more detail, a fan-out bonder can perform the following steps in order to build up a consistent matrix of die onto a carrier with an adhesive layer:

1) Carrier loading and clamping with vacuum; confirmation of successful clamping via vacuum monitoring.
2) Thermal calibration of left bond head coordinate system on N machine reference marks.
3) Thermal calibration of right bond head coordinate system on same N machine reference marks.
4) Global alignment against two global carrier fiducials (e.g. drilled holes) with (only) one bond head (the second bond head "knows" the global alignment marks via matching coordinate systems based on thermal calibration).
5) Parallel, asynchronous flip P&P of die with two bond heads performing picking from flipper, die alignment (sometimes called "upward alignment," measuring die position on bond head using an upward camera) and placing, until programmed number of die, or programmed time interval, or carrier fully populated, or end of production lot is reached, whichever occurs first.
6) If the programmed number of die or programmed time interval has been reached, the machine performs an intermediate thermal calibration by calibrating again on the same N machine reference marks with both bond heads (same procedure as in 2 and 3).
7) If carrier is not fully populated and end of production lot has not been reached, process control continues at 5.
8) Otherwise, if carrier is fully populated, carrier is unloaded and the whole procedure starting at 1 is repeated.
9) Otherwise the end of production lot is reached and machine stops the production run.

The above procedure is very effective to implement a stable global accuracy process. In an extreme case of 40 000 die per carrier with a machine throughput of 5 000 UPH and a programmed number of 20 000 die between thermal calibrations, the production run would result in a single thermal calibration and global alignment at the beginning of production, and the subsequent eight hours would be just picking, upward alignment, and placing (as displayed in Figure 16.4b), without having another camera operation that gives feedback about an actual substrate location.

Decreasing the number of die between thermal calibrations, e.g. from 20 000 to 500, would result in a procedure where every six minutes a thermal calibration would be performed, which simplifies the task from an accuracy perspective. The downside of achieving better accuracy through more frequent thermal calibrations would be a significant reduction in machine throughput due to the increased calibration operations performed during the production run. Furthermore, it should be noted that significant thermal drifts occurring during the six-minute interval between calibrations would still impact the P&P accuracy since these drifts are not captured by the machine.

In contrast, for a local alignment process, each actual substrate target location would be captured before each die placement, and the time for machine drift between local alignment and placement would usually be less than a second. Therefore, drift in terms of a bond head's coordinate system shift has almost no influence on local accuracy, since the drift error that is made by measuring a substrate location and the drift error that is made by moving the bond head to the target location would more or less completely cancel, which makes the challenge for engineers to implement machine capability for a local accuracy process tremendously easier.

Based on the experience of building fast and accurate bonders, it can be concluded that the deformation of a bond head's coordinate system caused by thermal drifts can be thermally calibrated in an easy way if the deformations can be modeled by linear mapping, and in contrast it will be very hard to perform periodic thermal calibrations if the coordinate system deformations are nonlinear in nature. The ideal situation is, of course, if the deformations of the bond head coordinate system are negligible and thus do not need to be considered at all. Figure 16.5 shows the coordinate system deformations of a current handling system, which are compensated by application of a thermal calibration procedure, and the negligible deformations of a next-generation gantry system, which is designed to support a 600 mm × 600 mm panel-level fan-out P&P process. The coordinate drifts of the current handling system are in the range of 4 μm, which, however, can be easily compensated since the deformations can be modeled more or less by linear mapping. The coordinate system deformations of the next-generation system are more or less negligible, and it is expected that no thermal calibrations are necessary for a 5 μm at 3σ global accuracy process.

All considerations made so far were regarding thermal drifts. In a dual bond head machine, there is also the important effect of mechanical cross talk

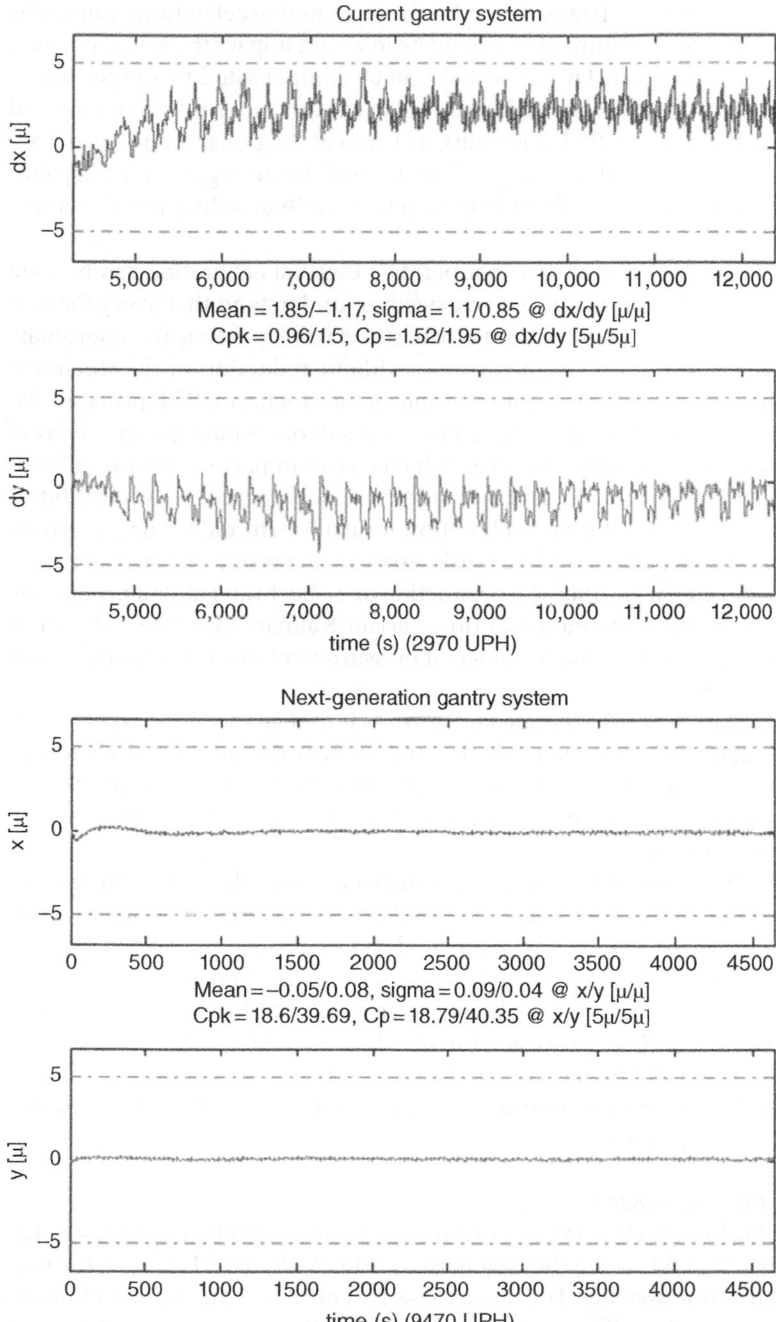

Figure 16.5 Thermal drift of two handling systems based on repeated position measurement of a fiducial matrix.

between the two bond heads, whereby the jerks and accelerations caused by rapid change of momentum of one bond head would impact the actual position of the other bond head. This cross talk can be compensated by proper multivariable control. Such multivariable control has been successfully implemented to increase the original P&P capability of 10 μm at 3σ global accuracy at 5000 UPH to 5 μm at 3σ global accuracy at 7000 UPH of the next-generation capability, resulting in 40% throughput improvement while doubling the placement accuracy.

It has been mentioned earlier that periodic thermal calibration can be used to compensate thermal drifts, but one needs to understand that every thermal calibration returns a result with a stochastic error. Updating the coordinate system calibration based on such results without reduction of the stochastic error would cause a shift of the mean value of the stochastic P&P process (discontinuities of the Gaussian center), and as a result one would get an overlay of differently centered Gaussian curves, which has an impact on the overall process standard deviation. In order to minimize such centering errors, filters could be used for the thermal calibration measurement data. Time-invariant filters, however, would show their weaknesses during restart phases of production (either in the beginning of a production or a short time after a production stop). With the implementation of time-variant Kalman observers [12], which can be designed for minimal variance of measurement noise, such weaknesses can be overcome.

In conclusion, a global accurate P&P process is crucial for chip-first FO-WLP and PLP packaging processes, which typically provide only global alignment marks on the carrier. Verification of a stable and robust global accuracy process according to the process specifications should be the first step in the selection of a fan-out bonder.

On the other hand, if local alignment marks are provided, a fan-out bonder needs to support also a local accurate P&P process, as shown in Figure 16.4a. This is usually the case for face-down chip-last processes like SWIFT and SLIM. Since the RDL is processed first, the bonder can find local structures in order to utilize a local accurate placement process. A SLIM process that incorporates a foundry RDL process with very fine structures can easily include local features for alignment. For such a process the next-generation P&P tool can utilize 3 μm at 3σ local accuracy, while a global accuracy process would run by standard at 5 μm at 3σ.

16.2.1.8 Clean Capability

The standard clean class for die attach equipment is ISO 6 [13]. Industry has proven that ISO 6 is a suitable standard for FO-WLP and PLP for RDLs with 10 μm/10 μm line spacing. For denser line spacing like 5 μm/5 μm and below, the industry prefers ISO 5 clean capability [13], which advanced fan-out bonders can provide as an option.

16.3 Avoiding Fan-Out Bonding Pitfalls

16.3.1 Die Tilt

Die tilt, which comes from a lack of coplanarity between bond tool and substrate, has three major impacts to the fan-out P&P process: flying die (extreme die shift during the molding process due to insufficient adhesion between die and carrier), P&P offset, and a larger spread (standard deviation) in the *x*, *y*, and theta placement accuracy. All these effects negatively impact the process yield; thus die tilt is a serious issue in a fan-out bonding process.

Usually, a coplanarity setup is established between bond head and work holder (Figure 16.6). Since the goal is to have a coplanar bond tool and substrate, the bond tool plane is perpendicular to the bond head axis, the bond head axis is perpendicular to the work holder plane, and the carrier surface plane is coplanar to the work holder plane (Figure 16.7). To guarantee the first requirement, special P&P tools that are manufactured from one part are used for fan-out P&P processes (Figure 16.7). This is in contrast to usual die bonder setups where the P&P head interfaces to a (configurable) tool holder that holds a tool (sometimes called nozzle) that is optionally covered with a tool tip. Such a complex tool stack would add too many manufacturing tolerances; thus tool holder, nozzle, and tip need to be manufactured as a single part. Such special P&P tools guarantee coplanarity between tool surface and bond head interface of ±2 μm at 10 mm and avoid the user influence of mounting a bond tool onto a tool holder with inadequate perpendicularity.

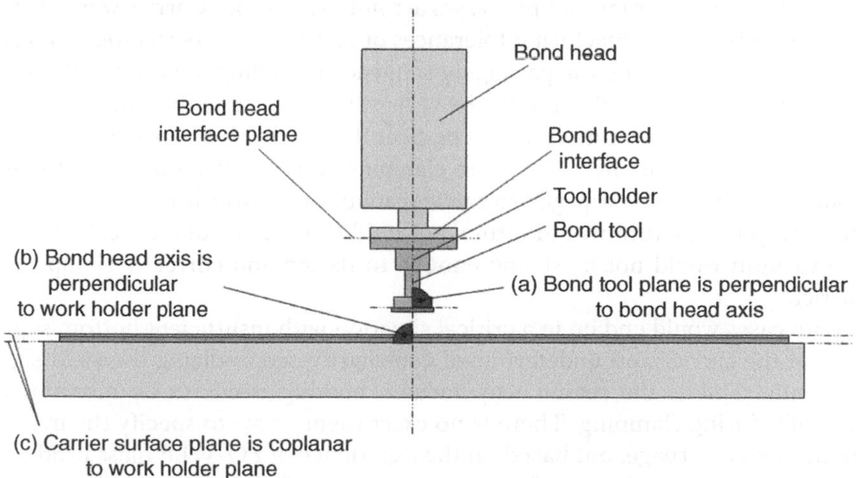

Figure 16.6 Coplanarity setup. (a) – (c) indicate three important alignment angles.

Figure 16.7 Pick and place tool manufactured from a single part.

The maximum coplanarity mismatch between bond tool and work holder has been initially specified to be max. ±10 μm at 10 mm for previous P&P systems. Since die tilt in this range has still shown recognizable yield impact, the specification for the next-generation P&P systems has been tightened to ±5 μm at 10 mm, which, at least for eWLB applications, has demonstrated non-recognizable yield impact in mass production.

16.3.2 Warped Carriers

The carriers used for fan-out processes are not exactly flat. Carrier warpage is either caused by manufacturing tolerances or results from thermomechanical stresses induced during the packaging sequence (handling, molding, debonding). These carriers are designed to be very stiff to withstand the handling during the assembly flow; thus it should be clear that the intrinsic spring force of a warped carrier is acting against the clamping force induced by the fan-out bonder vacuum. The warpage of a carrier cannot be arbitrarily large; otherwise the clamping vacuum would either not hold the carrier due to leakage, or the vacuum would not be strong enough to flatten and correctly clamp the carrier.

Both cases would end up in a critical situation with insufficient bottom support of the carrier and undetermined coplanarity, jeopardizing the yield significantly. This is the reason why vacuum buildup needs to be monitored carefully during clamping. There is no exact theory how to specify the maximum carrier warpage, but based on the experience of FO-WLP mass production, a maximum tolerance for carrier warpage of ±500 μm is sufficient to prevent a negative yield impact.

16.3.3 Influence of Motion Parameters

For advanced die attach systems like fan-out bonders, a P&P head movement is servo controlled on the basis of reference trajectories characterized by motion parameters such as travel distance, maximum jerk, maximum acceleration, and maximum velocity. The choice of motion parameters should usually not influence the center of the stochastic P&P process. The practical experience, however, shows that the P&P process center shifts with a change of the motion parameters, while in contrast, for a fixed set of motion parameters, the P&P process is highly repeatable. The reason is that with fast x/y movements of the bond head, transient oscillations are induced at the tool center point, which are still present when the die touches the substrate. Thus there are two options to set up the bond process:

a) Operating the bonder with very slow (nonaggressive) motion parameters, optionally combined with sufficiently large settling delays, in order to minimize or eliminate the residual oscillations at the tool center point. This would more or less eliminate the influence of the motion parameters on the bond position offset but would compromise the throughput due to slower movements.
b) Alternatively, one can use an aggressive parameter set for the motion profiles, combined with minimized settling delays, and keep in mind that the existing bond head offset will differ if one of the motion profile parameters is changed.

16.3.4 Lack of Placement Repeatability

Placement accuracy issues can be due to repeatability issues or drift issues. If the problem is a lack of repeatability, first the theta (rotational) repeatability data should be reviewed. An issue on theta repeatability has a direct impact on the x/y repeatability; thus this issue has to be fixed first. Low theta repeatability can be an indicator of insufficient coplanarity; thus in such a case it is well worth checking whether die is being bonded with tilt, and if so, first fixing the coplanarity issue.

16.3.5 Flying Die

The term flying die is used in the context of chip-first processes for die misaligned significantly more (e.g. more than three times) than the accuracy specification after the molding step. The misalignment of flying die is not a direct outcome of the P&P process; rather it is caused by the lateral forces induced on the die by the flowing mold compound during the molding process, in combination with insufficient adhesion between the die and the carrier. A flying die can thus have its root causes both in the P&P process and in the molding process. With focus on P&P, flying die can always be pinpointed to low adhesion to the carrier. One root cause for this is die tilt, where only a partial area of the die adheres well to the temporary adhesive, again emphasizing the seriousness of die tilt during P&P. Other root causes could be excessively low bond forces or short bond delays.

In order to avoid flying die, both bond force and bond delay need to be sufficiently large. On the other hand, there is an upper limit for the bond force due to the fact that the die would be pressed too deep into the adhesive layer of the carrier (e.g. adhesive tape), which would result in higher variance of the die surface z-location, causing subsequent process issues during molding and RDL buildup. Furthermore, an excessive increase of the bond delay would again compromise throughput. For a productive, high yield fan-out process, the parameters for bond force and bond delay should be chosen with regard to die geometry and the properties of the adhesive tape, examined on the base of design of experiments (DoEs).

Finally it should be noted that an additional post-bond inspection step in the fan-out bonder cannot be effective to detect a flying die, since the die fly occurs in the molding equipment, not in the fan-out bonder.

16.3.6 Process Margins

In mass reflow flip-chip bonding and thermo-compression bonding, the substrate has local fiducials, and the bumps are solder based or have solder caps; thus these can be recognized in an X-ray microscope to verify correct placement accuracy. The situation for chip-first face-down fan-out bonding (like FO-WLP) is, however, more complex. It is not easy to verify a ±10 μm global placement accuracy due to missing substrate alignment marks and the fact that the die can only be inspected from the backside, which is blank and also does not present any alignment marks. At most, a post-bond inspection (PBI) can recognize the die edges, but these are not sufficiently accurate since shifts from the sawing process and detection quality issues of the chipped die edges introduce large errors into this measurement. Therefore, an accurate die placement inspection step usually happens after molding. This means that a whole batch of carriers has to be first populated by the fan-out bonder, containing up to hundreds of thousands of die placements, before the batch goes to molding and later on to placement inspection. Thus large placement errors are detected too late in the assembly process, possibly jeopardizing hundreds of thousands of parts, and thus it is desirable to have sufficiently large process margins. This is the reason why fan-out bonders with 5 μm at 3σ global accuracy capability, supporting more than twice the process margin, remove issues in mass production, even when 10 μm at 3σ is the formal global accuracy specification for the P&P process.

16.4 Equipment Qualification for Fan-Out Pick and Place

16.4.1 Step-by-Step Qualification

Assuming that a new fan-out bonder has been installed, and the bonder needs to be qualified for FO-WLP or PLP mass production, a full batch of reference production lots would have to be run through the fan-out bonder to assess the bonder performance. Since it has been mentioned already that there is no suitable method to verify the P&P capability directly by PBI of the populated die on the carrier (due to lack of alignment marks), the populated carriers would also have to run through the molding process before being able to do a precise placement inspection after debonding of the carrier.

Such a scenario, however, would not be effective, since the die locations would also be heavily influenced by the molding process. While in production such a process flow makes sense (inspection of the die locations before start of RDL building is a proper inspection step) [1], for a fan-out bonder qualification such a procedure would not be smart.

For this reason a four-step procedure is recommended to qualify a fan-out bonder for production:

1) Auto-diagnostic of the individual core capabilities of the fan-out bonder.
2) Running a large area (wafer- or panel-level) glass-on-glass (GoG) process.
3) Running a glass-on-carrier (GoC) process.
4) Running a reference production lot with silicon test die.

Placement of glass die onto glass substrate has the great benefit of being able to measure die misalignment directly and very accurately with a high resolution camera by evaluating the distance of die and substrate alignment marks in the same camera picture. By using concentric chrome circles on glass material with ≥200 μm diameter for the alignment marks, a measurement accuracy of 0.1 μm at 3σ can be achieved with a camera supporting ≤2 μm pixel resolutions (see Figure 16.8). Such high inspection accuracy might be surprising but can be explained by the fact that the image processing algorithms calculate pattern locations based on a very large amount of pixel information contained in the images.

16.4.2 Auto-Diagnostic of Core Capabilities

A powerful fan-out bonder provides semiautomatic or fully automatic diagnostic procedures to qualify the individual core capabilities, which are required to run a tool-to-tool repeatable mass production process. Tool-to-tool repeatability ensures that all machines operate almost identically (within prescribed machine tolerances) and thus allows the loading of one single “master” recipe

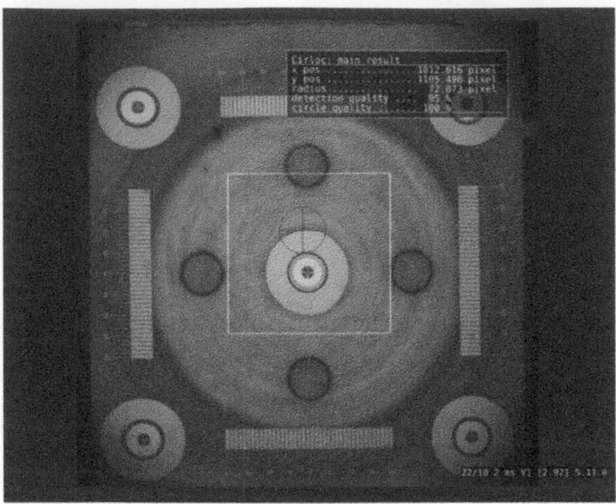

Figure 16.8 Measuring the misalignment of a glass die and glass substrate based on the center distance of concentric circular alignment marks in a single camera picture.

on any qualified production machine while yielding the same production output and quality. Fully automatic diagnostic procedures being supported on next-generation tools allow the machine to fetch required calibration or reference tools from a tool magazine, run the necessary diagnostic test, and show the test result at the procedure end. The test result provides information on whether the core capability being tested passed the repeatability criteria and thus is acceptable for production. Here is a list of the most important diagnostic procedures of such kind of system:

- Calibration of camera coordinate systems and illumination.
- Verification of PBI capability on glass die.
- Verification of flip tool alignment with ejection needles.
- Verification of work holder and bond head coplanarity.
- Verification of dipping station coplanarity.
- Verification of proper bond force calibration.
- Basic machine capability test (BMC test) (a repeated GoG process with a single glass die placed on a single glass substrate location)
- Verification of thermal capabilities (in the case of bond heat).

Special attention should be paid to the BMC test. This test verifies local placement capability with reference glass material on a single placement location. State-of-the-art fan-out bonders are equipped with an integrated standard BMC kit. Such a kit comprises a small glass substrate and a glass reference die located in a cavity. It allows the fan-out bonder to perform a BMC test with one button press. Once the BMC test is started, the machine fetches a reference P&P tool

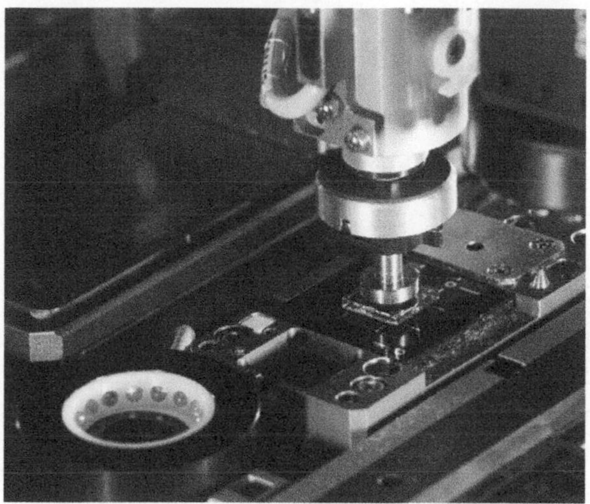

Figure 16.9 Integrated standard BMC kit with small glass substrate and reference glass die. The glass die can be temporarily fixed on the glass substrate by vacuum.

from a toolbox and picks the glass reference die from the cavity, aligns the die over the upward camera, performs an alignment on the glass substrate, and places the die onto the glass substrate where it is temporarily held in place by vacuum (see Figure 16.9). Subsequently the machine performs a PBI to measure and record the actual misalignment between glass die and glass substrate using the approach described earlier and shown in Figure 16.9. The procedure is then repeated a number of times, in these cases with the glass die being picked from the bond location instead of the cavity. After repeating this procedure of P&P and PBI 50–100 times, the glass die is placed back into the cavity, and statistical values like mean, standard deviation, and process capability number (C_{pk}) are calculated to assess the P&P process performance.

Based on the discussion so far, it should be clear that a single bond head-based BMC test with reference glass die and substrates with local alignment marks can reach a much higher placement accuracy specification than populating a large matrix of dies using two bond heads simultaneously, real production material, and only a few global alignment marks. Experience shows that the BMC accuracy should typically be two to three times higher than the accuracy expected in production. Thus it is a good empirical rule to demand a 5 μm at $C_{pk} \geq 2.5$ process with single bond head (local) BMC accuracy for a dual bond head fan-out bonder running a 5 μm at $C_{pk} \geq 1$ global accuracy process in production.

Successful performance of a BMC test, as is shown in Figure 16.10, gives confidence that all camera systems are installed correctly and performing well, the bond head is in good condition, and the machine is mounted properly on the production floor.

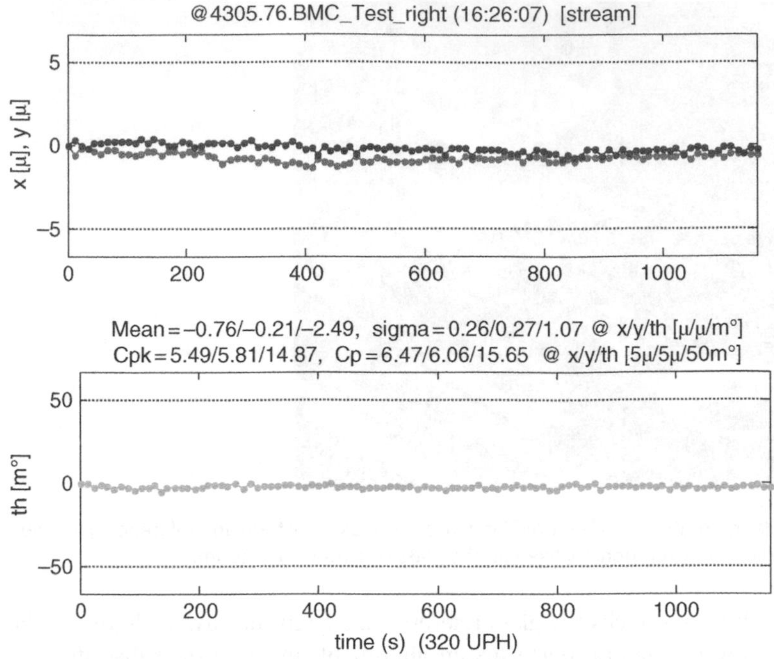

Figure 16.10 Results of a BMC test for a next-generation dual head fan-out bonder with 5μ at 3σ global accuracy and 3μ at 3σ local accuracy specification.

16.5 Running a Large Area Glass-on-Glass Process

Next, the P&P accuracy (x/y/theta) needs to be verified on the whole working area. In the first step, a qualification procedure needs to be performed to solely assess the placement capability of the bonder, keeping the influence of subsequent mass reflow or molding processes separate. This kind of capability test is usually done as GoG process. In the case of FO-WLP, such standard GoG process is based on a special carrier consisting of a standard FO-WLP steel carrier covered with double-sided adhesive tape on which an un-diced glass wafer is mounted, thus acting as a glass substrate. Usually this glass substrate would not require local fiducials, since the process to be qualified is a global accurate placement procedure. Providing a precise matrix of local fiducials, however, offers the convenient opportunity to verify the correct positioning of a glass die immediately after placement using the bonder's PBI function. As described before this inspection is very accurate since the misalignment between die and substrate alignment marks can be examined by pure vision metrology, without being influenced by inaccuracies from other subsystems such as the $x/y/z$/theta handling system. Note that the local alignment marks are not used for the actual P&P process, but only for PBI; thus the test verifies the global accuracy of the bonder.

Compared with the BMC test, the large area GoG test additionally verifies the ejection and flip process (the glass die is ejected from a wafer), as well as the important capability to perform a global accurate P&P of glass die on the whole working area. To avoid misalignment of the glass die between placement and PBI, an adhesive agent is used, such as spray glue applied to the glass carrier. For 5 or 3 μm accurate processes, it has been experienced that spray glue adds significant placement spread, which can be avoided if transparent sticky flux, applied with a roller onto the glass substrate, is used as an adhesive agent.

16.6 Running a Glass-on-Carrier Process

After successful qualification of the large area GoG process, which is purely a machine capability qualification, the next step is a GoC process qualification (see Figure 16.11). This cannot be considered as a machine capability verification since process materials like the temporary adhesive on the FO-WLP carrier (which might also require process heat) are used and have a measurable effect on the final placement accuracy. Using glass die for the GoC process is advantageous since the die locations can still be measured very accurately with external metrology equipment that supports global accurate measurements.

For an actual fan-out carrier, there are typically no local fiducials (except for die last processes), which is the reason why external metrology must be used to verify global accuracy. The global accuracy of the external metrology equipment itself should typically be a factor of 5–10 times greater than the global accuracy specification for the bonder in order to determine the global accuracy capability of the bonder with negligible influence from the metrology equipment. Usually, if the large area GoG process is capable, the GoC process should

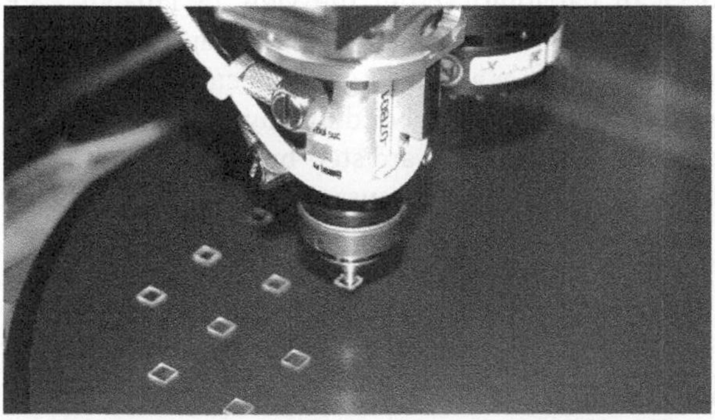

Figure 16.11 Large area glass-on-carrier process with 5 mm × 5 mm glass die on standard eWLB carrier.

also be capable. If this is not the case, then it is recommended to consider all pitfalls being discussed in the previous section.

16.7 Running a Reference Production Lot with Test Die

After successful qualification of large area GoG capability and GoC capability, the qualification of the fan-out bonder is usually complete. One can, however, also run an additional qualification based on silicon test die placement on a process carrier, which also has to run through the molding process. Note that the location of face-down silicon die can neither be inspected accurately on the carrier inside the fan-out bonder, nor is it possible to do this task on external optical metrology equipment based on a wavelength that cannot pass through silicon. After molding, the carrier is debonded, and the placement of the embedded die in the mold compound is finally inspected with external metrology equipment. The pitch of the P&P process typically has to compensate for the shrinkage in the molding process, and thus the fan-out bonder provides a software feature to apply this compensation. For high volume production this is usually an iterative procedure that is automated based on a systematic feedback loop [1]. Once this reference production lot has been qualified, the entire qualification of the fan-out P&P process can be considered complete.

16.8 Conclusions

In this chapter, the requirements for a fan-out bonder have been introduced by going through the core capabilities of an advanced die attach machine. These are die feeding, substrate handling, die flipping, fluxing, constant bond heat, pulse heat, accuracy, and clean capability. From the view of fan-out P&P requirements, it is important to know whether the process is chip-first or chip-last, or face-up or face-down, and whether a local or global accuracy placement process is in use. Important aspects of fan-out bonding have been covered in order to avoid pitfalls, and finally a straightforward step-by-step procedure for the qualification of a fan-out bonder has been discussed.

References

1 Martins, A., Pinho, N., and Meixner, H. (Fall 2012). Systematic feedback loop for silicon die pick and place process in reconstituted fan-out FO-WLP wafers in high volume production. *International Symposium on Microelectronics* 2012.

2 Brunnbauer, M., Fürgut, E., Beer, G. et al. (2006). An embedded device technology based on a molded reconfigured wafer. *2006 IEEE 56th Electronic Components and Technology Conference.*

3 Brunnbauer, M., Meyer, T., Ofner, G. et al. (2008). Embedded wafer level ball grid array (eWLB). *33rd International Electronics Manufacturing Technology Conference 2008.*

4 Keser, B., Amrine, C., Duong, T., et al (2007). The redistributed chip package: a breakthrough for advanced packaging. *2007 IEEE 57th Electronic Components and Technology Conference*, pp. 286–291.

5 Tseng, C.-F., Liu, C.-S., Wu, C.-H., and Yu, D. (2016). InFO (Wafer Level Integrated Fan-Out) Technology. *2016 IEEE 66th Electronic Components and Technology Conference.*

6 Rogers, B., Sanchez, D., Bishop, C., et al. (2015). Chips Face-up Panelization Approach For Fan-out Packaging. *2015 International Wafer Level Packaging Conference.*

7 Amkor Technology. Silicon Wafer Integrated Fan-out Technology (SWIFT™) Packaging for Highly Integrated Products, Amkor SWIFT/SLIM; white paper. www.amkor.com

8 Kim, Y.R., Kim, Y., Bae, J. et al. (2017). SLIM™, high density wafer level fan-out package development with submicron RDL. *2017 IEEE 67th Electronic Components and Technology Conference.*

9 Mahajan, R., Sankman, R., Patel, N., et al. (2016). Embedded multi-die interconnect bridge (EMIB) – a high density, high bandwidth packaging interconnect. *2016 IEEE 66th Electronic Components and Technology Conference*, pp. 557–565.

10 Kröhnert, S. (2016). High density package integration for wearables and IoT applications by WLFO based WLSiP and WLPoP. *European 3D Summit 2016.*

11 Pristauz, H. (2016). Advanced die attach platform for advanced packages. *Advanced Packaging Conference*, Grenoble (October 2016).

12 Zarchan, P., Musoff, H., and Lu, F.K. (2015). *Fundamentals of Kalman Filtering: A Practical Approach*, 4e. American Institute of Aeronautics.

13 ISO 14644-1 (2015). *Cleanrooms and Associated Controlled Environments – Part 1: Classification of Air Cleanliness by Particle Concentration.* International Organization for Standardization.

17

Process and Equipment for eWLB

Chip Embedding by Molding

Edward Fürgut[1], Hirohito Oshimori[2], and Hiroaki Yamagishi[2]

[1] *Infineon Technologies AG*
[2] *Apic Yamada Corporation*

17.1 Introduction

Today, semiconductor devices with rapidly changing features are the basis for applications that deliver comfort, communication, mobility, environmental, and medical care. New requirements for reliability or environmental constraints are the driving factors to develop new materials, processes, and equipment. During recent years, molding process development has became increasingly important. The development of new encapsulation methods including processes and materials allowed manufacturing of high performance, reliable, and cost-efficient microelectronic devices.

In this chapter a short history of molding is provided, which highlights that the introduction of compression molding was a major change for semiconductor assembly and packaging. Why compression molding was the key for the implementation of the innovative extended wafer-level ball grid array (eWLB) technology in 2006 is described. Next, the compression molding process for the new fan-out technology and the principle challenges are introduced. Finally, processing solutions that enable fully automatic compression molding with newly developed equipment are described. The new compression molding system was a significant breakthrough in semiconductor assembly and packaging technology. Conclusions and next steps complete this chapter.

Advances in Embedded and Fan-Out Wafer-Level Packaging Technologies, First Edition.
Edited by Beth Keser and Steffen Kröhnert.

17.2 Historical Background Molding

Up until the 1960s, hermetically sealed metal and glass packages were standard. In these early years of microelectronics, encapsulation by liquid glob top material was used to protect wire-bonded semiconductor devices from environmental influences [1]. Encapsulation material improvements, especially in ionic purity, mechanical properties, thermal stability, and minimization of hazardous contaminants, resulted in enhanced long-term device reliability and performance.

In the 1980s, transfer molding was introduced for microelectronic packaging (see Figure 17.1). During transfer molding, the encapsulation material is transferred from the pot into the cavity by plunger pressure. The material is usually in the form of a pellet. In the early days of transfer molding, cost-efficient manufacturing was considered a higher priority than component reliability. The transfer molding method was first possible with mono plunger systems based on a hydraulic press. These mono plunger mold tools were designed and manufactured mainly by semiconductor companies.

In the late 1980s, new requirements in the semiconductor industry, e.g. higher quality and yield combined with the request for miniaturization and increased functionalities, caused the need for the development of multi-plunger mold tools with a toggle press. These multi-plunger mold tools were designed and produced at pure toolmakers because of their high complexity. The mold tool is the main part of the process; therefore some of the former pure toolmakers are today's full equipment suppliers. Because of the proven advantages in terms of productivity, material waste minimization, and proper maintenance efforts, the transfer molding process with multi-plunger systems became a standard [1]. During the 1990s, the need to encapsulate larger and thinner chips showed the limit of transfer molding for long material flow distances. Thus, new encapsulation methods were required.

In 1997, Apic Yamada introduced compression molding for encapsulation of large and thin cavities using liquid molding compound (LMC). This latest innovation of introducing LMC and compression molding in semiconductor

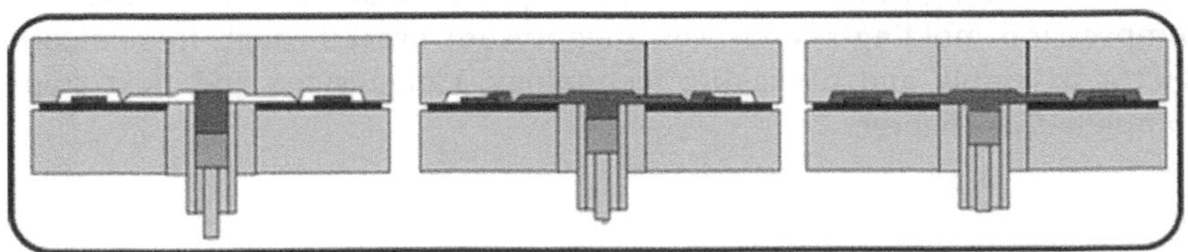

Figure 17.1 Transfer molding process schemes for single cavity lead-frame packages showing three process steps: (left) pellet melting inside pot, (middle) mold flow through runners and gates, and (right) complete mold filling of all mold tool cavities.

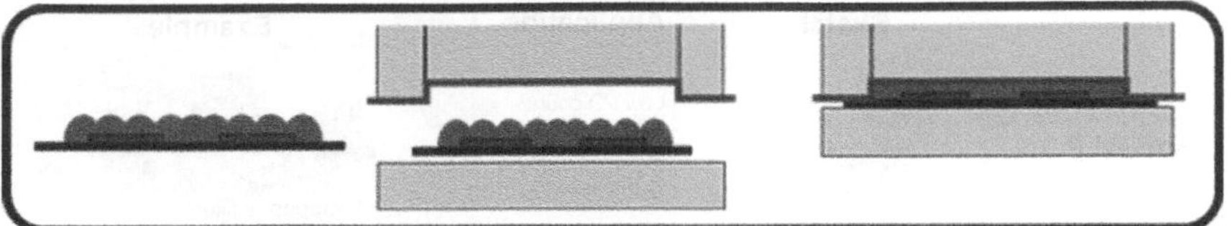

Figure 17.2 Compression molding scheme of a molded array package (MAP) based on substrate: (left) the molding compound is deposited on the chips that are fixed to a carrier, (middle) the unmolded device is placed on the bottom mold tool, and (right) the liquid molding compound fills the space between the chips fixed to the carrier and the cavity to shape the MAP.

encapsulation was first used for wafer backside coating. The introduction of compression molding was a major change in assembly and packaging technology. The origin of compression molding was the bakery industry followed by the general plastic industry. The plastic industry used compression molding first to replace metal with plastic parts. The typical applications were to manufacture parts such as rubber boots, scoops, fenders, and spoilers, as well as smaller and more complex parts.

The compression molding concept combines compound forming and pressure in a mold cavity. The compound is placed directly into a heated mold cavity. During closing and applying force, the compound is softened and formed by the mold cavity [2]. Figure 17.2 schematically shows the compression molding process.

In the next chapter the importance of compression molding for the eWLB technology (fan-out WLP) is demonstrated.

17.3 The Molded Wafer Idea: Key for the Fan-Out eWLB Technology

For standard wafer-level packaging (WLP), solder balls are mounted on semiconductor wafers. These wafers with solder balls are then diced and mounted on a printed circuit board. Before solder ball mount, an additional thin film redistribution layer (RDL) can be added (see Figure 17.3). These technologies successfully entered into the semiconductor industry mainly for small chips (< 5 mm × 5 mm). The major obstacle for WLP was the transfer of the high density pin count (I/O) on chip level to the printed circuit board (PCB). Due to the large difference in coefficient of thermal expansion (CTE) between silicon (3 ppm K^{-1}) and the PCB (typically 14–20 ppm K^{-1}), chips with an area larger than 5 mm × 5 mm as a rule of thumb require underfill to achieve sufficient board reliability. Thus, we need a fan-out area for small and high pin count chips. This leads to the fan-out eWLB package as shown in Figures 17.3 and 17.4.

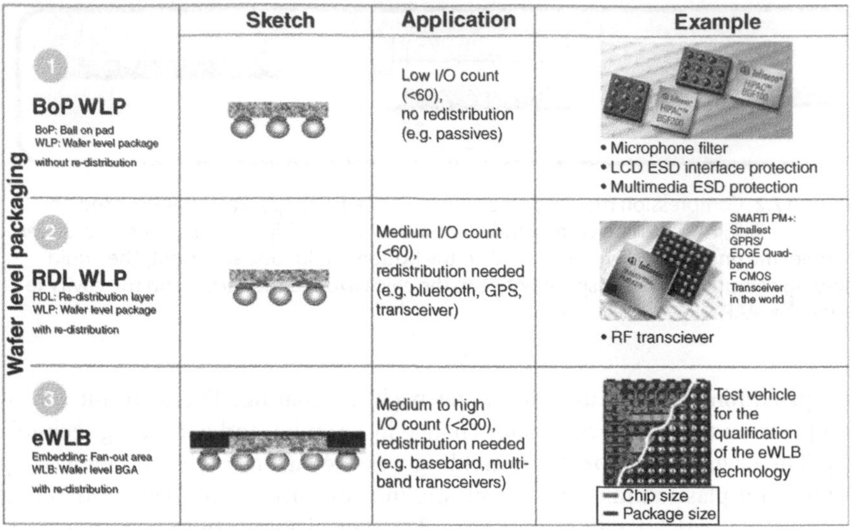

Wafer level packaging	Sketch	Application	Example
1 **BoP WLP** BoP: Ball on pad WLP: Wafer level package without re-distribution		Low I/O count (<60), no redistribution (e.g. passives)	• Microphone filter • LCD ESD interface protection • Multimedia ESD protection
2 **RDL WLP** RDL: Re-distribution layer WLP: Wafer level package with re-distribution		Medium I/O count (<60), redistribution needed (e.g. bluetooth, GPS, transceiver)	SMARTi PM+: Smallest GPRS/ EDGE Quad-band F CMOS Transceiver in the world • RF transciever
3 **eWLB** Embedding: Fan-out area WLB: Wafer level BGA with re-distribution		Medium to high I/O count (<200), redistribution needed (e.g. baseband, multi-band transceivers)	Test vehicle for the qualification of the eWLB technology — Chip size — Package size

Figure 17.3 Comparison of ball on passivation (no RDL) WLP, RDL WLP, and fan-out eWLB technologies.

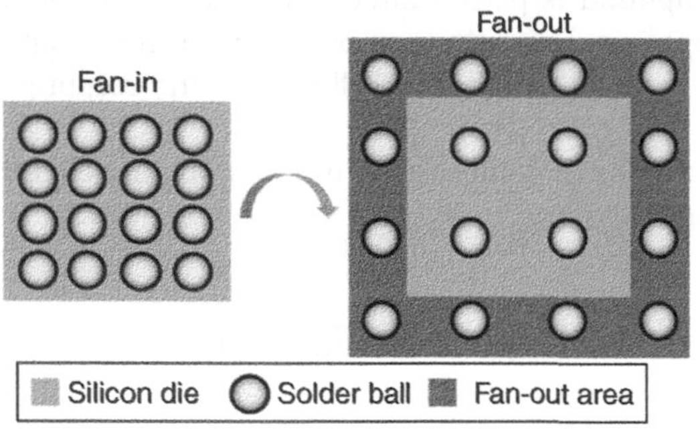

Figure 17.4 From fan-in to fan-out technology.

The fan-out eWLB technology fills the interconnect gap, allowing for the ongoing miniaturization of the pitch on the chip and the typically non-changing pitch on the application board [3].

Infineon was the first company to successfully investigate and introduce a fan-out eWLB package into the market. The basic idea of this new package concept was to dice the wafer into single chips and embed these chips in molding compound to form a reconstituted molded wafer. Figure 17.5 shows on the left the four main processes for creating the reconfigured wafer: tape lamination on carrier, chip placement, encapsulation (molding), and debonding. On the right of Figure 17.5, an image of the final reconfigured wafer with the die embedded in mold compound is demonstrated [4].

Figure 17.5 The four main processes for creating the reconfigured wafer via compression molding: tape lamination on carrier, chip placement, encapsulation and debonding (left) and image of the final reconfigured wafer with the die embedded in molding compound (right).

This molded wafer (reconfigured wafer) is the core of the new fan-out eWLB technology. The introduction of this molded wafer applying compression molding led to the launch of fan-out WLP. Further details on compression molding to manufacture the molded wafer are presented in the next section.

The new eWLB packaging concept based on a fan-out molded wafer enables new design opportunities in comparison with WLP. The concept offers the following main advantages:

- Integration of chips with different sizes in a standardized predefined package (example: memory chips of different generations).
- Embedding of different chips side-by-side into one system in a package (SiP).
- Elimination of bump and wire bond to improve the electrical and thermal performance of the package.

SiP using eWLB technology provides products with high integration potential capability, design flexibility, and high interconnection density. Figure 17.6 shows a plot of the chip size versus package size [5]. The advantage of the molding process is the capability to cover a large range of chip and package sizes. A new package platform is generated, which also allows multi-chip

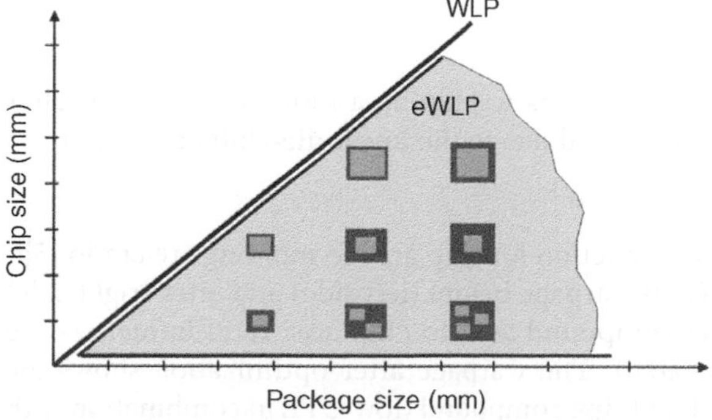

Figure 17.6 Flexibility of the eWLB package platform in comparison with the limitation of the WLP package platform.

integration. The following Figure 17.6 shows the dimension restriction of WLP in comparison with the dimension flexibility of eWLB. The package and chip have the same size (see the WLP line with slope 1) for the WLP package platform. The eWLB package platform offers improved flexibility with respect to chip size and multi-chip side-by-side integration (see the application below the WLB line, e.g. with slope <1). A multi-chip solution with 3D integration is required to address the slope >1.

Wafer front-end manufacturing has a significantly higher degree of automation compared with assembly and packaging manufacturing (back end). The development target was an encapsulation process with the highest possible automation to enable a manufacturing environment similar to front-end production. The high degree of automation for the eWLB package enables a high manufacturing yield in spite of the significantly increased number of process steps compared with typical back-end manufacturing.

The existing manufacturing methods within the semiconductor industry did not provide the following required features to fabricate a molded wafer:

- Temporary bonding of chips.
- Molded wafer with thermal stability.
- Molded wafer with identical dimension to SEMI specification [6].

Thus, a complete new manufacturing concept and encapsulation process had to be developed. Different encapsulation methods as shown in Figure 17.7 were considered.

The investigation and evaluation of the complex and new manufacturing process required many years of systematic feasibility studies in materials, processes, and automation. This included, for example, careful studies of die shift, wafer warpage, large and thin wafers, etc. Figure 17.8 shows an example of a die-shift optimization; Figure 17.9 shows an example of a warpage optimization [5].

Figure 17.8 demonstrates the eWLB die shift before and after optimization. The die shift before optimization (left side) increases from center to wafer edge:

$$X_{\text{offset}} = X^{*} C_{\text{correction}}$$

The die shift is related to mold area to chip area ratio. The die shift after optimization (right side) is improved due to the linear die-shift correction:

$$X_{\text{corrected}} = X_{\text{ideal}}{}^{*}\left(1 - C_{\text{correction}}\right)$$

The model for $C_{\text{correction}}$ is a function for chip area to molding area ratio [5].

Figure 17.9 shows the eWLB warpage before (left side) and after (right side) optimization. The molding compound area to chip area ratio influences the wafer warpage (CTE mismatch). The warpage after optimization shows the wafer warpage of improved molding compound (low CTE in combination with low E-modulus). The molding and debonding is also optimized to achieve a low warpage [5].

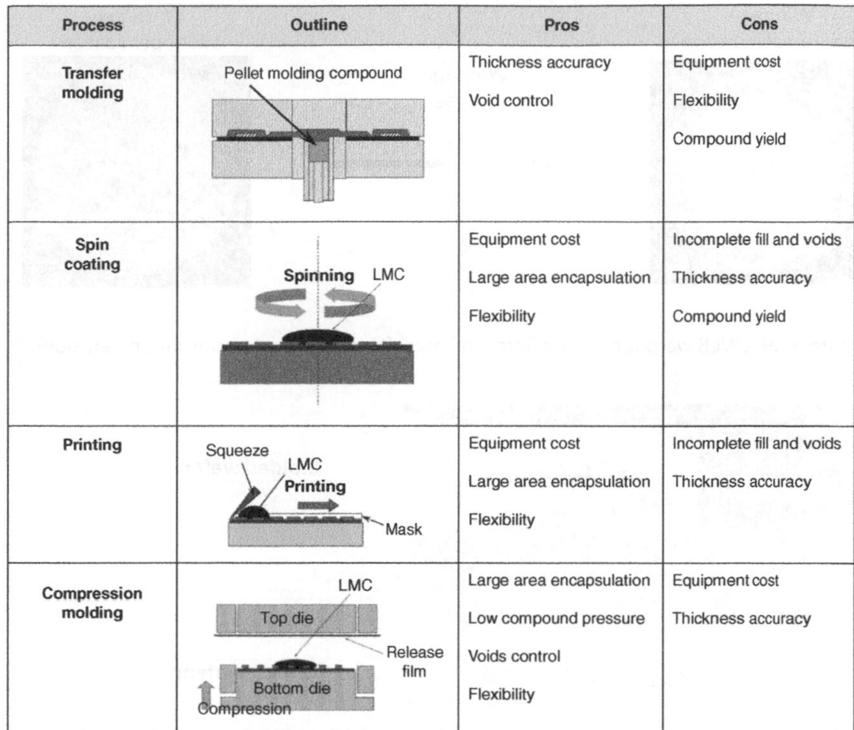

Process	Outline	Pros	Cons
Transfer molding	Pellet molding compound	Thickness accuracy Void control	Equipment cost Flexibility Compound yield
Spin coating	Spinning LMC	Equipment cost Large area encapsulation Flexibility	Incomplete fill and voids Thickness accuracy Compound yield
Printing	Squeeze LMC Printing Mask	Equipment cost Large area encapsulation Flexibility	Incomplete fill and voids Thickness accuracy
Compression molding	LMC Top die Release film Bottom die Compression	Large area encapsulation Low compound pressure Voids control Flexibility	Equipment cost Thickness accuracy

Figure 17.7 Molded wafer encapsulation methods with their pros and cons.

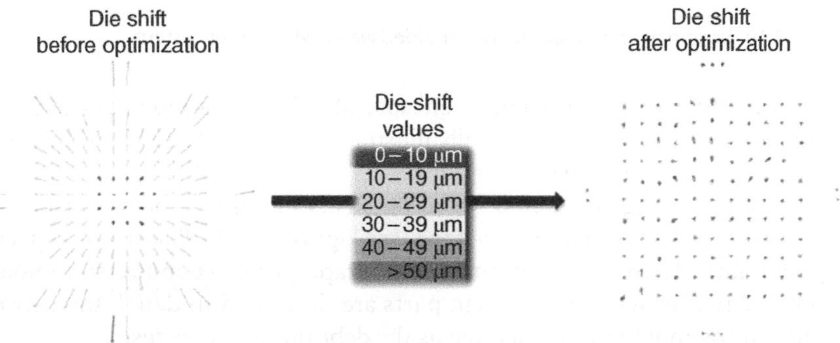

Figure 17.8 eWLB die shift before (left side) and after (right side) optimization.

Based on the research, the following main topics were identified as crucial:

- Temporary rigid carrier (carrier system)
- LMC
- Compression molding process

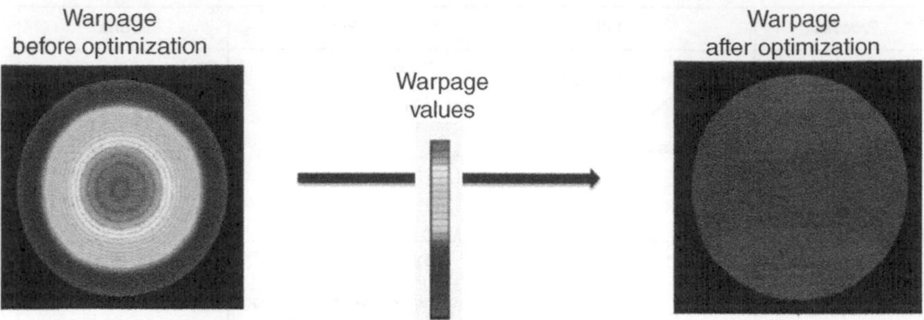

Figure 17.9 eWLB warpage before (left side) and after (right side) optimization (left side).

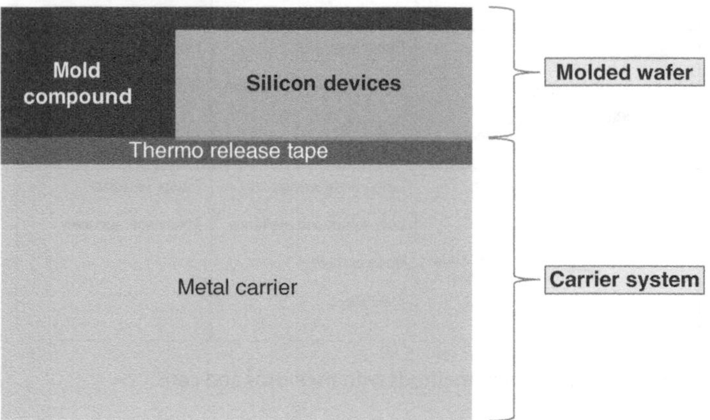

Figure 17.10 Schematic cross section of a molded wafer with carrier system.

A new manufacturing concept to encapsulate chips in a molded wafer was developed. Figure 17.10 schematically illustrates the cross-sectional scheme of a molded wafer and the carrier system [5].

The manufacturing concept is based on a carrier system that consists of a metal carrier and a thermal release tape (see Figure 17.10). The silicon devices are placed face-down on the thermal release tape prior to compression molding with LMC. Below, the two main parts are described in detail, the carrier system and the molded wafer, as well as the debonding processes.

17.3.1 Carrier System

The carrier system consists of two components, the metal carrier and the thermal release tape. The CTE of the metal carrier and molded wafer, which consists of the LMC and silicon die, needs to be properly adjusted to each other. The CTE of LMC is in the range of 6–14 ppm K^{-1}, whereas the CTE of the

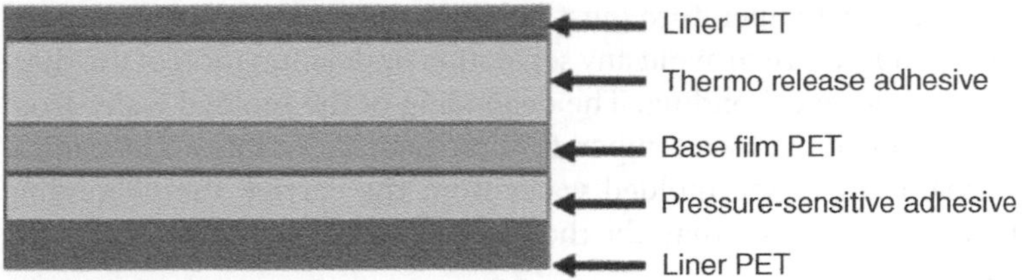

Figure 17.11 Cross-sectional scheme of the thermal release tape; multiple layers are required for the tape processing (lamination).

silicon die is about 3 ppm K^{-1}. Therefore, thermal expansion of the molded wafer may be described by an effective CTE of 4–12 ppm K^{-1}. Additional properties required of the metal carrier are handling robustness and low warpage.

The thermal release tape fixes the chips during molding and defines the active side of the molded wafer. The thermal release tape is a double-sided adhesion tape with different properties on both sides. On one side, a thin layer with pressure-sensitive adhesive glue is applied. On the opposite side, a thin layer with thermal release adhesion glue is applied. Figure 17.11 shows a cross-sectional scheme of a thermal release tape that is a key material for the eWLB manufacturing concept.

The unique tape function is called thermal release, which means a high adhesive adhesion (fixing of the die during molding) is changed to no or very low adhesive adhesion (debonding of the carrier system) at a defined temperature. Figure 17.12 shows the adhesion force of a thermal release adhesive glue layer in relation to the heat treatment temperature.

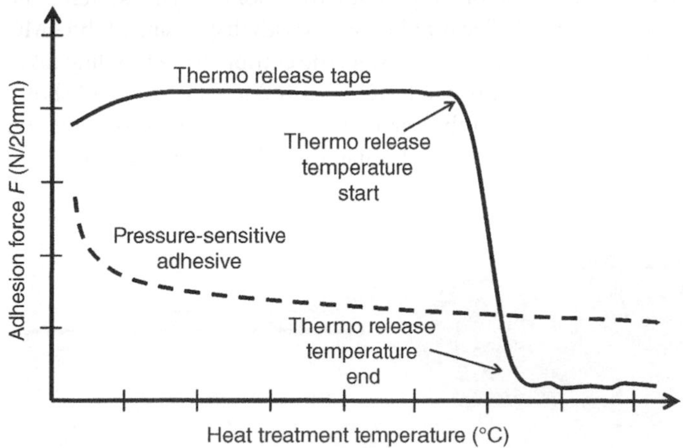

Figure 17.12 Adhesion force of a thermal release glue and a pressure-sensitive glue in relation to the heat treatment temperature.

The thermal release temperature must be above the molding and post mold curing (PMC) temperature to avoid any separation or delamination of the thermal release tape before debonding. The debonding of the molded wafer from the carrier system is done by a temperature process. To enable a mechanical stress-free separation of the molded wafer from the carrier, the debonding temperature has to be higher than the thermal release temperature.

17.3.2 Molded Wafer

The molded wafer consists of LMC and silicon die. The semiconductor chips are placed either face-down or face-up on the carrier system (metal carrier and thermal release tape). The opportunity to place only the known good die (KGD) enables an increased package yield especially in multi-chip configurations. Key factors are position accuracy between the chips and accuracy of the whole chip array. The carrier system consists of very accurate global alignment marks for pick and place and location holes to assure an accurate compression molding (see the Chapter 16 on pick and place by BESI). Figure 17.13 shows a 3D sketch (left) of the carrier system with the die after pick and place. On the right, the corresponding scheme of the cross section is shown.

The LMC is the key element of the molded wafer. The molded wafer consists of LMC and chips. Therefore the LMC properties define the molded wafer features like warpage and temperature stability. Compared with solid molding compounds, the LMC allows molding at lower temperatures. Details about the LMC material are described in the separate chapter of this book called "The Role of Liquid Molding Compounds in the Success of Fan-Out Wafer Level Packaging Technology" by Nagase.

After placement of semiconductor chips onto the mold carrier system, the encapsulation process takes place. The molding starts with dispensing of the LMC in the center of the carrier system. In the next step, the compression molding takes place. This is followed by PMC to enhance LMC cross-linking. Figure 17.14 shows the final molded wafer together with the carrier system below. The compression molding process will be explained in more detail in the subsequent sections.

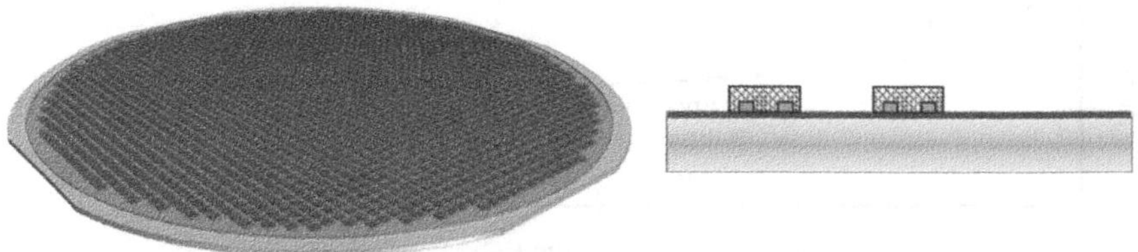

Figure 17.13 Carrier system with chips after pick and place (left) and the corresponding cross-sectional scheme of chips mounted face-down on the carrier system (right).

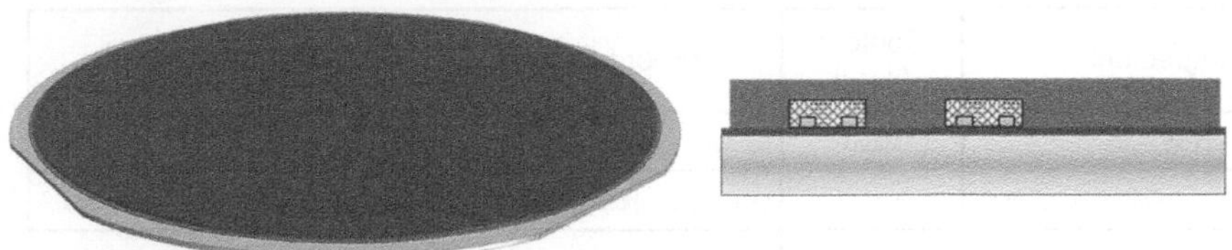

Figure 17.14 3D sketch of the molded wafer together with the carrier system below (left) and a scheme of the cross section (right).

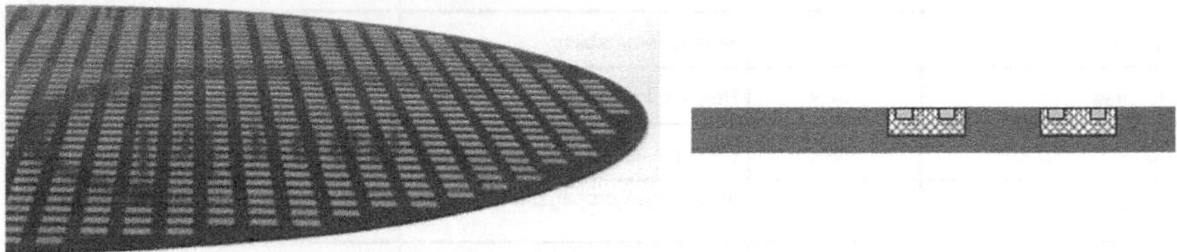

Figure 17.15 Photo of molded wafer after debonding (left) and the corresponding scheme of a molded wafer after debonding (right).

17.3.3 Debonding

After compression molding, the molded wafer will be separated from the carrier system using high temperature. The thermal release tape is removed at a defined high temperature. Figure 17.15 shows the molded wafer after debonding.

17.4 The Compression Molding Process

In this chapter, first the molding compound ingredients are introduced. Then, an overview of molding compound states of aggregation, cavity-down and cavity-up processing methods, molding compound preparation, temperature of compression molding, cavity filling, and mold release is given. This chapter concludes with a discussion of a comparison between transfer molding and compression molding capabilities.

17.4.1 Molding Compound Ingredients

Generally, the compression molding process can be performed with different polymer types like thermoplastic and thermoset materials. In the following, thermoset materials (molding compound) are the focus, which are most important for semiconductor packaging. The reasons are the required high temperatures (>175 °C) during package soldering and the high resistance to the

Ingredient	Content (wt%)	Major function	Typical agents
Epoxy resin	Matrix	Binder	Cresol – novolac
Hardener (curing agents)	up to 60	Linear/cross-polymerization	Amines, phenols, acid anhydrides
Accelerators	<1	Enhance rate of polymerization	
Fillers	up to 90	Reduce CTE and chemical shrinkage, higher E-modulus, reduce resin bleed, reduce moisture uptake	Fused silica, crunched silica
Flame retardants	<10	Retard flammability	Brominated epoxies, antimony trioxide
Release agent	Trace	Release form mold surface	Silicone, hydrocarbon waxes
Adhesion promoter	Trace	Enhance adhesion	Silane, titanates
Stress relief additives	<10	Inhibit crack propagation, lower CTE	Silicone, rubber, polybutyl acrylate,
Coloring agent	<1	Reduce device visibility, improve device marking	Carbon black

Figure 17.16 Ingredients of molding compounds that are used for encapsulation of semiconductor devices.

environment. The molding compound consists of a multitude of different ingredients. The most important ingredients are the filler and resin, including hardener. In addition, optional ingredients are flame retardant, wax, adhesion promoter, catalyst, stress absorber, coloring agent, and others. Figure 17.16 shows typical ingredients of a molding compound, content (wt%), major functions, and typical agents.

The mechanical properties like CTE and chemical shrinkage are influenced mainly by the filler content, which can be typically up to 90 wt%. The lowest CTE and chemical shrinkage requires the highest filler loading. The highest filler loading can be achieved by, for example, a tribology of spherical fillers. Figure 17.17 shows a cross section of a typical molding compound used for semiconductor packaging.

17.4.2 Molding Compound State of Aggregation

Compression molding is able to process molding compound with different states of aggregation. The mainly used states of aggregation for molding compounds are:

- Granulate molding compound (GMC)
- LMC
- Sheet molding compound (SMC)

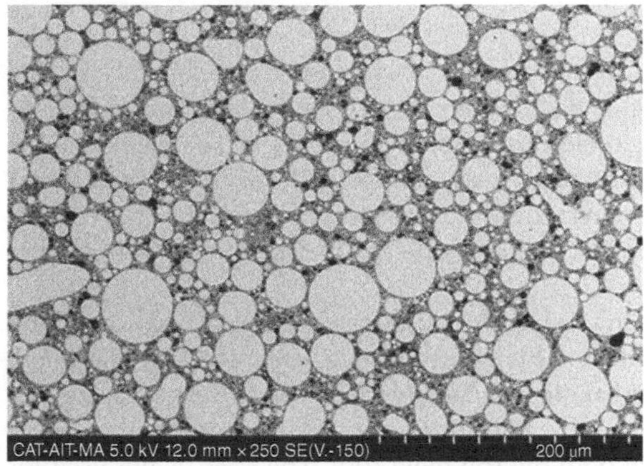

Figure 17.17 Cross section of a typical molding compound for semiconductor device application.

The most commonly used material for compression molding in the back end is the GMC. This aggregation state is based on a solid molding compound. The liquid state of aggregation is used mainly in front-end production for photomasks or polyimide passivation via spin coating. The sheet molding compound is well known for lamination in the PCB industry. Figure 17.18 shows the pros and cons of the three states of aggregation used for compression molding.

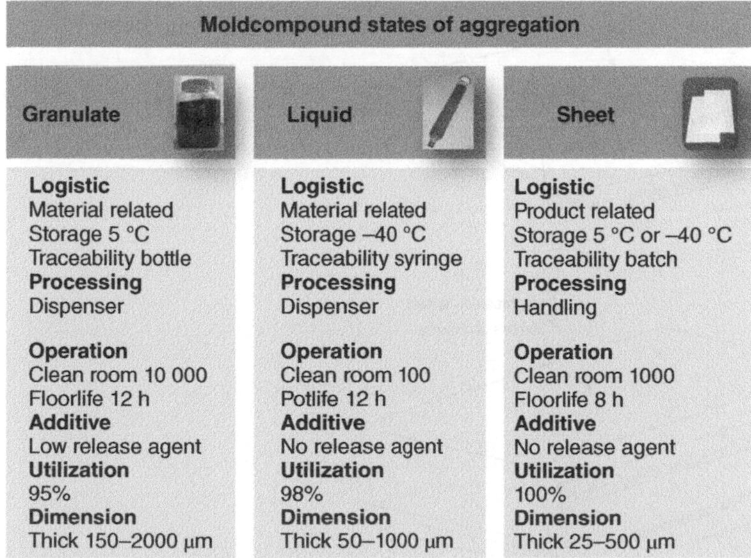

Moldcompound states of aggregation
Granulate
Liquid
Sheet
Logistic
Material related
Storage 5 °C
Traceability bottle
Processing
Dispenser
Operation
Clean room 10 000
Floorlife 12 h
Additive
Low release agent
Utilization
95%
Dimension
Thick 150–2000 μm
Logistic
Material related
Storage –40 °C
Traceability syringe
Processing
Dispenser
Operation
Clean room 100
Potlife 12 h
Additive
No release agent
Utilization
98%
Dimension
Thick 50–1000 μm
Logistic
Product related
Storage 5 °C or –40 °C
Traceability batch
Processing
Handling
Operation
Clean room 1000
Floorlife 8 h
Additive
No release agent
Utilization
100%
Dimension
Thick 25–500 μm

Figure 17.18 Molding compound states of aggregation for the compression molding process.

17.4.3 Processing Methods

Compression molding is a forming process, in which a plastic material is placed directly into a heated metal mold, then is softened by the heat, and is pressed into the mold cavity during mold closing (see Figure 17.19). The melting of the mold compound, forming, and final pressure is formed within the mold cavity. There are two different relevant methods for compression molding. Figure 17.19 shows the two different cavity molding process methods: cavity-down and cavity-up [2].

The cavity-down approach (Figure 17.19 left) places the molding compound directly into the bottom cavity. This approach is usually used for GMC and SMC. The reason is the fast handling capability that allows a direct placing into the bottom mold cavity. For this cavity-down processing method, the carrier system is placed face-down onto the top mold tool. The LMC is not used for the cavity-down approach because of the tackiness of the uncured LMC and the relative long dispense time compared with GMC and SMC. The advantage of the cavity-up approach is the possibility to distribute the molding compound over the area to minimize the mold flow. This method is also called compression molding by dipping. For the cavity-up approach (Figure 17.19 right), the molding compound is placed onto the carrier system prior to compression, usually using LMC.

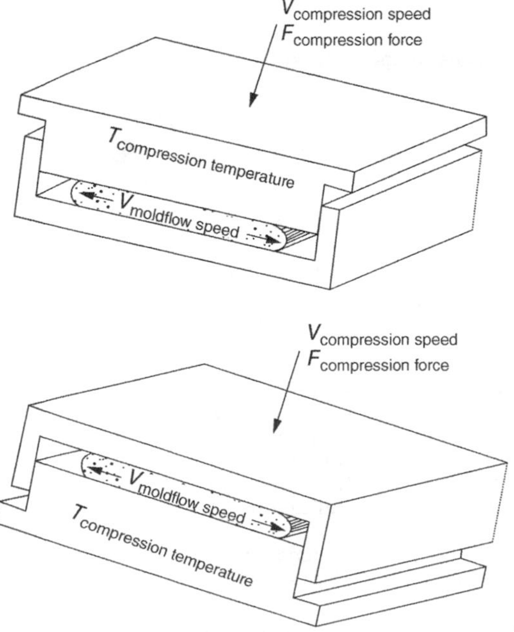

Figure 17.19 Compression molding methods cavity-down (top) vs. cavity-up (bottom).

17.4.4 Molding Compound Preparation

Before the cavity filling process can start, certain procedures must be accomplished to prepare the molding compound to be molded. This typically involves dispensing, weighing, and placement of the molding compound [4]. In the past, these processes were mainly done manually. To ensure the required quality and throughput for the semiconductor market, the processing must be automated. Slight changes in the amount of the molding compound or displacement can lead to unacceptable results.

GMC is usually dispensed by a vibration mechanism. GMC can be easily distributed over the area to reduce mold-filling issues. The GMC dust generation has to be managed by the molding equipment – especially by the dispenser. LMC is usually applied by a volume dispensing mechanism. Dispensing of GMC and LMC is time consuming. Therefore it is performed under room temperature before starting the molding process. A weight control loop is recommended to ensure accurate dispensing. The required molding compound weight (g) is implemented very precisely via software control. This is a main advantage compared with molding compound pellets that are used for transfer molding. The *in situ* weight control reduces the molding compound logistic complexity.

SMC can be tailored at the supplier or directly prior to the molding. Fully automatic handling is critical for SMC because of the flexibility and tackiness of the uncured sheet. The logistic complexity is increased due to product-related variants, e.g. package thickness, chip dimension, etc. A variation of packages end up in several different SMC dimensions, for example, sheet thickness and sheet size. The molding compound volume in combination with the mold tool dimension defines the thickness of the compression molded part. The volume (cm^3) is calculated by the product of the molding compound weight (g) and specific gravity ($g\,cm^{-3}$). The deviation of molding compound volume results directly in a thickness deviation. In addition, the placement accuracy has a strong impact on the cavity filling behavior of the molding compound. Displacement of the SMC can result in incomplete fill (mold flow too long) or mold flash (mold flow too short).

Although the molding compound preparation step may seem trivial, it has the potential to be the root cause of many problems that appear throughout the entire molding process. This eventually leads to flaws in the molded part, for example, incomplete fill or internal voids [4]. Careful analysis and control of this step must be taken to assure that it does not create unnecessary problems like flow marks or molding compound segmentation. The molding compound preparation process has to be developed to have no impact to the cycle time of the overall compression molding process [2].

17.4.5 Compression Molding Temperature

Compression molding of thermosets uses a constant and homogeneous temperature within the top and bottom mold tool. The automatic temperature control technique of the mold tool ensures a surface temperature with low deviation. The surface temperature has to be controlled within ±5 °C or even ±3 °C to ensure a constant process quality. The molding temperature is related to the molding compound used. LMCs are usually processed in the range of 120–150 °C. GMCs are usually processed in the range of 150–190 °C.

The mold tool temperature combines two different functions. First, the mold tool temperature heats up the molded part and molding compound. The GMC requires melting (molding compound condition changes from solid to liquid) before it can be processed. The LMC can be processed at room temperature or between room temperature and the molding temperature. The viscosity of the molding compound is related to the temperature. Higher molding temperatures reduce the viscosity of the molding compound in the beginning and shortens the processing time due to the high degree of molding compound cross-linking. A lower molding temperature enlarges the processing time and has a negative impact on the molding compound viscosity during mold filling. Figure 17.20 shows the viscosity of GMC and LMC materials vs. molding process time.

Second, the mold tool temperature drives cross-linking of the molding compound. The molding compound has to be sufficiently cross-linked to enable proper release and robust handling. A higher molding temperature increases the cross-linking speed of the molding compound. Thus, it reduces the required curing time. The mold tool curing time mainly determines the throughput of the whole molding process.

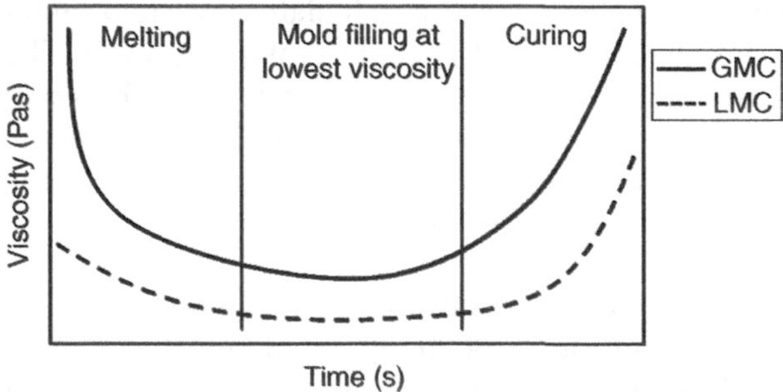

Figure 17.20 Viscosity of GMC and LMC material vs. molding process time.

17.4.6 Cavity Filling During the Compression Molding Process

The solid molding compound (e.g. GMC) curing reaction and its mechanism of solidification create a network of tightly connected molecules. Molding compounds are typically placed on the bottom mold tool that is 175 °C at least and starts melting during the heating-up process. The temperature can be seen as a catalyst for the curing process of any epoxy molding compounds. Once the molding compound is in full contact with the molding tool, the temperature transfer into the molding compound takes place immediately. The molding compound is a bi-stage material. This means it includes a solid and a fluid state of aggregation. As a result of heating up the molding compound, the fluidic stage of the material ratio is drastically increased. This means the molding compound can be considered as fluid. This stage of aggregation is most important to be used for filling the cavity. The curing of the molding compound is continuously ongoing and strongly temperature dependent: commonly an increase of 10 °C will double the reaction speed, whereas a decrease of 10 °C will halve the reaction speed. As an effect, the molding compound changes its stage of aggregation back to solid where the ratio from fluid stage declines and the solid stage increases again. The molding compound viscosity curve is shown in Figure 17.20. The compression is related to the viscosity of the material and the molded area. A high-end pressure is required to obtain a solid bulk molding compound with as little air entrapment as possible. LMC behaves differently from the GMC because of its low viscosity, and its workability at lower temperatures (even at room temperature) is already given. In addition, the LMC enables the whole encapsulation process to be driven at lower temperatures.

The flow speed of the molding compound has an important role for proper cavity filling. The optimized flowing speed of the material should be in the range of 1–10 mm s^{-1}. The molding compound distribution prior to the filling of the cavity reduces the impact of molding compound flow speed significantly. This is mainly used for GMC and SMC with the cavity-down concept. This is the so-called "dipping" compression molding process concept.

The GMC distribution prior to material filling for cavity-up has some limitations. Figure 17.21 shows the relation of compression speed versus molding compound flow speed. Especially for large and thin cavities, the compression needs to be specifically low to ensure an optimized molding compound flowability. The mold flow can be calculated as seen in Figure 17.21.

Another important parameter to ensure proper cavity filling is the end pressure. The final pressure reduces the internal and external voids and brings the molding compound into the final shape. In the compression molding process, the final pressure is in the range of 20–50 bar. Therefore it is significantly lower than in the transfer molding process where it is in the range of 50–100 bar. The lower final pressure results because the pressure is inserted directly into the cavity. The final pressure in the compression molding process p (N mm^{-2}) depends on the clamp force F (N) and the molding compound area A (mm^2):

$$p = F/A$$

The required clamp force for a cavity area with a diameter of 300 mm is, for example, in the range of about 300 kN.

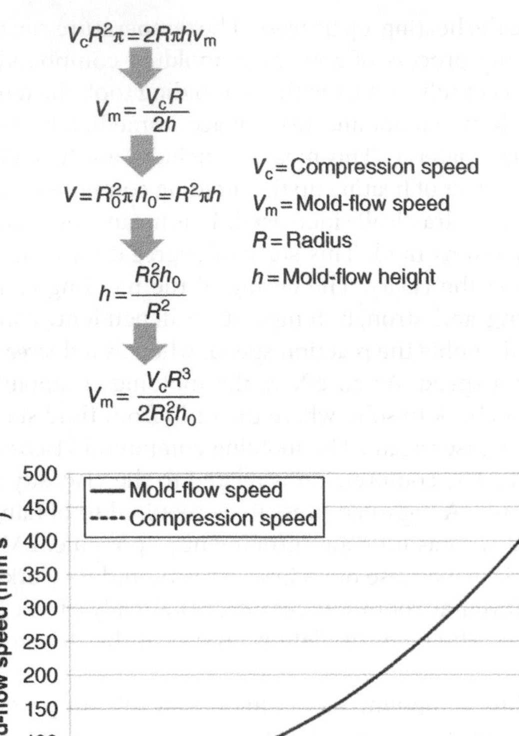

Figure 17.21 Mold flow speed vs. compression speed.

The mold tool design can be separated into two different concepts as shown in Figures 17.22 and 17.23.

17.4.7 Mold Tool Release During Compression Molding

The solidification process of thermoset materials is dominated by an exothermic chemical reaction – the curing reaction. The curing reaction is an irreversible process that results in a cross-linked structure of molecules. Thermoset materials are mainly used for semiconductor packaging. As known from the previous explanation, curing of thermoset materials is activated by temperature. The release from the molding tool is only possible when the molding compound achieves a certain curing stage. The cross-linking also determines the hot hardness of the material used. As the temperature of a molding compound

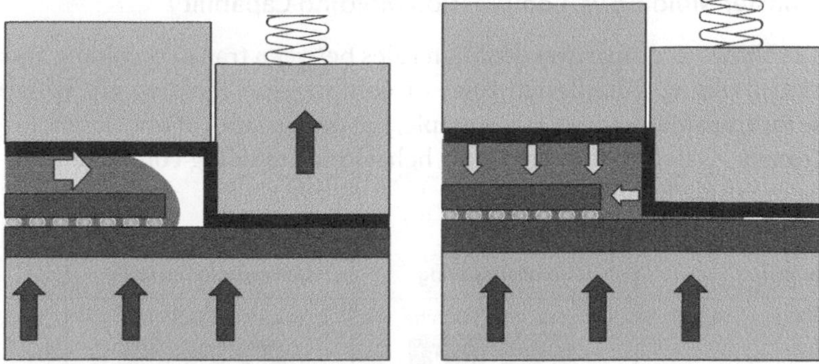

Figure 17.22 Mold tool design for spring-loaded cavity: during clamping the ring is touching the carrier system (left) and during final pressure (right).

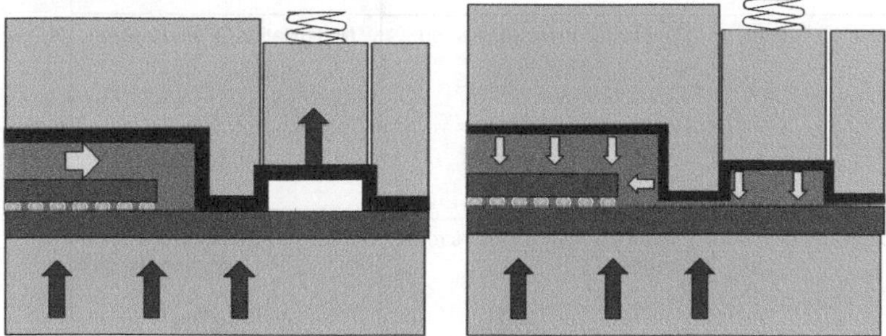

Figure 17.23 Mold tool design with overflow mechanism: (left) during clamping the ring is touching the carrier system and (right) during final pressure.

increases, hardness decreases, and at the glass transition temperature (T_g), a drastic change in hardness occurs. The hot hardness of a molding compound is the hardness at the molding temperature. Two different methods are used in semiconductor packaging to separate the package from the mold tool.

Separation from the mold tool can be done by ejector pins. It is mainly used for standard packages, for example, BGA, QFP, QFN, etc. The limitations are very thin and large cavities or high adhesion molding compounds like LMCs. The reason for this limitation is that the molded device has to withstand the release force and bending during ejection from the mold tool cavity.

Separation by release film is used for dedicated packages like for LED packaging, MEMS packaging, or WLP. The release film advantages are a low separation or release force even for high adhesion molding compounds like LMCs and that the process can be controlled much more easily with repeatable stability.

17.4.8 Transfer Molding vs. Compression Molding Capability

Figure 17.24 shows a comparison of design rules between transfer molding and compression molding. The design rules of compression molding are better than those for transfer molding. For example, the design rules of transfer molding are strongly related to the mold-flow behavior of molding compound due to the long molding distances.

Design rule item	Transfer molding	Compression molding
Molding area	up to 100 × 300 mm^2 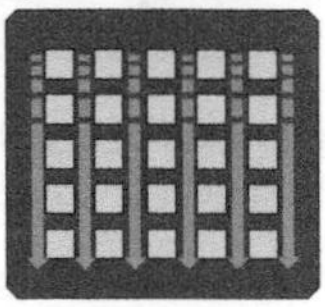	up to 600 × 600 mm^2 
Wire diameter	Limitation by wire sweep 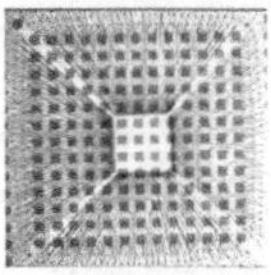	No limitation by wire sweep 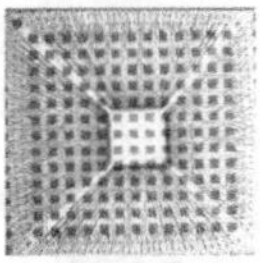
Molding compound utilization	Approximately 40–60 % resin usage 	Almost 100% resin usage due to no cull, no runner and no gate 
Molding gap above chip	3x filler cut minimum 60 μm 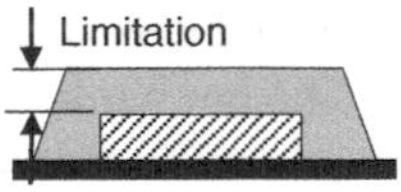 Limitation	1.5x filler cut minimum 30 μm 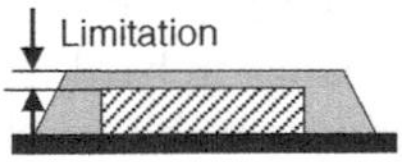 Limitation
Molding cavity thickness	Minimum 150 μm	Minimum 50 μm
Special features	1) Leaded packages 2) Double sided exposed 3) Retractable pin	1) Molded array package 2) Low pressure encapsulation 3) Low warpage
Molding compound state of aggregation	Pellet (package related)	Liquid and granulate (material related)
Cleaning interval	100–2000 shots	No cleaning required due to film assisted molding

Figure 17.24 Comparison of design rules for transfer molding and compression molding.

17.5 Principle Challenges for Chip Embedding with Compression Molding

17.5.1 Cavity Filling of Large Area

The moldability of large area compression molding like for 300 mm eWLB strongly depends on the mold-flow speed. This mold-flow speed has to be controlled in the range of 3–7 mm s^{-1}. If the mold-flow speed is too low, the mold compound viscosity increases during mold filling, which leads to incomplete cavity filling. If the mold-flow speed is too high, the risk of mold compound bleeding or internal voids can occur. Therefore, the mold-flow speed has to be controlled very accurately and must be very slow in order to achieve optimized cavity filling especially for increasing wafer diameter or distances. The mold-flow speed is related to the compression speed. Figure 17.25 shows a carrier system with chips, the dispensing of LMC, mold filling by compression molding, final pressure, and the molded wafer with mold carrier.

To ensure the optimized mold flow in the range of 3–7 mm s^{-1}, a compression speed of 0.1 mm s^{-1} has to be applied, for example, for the 300 mm eWLB. Figure 17.26 shows the optimized compression speed depending on the diameter of the molded wafer.

The original toggle press could not achieve the required compression speed. Toggle presses use the so-called toggle effect: the farther the toggle is stretched, the slower but more powerful the movement of the press becomes. The toggle press closing speed does not follow the servo motor target speed during the

Figure 17.25 . Carrier system with chips (1), the dispensing of LMC (2), the mold filling by compression molding (3 and 4), final pressure (5), and the molded wafer with carrier system (6).

Compression speed (mm s^{-1})	Molded wafer (mm)
0.20–0.30	100
0.15–0.25	200
0.10–0.20	300
0.05–0.15	400

Figure 17.26 The optimized compression speed (left) in relation to the diameter of the molded wafer (right).

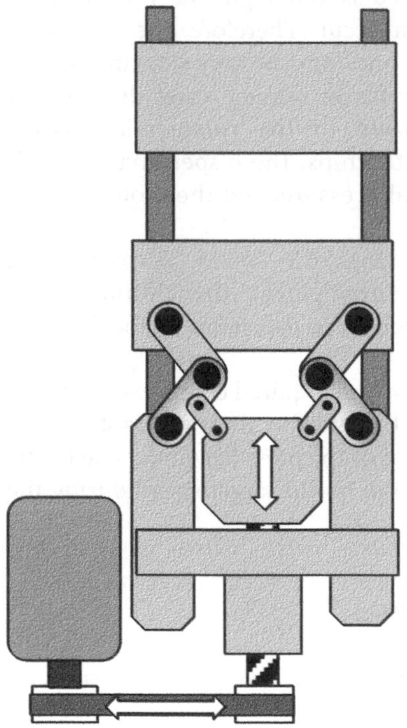

Figure 17.27 The toggle press mechanism was optimized on press clamping force.

final closing. The reason was that the toggle mechanism focuses on the press clamp force. Figure 17.27 shows a sketch of a state-of-the-art toggle press mechanism.

The challenge was to develop a very accurate direct drive, which can control very low speeds.

17.5.2 Planarity of Large Area

The planarity of transfer molding is related to the mechanical accuracy of the mold tool. The transfer mold press has an influence on molding responses like mold flash or mold bleed and not on the dimensional accuracy of the molded part. Thus, the molding tool accuracy is a key issue for the transfer molding process. Figure 17.28 shows a sketch of a transfer mold tool concept focused on the package thickness.

The planarity of compression molding is related to the mechanical accuracy of the mold tool in the case of using the compression molding overflow concept. The overflow concept is pressing out the excess mold compound in a separate overflow cavity outside of the device cavity. This concept requires de-gating after molding. The de-gating process breaks the overflow cavity away from the device cavity. This is usually done directly after molding within automatic molding equipment. The mold tool cavity has to be spring loaded to avoid such an overflow concept and therefore the de-gating process. The planarity of the spring-loaded mold

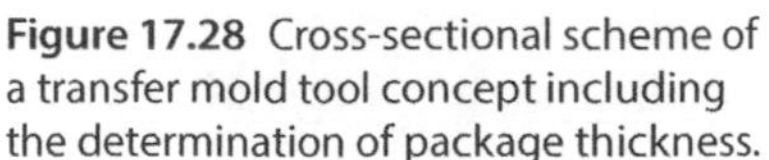

Figure 17.28 Cross-sectional scheme of a transfer mold tool concept including the determination of package thickness.

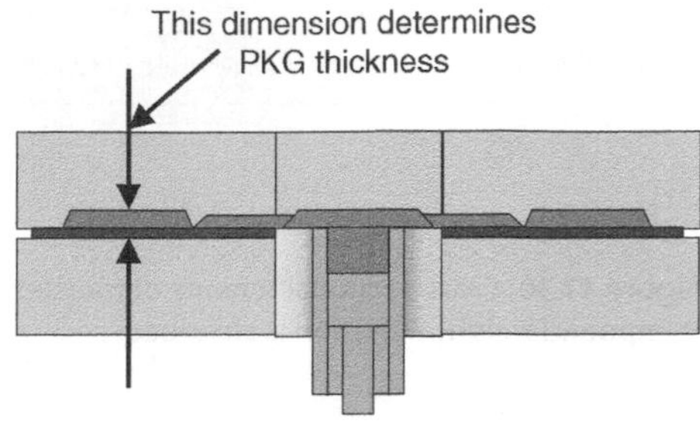
This dimension determines
PKG thickness

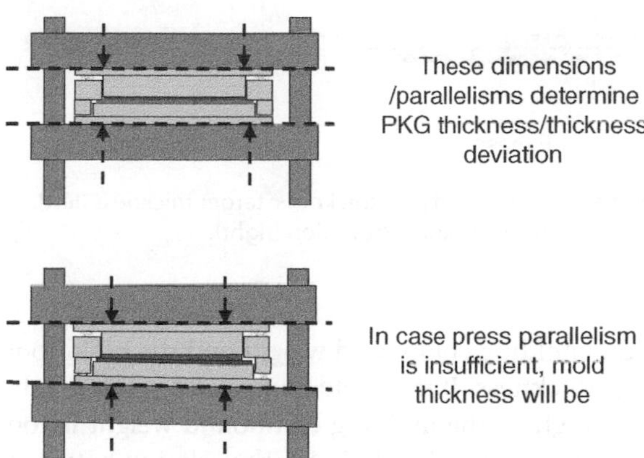
These dimensions
/parallelisms determine
PKG thickness/thickness
deviation
In case press parallelism
is insufficient, mold
thickness will be

Figure 17.29 Compression molding cross-sectional scheme of a parallel press (top) and insufficiently parallel press (bottom).

tool is related to the mechanical accuracy of the mold tool and the planarity of the molding press. Figure 17.29 shows the cross-sectional scheme of a compression molding press including a mold tool and the impact of the molded part parallelism.

The challenge was to develop a very highly parallel press mechanism.

17.5.3 Dimensional Accuracy

For transfer molding, the mold tool dimension determines the mold thickness and accuracy. Compound weight deviation is absorbed by the cull thickness. The cull thickness is defined by the runner thickness and the remaining mold compound generated by the plunger stroke. Figure 17.30 shows cross-sectional scheme of transfer mold tool including the molding compound weight deviation absorbed by the cull thickness.

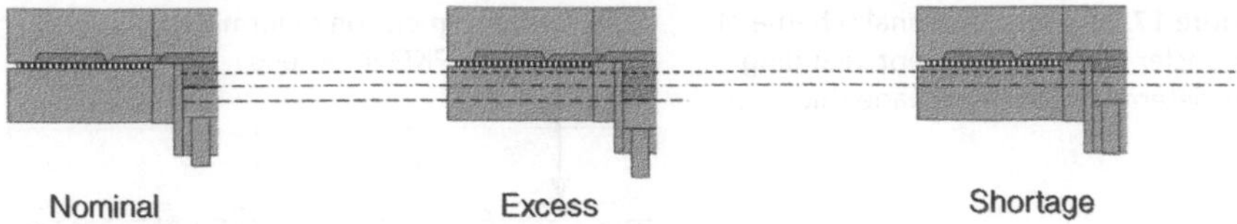

Figure 17.30 Cross-sectional scheme of transfer mold tool and how the molding compound weight deviation is absorbed.

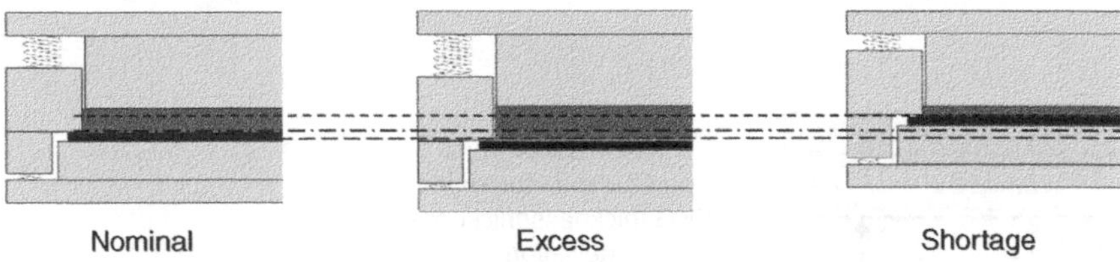

Figure 17.31 Cross-sectional scheme of the molded part thickness: target thickness (left), upper thickness deviation (middle), and lower thickness deviation (right).

For compression molding the molding compound weight and the mold tool volume determine the package thickness. If the mold compound is too much, the mold thickness will be too high. If the molding compound weight is too low, the mold thickness is too thin. Figure 17.31 shows the relation between the molding compound weight and the molded part thickness for a spring-loaded mold tool by compression molding.

The challenge was to develop a dispense method that very precisely controls the molding compound weight, even at high amounts.

17.5.4 Molded Wafer Identical to SEMI Specification

The molded wafer must be identical to the SEMI specification of a silicon wafer, because this allows the use of standard front-end tools. This means the wafer edge should not have any irregularities like an air vent or an ejector pin. The air vents are required to bring the internal and external mold voids under control. Molding compounds with very low viscosity or with fine filler have a limitation on the air vent height due to the mold-flow behavior. The ejector pins are required to have a proper mold release after molding compound curing. Large and thin molded areas are very critical to release due to the high mold release force and high bending properties under molding temperature. Figure 17.32 shows typical irregularities of molded parts.

The challenge was to develop a molding process to achieve an outer dimension identical to the SEMI specification of a silicon wafer.

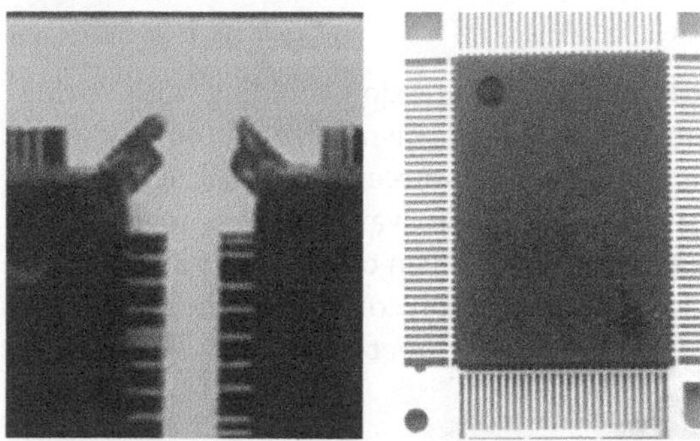

Figure 17.32 Transfer molded package with typical irregularities: (left) required air vents and (right) required ejector pins.

17.6 Process Development Solutions for Principle Challenges

17.6.1 Cavity Filling of Large Area

The mold filling of a large area requires a very low compression closing speed down to $0.1\,\mathrm{mm\,s^{-1}}$. This is solved by the change of the molding press mechanism from toggle to direct-drive ball screw. The direct-drive ball screw press mechanism is able to achieve a very high position accuracy. This is required especially for the spring-loaded mold tool concept. Figure 17.33 shows the construction of compression molding press with direct-drive ball screw mechanism.

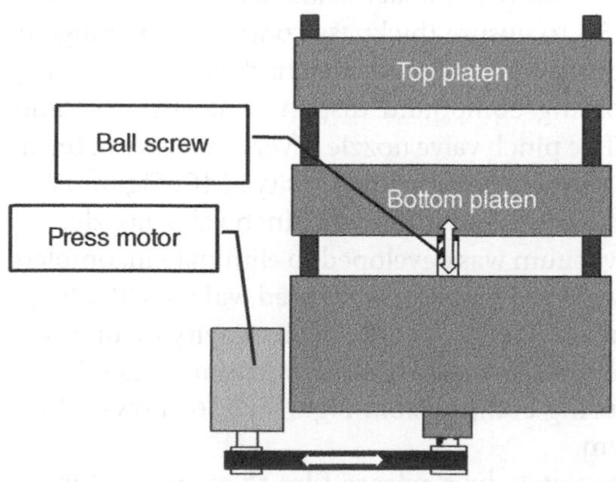

Figure 17.33 Sketch of a direct-drive ball screw press mechanism.

17.6.2 Planarity of Large Area

The planarity of large molded areas by compression molding with spring-loaded mold tool concept requires, a high parallelism of mold tool and mold press. To overcome this issue, a four-corner direct-drive ball screw press mechanism was developed instead of using the available press toggle drive. This four-corner direct-drive ball screw mechanism can be easily adjusted on-site by software and enables a closed-loop position control. Figure 17.34 shows a cross-sectional scheme of a four-corner direct-drive ball screw press mechanism.

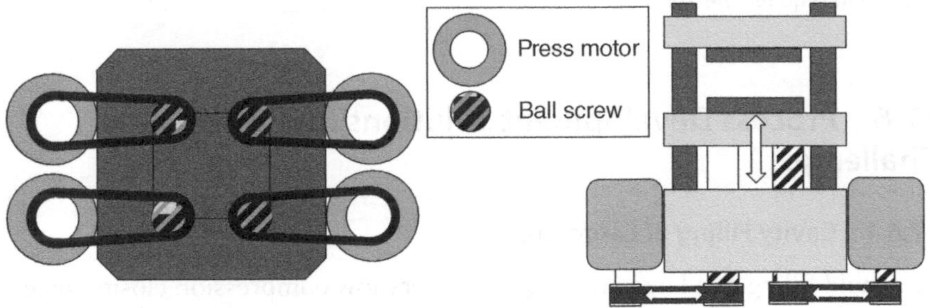

Figure 17.34 Four-ball screw drive press mechanism that is optimized for precise press planarity during compression speed and final pressure.

17.6.3 Dimensional Accuracy

A very accurate LMC dispenser for high viscosity and for high amounts of mold compound was developed to ensure minimal thickness deviation. The dispensing accuracy is around ±0.3 g to ensure thickness control in the range of 10 μm. Due to the fact that the mold carrier system weight deviation is already more than few grams, the molding compound dispenser needs a weight-controlled closed-loop system. The pinch valve nozzle is very effective in terms of dispense speed improvement, especially for high viscosity LMC. Figure 17.35 shows an LMC dispenser with closed-loop system and pinch valve nozzle.

The high performance cavity vacuum was developed to eliminate incomplete fill and air vents of the molded wafers. This ensures molded wafers with identical dimension to SEMI specification. The high performance cavity vacuum will evacuate the air completely before closing. Absolute vacuum values below 20 mbar have to be achieved during compression. Figure 17.36 shows a high performance vacuum pipe system.

The mold tool is covered completely by a release film to avoid ejector pin imprints. In addition, the release film enables a smooth releasing of the molded

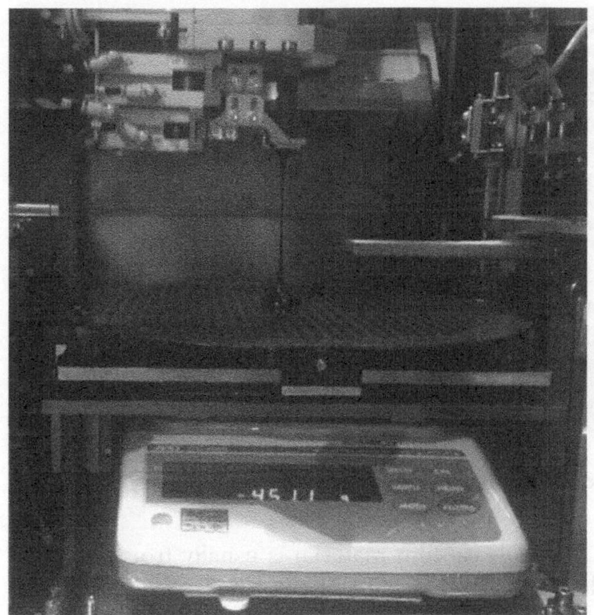

Figure 17.35 Liquid compound dispenser that is optimized for accurate dispensing and dispensing speed.

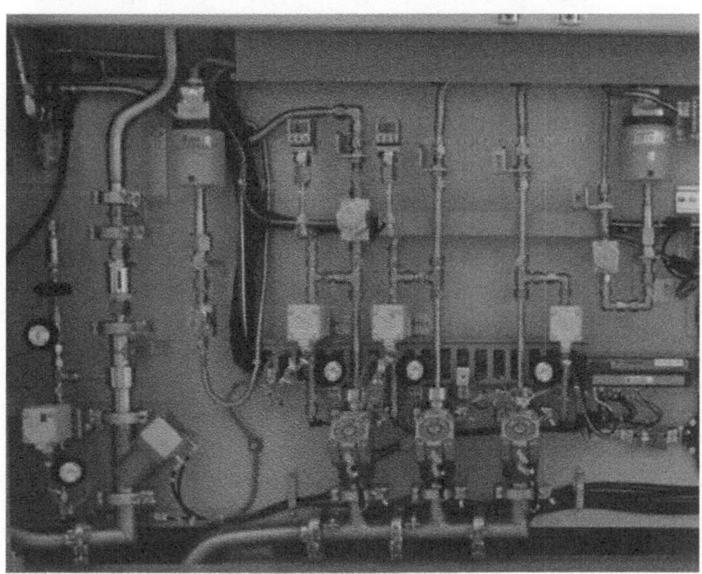

Figure 17.36 High performance vacuum pipe system.

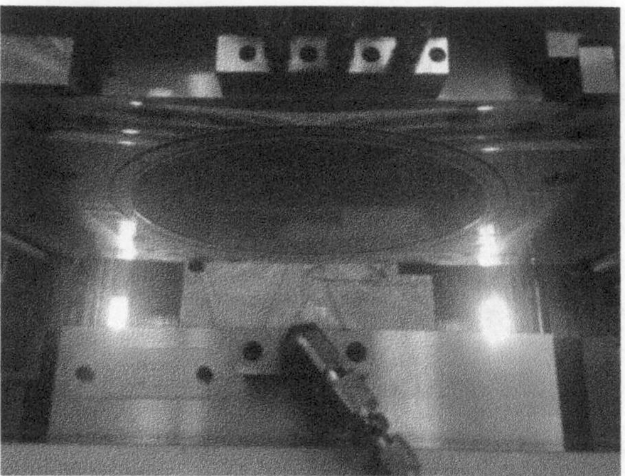

Figure 17.37 Top mold tool covered completely with release film.

part from the mold tool. The release film material is usually from the resin family of ethylene tetrafluoroethylene (ETFE) or polyethylene terephthalate (PET). The important release film properties are the releasability, elongation to break, and expansion during heating up. The film-assisted unit handles the release film automatically, and the suction into the cavity is done by a vacuum system. The mold tool is designed to ensure fully automatic wrinkle-free processing. Figure 17.37 shows a cavity-up mold tool covered completely with a release film.

17.7 Compression Molding Equipment for Chip Embedding

The compression molding for fan-out eWLB technology required a complete new equipment development. All required challenges necessary to achieve a robust process have been addressed. In addition, the processing of bare silicon wafers (e.g. TSV, wafer bump protection, backside protection) has also been considered. The equipment consists of two newly developed compression molding presses. These presses are equipped with a four-corner direct-drive ball screw press mechanism to ensure the highest planarity in combination with optimized mold-flow speed. The liquid dispensing unit is able to run very high viscosity LMCs. A granular dispensing unit can be selected as an option. The equipment has integrated PMC and post mold inspection. The handling is based on a commercial flexible robotic unit. Thus, the conversion effort is minimized. Figure 17.38 shows the fully automated compression molding system optimized for fan-out eWLB and fan-in WLP (bare silicon wafer).

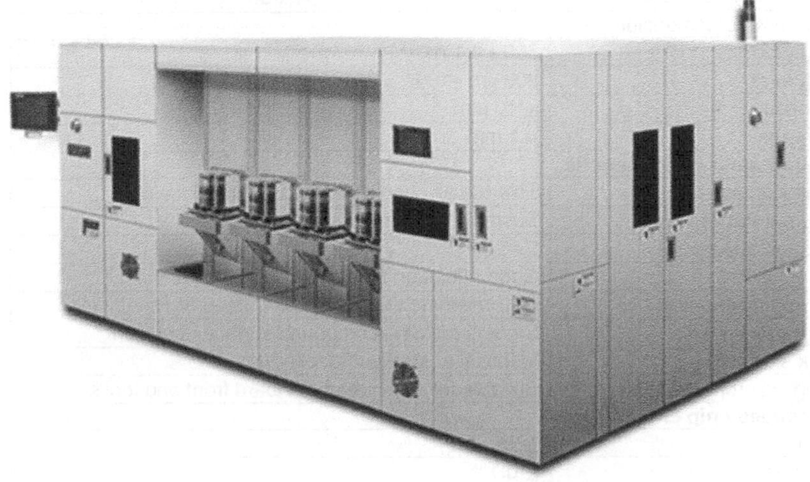

Figure 17.38 Compression molding equipment optimized for chip embedding (WCM-300 from Apic Yamada).

This equipment, named WCM-300, is able to process fan-out and fan-in wafer-level packages with a diameter up to 300 mm (round shape). The equipment is able to operate in a clean room environment down to class 1000. Automatic magazine loading and unloading, which is required by front-end production, can be applied. In addition, panels with a length up to 300 mm (square shape) can also be encapsulated. For larger panel dimensions, further equipment is under development.

17.8 Chip Embedding Features Achieved by Compression Molding

Applying compression molding, the results shown in Figure 17.39 for features of eWLB molded wafers with a diameter of 300 mm could be achieved.

The data in Figure 17.40 compare package type, total thickness variation, wafer diameter, and the overall equipment efficiency (OEE) for volume production.

17.9 Conclusions and Next Steps

This chapter demonstrated the importance of compression molding for developing the fan-out eWLB technology. The development and setup of the proper encapsulation equipment and process was of outstanding importance. The

eWLB features	Result
Molded wafer dimension	
Diameter	300 ± 0.1 mm
Thickness	300 μm – 1000 ± 20 μm
Flatness	± 10 μm
Warpage	< 1 mm
Molded wafer characteristic	
Chemical resistant	Tolerate all chemical used in RDL processing
Thermal stability	Sustains temperatures up to 260 °C
Moisture absorption	0.15 %
Adhesion	Good adhesion to all interfaces
Thermo-mechanical	CTE mismatch optimized to Cu and Si
Front side surface	Smooth molding compound surface condition
Back side surface	Identical to standard packages
Handling properties	Suitable for automated standard front end tool's
Embedded chip characteristic	
Position	± 15 μm
Rotation	± 0.1 °
Stand-off	5 ± 3 μm

Figure 17.39 eWLB features with corresponding results for 300 mm molded wafer technology.

Package type	Total thickness variation (TTV)		Wafer diameter tolerance		Overall equipment efficiency (OEE)
eWLB	Spec	40 μm	Spec	±200 μm	10 – 20 WPH
	Achieved	10 μm	Achieved	±150 μm	
2.5D FOWLB	Spec	15 μm	Spec	±200 μm	10 – 20 WPH
	Achieved	10 μm	Achieved	±150 μm	
2.5D Full Mold	Spec	40 μm	Spec	±200 μm	8 – 12 WPH
	Achieved	8 μm	Achieved	±100 μm	
3D FOWLB	Spec	40 um	Spec	±500 μm	8 – 12 WPH
	Achieved	10 μm	Achieved	±200 μm	

Figure 17.40 Total thickness variation, wafer diameter tolerance, and overall equipment efficiency for different package types.

importance of the right molding material was also highlighted. The temporary rigid carrier together with the thermal release tape, the liquid mold compound, and the compression molding process were identified as the most critical parts that had to be solved in the first years of the fan-out eWLB technology to construct the molded wafer. Special challenges that had to be solved for compression molding equipment were large area cavity filling, planarity of the large area, dimensional accuracy, and SEMI specification compliance of molded wafers.

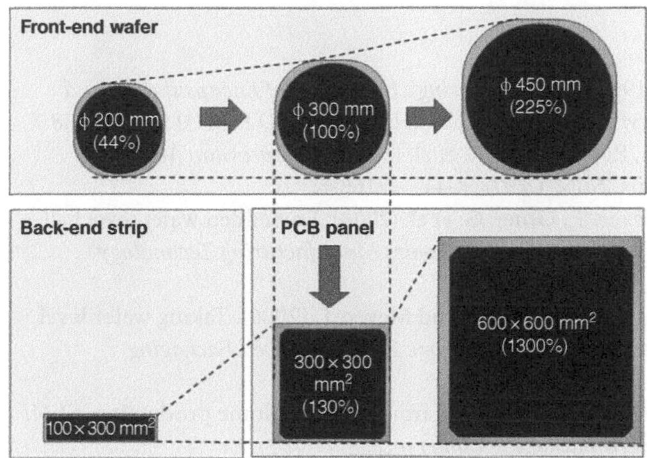

Figure 17.41 Front-end-like round wafers compared with square back-end strip and PCB panels.

In the future, main trends will focus on cost requirements, thinner devices, and higher system integration. System integration, i.e. the integration of more functionality into smaller volume, will be a major trend. This will include technologies like side-by-side, 3D integration, and integration of passives and antennas into one package. The encapsulation equipment combined with new mold material must be designed properly for this trend. One example is equipment that can position chips side-by-side with less than 50 μm spacing in between. Another example is preparing equipment that offers 3D stacking capabilities. For further cost reduction, the large size molding process is investigated with respect to introducing square panel format. The development of fan-out panel-level packaging (FO-PLP) up to 600 mm × 600 mm or even more (see Figure 17.41) is expected to provide a significant cost reduction. In addition, material improvements are under investigation, e.g. in respect to molded wafer warpage. The industry is presently introducing new types of molding compound aggregations such as granulate (GMC) or sheet (SMC). Figure 17.41 compares front-end-like round wafer with square back-end strip and PCB panels.

Acknowledgments

Infineon Technologies gratefully acknowledges the partial funding of molding processes by the project SIPHA, funded by the German BMBF (grant number 01M3177A), and by the project ESIP, funded by the ENIAC JU (grant number 120227) and the German BMBF (grant number 16 N10971).

References

1 Bartholomew, M. (1999). *An Engineering's Handbook of Encapsulation and Underfill Technology*. Electrochemical Publications LTD ISBN 0 9011150 38 X.
2 Davis, B., Gramann, P., Oswald, T.A. et al. (2013). *Compression Molding*. Gardner Publications ISBN-13 978-3-446-22166-6.
3 Brunnbauer, M., Meyer, T., Ofner, G. et al. (2008). Embedded wafer level ball grid array (eWLB). *International Electronics Manufacturing Technology Conference 2008*.
4 Fürgut, E., Beer, G., Brunnbauer, M., and Meyer, T. (2006). Taking wafer level packaging to the next Stage. *SEMI Europe 2006 Advanced Packaging Conference*.
5 Fürgut, E. (2010). eWLB reconstitution from idea to volume production. *SEMI Europe 2010 Advanced Packaging Conference*.
6 SEMI M1-1016 (2016). Specification for polished single crystal silicon wafers; latest publication April 2016.

18

Tools for Fan-Out Wafer-Level Package Processing

Nelson Fan, Eric Kuah, Eric Ng, and Otto Cheung

ASM Pacific Technology

18.1 Turnkey Solution for Fan-Out Wafer-Level Packaging

In this chapter, we discuss three critical processes for fan-out wafer-level packaging (FO-WLP) technology. They are die placement, large format encapsulation, and handling of finished packages after singulation. Unlike traditional packaging platforms, the majority of FO-WLP is being manufactured in 12 in. wafer formats rather than small strip formats. Moreover, the FO-WLP technology platform is flexible as it can address different device design requirements with densely routed redistribution layers (RDL) and by deploying many different die placement approaches, such as chip-first and chip-last coupling with either a local or global alignment method. Depending on which approach is being used, the die placement can be done at room temperature or at an elevated temperature with die facing up or facing down. During molding, the reconstructed wafer can also be molded with either die face-up or face-down orientation, depending on the package structure. Specifically, the latter is deployed for process designs comprising molded underfill (MUF) processes. Often, packaging subcontract manufacturers offer multiple fan-out approaches to their customers; therefore, the required die placement tool and encapsulation tool both have to be able to handle large format material and be flexible and robust enough to cater to different process requirements. After the package singulation process, the individual units will be handled by post-singulation handling systems designed for testing, laser marking, final all-sided inspection, and tape and reel packing. In order to ensure good quality output for downstream surface-mount technology (SMT) assembly, the test handling system needs to have a reliable optics system for inspection and a chip component

Advances in Embedded and Fan-Out Wafer-Level Packaging Technologies, First Edition.
Edited by Beth Keser and Steffen Kröhnert.

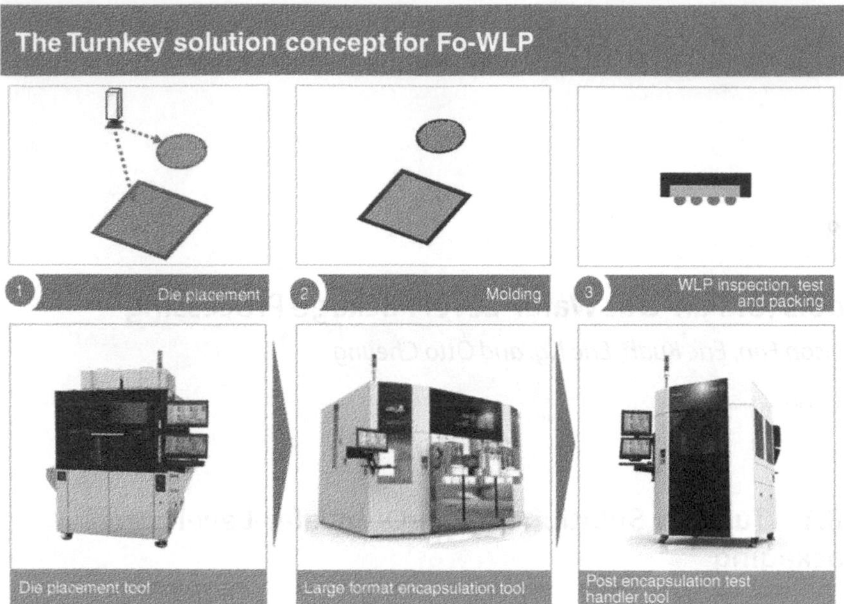

Figure 18.1 Turnkey solution for FO-WLP.

handling system to avoid inducing further defects to the tested objects. The details will be discussed in the subsequent sections. Figure 18.1 shows the tools developed for the manufacturing of FO-WLP.

18.2 Die Placement Process and Tools for FO-WLP

Today, advanced packaging technology innovation is increasingly important for driving the enhancement of package performance, lowering cost, and achieving a small form factor. Enabled by thin film-based interconnects, various FO-WLP packaging technologies such as embedded wafer-level ball grid array (eWLB) [1], integrated fan-out (InFO) [2], and fan-out chip on substrate (FOCOS) [3] have emerged and evolved to potentially replace many current packaging technologies from low and high density 2D packages like flip-chip chip-scale packages (FCCSP) to high-end 2.5D packages like chip on wafer on substrate (CoWoS) and even 3D packages like molded core embedded packages (MCeP) and high bandwidth package on package (HBPOP) for different applications. eWLB, as an example of a low density 2D package, has been a good replacement for conventional FCCSP with a cost-effective process and a thin package profile due to its substrate-less nature [1].

Normally, in FCCSP manufacturing, flip-chip bonding of the chip with solder bumps occurs onto a laminated substrate in a small strip format followed by a reflow process and MUF. In eWLB manufacturing, which is a chip-first approach, meaning the chip is bonded prior to an RDL process, the chip is first bonded onto a double-sided adhesive laminated onto a 300 mm diameter carrier followed by wafer-level compression molding. After molding, the RDL circuitry is deposited onto the active chip area and the mold compound surface at the periphery of the die by means of thin film technology. Unlike FCCSP, without the use of substrate, the overall package thickness of eWLB is reduced. It is also a more cost-effective process by making use of batch processing in large format manufacturing [1].

For higher density package designs, the InFO package, which is also a 3D package, sets another reference for using FO-WLP technology as an alternative to package-on-package (PoP) structures for mobile application processor devices [2, 4]. FOCOS is yet another example of even higher density FO-WLP with heterogeneous integration capability to substitute 2.5D packages with TSV Si interposers for high-end applications such as networking servers and high performance computing [3]. It is achieved by high density RDL technology with tight line width and spacing (L/S) scaling at 2 μm and below.

Another way to achieve 2.5D substitution can be done by the chip-last FO-WLP approach, in which the chip is flip chip bonded followed by reflow onto a prefabricated high density RDL on top of the temporary carrier. Each of these fan-out packaging technologies has unique features and merits in terms of form factor, electrical and thermal performance, package reliability, and cost-effectiveness. Figure 18.2 summarizes the various fan-out packaging

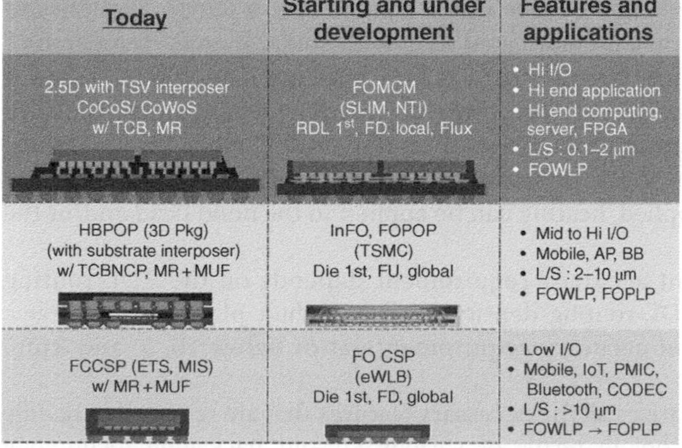

Today	Starting and under development	Features and applications
2.5D with TSV interposer CoCoS/ CoWoS w/ TCB, MR	FOMCM (SLIM, NTI) RDL 1st, FD, local, Flux	• Hi I/O • Hi end application • Hi end computing, server, FPGA • L/S : 0.1–2 μm • FOWLP
HBPOP (3D Pkg) (with substrate interposer) w/ TCBNCP, MR + MUF	InFO, FOPOP (TSMC) Die 1st, FU, global	• Mid to Hi I/O • Mobile, AP, BB • L/S : 2–10 μm • FOWLP, FOPLP
FCCSP (ETS, MIS) w/ MR + MUF	FO CSP (eWLB) Die 1st, FD, global	• Low I/O • Mobile, IoT, PMIC, Bluetooth, CODEC • L/S : >10 μm • FOWLP → FOPLP

Figure 18.2 Use of various thin film fan-out packaging technologies.

technologies that are alternative solutions to the current packaging technologies described above.

Different FO-WLP approaches are achievable by different process steps and sequences. Figure 18.3 illustrates the major processing steps of various fan-out packaging technologies. It shows the different ways for chip placement. In the first approach, the chip is bonded on a 200 or 300 mm diameter mother wafer by means of die-attach film (DAF) bonding. Chip orientation, which is defined as the orientation of the active side of the chip when it is bonded, is face-up. The second approach shows an example of achieving heterogeneous integration by FO-WLP in a multi-chip module package configuration. Unlike the first approach, a chip is bonded on the wafer carrier with a prefabricated RDL with the chip orientation facing down. Here, the interconnect can be achieved by either mass reflow soldering or the more advanced thermal compression bonding (TCB) process.

The die placement process for each type of these fan-out packaging technologies can be characterized in different ways: (i) die placement orientation, (ii) die alignment mode, (iii) die placement process condition, and (iv) die placement accuracy.

Die placement orientation can be classified into two modes. They are the face-up and face-down modes. The former has the die active side face-up during die placement, and the latter has the die active side face-down. Die alignment modes include local alignment and global alignment. Local alignment mode uses a local alignment mark near the bond position locally as a reference. Global alignment mode, on the other hand, uses a common alignment mark on the carrier or substrate carrier as a reference.

Die placement process conditions can be classified in two common ways: (i) at room temperature and (ii) at elevated temperature and force. For the room temperature process, it can be face-down bonded onto a temporary adhesive or flip chip bonded on a prefabricated RDL with solder bumps. Temperature and force processes may be referred to as DAF bonding. The chip is bonded at an elevated temperature with force onto the carrier. Depending on the DAF material, the bond temperature ranges from 80 to 150 °C, while the bond force is usually below 200 N, which also depends on the die size. Depending on where the DAF is applied, heating can be applied to the bond head and/or the bond stage.

The die placement accuracy requirement depends on the RDL routing density. A higher RDL routing density requires tighter placement accuracy. There are placement accuracy requirements at or below 10, 5, and 3 μm, respectively.

Figure 18.4 summarizes all the necessary features that are required to handle the die placement tool in terms of various fan-out packaging technology requirements.

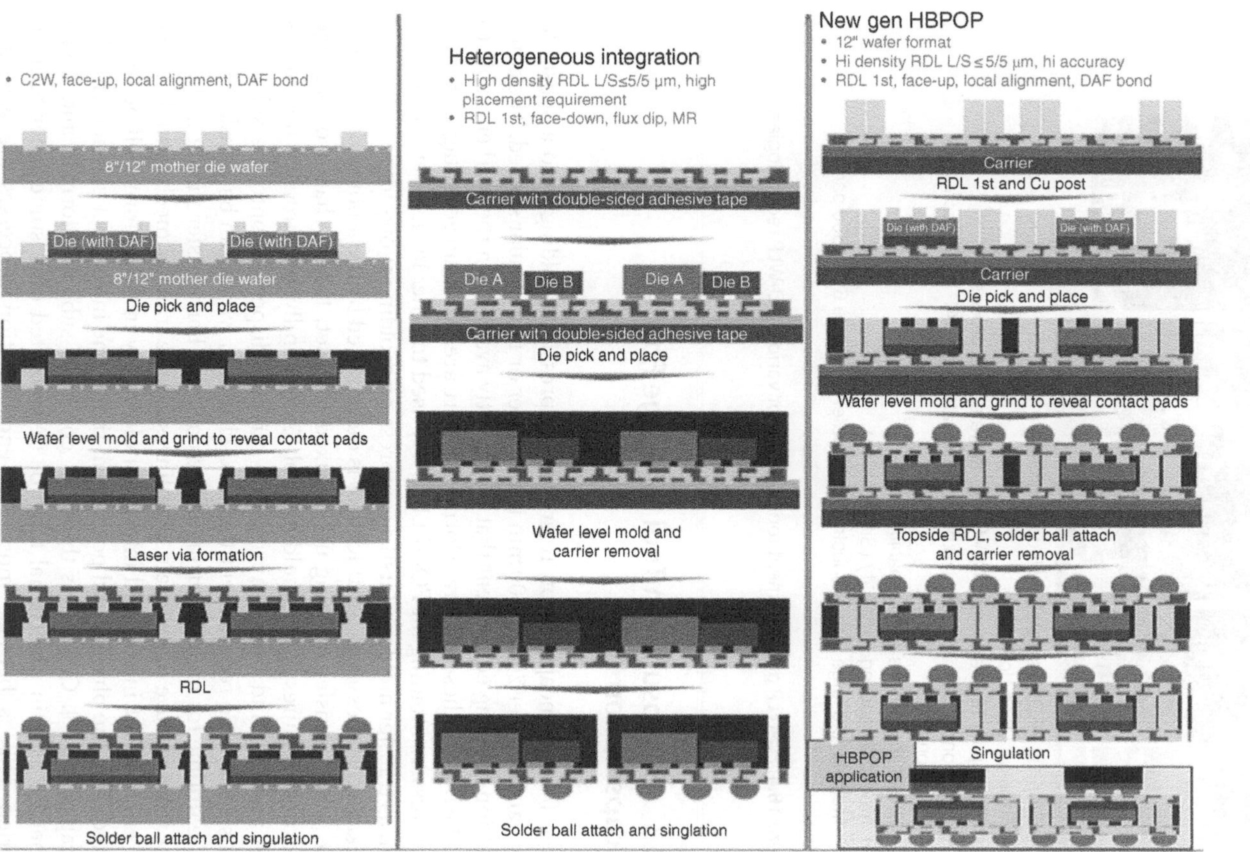

• C2W, face-up, local alignment, DAF bond
8"/12" mother die wafer
Die (with DAF)
Die (with DAF)
8"/12" mother die wafer
Die pick and place
Wafer level mold and grind to reveal contact pads
Laser via formation
RDL
Solder ball attach and singulation
Heterogeneous integration
• High density RDL L/S≤5/5 μm, high placement requirement
• RDL 1st, face-down, flux dip, MR
Carrier with double-sided adhesive tape
Die A
Die B
Die A
Die B
Carrier with double-sided adhesive tape
Die pick and place
Wafer level mold and carrier removal
Solder ball attach and singlation
New gen HBPOP
• 12" wafer format
• Hi density RDL L/S ≤ 5/5 μm, hi accuracy
• RDL 1st, face-up, local alignment, DAF bond
Carrier
RDL 1st and Cu post
Die (with DAF)
Die (with DAF)
Carrier
Die pick and place
Wafer level mold and grind to reveal contact pads
Topside RDL, solder ball attach and carrier removal
Singulation
HBPOP application

Figure 18.3 Illustrations of the major processing steps of various FO-WLP technologies.

- Chip 1st / RDL 1st (global / local aligment)
- Face-up / face-down
- Flux dip for high accuracy C2W (chip-to-wafer) FC process
- High bond force with elevated bond temperature for DAF (die attach firm) bonding
- Die-shift compensation capability
- Multi-die handling capability (SiP solution)
- Multiple and large format substrate (12" wafer carrier to 670 × 600 panel)

Figure 18.4 FO-WLP die placement tool suitable for various FO-WLP technologies.

18.3 Encapsulation Tool for Large Format Encapsulation

Large format encapsulation refers to compression molding onto a substrate that can be either 200 or 300 mm in diameter or rectangular shaped. The compression molding equipment that is currently available in the market is able to handle both of these substrate formats. There are two main advantages of using compression molding for FO-WLP as opposed to other forms of encapsulation such as transfer molding. Firstly, compression molding allows for more flexible package thicknesses as the thickness is determined by the amount of encapsulant, while transfer molding uses a fixed tool set; thus a change of package thickness requires a new molding tool [5]. Secondly, compression molding is film assisted, and the encapsulant is directly dispensed onto the substrate. Thus there is no requirement for mold cleaning, which can be translated into more uptime for manufacturing of the package. For transfer molding, the end user will still have to maintain cleanliness for the transfer mechanism because the encapsulant introduced into the mold cavity in pellet form will be transferred from the plunger and pot system that come into direct contact with the encapsulant [7]. Circular is the most common substrate format, and it can either be used in its original form or be mounted onto some other form of carrier. This is because most post-encapsulation processes are generally

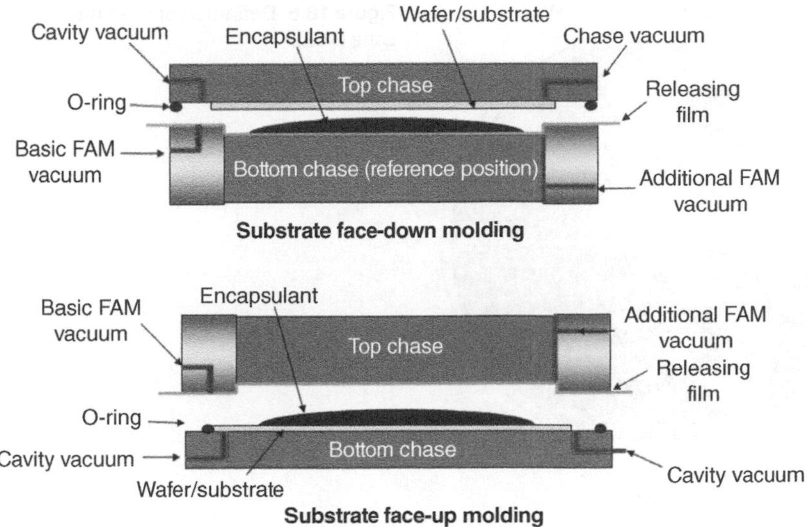

Figure 18.5 Direction of molding.

designed to handle circular formats. Metal and glass are most commonly used for the carrier [6]. However, such carriers are generally custom manufactured to have the coefficient of thermal expansion (CTE) that is as close as possible to the encapsulant (epoxy based) and the silicon die [6].

Compression molding can be performed with the active side of the silicon die face-up or face-down depending on the package configuration. The substrate is either substrate face-up or substrate face-down with reference to the bottom chase (see Figure 18.5). Substrate face-down placement is used when there is a challenge to resolve mold voids. Placing the substrate in a face-down direction allows it to have more time for the evacuation of air from the molding tool, and the encapsulant can be preheated without contact with the silicon die attached to the carrier before compression molding [6, 7].

Generally, the compression molded package is molded with a keep-out-zone (KoZ) outline, which means that the package outline is smaller than the dimension of the carrier (see Figure 18.6). Table 18.1 is a comparison of three different manufacturers of compression molding equipment that can be used for encapsulating FO-WLP. Generally, all three manufacturers have the required tonnage and necessary features for molding FO-WLP packages. However, one key requirement for molding multi-die FO-WLP when using liquid encapsulant, the most commonly used form of encapsulant [5], is the ability to dispense patterns beyond glob tops (see Figure 18.7). In Table 18.1, manufacturer C does not have the capability to dispense more than one dispensing pattern for liquid encapsulants.

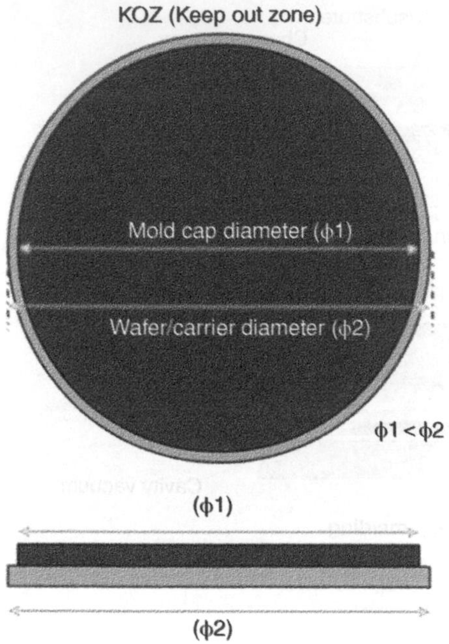

Figure 18.6 Definition of keep out zone (KoZ).

Braun [5] and Kuah [7] have shown that in order to overcome moldability challenges such as flow mark and incomplete fill, glob top dispensing patterns at the center should not be used. Dispensing patterns such as maze, star, and serpentine should be used instead, as they allow the encapsulant to cover a wide area on the substrate and result in short mold flow paths [7] (see Figure 18.8). This ability to dispense multiple types of dispensing patterns will help to solve moldability challenges such as flow mark and incomplete fill [5, 7]. However, it can be overcome by designing a dispensing table that can travel in x–y directions with the dispenser mounted in a fixed position (see Figure 18.8). Although manufacturer A has the capability to mold very thin mold caps, this is not required for FO-WLP; however, five- and six-sided wafer-level packaging (WLP) does require thin mold caps [6].

A typical compression encapsulation tool is made up of input/output modules, mold press, mold chase, aligner, dispenser, and automatic handlers. It is noted here that there will be some variations among the different vendors shown in Table 18.1, but the basic compression encapsulation system is still very similar in configuration.

Figure 18.9 shows the input/output modules that can be configured to accept wafers of up to a maximum of 300 mm circular format (shown at the left side of Figure 18.9). For other formats, such as metal carriers larger than 300 mm, a cassette will be used instead (shown at the right side of Figure 18.9).

Table 18.1 Comparison of compression system from three manufacturers.

Firm	Expandable/ press	Mode of molding	Tonnage (T)	Handler	Mold clean	5S/6S in a single system	Liquid dispensing capability	Overmold capabilities	Thinnest mold cap (μm)	Largest substrate (mm^2)
A	Yes/2	Die-up, die-down	60	Customize robot for face-up/ face-down molding	Yes	Yes	Multiple patterns	Yes, without FAM	35	SQ 340 mm
B	Yes/2	Die-down	80	Normal robot and loader	No	No	Glob via spinning	No	300	SQ 320 mm
C	Yes/2	Die-up	36	Normal robot and loader	No	No	Glob	Yes with FAM	200	SQ 300 mm

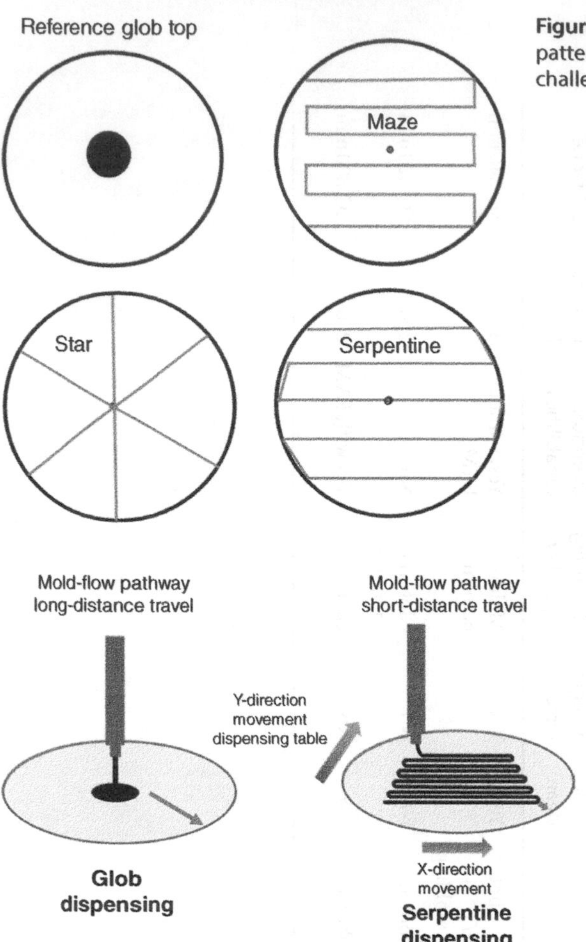

Figure 18.7 Various dispensing patterns to resolve moldability challenges.

Figure 18.8 Dispensing with x–y table movement.

Figure 18.9 Input and output load ports for different substrate formats.

Figure 18.10 shows the molding press installed with the mold chase. It is not notably special, except that the design could be configured to accept both types of mold tooling for either substrate face-up or substrate face-down molding.

Above is a dispenser that can be configured for both liquid (Figure 18.11a) and granular (Figure 18.11b) encapsulants.

Lastly, Figure 18.12 shows a SCARA robot handler transporting the substrate from the input module to the aligner and then to the dispenser, followed by the

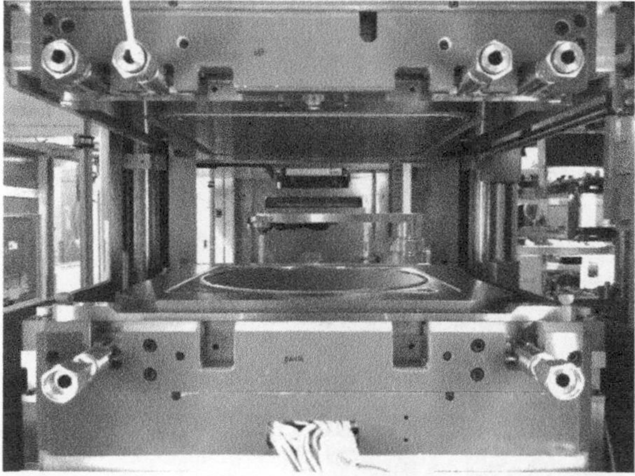

Figure 18.10 Molding press system with convertible mold chase.

(a)

(b)

Figure 18.11 (a) Dispenser for liquid. (b) Dispenser for granules.

Figure 18.12 SCARA robot handler.

mold press. Once molding is completed, it will transport the molded substrate back and place it into the output module.

18.4 The Test Handling and Packing Solution for Wafer-Level Packaging and FO-WLP

Final packing is an important process that requires good quality output before conducting SMT placement onto a printed circuit board. Detecting micro-defects is extremely important to ensure the final quality. A system comprising a series of processes including final testing, laser marking, and final all-sided inspection is needed to achieve the goal of good quality output for FO-WLP and fan-in WLP today. A few types of equipment can be applied to work on this procedure, though the turret platform is definitely the only way to provide an all-in-one solution in one system (see Figure 18.13).

The turret platform provides a total solution for both fan-in and FO-WLP final packing applications with the most advanced technologies to address today's high quality requirements. Contactless précising is a noncontact solution to ensure die placement accuracy without the risk of creating die defects during handling. A-Eye inspection is the latest technology used for WLP micro-crack detection. Additionally, the Smart iFlip solution for handling

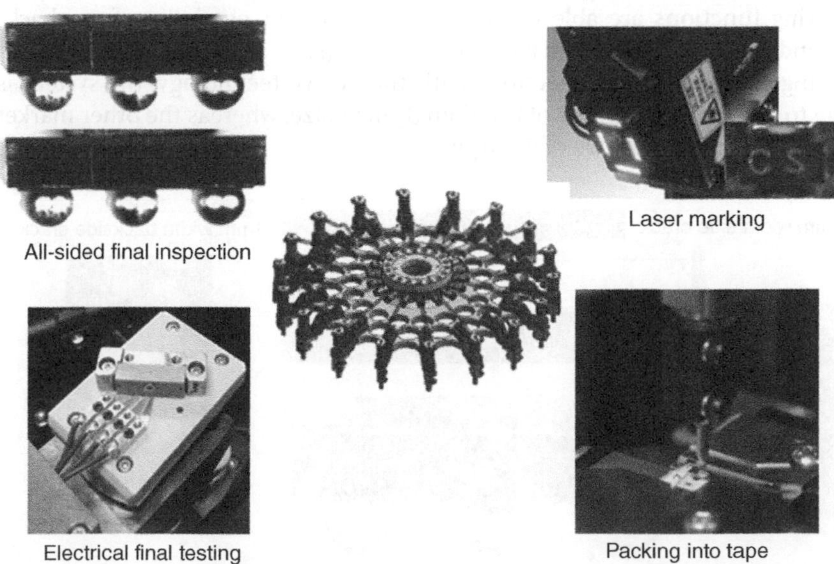

Figure 18.13 All-in-one total solution using turret platform with final testing, laser marking, all-sided inspection, and packing.

Figure 18.14 Contactless précising for aligning package with modules in the x–y direction.

ultrasmall WLP especially during system conversion is all essential to overcome today's challenges.

Contactless précising refers to the handling of packages without any physical contact. The system uses vision to locate the package position, and the pick head is built with an individual rotary motor for angular rotation. Submodules like test modules or offload modules will move in x–y directions to align with the package (see Figure 18.14).

A-Eye technology provides smart optical and algorithm solutions for WLP defect detection. Systems equipped with microscopic grades of lenses with autofocusing functions provide the best image quality. Algorithms with smart filtering functions are able to locate cracks without overkilling from background noise such as saw marks and step cuts (see Figure 18.15) while maintaining yield in mass production. With this A-Eye technology, the system is able to detect micro-cracks of less than 3 μm in size, whereas the other market solutions are only capable of 5–10 μm.

Figure 18.15 *A-Eye inspection* detecting micro-crack width with autofocusing function and smart noise filtering algorithm.

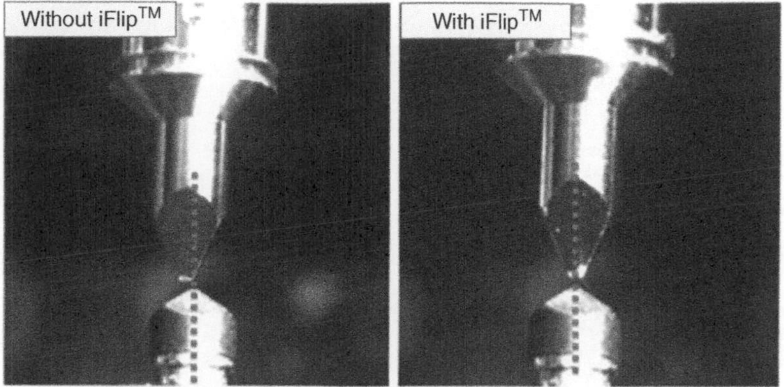

Figure 18.16 *iFlip* function reduces the setup time for ultrasmall-size WLP.

Smart iFlip solution, as a fast conversion tool, automatically locates all the combinations of turret heads and flipping heads by vision inspection. The flipper module will then index accordingly based on the vision inspection results (see Figure 18.16), which greatly reduces the conversion time for ultrasmall WLP without any need for manual adjustment.

References

1 Brunnbauer, M., Fürgut, E., Beer, G. et al. (2006). An embedded device technology based on a molded reconfigured wafer. *IEEE/ECTC Proceedings*, 2006, pp. 547–551.

2 Tseng, C.-F., Liu, C.-S., Wu, C.-H., and Yu, D. (2016). InFO (wafer level integrated fan-out) technology. *IEEE/ECTC Proceedings*, 2016, pp. 1–6.

3 Lin, Y.-T., Lai, W.-H., Kao, C.-L. et al. (2016). Wafer warpage experiments and simulation for fan-out chip on substrate. *IEEE/ECTC Proceedings*, 2016, pp. 13–18.

4 Hsieh, C.-C., Wu, C.-H., and Yu, D. (2016). Analysis and comparison of thermal performance of advanced packaging technologies for state-of-the-art mobile applications. *IEEE/ECTC Proceedings*, 2016, pp. 1430–1438.

5 Braun, T., Voges, S., Kahle, R. et al. (2015). Large area compression molding for fan-out panel level packing. *IEEE/ECTC Proceedings*, 2015, pp. 1077–1083.

6 Che, F., Ho, D., Ding, M., and Zhang, X. (2015). Modeling and design solutions to overcome warpage challenge for fan-out wafer level packaging (FO-WLP) technology. *IEEE/EPTC Proceedings*, 2015, pp. 2–4.

7 Kuah, E., Hao J.Y., and Chan, W.L. (2016). Wafer level encapsulation – an alternative format for discrete packaging: its challenges and solutions. *Proceeding of International Wafer-Level Packaging Conference*, 2016, pp. 1–8.

19

Equipment and Process for eWLB

Required PVD/Sputter Solutions

Chris Jones[1]*, Ricardo Gaio*[2]*, and José Castro*[2]

[1] *SPTS Technologies Ltd, Newport, UK*
[2] *Amkor Technology Portugal S.A., Portugal*

19.1 Background

In a fan-out WLP (FO-WLP) structure, conducting redistribution layers (RDL) are formed in dielectric material to relocate device input and output (I/O) electrical connection points to different areas of a die surface, to areas of mold real estate around the periphery, or even to link adjacent die embedded in mold, depending on the packaging structure and application. Many layers of RDL with vias connecting each layer are used when I/O density is high to avoid interconnect congestion. They are formed in photosensitive resist materials following the basic process flow illustrated in Figure 19.1.

For multilevel RDL structures, the RDL formation process is repeated, depending on the number of conductive routing layers required, with the final plating pass acting as an under-bump metal (UBM) layer prior to solder or ball drop connection.

RDL conductor metal is predominantly copper (Cu) deposited using electrochemical deposition (ECD). ECD requires a thin conducting seed layer to initiate the ECD process. Cu is typically used in combination with an adhesion layer underneath, often titanium (Ti) or titanium–tungsten (TiW). The industry standard technique for depositing these materials is physical vapor deposition (PVD) or, to be more precise, sputtering. For brevity, PVD will be the term used here. An example of a PVD production system is shown in Figure 19.2.

During early FO-WLP, when Infineon was developing their embedded wafer-level BGA (eWLB) technology, 200 mm PVD systems were used for

Advances in Embedded and Fan-Out Wafer-Level Packaging Technologies, First Edition.
Edited by Beth Keser and Steffen Kröhnert.

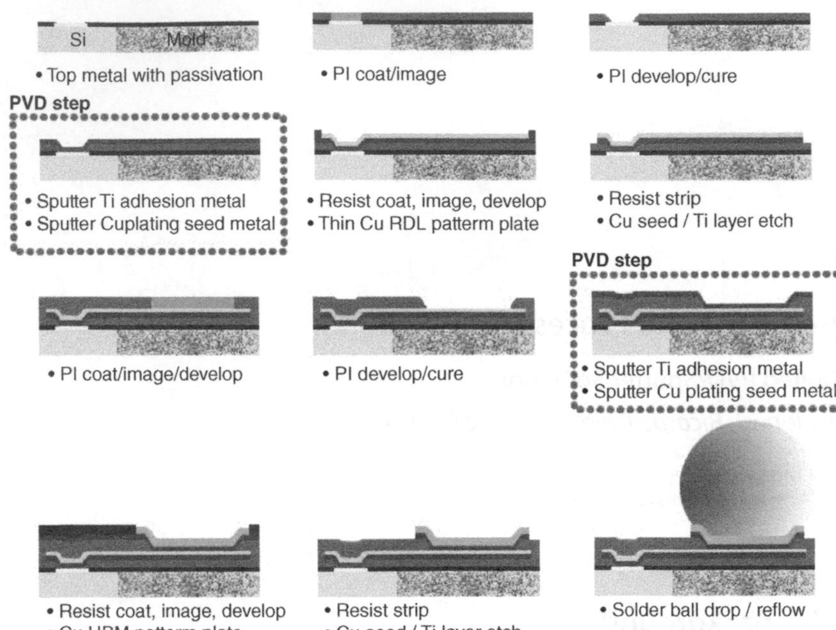

Figure 19.1 RDL process flow.

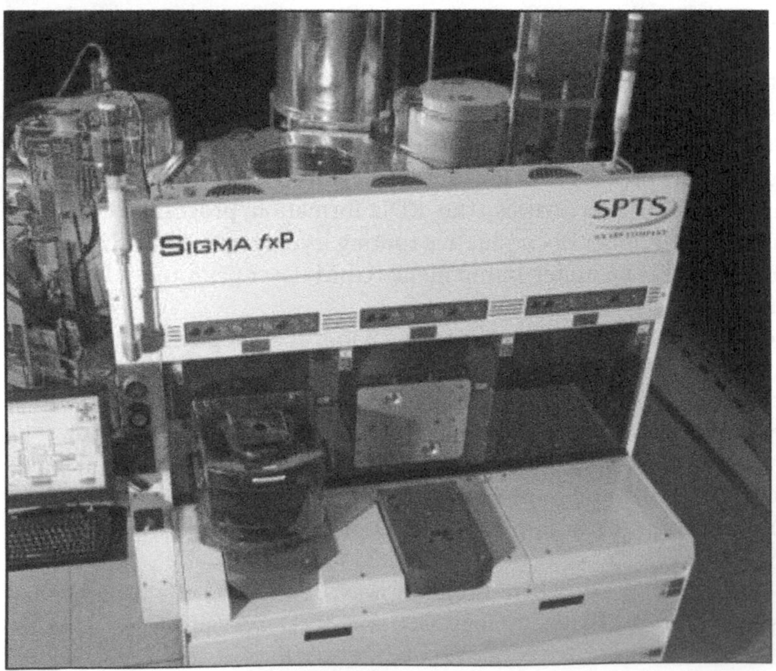

Figure 19.2 SPTS 300 mm Sigma $f \times P$ system for FO-WLP PVD.

development and initial production [1, 2]. This early research alerted engineers to the challenges of sputtering films on molded wafers at a relatively early stage. In particular, it enabled them to plan solutions for issues related to the inevitable scaling to larger wafer sizes.

Since reconstituted mold wafers do not necessarily need to follow Si diameter standards, PVD systems are used today in volume production on a variety of mold wafer formats: 200, 300 mm, and larger. Device types ranges from low I/O packages on molded substrates (also called chips-first FO-WLP) to high density I/O packages where multilayer RDL, die, and mold encapsulation takes place on a temporary carrier wafer (avoiding wafer warpage problems and allowing tighter patterning L/S resolution), also called RDL-first.

A typical PVD system for FO-WLP is composed of degas, pre-clean, and PVD module hardware clustered around a central wafer handler (Figure 19.3). Degas and pre-clean modules are used to remove contaminants and clean electrical contacts on the die. PVD modules are used to deposit the adhesion and seed metals, one chamber for each material (delivering optimal "on-wafer" film properties). All modules are high vacuum, which is an industry standard practice for producing best quality films in a PVD reactor, more so for applications involving organic materials such as FO-WLP [3, 4].

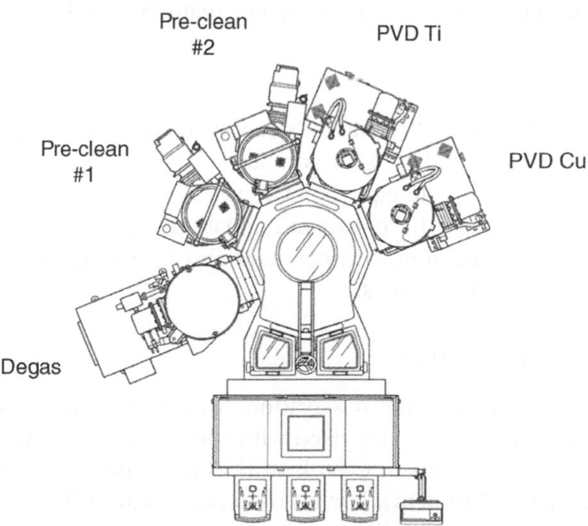

Figure 19.3 Example PVD cluster system for FO-WLP.

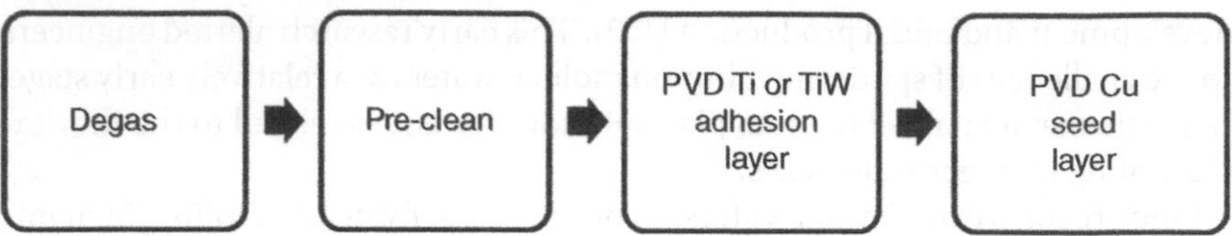

Figure 19.4 PVD RDL process flow.

19.2 Process Flow

The FO-WLP PVD process flow involves the following stages (Figure 19.4):

19.2.1 Degas

Contamination during sputtering can impact film properties and subsequent electrical performance, so it is important to remove contaminants such as moisture prior to PVD module visits. In the degas chamber, incoming wafers are raised in temperature, ideally under high vacuum conditions, to cause absorbed moisture and other contaminants to outgas from the substrate. For normal UBM and RDL processes on Si substrates, degas times from 30 to 300 seconds are typical depending on the amount and type of passivation material involved, with temperatures operating in the 150–250 °C range, again depending on the tolerances of the materials involved or other integration constraints. With FO-WLP structures the temperature limitations are lower, typically 120–130 °C. Degas times to minimize contact resistance (R_c) are typically on the order of 20–40 minutes depending on device R_c requirements and the organic materials used.

19.2.2 Pre-clean

Native oxide growth on exposed metal contacts is removed using a sputter etch process to ensure good electrical contact is made when the first PVD metal layer is deposited. Thermal oxide equivalent removal amounts in the 150–300 Å range are typical and no different from mainstream Si UBM and RDL. The challenges come when dealing with particles and R_c management issues generated from working with organics.

19.2.3 PVD Adhesion Layer Deposition

Since Cu does not adhere well to organic passivation, an adhesion layer is required beforehand. Ti or TiW is commonly used with the material selected either for patterning integration reasons or even historical reasons (e.g. what was available at the time of development). For automotive products TiW is generally preferred because of its superior diffusion barrier properties and subsequent impact on long-term device reliability. From a production perspective,

Ti is preferred over TiW. TiW is a brittle material, and chambers can suffer from particle problems and relatively short mean wafers between cleans (MWBC) if thermal cycling and chamber furniture surface finish is not managed carefully. As only a flash layer is required for adhesion, films are relatively thin, typically in the 500 Å–1 kÅ range.

19.2.4 Cu Seed Layer Deposition

This is the final step in the process flow, the requirement being to deposit a relatively thin layer of Cu on the wafer to act as a nucleation layer for Cu ECD. Cu seeds are typically in the 1.5–3 kÅ range. Layers are thicker for structures with thicker passivation or narrower critical dimensions (CDs) where step coverage is reduced. With Cu ECD, minimum thicknesses of 20–50 nm on surfaces are required.

With the Cu seed layer deposited, wafers are passed back from the vacuum back end of the PVD system into their FOUPs at atmosphere via the equipment front-end module (EFEM) for transfer on to the next process stage, namely, resist coating and then patterning and selective plating of the exposed Cu seed material to form the RDL.

19.3 Equipment Challenges for FO-WLP

Initially, the requirements for PVD may look no different from mainstream UBM and RDL processing, but there are several aspects to the process flow that present significant challenges to the user. It cannot be assumed that regular UBM and RDL sputter equipment can be used for FO-WLP, at least not in a productive manner with high yields.

19.3.1 Contamination

The epoxy mold and organic passivation materials used in FO-WLP structures suffer badly from moisture absorption at atmosphere. Moisture and other organic-based contaminants outgas when elevated to higher temperatures or placed under vacuum. If not addressed, the presence of contamination at the adhesion metal deposition stage can create electrical contact resistance and adhesion issues. Figure 19.5 shows typical residual gas analysis (RGA) of the outgassing characteristics of a 300 mm molded wafer when placed in a vacuum degas module, with wafer temperatures driven to 120 °C. Mass 18 [H_2O] and Mass 28 [CO/N_2] dominate. From background O_2 levels (and its proportion to N_2 in air), it is clear that the Mass 28 levels are predominantly CO [5].

Whereas conventional circuits built on silicon can withstand temperatures in excess of 400 °C and can be degassed rapidly without impacting system throughput, the mold and dielectrics used in FO-WLP have heat tolerances

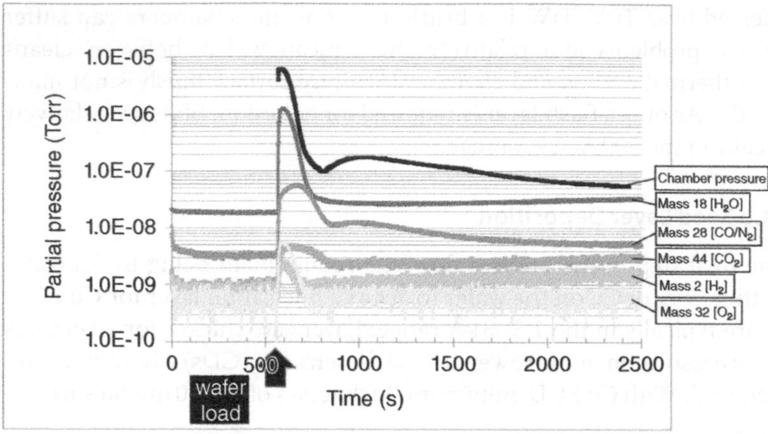

Figure 19.5 RGA mass scan of a 300 mm molded wafer with PI passivation in a vacuum degas module.

closer to 120 °C for production, largely due to the T_g of the epoxy mold compound, which is approximately 150 °C. Temperatures exceeding this threshold can cause material decomposition and excessive wafer warping.

Degassing wafers at such low temperatures takes a longer amount of time to fully remove contaminants and can drastically reduce the throughput of a conventional sputter system that uses single-wafer degas chamber technology. The data in Figure 19.5 shows that for a typical 300 mm molded wafer with polyimide (PI) organic passivation, it can take up to 30 minutes (1800 seconds) for contamination levels to return to initial values.

An *ex situ* oven bake prior to PVD can be used to reduce degas times, but depending on device performance requirements, this may not always be sufficient: moisture reabsorption between the oven and the PVD system has the potential to impact both plasma process stability in the PVD system and overall device electrical performance. Oven bake also requires additional investment and floor space and increases overall manufacturing process flow time.

For a mainstream UBM and RDL sputter system with single or even dual wafer degas configuration and no *ex situ* oven bake, throughput can be limited to <5 wafers per hour (wph), which is low compared with >30 wph throughputs usually achieved with traditional Si UBM and RDL processes based on inorganic passivation material such as silicon nitride (SiN). It is worth noting that more advanced passivation schemes for Si UBM and RDL involving organics such as low temperature cure PI or polybenzoxazole (PBO) can also suffer similar productivity problems, because of similar thermal budget restrictions. Compromises in throughput for both Si and FO-WLP cases can be made (i.e. shorter degas times) but potentially at the expense of device electrical performance. It depends on the performance specifications for the packaged device.

19.3.2 Increased I/O Density

After successful degas, the molded wafer needs to be pre-cleaned in a plasma etch module. This facilitates the removal of trace native oxide layers from exposed metal that the RDL will connect to. With the increase in density of interconnects as more complex dies are packaged using FO-WLP techniques, there is a need for multiple levels of RDL to overcome interconnect congestion. At the uppermost level >70% of the wafer surface can be exposed metal contacts, areas much larger than those encountered in mainstream Si UBM and RDL (typically <20%).

For PVD systems with pre-clean modules based on inductively coupled plasma (ICP) technology with ceramic chamber walls, this increase in exposed metal contact area presents a problem for RF stability. The pre-clean process will involve an "over etch" to accommodate any cross-wafer variation in the etch process. During this stage metal contacts on certain parts of the wafer will gradually be revealed as native oxide is cleared. As etching continues, exposed metal will be sputtered onto the ICP ceramic sidewalls. As metal coats up the chamber walls, the RF will couple through to the metal rather than the plasma, with the process rapidly falling out of control. Figure 19.6 shows an accelerated failure case, using wafers with 60% pure repeatedly etched Cu – i.e. no native oxide present after the wafer has been etched the first time.

PVD systems using diode etch technology, otherwise known as capacitively coupled plasma (CCP), benefit from operating with a metallic chamber wall and therefore avoid this coupling issue, but diode etch technology has its own drawbacks for use in FO-WLP that will be explained later in this chapter.

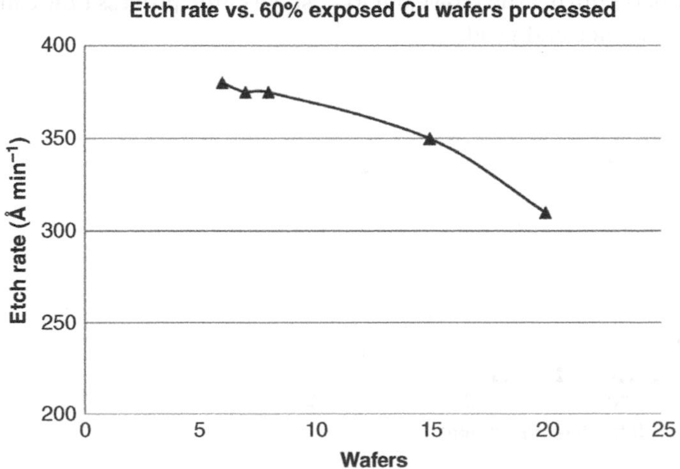

Figure 19.6 ICP etch rate deterioration sputter etching pure metal.

19.3.3 Organic Passivation Particle Management

Due to the presence of organic passivation such as PI or PBO on FO-WLP wafers, carbon-based material buildup on the ICP chamber wall will take place during pre-clean, as organic material is sputtered off the wafer surface in parallel to the native oxide on the contacts being removed. Organic material does not adhere well to ceramic surfaces and, if not carefully managed, can result in premature particle failure (Figure 19.7). Particle management in general becomes more critical as L/S dimensions are reduced for higher density packages.

19.3.4 Contact Resistance (R_c) Management

The use of organic passivation in FO-WLP structures also introduces challenges for R_c during the pre-clean step that intensify as the CDs of the features reduce. As the surface of the wafer is etched, native oxide is removed from the exposed metal contact, but the organic passivation is also physically bombarded. This ion bombardment damages the surface of the passivation, releasing carbon-based (C-based) volatiles, which, in turn, recontaminate the metal contacts the process is attempting to clean (Figure 19.8).

Aside from advanced technology node devices, power management-based devices (PMIC, PMU) are particularly sensitive to this parasitic resistance issue because of the relatively high currents involved with their operation. High R_c can lead to increased power consumption, increased temperature, and resistance–capacitance (RC) time delays on certain device types, so management of this parameter during the UBM and RDL stage of the process flow can be critical to performance and yield.

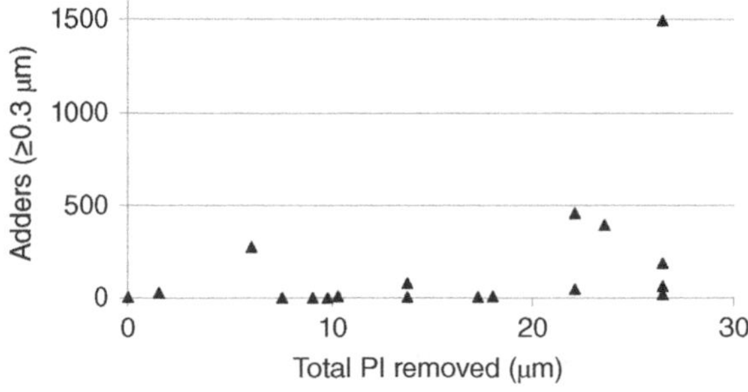

Figure 19.7 Particle performance in non-optimized ICP pre-clean etching wafers with PI passivation.

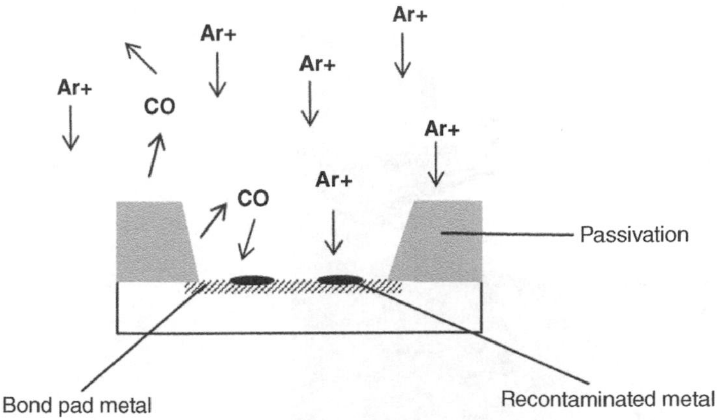

Figure 19.8 Contamination of pads during pre-clean.

19.3.5 Wafer Warpage

Epoxy-molded wafers can be warped after curing, and the size and shape of the warpage hinge on the different size, density, and placement of the die. An FO-WLP PVD system must therefore be able to minimize temperature-induced shape shifting and accommodate wafers with high incoming bow. The majority threshold for bow is on average less than 6 mm for high density packages, as it is not easy to make uniformly thick, fine resolution conductors on a substrate exhibiting high warpage. This is one of the major challenges facing FO-WLP packages on a large area panel format.

Although FO-WLP 300 mm substrates involving glass or silicon carriers with RDL-first typically exhibit bow in the 2–3 mm range, unsupported chips-first 300 mm wafers can experience warpage up to 6 mm (Figure 19.9). For mold substrates >300 mm, 7–8 mm warpage is experienced. Traditional PVD systems for UBM and RDL have never had to process wafers with such large bow. Processing SEMI standard thickness Si meant that wafer bows were never more than 1 mm at most.

Often wafer handling clearances, robot arm velocities, and wafer position sensors are not compatible with wafers warped to this degree, preventing many existing PVD toolsets from being utilized. Temperature management techniques such cold electrostatic chuck (ESC) pedestals on PVD systems can also have wafer bow limitations, ruling some out of FO-WLP process flows.

19.3.6 Capital Cost

Always on the radar of any packaging house but never more so than with FO-WLP lines, cost-effective sputtering must form an integral part of the PVD solution. On that basis, the PVD chamber design used must not be

Figure 19.9 Bowed 300 mm FO-WLP molded wafer in equipment front-end module (EFEM).

over-engineered. Unlike mainstream Si interconnect PVD, topography is relatively low so high cost ionized PVD capability is not required. Neither are thick metal layers such as those used for the bond pad layers of die on advanced node devices. So high power, high deposition rate capability is not required either. PVD modules for mainstream Si applications are effectively over-specified for UBM and RDL applications, with high capital equipment cost. This makes the overall performance benefits for adopting an FO-WLP structure difficult to justify.

19.4 Equipment Developed to Overcome Challenges

19.4.1 Solution for Contamination

Multi-wafer degas (MWD) technology has emerged as a compelling solution to the degas throughput problem, enabling multiple wafers to be degassed in parallel before being individually transferred to subsequent process steps, without breaking vacuum. With this approach, wafers are dynamically pumped under clean, high vacuum conditions, with radiation heat transfer warming wafers directly to temperatures within the operating budget. The use of high vacuum radiative heating prevents contamination cross talk.

Each wafer can spend up to an hour inside the MWD, but because they are processed in parallel, a "dry" wafer is output for metal deposition every 60–90 seconds, at a rate of between 30 and 50 wafers per hour (depending on the wafer bows and film thicknesses involved). This approach increases PVD

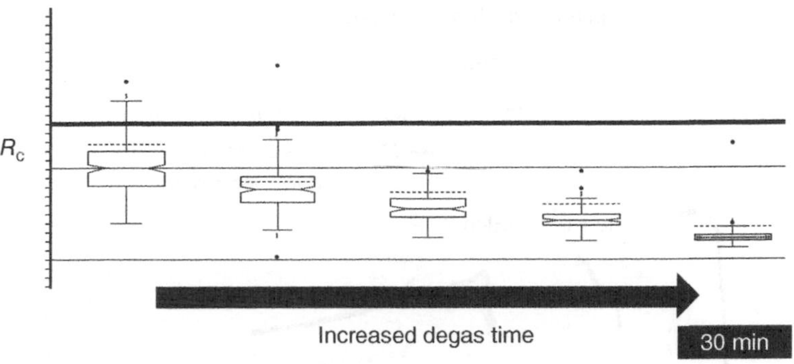

Figure 19.10 R_c performance with increased degas time.

system throughput significantly compared with single-wafer degas processing technology, and as lower cost materials emerge with even lower thermal budgets or worse outgassing behavior, longer degas times can be accommodated with no impact on throughput.

MWD technology has also been shown to eliminate the need for an *ex situ* bake prior to PVD, saving capital expenditure, floor space, and overall flow time for the manufacturer.

Figure 19.10 shows how normalized mean R_c and R_c spread across a 300 mm molded test vehicle wafer reduces with increased degas time (T_{MAX} = 120 °C), illustrating the time required for FO substrates to be left in vacuum degas conditions to achieve the best electrical performance [6].

19.4.2 Atmospheric Degas vs. Vacuum Degas

If a substrate involving organic material is not degassed sufficiently prior to pre-clean, it produces high levels of outgassing that can affect plasma stability during etch and film quality (R_c) during subsequent sputter deposition. Locating the MWD station in the atmospheric front end of the PVD system to minimize complexity can impact the efficacy of the process. To demonstrate this, two identical PBO-coated wafers were monitored using an RGA in a pre-clean module during process: one originating from an atmospheric degas station and the other originating from a vacuum degas station – both wafers given the same 30 minute degas time beforehand at 120 °C. The data obtained (Figure 19.10) shows increased levels of outgassing (H_2O partial pressure) from the wafer that received an atmospheric degas, demonstrating that a vacuum-based degas is more effective at removing moisture.

As previously explained, this increase in contamination during pre-clean has implications for process plasma stability and potentially device electrical performance. Therefore, the decision to locate the MWD station in a high vacuum environment is validated. It should be noted that a vacuum-based

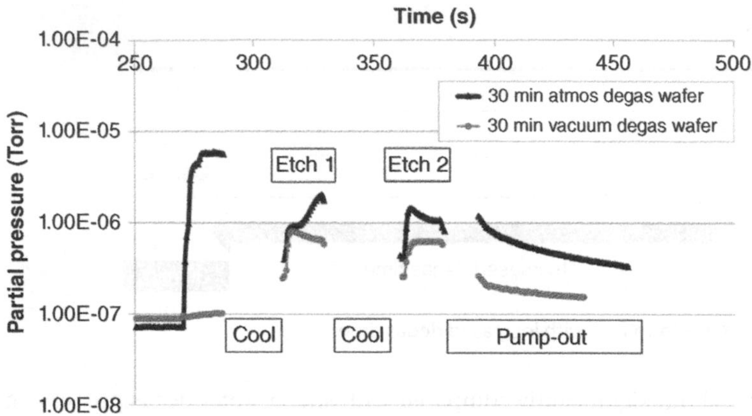

Figure 19.11 Outgassing during pre-clean: atmospheric vs. vacuum degas pretreatment.

MWD approach is only feasible for systems based on pick/place robot wafer handling in the vacuum back end. Wafer handlers based on a carousel or "indexed" approach where wafers are supported by the handling system throughout the process flow are prevented from loading wafers into a multi-slot chamber by the fundamental nature of their design. In those circumstances, multi-slot degas hardware can only be located in the atmospheric side of the system and potentially suffer the problems with moisture reabsorption indicated by the data in Figure 19.11. Degas at atmosphere may also produce relatively poor temperature uniformity and risk of particulates if high flow purge gases are used in an attempt to resolve temperature uniformity issues.

In contrast, the MWD degasses wafers under high vacuum and uses radiative heat transfer, producing excellent WIW temperature uniformity (<5 °C), low particle levels as there are no purge gases, and no risk of moisture reabsorption or cross-contamination because of the low vacuum conditions employed [7].

19.4.3 Solutions for Increased I/O Density

By inserting a discontinuous metal liner in the ICP chamber between the wafer and the ceramic chamber wall (i.e. a Faraday cage), it is possible to maintain the ICP effect while sputtering metal onto the sidewalls; the design of liner and RF antenna circuit ensures that a continuous band of metal is prevented from forming. With the liner installed, etch rate stability can now be maintained at levels previously experienced with low exposed Cu contact structures (Figure 19.12).

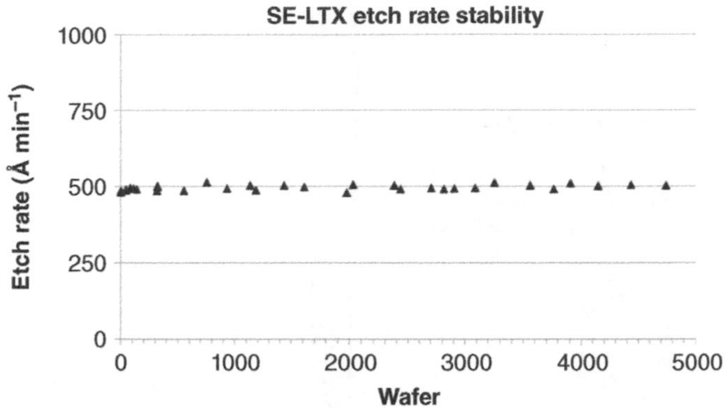

Figure 19.12 Pre-clean etch rate stability.

19.4.4 Solution for Particle Management

The presence of the metal liner on the chamber sidewall for exposed metal compatibility now also allows *in situ* metallic pasting to be employed to adhere problematic organic material to the sidewalls. Al-coated silicon or Al-coated metal disks are commonly used for this purpose. Preventative maintenance intervals can be extended to match adjacent PVD module maintenance activity timelines, with no detrimental impact on system uptime.

In situ coaxial pasting (co-pasting) during the etch process significantly improves productivity further by reducing the frequency of wafer-based pasting (typically every 12 wafers). With the co-pasting approach, the "RF-live" back sputter shield located around the platen is deliberately designed to be etched, sputtering metal away from the wafer, onto the sidewalls of the chamber, pasting down organics as the wafer etch process takes place (Figure 19.13).

Vapor phase decomposition (VPD) analysis during a 5000 wafer marathon showed that the use of this co-pasting design does not adversely impact contamination levels detected on wafer (Figure 19.14), proving that the concept of sputtering metal away from the wafer is working successfully.

The co-pasting concept is compatible with bowed molded wafers and has been optimized to prevent arcing. This solution is now running FO-WLP production wafers at multiple sites. Co-pasting increases the wafer-based pasting interval by >100×, maintaining low particle levels throughout, even at small bin sizes (>0.2 μm) (Figure 19.15).

19.4.5 Solution for R_c Management

Common approaches to tackle this problem involve limiting wafer temperature during process, thus reducing the volatility of the C by-products. Chilled pedestals and chamber furniture are commonplace as a result. In addition to

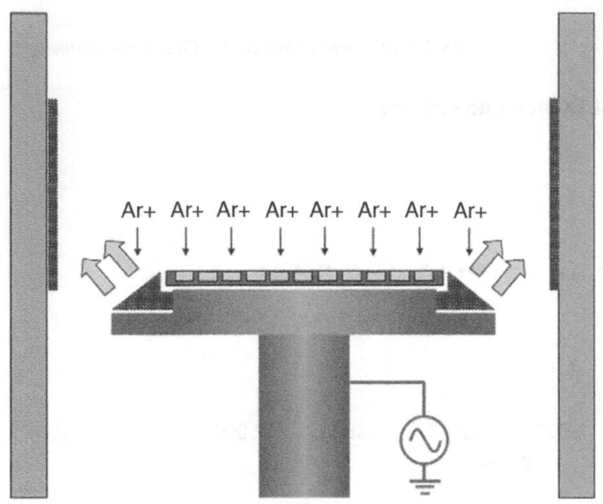

Figure 19.13 *In situ* pasting during pre-clean.

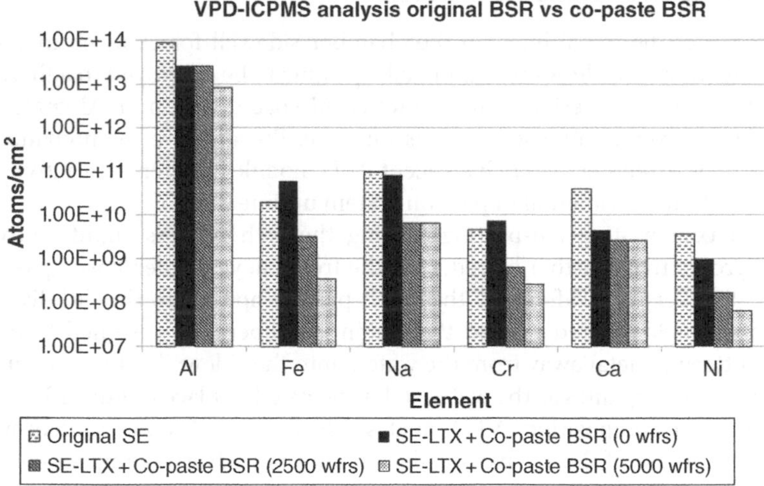

Figure 19.14 VPD metallic contaminant levels using co-pasting technique.

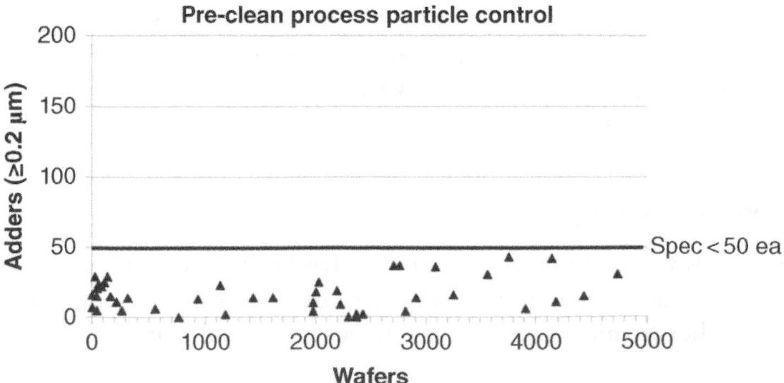

Figure 19.15 SE-LTX pre-clean particle performance with co-pasting technique employed.

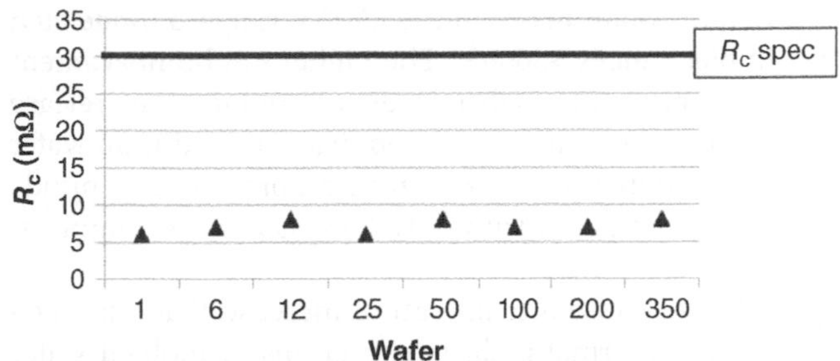

Figure 19.16 RC stability using co-pasting over 350 wafers with no wafer pasting.

reducing contaminant volatility, vacuum pumping speed is increased to pump away contaminants before they have an opportunity to react with the exposed metal contacts. The use of *in situ* pasting for particle control reasons can also be used for R_c management purposes, with the co-pasting technique producing stable R_c performance without the need for frequent wafer pasting (Figure 19.16).

19.4.6 Why Soft ICP Etch Is Best

Given the challenges of large exposed metal and organic material adhesion with ICP ceramic walls, it would be tempting to employ a CCP or diode etch approach to pre-clean, as the single source concept in the form of a RF-driven platen allows the chamber walls to be of metal construction. The diode etch approach has several disadvantages however. First, the diode etch concept, being single source, requires relatively high bias voltages to initiate and sustain a plasma discharge (Figure 19.17).

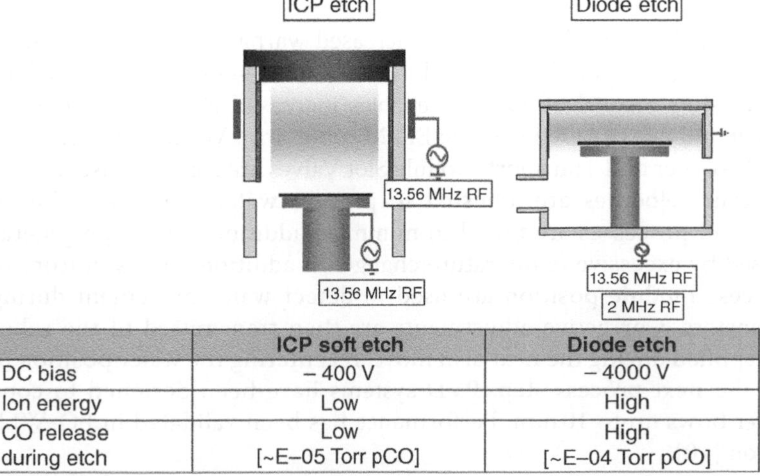

	ICP soft etch	Diode etch
DC bias	– 400 V	– 4000 V
Ion energy	Low	High
CO release during etch	Low [~E–05 Torr pCO]	High [~E–04 Torr pCO]

Figure 19.17 ICP vs. diode etch comparison.

Where an ICP etch generates wafer biases in the <500 V range, a diode etch module typically works in bias ranges >3000 V. The higher ion bombardment energy involved produces significantly higher levels of contaminant release from the organic passivation on the wafer. For an equivalent maximum wafer temperature during etch, CO partial pressures can be an order of magnitude higher in a diode etch system compared with ICP. This has obvious implications for R_c.

The high ion energy will also cause a temperature increase that will need additional cool steps given the thermal budget restrictions of molded wafer processing. The cool steps will in turn impact module throughput, assuming the degas throughput bottleneck with an MWD approach has already been solved. With mainstream Si UBM and RDL PVD systems, these problems can be overcome through the use of ESC clamping combined with backside cooling gas to maintain wafer temperatures at low levels, minimizing contaminant volatility. With molded wafer processing, however, ESCs are not reliably compatible because of the excessive wafer bows that can be encountered.

Another disadvantage of diode etch is that etch rate is typically 30–50% of an ICP equivalent, limiting the throughput of the system after the degas bottleneck is addressed.

Finally, the diode etch process is less directional than ICP due to its single source and higher discharge pressures. As CDs reduce the aspect ratio of contact, topography will inevitably increase. For a diode configuration this will mean longer etches are needed to remove native oxide in the base of the contact. Conversely, the ICP etch, being dual source and low pressure (<2 mT), produces an anisotropic etch, directional in nature. As aspect ratios increase, the ICP etch is less affected, and throughputs can be maintained.

19.4.7 Solution for Wafer Warpage

To successfully handle substrates with increased warpages, several aspects of the system design need to be modified. At the atmospheric front end, FOUPs will use 13 slots instead of 25. This enables increased clearances for wafer transfers in and out of FOUPs by the EFEM robot arm. Vertical clearances at the EFEM Aligner and Transport Module Slot Valves are also increased.

Robot arm velocities are adjusted to prevent wafer movement during transfers, and processes are tuned to minimize additional warpage generation caused by excessive temperature change. In addition, optics in front of each process module position are used to detect wafer movement during wafer transfers. Corrective adjustments are then transmitted to the robot arm and applied during the next arm move, recentering the wafer position in time for the next process step. PVD systems have been designed to cope with wafer bows up to 10 mm. Performance has been validated in FO-WLP production [10].

19.4.8 Solution for Capital Cost Reduction

PVD RDL seed depositions are relatively simple, and for that reason it is important to use hardware that does not exceed requirements. Traditional front-end-of-line (FEOL) PVD chambers used in semiconductor device fabrication facilities are designed to offer features such as high deposition rates, directionality, and advanced uniformity control, which are not critical for FO-WLP. A basic conventional PVD module will meet the requirements. A picture of a conventional magnetron sputter PVD module is shown in Figure 19.18 and is specifically designed for BEOL processing. The simplicity of the design allows cost savings to be passed on to the user, contributing to the overall task of minimizing the costs to run the FO-WLP application.

Figure 19.18 SPTS Inspira PVD module for UBM and RDL.

19.5 Additional Equipment Features

In addition to the features already described to deliver productive PVD processing capability, it is worth mentioning an additional technique employed to manage the heat generated by the sputtering process (condensation energy). Si-based UBM and RDL platforms often take advantage of forced clamping systems with backside gas cooling to limit wafer temperatures, but as explained previously, with highly bowed wafers forced clamping is not always possible. For most the alternative is to use time alone to allow wafers to cool after deposition, but heat transfer is poor at high vacuum so wait times are long, and this has obvious implications for throughput. A more efficient alternative is to use a backfill cool approach.

The backfill cool technique is similar to clamping in that argon is used to conduct heat away from the wafer to a relatively cold platen beneath, except with backfills; no clamping is involved (Figure 19.19). The procedure involves isolating the vacuum pump temporarily, filling the chamber with argon to reach ~1 T, pausing while heat is transferred away from the wafer, then opening the vacuum line, pumping the argon away, and commencing sputtering.

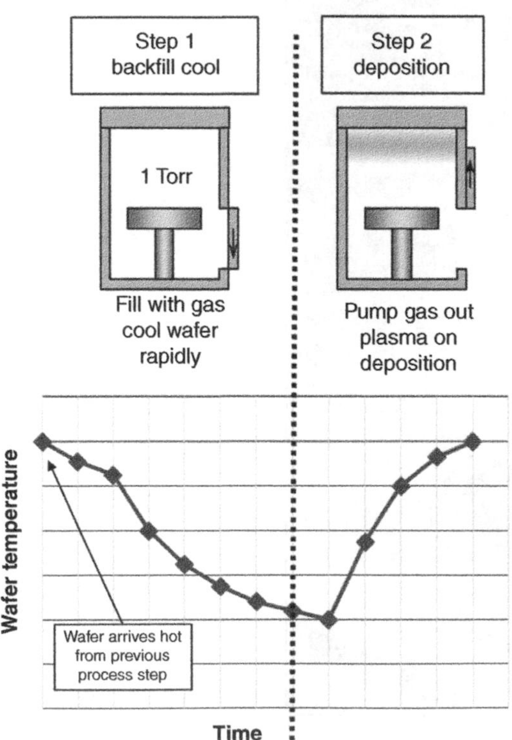

Figure 19.19 Backfill cool concept for bowed wafer cooling.

Backfill process times and insertion points in the process flow are coordinated to maximize system throughput, working within defined thermal budget restrictions of the process.

19.6 Design Rules Related to the Equipment

As previously described, PVD systems need to be able to cope with wafer bows up to the 10 mm range to fully accommodate the variety of FO-WLP process flows and die combinations in use. Mold thickness is generally in the 500–1000 μm range, but it is warpage that limits the PVD system capability rather than the thickness. Obviously the two can be interrelated.

Wafers in the 200–330 mm range can be processed. The use of conventional PVD chambers with relatively short target to wafer spacing (~50 mm) means that designs with low topography can be processed successfully (e.g. vias with aspect ratios up to the 1–1.5:1 range). Given that most structures have 20–50 μm CD and 5–10 μm passivation thickness, this means conventional PVD falls safely within the operating range.

For those with an interest in depositing PVD seeds for through-mold via (TMV) purposes, alternative PVD chamber hardware is available [8]. Ionized PVD is used whereby material not only is sputtered from the target but is also ionized. By applying a DC bias to the wafer, sputtered metal ions are attracted and accelerated toward the wafer, driving the material with a more vertical component of direction down into high aspect ratio vias. With additional DC bias applied, the material in the base of the vias can be re-sputtered onto the sidewalls, providing a conducting path for an ECD process to fill the TMV with conducting material such as Cu (Figure 19.20).

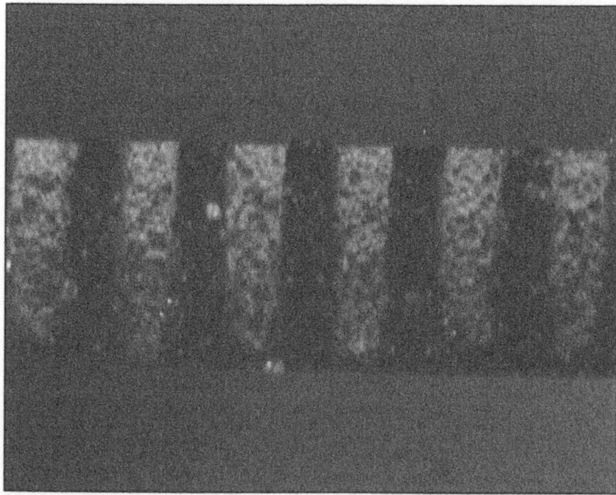

Figure 19.20 Continuous Cu in through-mold via test structures.

Depositing a metal liner into a TMV created by laser drilling is challenging because of the roughness of the via sidewalls – much rougher than anything produced with the more traditional Si-based TSV structures plasma etched using Bosch or equivalent etch techniques. Despite this, with enough material deposited in the field, ionized PVD has been shown to produce sufficient coverage for plating [9]. The challenge comes more from managing the thermal effects of working with an ionized metal plasma reactor when temperature-sensitive materials are involved.

19.7 Reliability

The data in Figure 19.21 illustrates the reliability of current generation PVD tools for FO-WLP production [10].

The data shows that systems processing warped FO-WLP wafers involving mixed die combinations (in some cases warping up to 7–8 mm for substrates >300 mm diameter) can operate with uptimes >90%. Breakage rates that are <1 in 70k have been reported, in line with mainstream silicon PVD system performance.

From a device performance perspective, Figure 19.22 shows R_c values collected over a six-month period (approximately 5000 measurements, 35 μm CD test structures, PI passivation). Low R_c values <2 mΩ are achieved and maintained throughout the monitoring period.

19.8 Next Steps

Aside from the potential introduction of ionized PVD for higher aspect ratio features, FO-WLP PVD developments will continue to focus on R_c management and cost reduction. For R_c management, new pre-clean chamber designs will emerge with continued emphasis on contamination prevention or rapid removal.

	Month 1		Month 2		Month 3	
Internal ID	Uptime	Utilization of uptime	Uptime	Utilization of uptime	Uptime	Utilization of uptime
Sigma #23	89%	89%	88%	90%	90%	88%
Sigma #25	90%	86%	92%	91%	90%	92%
Sigma #26	91%	90%	91%	89%	92%	90%
Sigma #82	Install to Qual		88%	75%	92%	85%

Figure 19.21 FO-WLP PVD system reliability data.

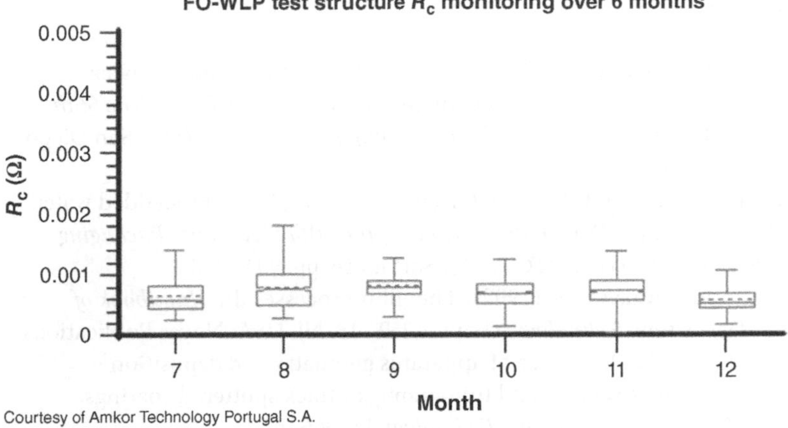

Figure 19.22 FO-WLP device test structure R_c repeatability.

Cost reduction is a continuous process, so it will be on any FO-WLP production roadmap. Improvements in mechanical throughput are anticipated with the development of faster robot arm systems from the robotics OEMs. This is most likely to impact high-end RDL-first process flows that run with reduced bow substrates and higher transfer speeds.

For degas performance it is anticipated that reduced cost passivation and mold materials will continue to emerge, some with higher temperature tolerances, but with potentially worse outgassing behavior. MWD temperature capability should be in place to address this.

For incoming wafer conditions, bow values are expected to increase as material reduction takes place from efforts to reduce costs as well as thinner molded wafers for low profile packages (<500 μm). This will put more demands on the PVD systems as well as equipment and materials for the entire FO-WLP process flow, especially those running on larger wafer sizes. Adoption of larger and more diverse combinations of SIP configurations will also lead to higher warpages – and with SIP being one of the most popular emerging package types for FO-WLP, this is sure to drive the average warpage values up.

Currently, there is a lot of research in the area of panel-level fan-out [11]. It is clear that there might be a place for panel-level fan-out with low density packages, provided identified technical roadblocks through the process flow are overcome, but the high performance packages using high density RDL will almost certainly remain wafer based largely because of the lithography limitations encountered with large bowed panel processing. Return on investment (RoI) arguments for developing a panel-based equivalent PVD system are not compelling today [12].

References

1 Brunnbauer, M., Furgut, E., Beer, G. et al. (2006). An embedded device technology based on a molded reconfigured wafer. *Proceedings of the 56th Electronics Components and Technology Conference (ECTC 2006)*, San Diego, USA, pp. 448–551.

2 Brunnbauer, M., Fürgut, E., Beer, G., and Meyer, T. (2008). Embedded wafer level ball grid array (eWLB). *Proceedings of the 10th Electronics Packaging Technology Conference (EPTC 2008)*, Singapore, pp. 994–998.

3 Wasa, K. and Hayakawa, S. (1992). Thin film processes. In: *Handbook of Sputter Deposition Technology*, chapter 2, 2–48. NJ, USA: Noyes Publications.

4 Thornton, J. (1974). Influence of apparatus geometry and deposition conditions on the structure and topography of thick sputtered coatings. *Journal of Vacuum Science and Technology* 11: 666–670.

5 Jones, C., Butler, D., and Burgess, S. (2016). High productivity UBM/RDL deposition by PVD for FO-WLP applications. *Proceedings of the International Wafer Level Packaging Conference (IWLPC 2016)*, San Jose, USA.

6 Jones, C. (2013). Cost-effective & productive PVD for Advanced Packaging UBM/RDL with Sigma fxP. Presented at SEMICON Taiwan 2013 Advanced Packaging Workshop, Taipei, Taiwan.

7 Knight, N., Burgess, S., Jones, C. et al. (2017). Rc management for next generation PVD UBM/RDL metallization schemes. *Proceedings of the International Wafer Level Packaging Conference (*IWLPC *2017)*, San Jose, USA.

8 Urbansky, N., Burgess, S.R., Schmidbauer, S. et al. (2002). Advanced hi-fill® for interconnect liner applications. *Microelectronic Engineering Vol.* 64: 99–105.

9 Butler, D. (2015). High density packaging on wafer level fan-out: deposition and via drilling solutions tailored for non-silicon substrates. Presented at SEMI European 3D Summit 2015, Grenoble, France.

10 Vishwanath, S. (2016). Achieving high quality and low cost PVD – critical for success in advanced packaging. Presented at SEMI Taiwan WLCSP Equipment Technology Forum, Taipei, Taiwan (June 2016).

11 Werbaneth, P. (October 2016), "First-Mover Advantage: Fan-Out Panel Level Packaging at IWLPC 2016", 3DInCites [Online]. https://www.3dincites.com/2016/10/first-mover-advantage-fan-out-panel-level-packaging-at-iwlpc-2016 (accessed 6 August 2018).

12 I-Micronews (July 2017). Inside PVD Technology for Fan-Out by SPTS, Yole Development [Online]. https://www.i-micronews.com/news/advanced-packaging/9596-inside-pvd-technology-for-fan-out-by-spts.html (accessed 6 August 2018).

20

Excimer Laser Ablation for the Patterning of Ultra-fine Routings

Habib Hichri, Markus Arendt, and Seongkuk Lee

SUSS MicroTec Photonic Systems, Inc., Corona, CA, USA

The semiconductor industry is driven by the miniaturization of transistors and scaling CMOS technology to smaller and more advanced technology nodes while at the same time reducing the cost. Photolithography has long been the dependable standard for structuring organic materials used in advanced packaging. For several reasons, however, this multistep process is beginning to limit package designs and performance. To overcome these limitations, new advanced packaging technologies have been developed, enabling more functionality to be integrated along with various types of devices in the same package. A significant level of activity is currently underway within the advanced packaging industry.

One of the most well-known examples of a fan-out wafer-level packaging (FO-WLP) structure is embedded wafer-level ball grid array (eWLB) technology by Infineon Technologies AG. This technology uses a combination of front- and back-end manufacturing techniques with parallel processing of all the chips on a wafer, which can greatly reduce manufacturing costs. Its benefits include a smaller package footprint compared with conventional lead frame or laminate packages, medium to high I/O count, maximum connection density, and desirable electrical and thermal performance. It also offers a high performance, power-efficient solution for the wireless market [1]. Infineon was the first company to commercialize its own eWLB packaging technology in an LG cellphone in 2009. In 2015, Taiwan Semiconductor Manufacturing Company (TSMC) established the "integrated fan-out" (InFO) line and multiplied the existing FO-WLP capacity by producing the application processor for the Apple smartphone through single chip packaging with multilayer routing [2]. This showed that the FO-WLP process flow could meet the continuous trend toward miniaturization, increasing the performance and

Advances in Embedded and Fan-Out Wafer-Level Packaging Technologies, First Edition.
Edited by Beth Keser and Steffen Kröhnert.

cost-effectiveness of electronic devices. While traditional organic flip-chip substrates using semi-additive processes (SAP) have not been able to scale to ultra-fine RDL pitches and via openings below 10 μm, photosensitive spin-on dielectrics and RDL processes used for wafer-level packaging do not sufficiently address the cost reduction need. To enable wafer-level packaging (WLP) to reduce cost for RDL and scale interconnect pitch to 40 μm and below, excimer laser ablation is introduced as a direct patterning process that uses proven industrialized excimer laser sources emitting high-energy pulses at short wavelengths to remove polymer materials with high precision and high throughput. The combination of a high power excimer laser source, large-field laser mask, and precision projection optics enables the accurate replication and placement of fine resolution circuit patterns without the need for any wet processing. In addition, with excimer laser patterning technology, the industry gains a much wider choice of dielectric materials (photo and non-photo) to help achieve further reductions in manufacturing costs as well as enhancements in interposer and package performance.

This chapter reviews the status of advanced packages and details an innovative and alternative patterning process. This new patterning solution uses excimer laser ablation to integrate via and RDL traces in one patterning process step, followed by seed layer deposition, plating, and planarization steps. It therefore uses the dual damascene process from the front end of line (FEOL) to create connections for advanced packaging in the back end.

20.1 Advanced Packaging Applications and Technology Trends

A wide variety of advanced packaging technologies exist to meet the current requirements of the semiconductor industry. However, the consumer demand for higher functionality in smaller and thinner devices, like smartphones or tablets, as well as high data bandwidth in cloud and edge computing, drives the need for next-generation interposers and packages with finer features, improved materials, and increased package reliability [3]. At the same time, continuous cost reductions are essential to enable pervasive high volume applications for any new technologies. These cost pressures are felt throughout the supply chain, from chip manufacturers to assembly and test houses to consumer device manufacturers. Consequently, the industry is seeking new innovative approaches and enabling technologies to lower manufacturing costs for devices and packages that will meet the next-generation bandwidth, power, and performance requirements.

Advanced packaging approaches including chip-on-chip stacking, high density fan-out wafer-level packages, and embedded ICs will need thinner substrates, redistribution layers, and other interconnects with finer traces and higher resolution vias. To increase package reliability, the industry is searching for better performing dielectric materials and process solutions that overcome the scaling challenges of current photolithography and metallization processes.

In response to these vital industry needs, we introduce a novel dual damascene process integration flow, combined with the direct patterning benefits of excimer laser ablation. Excimer ablation enables the use of a wider choice of dielectrics (photo and non-photo) with improved thermal, mechanical, and/or electrical properties. Note that better performing non-photosensitive materials have been available for quite some time, but could not be patterned at an acceptable cost level since it required costly etching methods.

Another approach to reduce cost in packaging is by use of larger panel substrate formats, which provide a 2.5 times larger usable field size compared with 300 mm round wafers. The dual damascene RDL process through excimer laser ablation is scalable for both wafer and panel processing.

20.2 The High Density Structuring Challenge

Advanced packages, including chip-on-chip wafer-level packages (WLP) (e.g. 2.5D), chip-on chip stacking, and embedded IC, all have a need to structure thin substrates, redistribution layers, and other package components at high resolution. A very common example is drilling vias down to contact pads for applications like fan-in WLP or FO-WLP. Another emerging example is trenching to form embedded connectors in thin substrates or interposers. In the case of organic dielectrics, these structuring needs have usually been met by lithography, using either a projection light source or photomask, or sometimes by laser direct imaging (LDI) [4]. The relentless, consumer-driven push for higher functionality from smaller products, however, is creating a fast-growing need for next-generation packages with smaller feature sizes and tighter pitches, which in turn means the use of smaller structures (e.g. vias with diameters as small as 5 μm). This situation presents several challenges for lithography. Specifically, lithography requires the use of photoimageable materials, such as polyimides (PIs), polybenzoxazoles (PBOs), and epoxies, which can cause thermal expansion (CTE) mismatch between these materials and the chips. Furthermore, higher reliability packages have bigger thermal loads, exacerbating this problem. As a result, this CTE mismatch can cause issues such as stress damage to low-κ dielectrics and wiring layers on the semiconductor die, as well as processing issues caused by increased warpage. In addition, many photopolymers are limited in the resolution and via wall angle they can support. In the case of vias, for example, this limits the minimum achievable diameter and interconnect density, as well as their practical aspect ratios.

Devices that are already seeing an impact are those involving larger dies, higher I/O densities, thinner Si chips and substrates, and, obviously, any high heat load applications. Moreover, lithography is a multistep process involving developers and other wet chemicals, making it increasingly unattractive for two reasons. First, there is the cost of ownership associated with these chemicals and their safe handling and disposal. Second, like every other industry, advanced packaging is under pressure to use greener manufacturing that is more energy efficient and less polluting. Eliminating multistep lithography and its associated chemicals would be very congruent with this goal. A very attractive alternative to photolithography would be a technique that can directly structure PIs, PBOs, and epoxies at even higher resolution than today's bleeding edge packages, and allows a wider material selection to contain the CTE issue, at reduced cost. Excimer laser ablation now provides that alternative.

20.3 Excimer Laser Ablation Technology

The laser ablation of thin film polymers is in principle not a new technology for wafer-level packaging. The laser ablation of polymers was first reported in 1982 [5]. Low speed and high cost were the major barriers for further developments, but the lithographic approach of excimer laser systems using quartz masks on stepping/scanning platforms has improved this technology to overcome the limited throughput.

Excimer laser ablation is a one-step dry-etch patterning process that differs from solid-state laser ablation in that the excimer process is based on photochemical (photolytic) bond breaking, while the solid-state process relies on heat (pyrolytic), melting, and evaporation to ablate polymers [6, 7]. The relatively nonthermal characteristics of "ablative photodecomposition" with excimer lasers were so notable that the term "cold ablation" was coined [5, 8].

The "photo" term in photochemical comes from "photon," the energy particle for laser ablation. One requirement for photochemical bond breaking is that the applied photon energy must exceed the binding energy of the material to be ablated. As an example, Kapton (a PI) is primarily made up of carbon (C), oxygen (O), and nitrogen (N) molecules that are covalently bonded together [6]. The photon energy level in electron volts (eV) to break C–C or C–N bonds is 3.0–3.5 eV, and for C–O or C–H bonds is 4.5–4.9 eV [9]. In comparison, the photon energy available in common industrial ultraviolet (UV) excimer lasers is 4.02 eV for 308 nm, 5.00 eV for 248 nm, and 6.42 eV for 193 nm wavelength, which are sufficient to break chemical bonds in polymers [10].

Most polymer materials used as dielectrics in advanced packaging strongly absorb in the UV spectrum. Organic materials, such as PI, have high absorption coefficients (α): 135 000 cm^{-1} at 248 nm and 35 000 cm^{-1} at 308 nm. In contrast, the absorption coefficient for 355 nm (a typical YAG solid-state laser wavelength) is only approximately 10 000 cm^{-1} [11]. Therefore, when an excimer laser pulse is focused onto a polymer material so that the intensity (fluence) is above a material-dependent threshold value, the high-energy UV photons directly excite electrons and break molecular bonds (bond scission) in the long polymer chains [5].

In addition, some of the smaller broken chains immediately react with other ablation products to create new bonds. If the number of broken bonds does not exceed the number of new molecular bonds by a certain amount, then only heating occurs and the ablation process does not progress [6]. This is the case if an insufficient fluence or energy density (number of photons per second, per unit area) is applied to the material. This condition is considered to be below the ablation threshold. The ablation threshold or flux threshold is the volume density of photons per second above which the number of bonds being broken are more than those being formed – above this threshold ablation starts [12]. The ablation threshold is an important characteristic because it can facilitate selective ablation, meaning ablate vias in polymer down to a metal pad without damaging the metal. Selective ablation is possible when there is a significant amount of difference between the two materials' ablation thresholds. Above the ablation threshold, the primary mechanism is direct bond breaking, which progresses to create smaller polymer chains, molecules, and gas vapor. In addition, the secondary mechanism called thermal scission assists in further breaking down the polymer chains, within the same volume. When the bond density in the ablated volume drops below a critical amount for the specific material, the small chains and carbon debris in powder and vapor form explode out of the top of the material, ejecting the ablation materials at high speed (Figure 20.1b) [6].

The result of excimer laser ablation is a well-defined feature that is mask defined and dry etched into the material to a certain depth and are material, fluence, and wavelength dependent. Since a typical excimer laser pulse duration

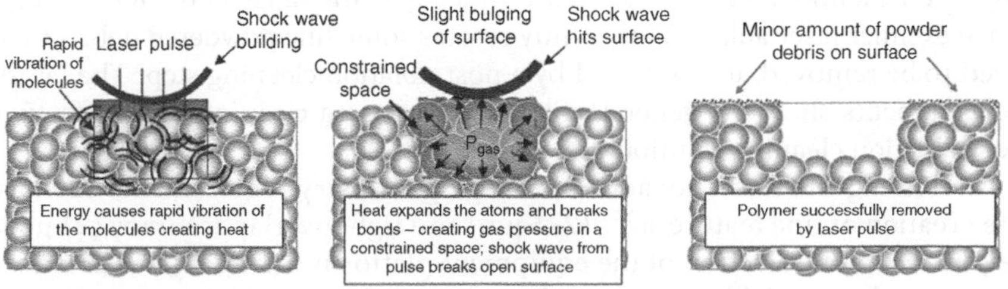

Figure 20.1 Schematic illustrations of excimer laser ablation of polymer in a one-step dry-etch process [6].

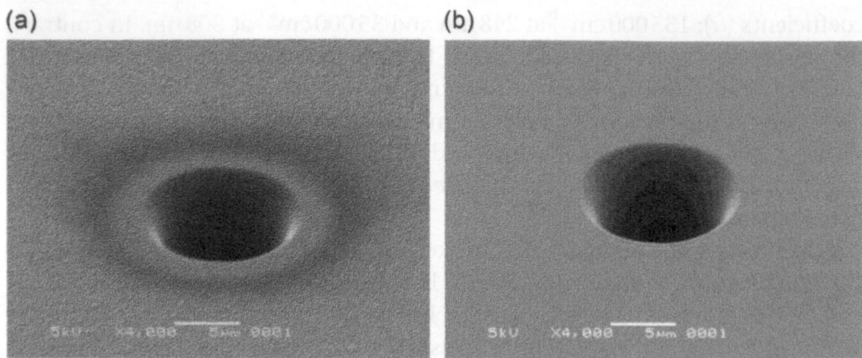

Figure 20.2 Top view of dielectric (PBO HD8820) surface post-ablation (a) and post-O2 plasma cleaning (b).

is around 20–30 ns, the interaction with the material occurs very rapidly, resulting in little time for thermal transfer in the material. Excimer ablation of polymers produces little to no heat-affected zone (HAZ) and typically only minor amounts of submicron carbon debris (Figure 20.1c). The removal of the loose carbon powder debris can be accomplished by common industry wet cleaning or by O_2 plasma cleaning (Figure 20.2) [13].

Excimer laser patterning technology uses the advantage of high power (i.e. 300 W) excimer laser sources to emit high-energy pulses at short wavelengths. Depending on the material's UV absorption properties and laser wavelength, each laser pulse removes a certain amount of material at specific laser fluence. Empirical results indicate the ablation rates for advanced packaging dielectrics are in the range of 0.1–0.8 μm per pulse and are material, fluence, and wavelength dependent (Figure 20.3). This inherent property enables accurate depth control by adjusting the number of pulses applied to the material. Furthermore, it allows the creation of blind holes. The technology also supports the fine-tuning of feature sidewall angles by adjusting the laser fluence. The sidewall angles for excimer ablated features are typically in the range of 65–83°. Finally, since excimer laser ablation inherently creates some fine powdered debris that need to be removed, it is followed by a post-ablation cleaning step. The latest developments show the general suitability of plasma cleaning as an effective post-ablation cleaning solution.

Addressing the latest advanced packaging industry requirements such as the creation of fine feature sizes (2–5 μm) and tight overlay accuracy (<1 μm) requires careful selection of the equipment platform. Excimer laser ablation tools are also available in step-and-repeat platforms, which are similar to UV lithography stepper systems. The combination of an industrialized excimer laser source, a quality beam delivery system (BDS), and precision

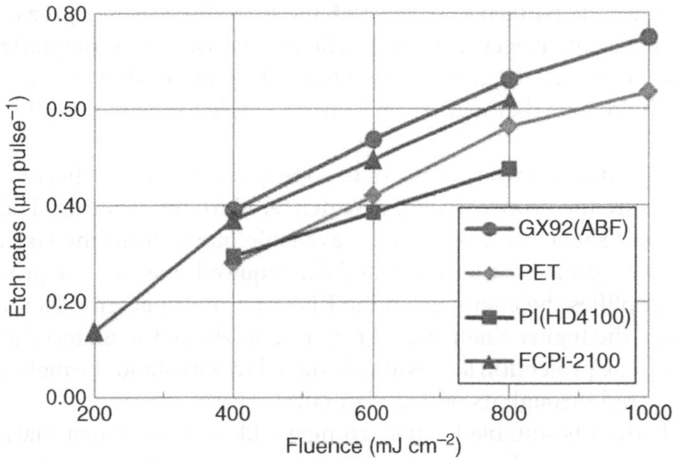

Figure 20.3 Example of ablation rates in ABF (GX92), PET, PI (HD4100), and FCPi-2100 (Fujifilm) as a function of fluence.

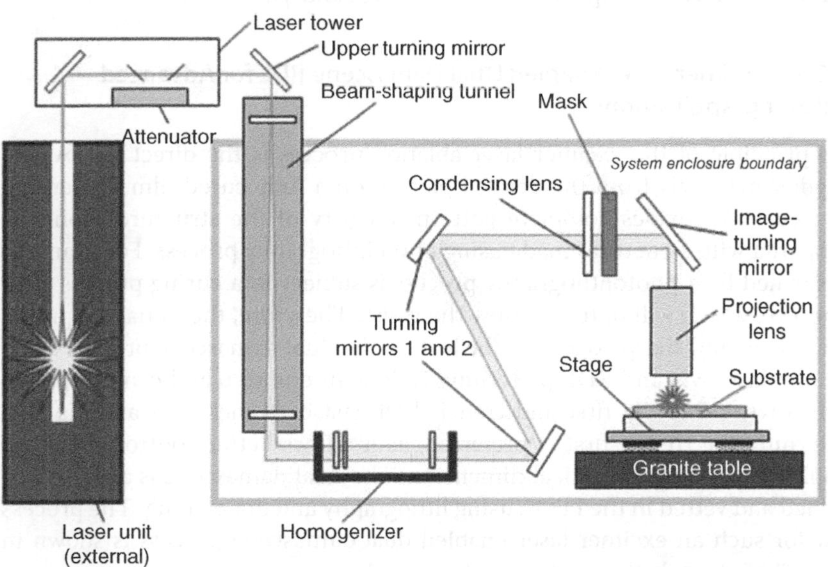

Figure 20.4 Example beam delivery system schematic for an excimer laser ablation tool.

reduction projection optics enable high resolution imaging and accurate pattern replication. In addition, with sufficient high energy output from the excimer laser (i.e. 300 W, 1 J), it is possible to pattern large areas at a time through quartz masks to maximize throughput. A schematic of the excimer laser stepper is shown in Figure 20.4.

A laser mask defines the pattern to be ablated and provides a high degree of pattern fidelity and placement accuracy. Typical laser masks are made of quartz with aluminum or chrome coating. Aluminum is mostly suitable due to a very high damage threshold level that makes the Al mask stable for a long production cycle time.

The beam delivery optics is typically customized to meet specific application requirements to ensure the proper fluence gets delivered to the substrate. The ablation field size and shape are based on the available power from the laser, the die or package size, projection lens size, and the required fluence. The projection lens then amplifies the energy from the BDS by a multiplication factor (i.e. 2.5×) to achieve the higher fluence level required to ablate the material at the substrate plane. The projection lens is also designed to withstand the higher fluence levels projected through its optical elements.

Excimer laser ablation is suitable for patterning a wide variety of materials, such as polymers, dielectrics, thin metals (<600 nm), epoxies, EMCs, nitrides, and other materials. It even allows the use of non-photosensitive materials, some of which offer better thermal and mechanical properties (i.e. lower CTE and residual stresses, higher T_g, and thermal stability).

20.3.1 Excimer Laser Enabled Dual Damascene RDL for Advanced Packaging Applications

The principle of the excimer laser ablation process is the direct removal of polymer materials from the desired pattern on a fully cured film. Patterning after curing provides complete pattern integrity of the structure profile as compared with structures made using photolithography process. For example, via formed by a photolithography process is subject to a curing process after exposure, which will in turn reflow the resist. Therefore, the initial via profile is modified and sharp edges may be lost. In the dual damascene process using excimer laser, via and RDL patterning is done in one step and can be applied both ways: either via first and trench RDL last or trench first and via last. Patterning the trench first is preferred, as it allows better control of the via profile and its bottom critical dimension. The dual damascene is already well defined and vetted in the FEOL using lithography and dry etching. The process flow for such an excimer laser enabled dual damascene process is shown in Figure 20.5 (trench-first and via-last process).

The first step is coating the polymer on the wafer and curing before the patterning process. The thin film polymer could incur shrinkage during cure of up to 40%, which limits the feature sizes patterned by photolithography. In the next step, the traces are ablated into the polymer layer with the excimer laser. The ablation depth can be controlled by the number of pulses. After trace ablation the mask is changed, and the vias are drilled by the laser. The excimer laser ablation tool allows high submicron alignment accuracy between the two mask

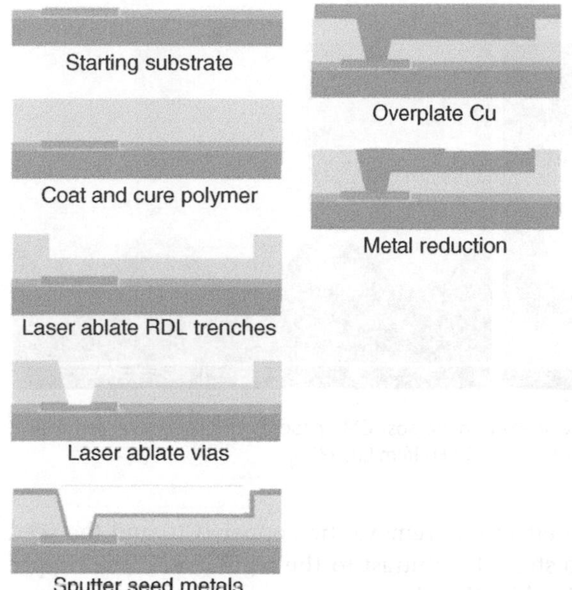

Figure 20.5 Schematic description of the dual damascene process for an RDL.

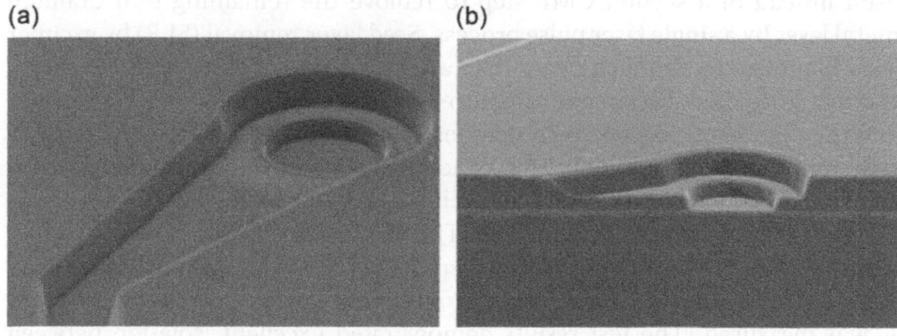

Figure 20.6 SEM image of an embedded RDL and via formed by excimer laser ablation in PBO film (HD8930, 13 μm thick). (a) Overhead view. (b) Cross section.

layers. It is therefore not necessary to provide a larger capture pad size. Figure 20.6 shows the dual damascene RDL and via structure ablated into a single dielectric layer using a 308 nm excimer laser in fully cured 13 μm thick PBO film. After formation of via and RDL, a seed layer is sputtered, which consists of a 100 nm titanium adhesion layer and a 400 nm copper layer. The structure is then filled by electroplated copper. The copper overburden is removed by a chemical and mechanical polishing (CMP) step or alternatively by lower cost methods such as fly cutting, which is diamond bit cutting. Such

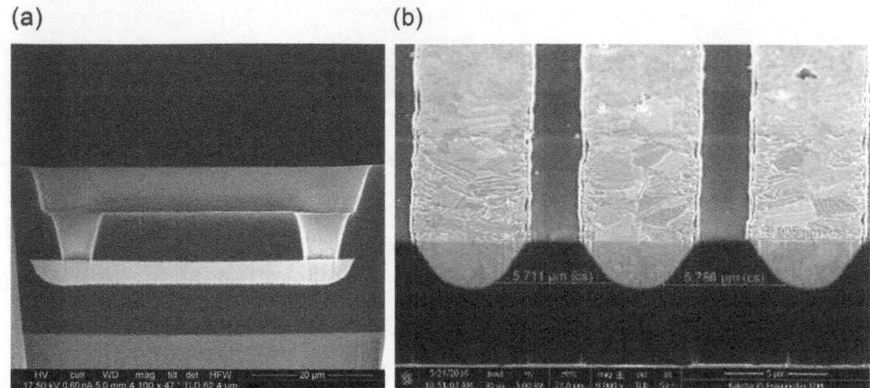

Figure 20.7 Cross-sectional view of line and via post-CMP process and seed layer removal using excimer laser. (a) Fujifilm FCPi-2100. (b) Fujifilm LTC9320.

a planarization process can effectively remove the copper (Cu) and the seed layer simultaneously in two steps. In contrast to the regular SAP, the copper structure height is not defined by the plating process. Therefore the copper thickness is independent of the local feature densities, feature size differences, and galvanic bath flow conditions over the wafer. The excimer laser can be used instead of a second CMP step to remove the remaining thin titanium metal layer by a single laser pulse process. Seed layer removal (SLR) by excimer laser is proven to be much faster than a CMP process. Figure 20.7a shows an excimer laser dual damascene structure of a 5 μm line and a 5 μm via without any misalignment. Figure 20.7b demonstrates a dense routing of a 5 μm line structure with a space of only 2 μm. The titanium removal by laser left behind no remaining metal parts, even at small line spacing, which would be very challenging to achieve for an SAP process. The SLR has no effect on thicker copper trace lines. Successful isolation between dense lines was demonstrated by an inter-digital electrode structure where only a low leakage current below 10e-9 A was measured. The test results demonstrated excellent isolation between dense lines using the excimer laser enabled dual damascene process described above.

20.3.2 Key Advantages of Excimer Laser Enabled Dual Damascene RDL

This integration scheme offers a shorter, faster, and more cost-effective process that overcomes the challenges of conventional SAP. The elimination of multiple photolithography steps such as the second exposure, development, and cleaning such as resist stripping and plasma ashing makes for a streamlined, lower cost process. Additionally, the ablation of trenches and vias in each

(a)

(b)

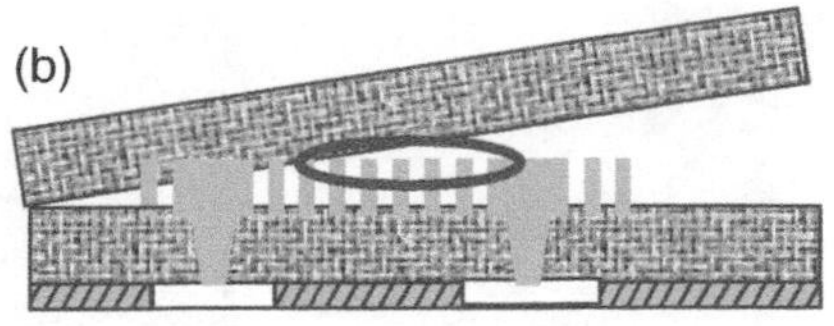

Figure 20.8 Process barriers for the POR lithography process flow: (a) topography formation from coating or lamination of dielectric material on top of RDL traces and (b) gap filling issue with small spacing between RDL traces.

RDL is performed in one step, lending to the reduction of more process steps and improved multilayer registration. Overall, the number of process steps is reduced by around 50% (Figure 20.12a and b). Photo or non-photo dielectrics can be used for this process as long as the dielectric absorbs UV light at 248 or 308 nm, allowing for a wider material range for CTE matching of the entire package. Excimer lasers inherently cannot ablate metal layers greater than 1 μm thick. This unique property is beneficial for RDL via drilling because the metal pad acts as a natural ablation stop layer.

The dual damascene process ends in a planarization step, as shown in the metal reduction step in Figure 20.5, for each RDL that facilitates single-layer and multilayer RDL buildup without the inherent problems reported in the lithography process of record (POR). One of the process challenges inherent to the current POR is applying a second dielectric layer over the 3D metalized RDL structures (Figure 20.8). The problem can be twofold: (i) undesirable surface topography caused by the underlying structures, and (ii) as the industry moves toward smaller lines and spaces (L/S), the POR techniques for filling the gaps between structures may hit their limits. Another POR challenge for generation of fine RDL structures is effective SLR. Wet-etching the tight spaces between fine-pitch RDL traces requires longer chemical residence time to fully remove the seed layers and reduce any impact on yield due to leakage and shorting. The increased chemical residence time will affect the integrity of the standing traces, causing undercutting and thickness erosion. Even excimer laser ablation as a potential alternative method to remove seed layers has limitations when the RDL width is smaller than 4 μm, since the pulses are no longer able to clear the bottom completely, causing a footing pattern. With the excimer laser enabled dual damascene process, the requirement to clear the seed layer in fine RDL trenches is completely eliminated.

20.3.3 Electrical and Reliability Data for Excimer Laser Enabled Dual Damascene

The following examples support the viability of using planarized RDL and patterning with excimer laser ablation technology. For the evaluation of the yield, a daisy chain design was used. The line width in the design is 5 μm and the via

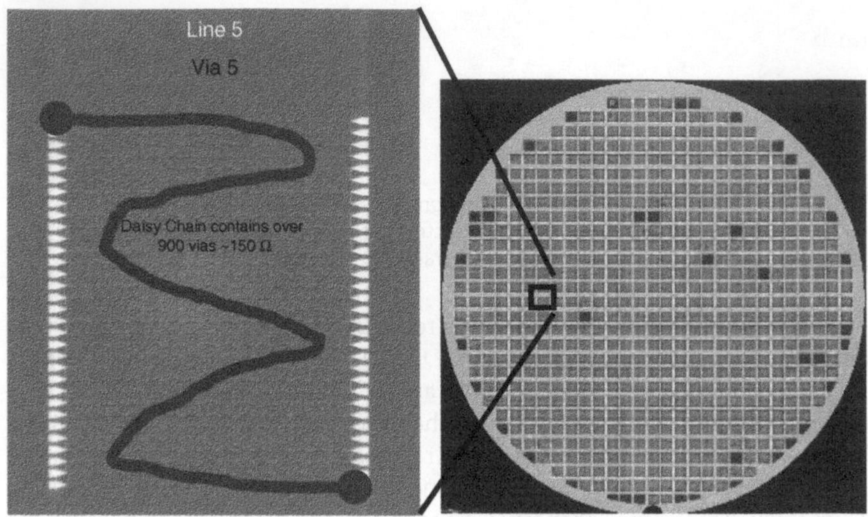

Figure 20.9 Daisy chain yield over a 200 mm wafer.

diameter is also 5 μm. Each chain contains of 960 vias. The average resistance of a complete chain is in the range of 150 Ω. Figure 20.9 shows a yield map of a 200 mm wafer where the daisy chains were ablated in a low temperature cured (230 °C) PI LTC9320 by Fujifilm Electronic Materials, which is commonly used for RDL generation on temperature-sensitive reconfigured wafers. Each green mark is a functional chain and a red mark indicates an open circuit in the chain. The higher numbers of defected chains at the edge are related to the edge bead removal of the polymer.

The excimer laser-based dual damascene process flow is currently under evaluation for other polymers that are used for fine line routing of organic interposers like the non-photo dry film Ajinomoto Build-up Film (ABF). The initial results are comparable with PI results. Figure 20.10 shows a crosscut of an embedded line in ABF.

Reliability data was collected for micro-via (8 μm) and trenches (5/5 μm line/space) in ABF material patterned by excimer laser on glass panel [14]. The micro-via passed successfully after thermal cycling for 4300 cycles from −55 °C for 15 minutes to 125 °C for 15 minutes and bHAST test (130 °C, 85% RH, 5.0 V) for two pitches of 20 and 40 μm (Figure 20.11). The 5/5 μm line/space traces enabled by excimer laser were also able to pass the thermal shock test from −55 °C for one minute to 125 °C for one minute after 1000 cycles.

20.3.4 Process Cost Comparison to Current POR

Using a wafer-level packaging (WLP) case study, cost estimates for each step were collected and compared between both integration schemes: current

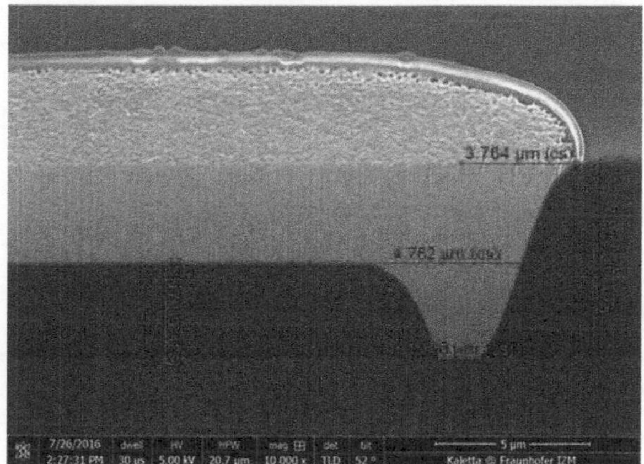

Figure 20.10 Excimer laser enabled dual damascene formation in ABF film GY50.

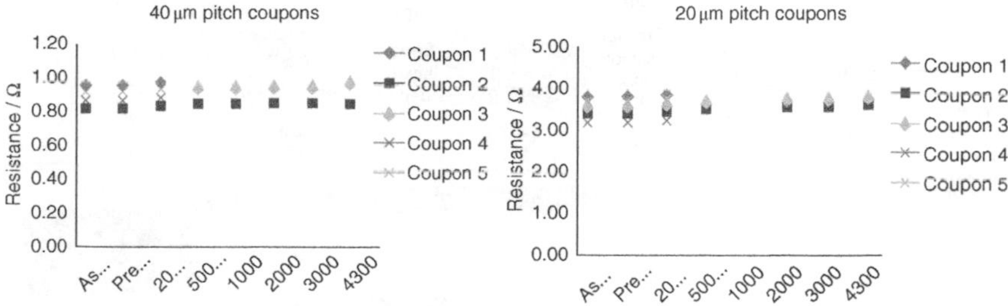

Figure 20.11 Reliability for 8 µm excimer laser ablated micro-via in ABF GX92 (thermal cycle: −55 °C (15 minutes) to 125 °C (15 minutes).

lithography POR (Figure 20.12a) and the proposed excimer laser enabled dual damascene process flow (Figure 20.12b). This cost comparison was made using the following design assumptions:

- 14 × 14 package.
- 10 × 10 die.
- 2000 I/Os.
- 10 µm via depth.
- Final cured dielectric thickness: 20 µm.
- Photo and non-photo dielectrics.
- RDL features: 10 µm/10 µm.
- RDL thickness: 10 µm.

The following cost assumptions were also included:

- Overhead and profit margins are not included in the results.

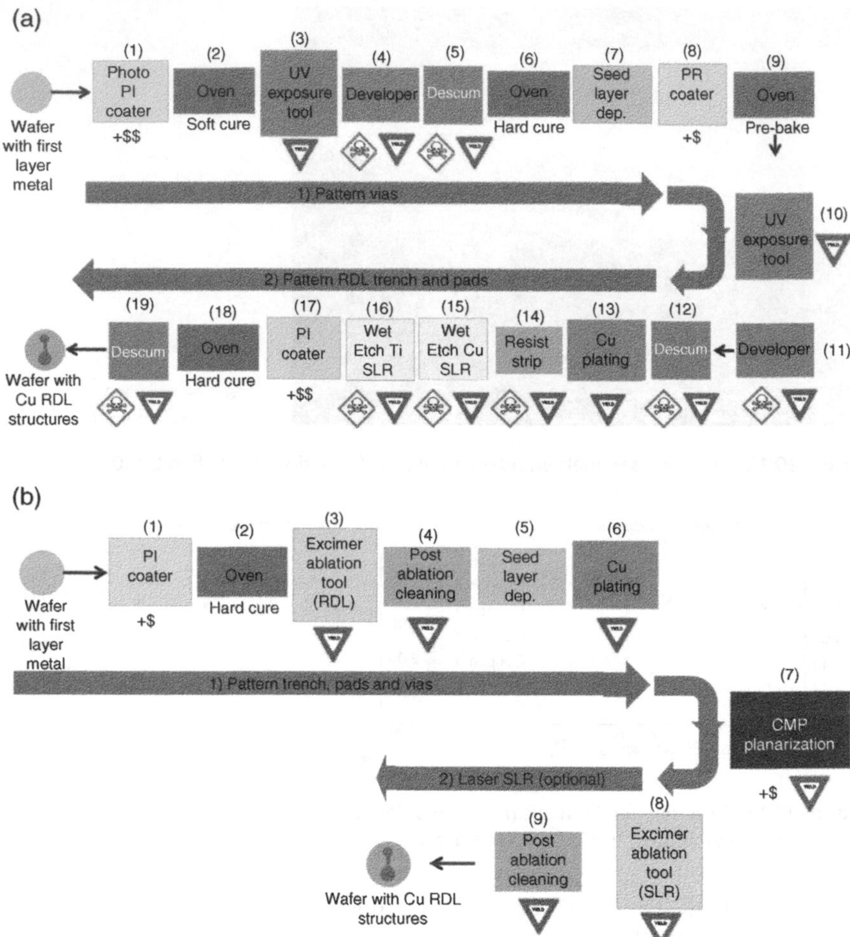

Figure 20.12 (a) Current POR photolithography process integration flow for one RDL formation. (b) Proposed excimer laser enabled dual damascene process flow for one RDL formation.

- Material costs reflect the material required for process, not material left on the wafer.
- Costs such as utilities for any given tool are generally accounted for in a factory's overhead rate.
- Disposal costs of material waste are typically incurred in the overhead rate as well, but for this cost comparison, a per-wafer estimated cost was used.

The cost comparison shows that using the excimer laser enabled dual damascene process flow provides cost savings of up to 30%, depending on the dielectric material chosen (photoimageable or non-photoimageable), with the largest

	Process cost comparison	
	Photolithography cost/wafer ($USD)	Dual damascene cost/wafer ($USD)
Capital	44.64	35.39
Labor	1.37	0.86
Material	27.93	*15.24
Total	**73.94**	**51.49**
*cost of non-photo dielectric		**30% savings**

Figure 20.13 Process cost comparison results for current POR and dual damascene process flow for one RDL formation.

	Cycle time comparison	
	Photolithography	Dual damascene
Cycle time	5.80 h	2.38 h
		59% savings

Figure 20.14 Cycle time comparison results for current POR and dual damascene process flow for one RDL formation.

savings expected with non-photoimageable dielectrics (Figure 20.13). With photo dielectrics, the cost saving is estimated to be in the 25% range. In addition, this process flow provides cycle time reductions of up to 59% as compared with the current POR process integration flow (Figure 20.14). The cost and time savings are the result of reducing the number of process steps and eliminating wet chemical waste-related costs. The biggest cost contributors in the dual damascene process scheme are the excimer laser patterning and the planarization process – the cost model assumes CMP planarization. With the significant reduction in yield-loss process steps, it is reasonable to expect an improvement in yield, although this was not factored into the cost assessment.

20.4 Summary and Conclusion

The application of PCB-like laser ablation technology to a mask-based lithographic process using a stepper platform allows the fabrication of ultra-fine structures with high throughput. Feature sizes of 5 μm and below for both vias and RDL are demonstrated and show an excellent yield. There are many benefits of using an excimer laser compared with conventional lithographic structuring: elimination of many process steps and consumables from the integration flow, the pattern integrity of the structuring, the ability to pattern ultra-fine vias and RDL, overcoming the obstacle to remove the seed layer in fine RDL trenches, and the use of non-photo-definable materials that enable a wider option for CTE matching and are often cheaper. The excimer laser enabled dual damascene process presents significant cost benefits compared with the current POR-based lithography process flow and enables users to reach the requirements for advanced packaging platforms where dense routing in combination with multilayer buildup is needed.

References

1 Pitcher, G. (2009). Good things in small packages. *Newelectronics* (23 June), pp. 18–19.
2 Fan-Out Wafer-level Packaging (FOWLP). http://www.3dic.org/FOWLP (accessed 6 August 2018).
3 Ahmad, M. (2017). Advanced packaging: five trends to watch in 2017. *Electronics Products* (25 January). https://www.electronicproducts.com/Packaging_and_Hardware/Device_Packaging/Advanced_packaging_five_trends_to_watch_in_2017.aspx (accessed 6 August 2018).
4 Souter, M. and Clark, D. (2014). Next-generation laser structuring method for high density packages. *Chip Scale Review* May–June: 42–45.
5 Srinivasan, R. and Mayne-Banton, V. (1982). *Applied Physics Letters* 41: 576.
6 Akhtar, S.N. (2015). Master's thesis, Department of Mechanical Engineering, Indian Institute of Technology Kanpur, India.
7 Garrison, B., Srinivasan, R., and Vac, J. (1985). *Science and Technology, A* 3 (3): 746.
8 Basting, D. and Marowsky, G. (2005). *Excimer Laser Technology*, 139. Springer Verlag Berlin-Heidelberg.
9 Meijer, J. (2004). *Journal of Material Processing Technology* 149 (1–3): 2–17.
10 R. Delmdahl, and R. Pätzel, (2008), Phys. *Status Solidi (c)*, 5: 3276–3279. doi: https://doi.org/10.1002/pssc.200779515.
11 Urech, L. and Lippert, T. (2010). Photoablation of polymer materials. In: *Photochemistry and Photophysics of Polymer Materials*, Chapter 14, 541–563. Wiley.
12 Sutcliffe, E. and Srinivasan, R. (1986). *Journal of Applied Physics* 60: 3315–3322.
13 Hichri, H. and Arendt, M. (2017). Excimer Laser Ablation for Microvia and fine RDL Routings for Advanced Packaging. *Chip Scale Review* Sep-Oct: 11–14.
14 Hichri, H., Arendt, M., Gingerella, M. (2016). Novel process of RDL formation for advanced packaging by excimer laser ablation. *Proceedings of the IEEE Electronic Components and Technology Conference* (2016).

21

Temporary Carrier Technologies for eWLB and RDL-First Fan-Out Wafer-Level Packages

Thomas Uhrmann and Boris Považay

EV Group – St. Florian am Inn, Austria

Fan-out wafer-level packaging (FO-WLP) process flows typically fall under two basic integration categories called chip-first and chip-last. With the chip-first approach, individual chips are embedded into epoxy mold, forming freestanding molded wafers as a basis for redistribution layers (RDL) and bumping, referred to as embedded wafer-level ball grid array (eWLB). In the chip-last approach, RDL are processed first, before the die are individually attached and overmolded. In both process flows as shown in Figure 21.1, temporary wafer carrier technologies play a crucial role. For chip-first (or RDL-last), temporary carriers are used for package-on-package (PoP) technology, where thinning device wafers below 400 μm does not allow for the use of freestanding molded wafer handling, as discussed by Campos et al. [1–3]. Slide-off debonding has been found to work reliably within the molded wafer process flow.

For eWLB, the warpage of the molded substrates is generally one of the biggest challenges for high volume manufacturing. The different coefficient of thermal expansion (CTE) for silicon and mold compound will cause each product to behave slightly differently in terms of intermediate thermal expansion. Especially, changes in terms of molded wafer thickness, die content, die location, die spacing, and height distribution introduce variation from product to product [3]. Furthermore, reduction of molded substrate thicknesses below 400 μm will cause the reconstituted wafers to acquire a flexible behavior that no longer allows self-support and freestanding handling. Also, reducing the wafer thickness often increases *z*-axis asymmetry, adding nonlinear behavior to the already flexible wafer. Processing such wafers makes temporary bonding of the thin molded wafer to a carrier an inevitable step, serving as pure mechanical

Advances in Embedded and Fan-Out Wafer-Level Packaging Technologies, First Edition.
Edited by Beth Keser and Steffen Kröhnert.

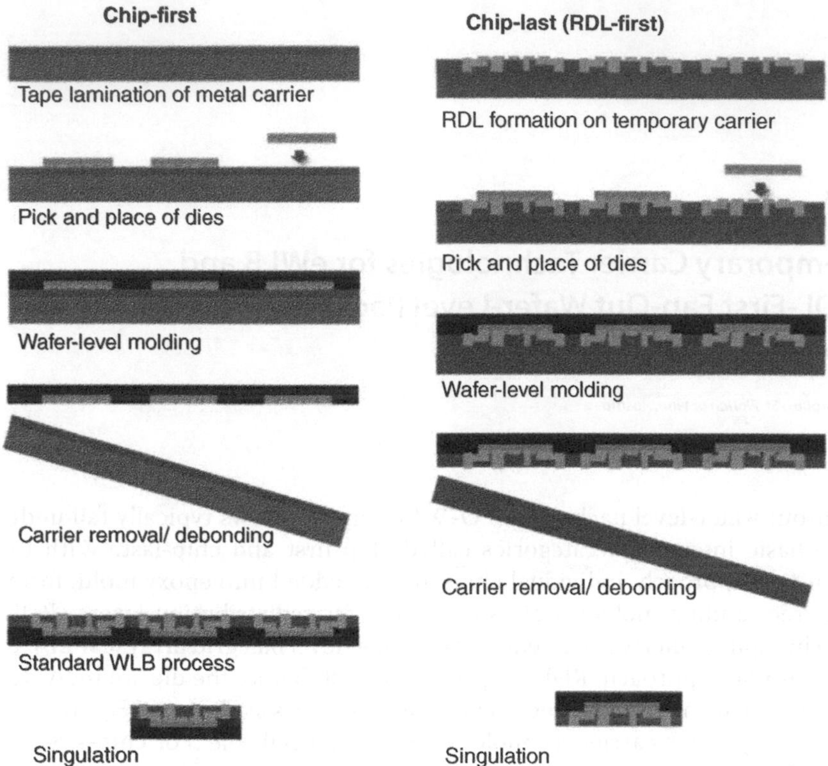

Figure 21.1 Schematic manufacturing process for chip-first and chip-last integration, including the use of temporary carriers.

support during processing. Depending on the carrier material, different adhesive types and respective debonding methods can be used.

Contrary to this, chip-last is a pure buildup process, where the temporary release film forms the base layer. The package is built successively on top of the carrier wafer, and UV laser debonding is the final process step separating the devices from the carrier; hence fundamental knowledge of critical process parameters for UV laser debonding is the key to ensuring a high overall process yield.

Temporary wafer bonding is classified by the mechanisms that break the temporary bond interface and their point of interaction: direct chemical bond release by laser radiation (laser debonding) via ablation or gas generation or elevated temperature to soften or at higher levels to thermally dissociate the adhesive layer (slide-off and lift-off debonding). Notably, the debonding temperature is only linked to the debonding mechanism for slide-off and lift-off debonding, while laser-initiated debonding decouples the debonding

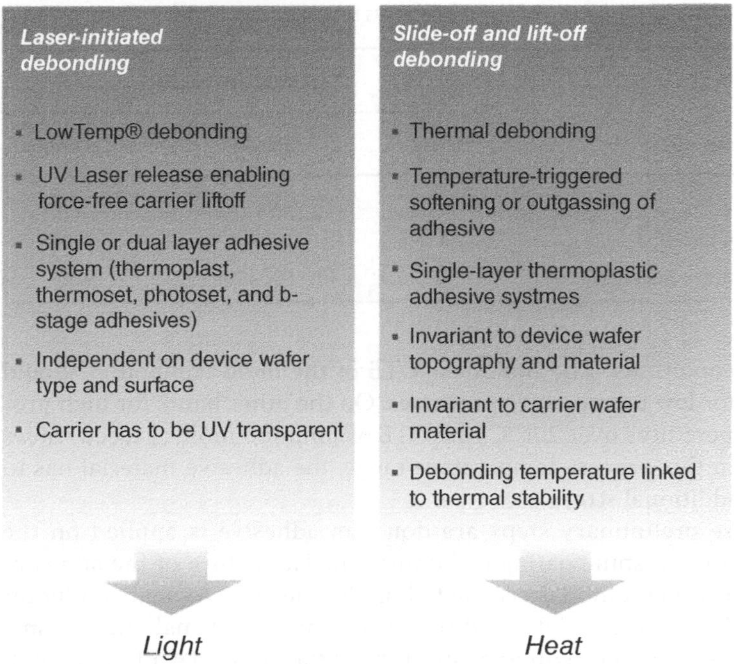

Laser-initiated debonding

- LowTemp® debonding
- UV Laser release enabling force-free carrier liftoff
- Single or dual layer adhesive system (thermoplast, thermoset, photoset, and b-stage adhesives)
- Independent on device wafer type and surface
- Carrier has to be UV transparent

Light

Slide-off and lift-off debonding

- Thermal debonding
- Temperature-triggered softening or outgassing of adhesive
- Single-layer thermoplastic adhesive systmes
- Invariant to device wafer topography and material
- Invariant to carrier wafer material
- Debonding temperature linked to thermal stability

Heat

Figure 21.2 Classification of temporary bonding and debonding processes.

mechanism from the global process temperature. Figure 21.2 provides more detail on each debonding process.

Depending on the integration process flows for FO-WLP, requirements for temporary bonding differ, separating into laser-initiated debonding of RDL-first buildup packages, while slide-off and lift-off debonding is used for chip-first packages to process molded substrates.

21.1 Slide-Off Debonding for FO-WLP

Temporarily attaching the molded reconstituted wafer to a suitable carrier was adopted to overcome the warpage and flexibility by stabilizing the molded wafer and reducing its topography variations. The selection of a carrier wafer and its dimensions, as well as of adhesive materials, is the key element to a successful temporary bonding and debonding technique. Epoxy-based mold compounds typically have high and nonlinear CTE. On account of these, several different carriers with different CTE, stiffness, and carrier thicknesses were evaluated, as shown in Table 21.1. Overall, the carrier CTE should closely match the molded wafer CTE under all different

Table 21.1 Mechanical properties of carrier materials for temporary bonding in FO-WLP.

	Carrier materials		
Mechanical properties	**Glass**	**Silicon**	**Ceramic**
CTE (ppm °C^{-1})	7.8	2.6	>8.0
Young's modulus (GPa)	74	130	>300
Carrier thickness range (μm)	700	600–700	550

processing conditions. The nonlinear CTE of the mold compound should match well for low temperature processes. On the other hand, for high processing temperatures over 200 °C, the CTE of common molds exceed values of 15 ppm. In this process temperature range, the adhesive material has to absorb the additional stress.

Once these preliminary steps are done, an adhesive is applied on the carrier or through spin coating or lamination. The nature of the adhesive and its deposition method is crucial to bonding success. As semiconductor processing involves a variety of different chemical, thermal, and plasma processes, selecting the right adhesive is essential, and very often material properties and functions are contradictory. Proficiency in temporary bonding and debonding involves several parameters, the first of which is the initial condition of the wafers. The final bonding quality (i.e. success in attachment and detachment) is dependent on a combination of parameters such as temperature, atmosphere, chemical exposure, and mechanical stress of the post-temporary bonding process, as well as total thickness variation (TTV), the presence of particles and voids, bow and warp, and the centricity of the wafers. These characteristics will help define the adhesive and process to be used.

Polymers are classified into two families: thermoplastics and thermosets. Thermoplastic adhesives have the ability to be solidified and re-melted, which is their key differentiator over thermosetting polymers, whose solidification is irreversible. In contrast thermoplastic residues can be chemically removed after debonding, which reduces contamination for the rest of the process. The maximum permitted operating temperature in the bond interface and the required rigidity are the most decisive parameters, as the adhesive layer has to remain stable throughout the process, especially for processes that involve heat and strong mechanical stress such as backgrinding.

TTV after temporary bonding is of central importance to bonding. Any local thickness variation of the temporary adhesive will be transferred to the device wafer. TTV variations mainly stem from spin coating and baking uniformity issues during the application of the adhesive film and pressure uniformity issues during bonding. Additionally, the adhesive must be compliant with the debonding process, which tends to require room temperature conditions.

The development of bonding processes for FO-WLP has been guided by restrictions of the molding and the RDL fabrication process steps. The correlation between increasing mold stress and rising temperature has fostered the adoption of low temperature adhesives, making thermoplastic adhesives an appropriate solution. Because of the initial warpage of the molded wafer, materials with high adhesion characteristics were considered to maintain a plain contact between the carrier and the mold and to prevent delamination. Finally, the RDL fabrication and its chemical environment have also influenced the choice of the adhesive material.

Besides the temporary adhesive, the carrier wafer properties determine the stability and process cornering of the temporary bond and debond step. These are primarily (i) thermal compatibility of the carrier substrate to the FO-WLP device wafer to minimize stress upon their interface (CTE mismatch across the whole operation temperature range), (ii) mechanical properties such as high strength and stiffness to ensure sound support and to enforce flatness to the set wafer carrier, and (iii) chemical stability in harsh chemical environments imposed by the FO-WLP process such as etching and electroplating.

Based on the results obtained during extensive experiments, some common carrier materials were found to be inadequate, and new types of carriers, together with the most suitable matching temporary bonding adhesives and process conditions, were found to meet the demand. After careful selection, the best combination of such parameters enabled a robust temporary bonding solution for FO-WLP. No visual or internal abnormality was seen on the optimized configuration, and minimum wafer warpage was measured during critical FO-WLP process steps [3].

As a result, the right carrier technology has been found as a combination of carrier material, carrier thickness, and adhesive properties. In terms of carrier material, the CTE as well as material stiffness has to be high. For these reasons, silicon and glass carriers do not fit well as carrier material for mold-first, RDL-last FO-WLP. In addition, the high stress level during processing demands an increased carrier thickness of 1.0–1.3 mm, in order to reduce bow and warp levels after processing. A last main determining factor is the adhesive. The layer thickness as well as viscosity under process conditions, or material rheology, determines the maximum process temperature. Thinner adhesives undergo less material flow during processing and hence lead to a larger process window.

Finally, process conditions have to be adapted to work with carrier-mounted FO-WLP. As the carrier system is much thicker, including several different materials with different thermal conductivity, temperature ramp up and ramp down, for example, in deposition processes, have to be lowered to provide strain relaxation in the carrier system.

21.2 Laser Debonding: Universal Carrier Release Process for Fan-Out Wafer Packages

UV laser debonding has several advantages over other debonding techniques that make it an extremely flexible debonding process. For slide-off debonding, material properties such as glass transition and material softening determine bonding, debonding, and maximum process temperature. In the case of laser-induced debonding, even permanent adhesive films with suitable photoactive response can be used. Hence, the debonding process is a decoupled function of the carrier system. Due to the well-controllable optical interaction, a wider range of chemical and physical parameters, such as spectral absorption and photosensitivity, scattering, or reflectivity can be utilized to separate the response of the different layers. This way, the individual process requirements, including high temperature, tension, or pressure, can be supported when selecting the optimal adhesive combination. Further flexibility can be achieved by adding functionality to the debond layer that also protects the substrate from side effects like thermal or mechanical stress, nonplanarity, or contamination. In contrast to mechanical or slide-off debonding where the adhesives are engineered for embedding of structures and adhesion to surfaces and also include the debonding trigger, this is not the case for laser debonding. The adhesive in laser debonding has no additional function for the debonding process, but is solely responsible for detachment after photoactivation. The only exceptions to this rule are polyimides. Their chemical structure allows for embedding and bonding of the materials and also acts as the photoactivated UV laser debonding layer. Meanwhile, their process latitude or temperature process window allows for very high thermal temperature stability (approaching 400 °C), and the material has high absorption at the laser debond wavelengths and also dissociates during UV exposure and so does not debond as easily as a material developed for very low levels of laser-initiated decomposition. During the other process steps, the material is almost inert.

The high independency of the laser debond process from the rest of the process flow, compared with mechanical or slide-off debonding, allows for a much more straightforward integration of temporary carrier technology. Another key benefit of UV laser debonding is its insensitivity to different substrate types, passivation layers, and metals. Laser debonding confines the temporary adhesion function to the carrier substrate side and does not impose additional limitations on the substrate, other than the ability to bond itself. Other debonding techniques rely on surface energy during adhesion or release, temperature behavior, solvent permeability properties, or defined release at specific stress levels of the bonded surfaces to ensure proper debonding. In contrast, UV debonding relies solely on the selective decomposition of the laser release layer. The substrate is kept at room temperature during debonding, meaning no thermal activation is needed. This allows for the use of a wide variety of temporary bonding adhesive types, which can be made with very low adhesion properties since the separation of the product wafer from the carrier wafer is force-free after debonding. The combination of this instantaneous bond release, avoidance of the thermal activation step, and high scanning speed associated with UV diode-pumped solid-state (DPSS) lasers allows increased debonding throughput compared with other debonding techniques. A drawback of laser debonding is that it requires a UV transparent glass carrier in order to allow the laser light to initiate the laser debond process at the bond interface. However, where sodium ion contamination or mismatch of glass properties have been issues with this approach in the past, glass manufacturers can now offer suitable glasses that eliminate these obstacles.

Table 21.2 gives an overview of available UV laser debonding technologies today. Predominantly two different types of lasers are found for laser debonding today, namely, gas-based excimer lasers and DPSS. Excimer lasers offer high power with a beam that can be formed into rectangular or square shapes with a uniform power density. However, they come with high maintenance and footprint. In the last couple of years, DPSS lasers have closed the power gap, are cheaper to maintain, and have a smaller footprint. However, they come with a Gaussian beam profile, as sketched in Table 21.2. In this case, the energy distribution leads to debonding around the full width at half maximum of the Gaussian beam shape. Around the maximum, the excess energy leads to burning of the debond layer. For this reason, the further discussed system features a DPSS laser with a flat-top beam profile, delivering a uniform debond at low system and process cost.

21.3 Parameters Influencing DPSS Laser Debonding

To limit the thermal input associated with laser debonding, UV DPSS lasers are favored for the debonding step, since they provide the benefit of rapid absorption of the UV light immediately at the glass release layer interface and also allow for direct photochemical interaction at chemical bonds provided by readily available materials. Ideally, release layer materials feature a high absorption coefficient in the relevant wavelength regime around the laser wavelength, typically situated around the well-known frequency-tripled Nd:YAG laser emission at $\lambda \approx 355$ nm. In this case, penetration depth of light is limited to less than 200 nm for most commercially available debond materials. Furthermore, due to the limited peak power of these lasers, it is essential that the transmittance toward the bond interface is high and their radiation can be concentrated well enough to reach the required instantaneous irradiance ($\mathrm{W\,m^{-2}}$) rather

Table 21.2 Comparison of UV laser sources for debonding in terms of their respective benefits and drawbacks as well as schematic beam profile.

	Solid-state laser		
	Gaussian beam profile	Flat-top beam profile	Excimer laser
Pro	• Easy optical setup • No halogen gas consumption • Less maintenance	• Quasi top-hat/line beam profile • Excellent spatial control • High beam reproducibility • Minimum consumables • Easy optical setup	• Top-hat beam profile for minimized heating • Debond processes studied

Con	• Gaussian beam profile with high excess energy leads to carbonization		• Complicated optical setup
			• Halogen gas
			  ○ Safety concern
			  ○ Must be refilled (halogen gas is consumable)
			• Consumables

Beam profile	• Gaussian beam profile Radiant exposure [a.U.] x [a.U.]	• Quasi top-hat beam profile Radiant exposure [a.U.] x [a.U.]	• Top-hat beam profile Radiant exposure [a.U.] x [a.U.]

than only the accumulative radiant exposure ($J\,m^{-2}$). Distinguishing the two is essential to the success of the debond process, which involves interconnected physical subprocesses at different time scales that are detailed below. The beam quality factor of single-mode DPSS lasers is typically close to $M^2 \approx 1$, which means that the beam cross section has a unidirectional Gaussian beam shape reaching the maximum possible power density at its peak. Since debonding necessitates a two-dimensionally homogenous distribution and the interwoven photochemical, photothermal, and mechanical processes already mentioned above that affect the process outcome, the spiky Gaussian power distribution has its drawbacks.

Photochemical dissociation is energy dependent and bleaches the formerly absorptive debond layer after irreversible breakage of the targeted bonds during exposure. Henceforth, the debond material locally increases penetration depth for multiple or longer low power exposures. However, too high intensities trigger different photochemical and thermal-initiated molecular and macroscopic changes such as carbonization, which only leaves a slim process window for scanning Gaussian spot ablation. Depending on the individual process, local heating or stress necessary for debonding can occur at medium and higher power settings. At the beam center, however, significant overexposure can lead to pronounced carbonization [4]. The widely absorbing carbon also locally increases heat buildup and scattering that nonlinearly increases the local temperature and results in even further carbonization as detailed in Figure 21.3.

Instead of generating high instantaneous irradiance with high power continuous wave lasers, pulsed irradiation can provide the required power densities by compressing the energy in smaller time scales. Like a spiky exposure profile, pulsed radiation also leads to inhomogeneous ablation; however, pulses have the advantage of confining local heating to the debond layer, as long as the thermal pulse propagation stays within the exposure field during optical energy deposition. This is typically the case for ns or shorter pulses. On the other hand the tails of the spatial Gaussian distribution are below the activation-threshold but still deposit energy, which leads to unnecessary global heating. Beam-shaping optics of the single-mode beam is a decisive way to flatten the central energy distribution and reduce the tail length to widen the confined process window again. Fine-tuning of the spot shape to a so-called flat-top beam profile together with exact relative laser spot positioning efficiently leads to homogenization and reduces overexposure effects. This flat-top beam profile facilitates a low maintenance and running cost of DPSS lasers and provides an optimal approach to laser debonding [5].

Due to the predetermined, but complex, effects during the debond process, the outcome is not only controlled by the laser beam shape but also by the deposited laser fluence or radiant exposure ($J\,m^{-2}$) per pulse. This important parameter affects the debond result by controlling the microscopic

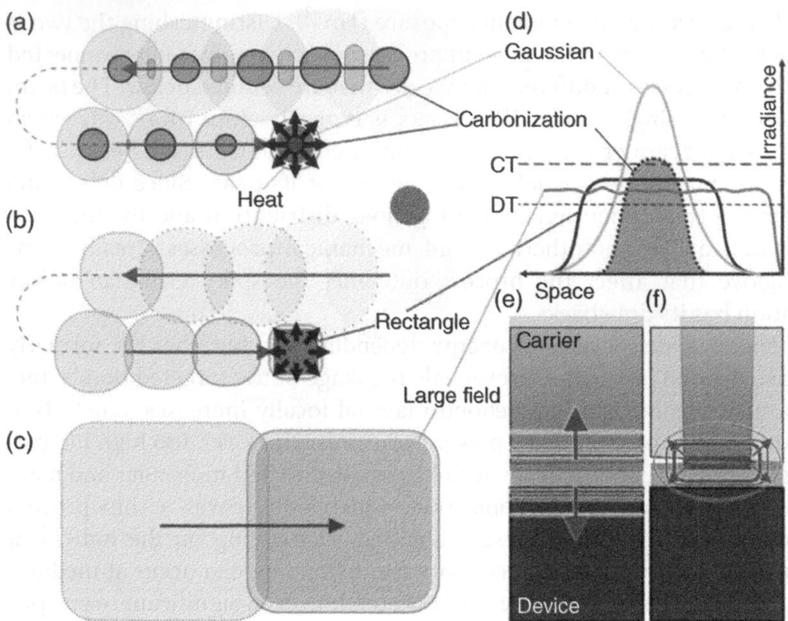

Figure 21.3 (a) Effects of a laser shot sequence with small Gaussian compared to a rectangular beam profile (b) and to large area exposure (c). (d) Depicts the individual irradiance profile shapes and the dissociation and carbonization thresholds (DT, CT) of irradiance. Carbonization is more likely under Gaussian exposure. The rectangular/"flat-top" profiles allow for more precise maneuvering between the two thresholds. Larger fields dissipate heat in a one-dimensional fashion (e), while smaller area exposure can make use of three-dimensional heat dissipation (f).

photochemical and photothermal dynamics as well as the debond forces that appear as a result of the gas generation and heating within the debond material. Both laser power and pulse length determine the deposited energy per pulse. While the laser power can be rather easily tuned, pulse length is typically restricted to a narrow frequency band by the specific laser source. Once these parameters are optimized and controlled, and the process window for the individual shot area is set, the laser scanner is fine-tuned to reach the optimum overlap between successive scans or even via multiple shifted exposures. This means that the patterning is key to the process outcome. Depending on the release layer and the mechanism of response, the full debond fluence can be applied in an individual shot, or multiple shots may be overlapped to deposit the required energy such that the debond kinetics allows for the right photochemical decomposition, outgassing, and successful separation while maintaining an optimum arrangement for homogeneity, residual ablation roughness, stress control, heat distribution or carbonization, or other particle generation. In comparison with longer wavelengths in the green or infrared spectrum,

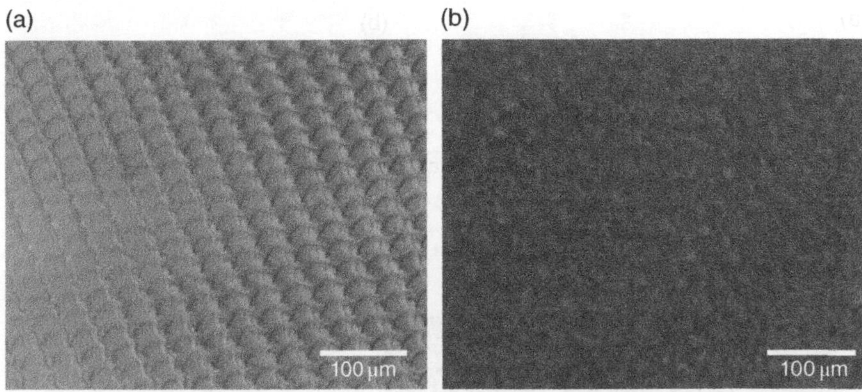

Figure 21.4 (a) Bright-field microscope images of UV debonding without overexposure showing successful debonding without thermal carbonization. (b) Laser fluence exceeding allowed carbonization limit when increasing overlap for better homogeneity without beam shaping.

where debonding is only related to photothermal decomposition, UV wavelengths decompose the material into gaseous states directly, i.e. photochemically as well as photothermally, and have a significantly larger process window without carbonizing the interface or inducing excess heat. A comparison of an optimally tuned UV laser debonding process versus an overexposed photothermal carbonization can be seen in Figure 21.4. The bright-field microscope images illustrate a similar surface roughness in either case; however, the absorption of carbon in Figure 21.4b reduces the brightness and also suppresses the shiny polished bubble-like appearance of the gas–solid interface seen as differently colored shades in Figure 21.4a.

Even though laser debonding is a room temperature process, it still introduces a thermal component that can vary with wafer parameters. The laser debonding process described above, utilizing both a release layer and bond layer, is schematically shown in Figure 21.5 in order to highlight the thermal component of the debonding process. Depicted in Figure 21.5a is a cross section with individual laser scan shots progressing from left to right. The absorption of the laser is shown in purple in Figure 21.5b, where the laser penetration depth is inversely proportional to the absorption coefficient of the release layer. It should be noted that the thermal spike that appears close to the interface is directly related to the energy deposited by the laser. As the laser is absorbed in the topmost portion of the debond layer, the heat is conveyed into the surrounding material, and temperature rapidly decays away from the glass/release layer interface as well as laterally from the heated spot. Figure 21.5c and d depicts the temperature evolution during debonding, including how the heat generated at the glass surface diffuses through the stack, increasing the

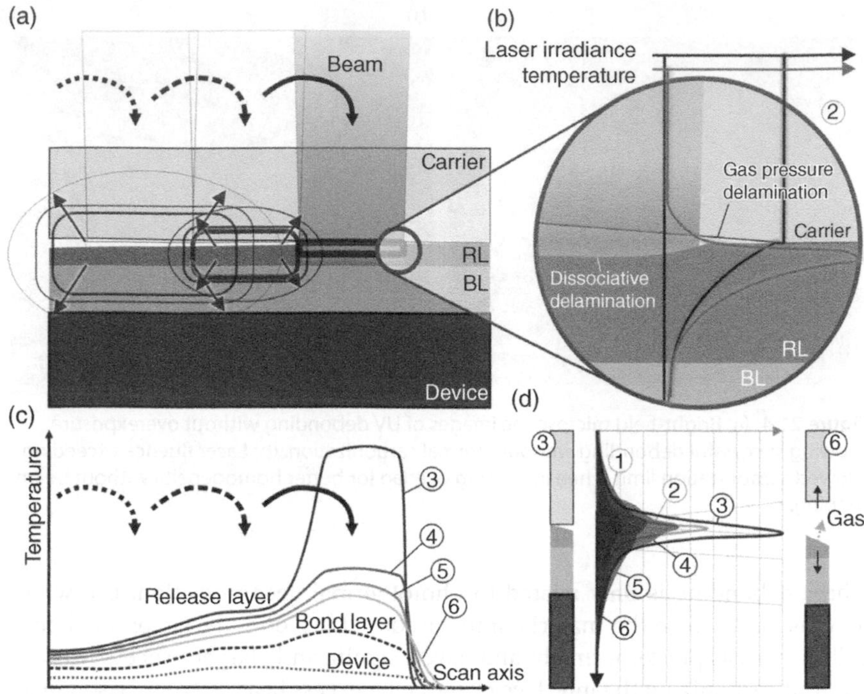

Figure 21.5 Schematic of the temperature evolution during UV DPSS laser debonding.

temperature via accumulation across several laser shots, thereby delaying the heat deposition and significantly reducing peak temperatures that might damage the device surface. Phases 1 and 2 are displayed at a different time interval (in the ns range) where the pulse still heats the release layer, while phases 3–6 are in the μs range when the heat dissipates.

The smaller size of DPSS laser spots is beneficial for this debonding case, as the field size is of the same order of magnitude as the layer thicknesses. Therefore, heat diffusion occurs in all directions in space, and the in-plane component reduces the temperature increase of the other layers considerably. Measurements have shown that this heating of the device surface is less than +30 °C owing to the fast heat dissipation into the surrounding regions. On the other hand, with large area exposure such as generated by excimer lasers, this 3D heat diffusion model is not applicable anymore. Caused by their large exposure area and low repetition rate, the generated heat can only diffuse orthogonally to the sample surface, as the perimeter of the debonding area is comparatively small. In other words, controlling the temperature impact is much easier in the small spot case. Large spots contrarily have the advantage of better homogeneity control within the exposure field and a relatively smaller border region as a source of inhomogeneity, although those stitching regions become more prominent than with the rather continuous small spot irregularities.

Concentrating the energy in time and space drastically improves efficiency, especially in thermal-dominated materials like polyimides, and improves the debond speed to more than 1000 $mm^2\ s^{-1}$ while covering large areas at lower laser energy. As a result, both throughput and thermal exposure of the sensitive components are positively affected, as are the overall costs of the system.

The debond gap does not open only because of the ablation process and the dissociative removal of material but also due to the conversion and heat-induced gas pressure buildup, which can introduce local debond forces of variable strength that are given by the material's chemical and physical properties and are controlled by laser settings such as pulse energy, spot size, spot overlap, and the timing. Again, the control of the laser parameters and their fine-tuning to the material is crucial for a successful and homogenous delamination across the whole substrate due to a precisely defined debond force on a microscopic level. In practical terms, a successful and stable laser debond implies fine and precise management of the laser process parameters, continuous monitoring, and detailed and fast strategies to map out the process window that is individually tailored to the product requirements.

From an engineering perspective, the debond process is not always completed after the laser treatment. Depending on the postexposure behavior of the individual bond material and the quality of the treatment residual, no or some regained holding force has to be overcome to separate the carrier from the substrate. Yet again, the temporary bond material's behavior and timing is decisive for successful completion. After separation, an optional removal of residue, cleaning of the substrate and the carrier with suitable solvents, is a quite common procedure, which has to be considered when choosing a particular material system.

In addition to maintaining exact control of the debonding process, the cost of debonding must be minimized. UV DPSS lasers for debonding feature a high pulse repetition frequency, which results in a high debonding throughput. Depending on the type of adhesive, required dose, and spot size, typical debonding can be achieved in less than one minute for 300 mm wafers. For panels, the debond time scales according to the panel size.

UV laser debonding is a universal debonding process that is applicable to a very wide range of devices, wafer types, and surfaces. It offers high process latitude and temperature stability to ease the integration of temporary carrier technology into the wafer production flow for stacked device applications. The high repetition frequency, low consumable costs, long lifetime and uptime, and high throughput of DPSS lasers with a small footprint of the source, coupled with precise and stable beam-shaping optics, provide a reliable and efficient combination that enables debonding for cost-sensitive advanced package types, such as FO-WLP.

Acknowledgments

We would like to acknowledge Steffen Kröhnert, Jose Campos, Andrea Cardoso, Mariana Pires, Eoin O'Toole, and Raquel Pinto of NANIUM S.A. This work was partially carried out in the Enhanced Power Pilot Line (EPPL) project, funded by EU and national grants under the ENIAC program.

References

1 Brunnbauer, M., Meyer, I., Ofner, G. et al. (2008), Embedded wafer level ball grid array (eWLB). *33rd International Electronics Manufacturing Technology Conference 2008* (2008).

2 Jin, Y., Baraton, X., Yoon, S.W. et al. (2010). Next generation EWLB (embedded wafer level BGA) packaging. *12th Electronics Packaging Technology Conference* (2010).

3 Jose Campos, A.C. (2015). Temporary wafer carrier solutions for thin FO-WLP and eWLB-based PoP. In: *iWLPC (International Wafer Level Packaging Conference)*. San Jose, CA: SMTA International.

4 Lippert, T. (2009). UV laser ablation of polymers: from structuring to thin film deposition. In: *Laser-surface Interactions for New Materials Production* (ed. A. Miotello and P.M. Ossi), 141–175. Berlin: Springer.

5 Andry, P., Budd, R., Polastre, R. et al. (2014). Advanced wafer bonding and laser debonding. *IEEE 64th Electronic Components and Technology Conference (ECTC)*, Orlando, FL (S. 883–887).

22

Encapsulated Wafer-Level Package Technology (eWLCSP)

Robust WLCSP Reliability with Sidewall Protection

S.W. Yoon

STATS ChipPAC, JCET Group

22.1 Improving the Conventional WLCSP Structure

The WLCSP was introduced in 1998 as a semiconductor package wherein all packaging operations were done in wafer form [1]. The resultant package has dielectric layers, thin film metals, and solder balls directly on the surface of the die with no additional packaging. The basic structure of the WLCSP has an active surface with polymer coatings and solder balls with bare silicon exposed on the remaining sides and back of the die. The WLCSP is the smallest possible package size since the final package is no larger than the required circuit area. Based on the small form factor and low cost, the number of WLCSP used in semiconductor packaging has experienced significant growth since its introduction. The growth has been driven aggressively by mobile consumer products because of the small form factor and high performance required in the package design. Although WLCSP is now a widely accepted package option, the initial acceptance of WLCSP was limited by concerns with the surface-mount technology (SMT) assembly process and the fragile nature of the exposed silicon inherent in the package design. Assembly skills and methods have improved since the introduction of the package; however damage to the exposed silicon remains a concern. This is particularly true for advanced node products with fragile extreme low dielectric constant (low-ELK) dielectric layers used in the silicon processing. One method commonly used to improve die strength and reduce silicon chipping during assembly is lamination of an epoxy film on the back of the die. The film is laminated and cured on the back of the wafer prior to singulation to strengthen the die, in spite of the fact it adds cost to the package. By the nature of the backside lamination process, the uncoated sides of the die continue to be exposed after

Advances in Embedded and Fan-Out Wafer-Level Packaging Technologies, First Edition.
Edited by Beth Keser and Steffen Kröhnert.

dicing the wafer, and the silicon continues to be at risk for chipping, cracking, and other handling damage during the assembly process.

Despite the indisputable benefits of WLCSP, there are a number of concerns that have continued to plague the adopters of this technology since its inception:

1) WLCSP is essentially a bare die with exposed Si surfaces. The package suffers mechanical damage in the form of chipping and cracking in the course of processing, shipping, and during SMT operations. This necessitates additional processing (e.g. backside coating or lamination) for partial die protection and inspection steps to ensure outgoing product quality, leaving the product still exposed to potential field failures due to the risk of marginally damaged parts being shipped that may not be "caught" by inspection. Also, WLCSP is typically tested at wafer level prior to saw, which is a quality risk.
2) The manufacturing infrastructure for WLCSP is entirely dependent on the incoming wafer diameter. As designs migrate to larger wafer diameters (as in the 200–300 mm transition or the future transition from 300 to 450 mm), new investments become necessary to support the capacity requirements, while investments in the existing infrastructure may be rendered obsolete.
3) From a design perspective, WLCSP is effectively a "fan-in only" package. The input and output signals (I/O) must be accommodated on the area of the die at the desired terminal pitch; hence, there is a threshold of I/O density above which the WLCSP package becomes unusable and a change to a completely different packaging solution becomes necessary. Such situations often arise with node transitions that result in die shrinks. For example, a change from WLCSP to wire-bond ball grid array (FBGA), flip-chip BGA (fcFBGA), or quad-flat no-lead (QFN) packaging is not uncommon, which entails a radical change in package footprint, form factor, performance, and cost structure.

Enter the eWLCSP that is a simple variation of the broader fan-out wafer-level packaging (FO-WLP) platform (trade named eWLB for embedded wafer-level ball grid array). eWLCSP retains the compelling benefits of WLCSP packaging while addressing many of the key concerns mentioned above. The structure of the package, the fabrication process, the unique advantages, and the preliminary product/reliability assessment are discussed here.

eWLCSP has been developed to provide five-sided protection for the exposed silicon in a WLCSP. The process starts with an existing high volume manufacturing flow developed for eWLB fan-out products. The implementation of this process flow into 300 mm diameter reconstituted wafers has been described in detail in previous presentations [2]. In this manufacturing method the wafer is diced at the start of the process and then reconstituted into a standardized

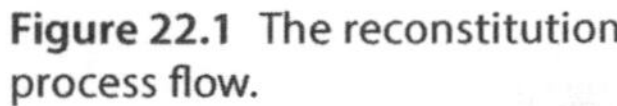

Figure 22.1 The reconstitution process flow.

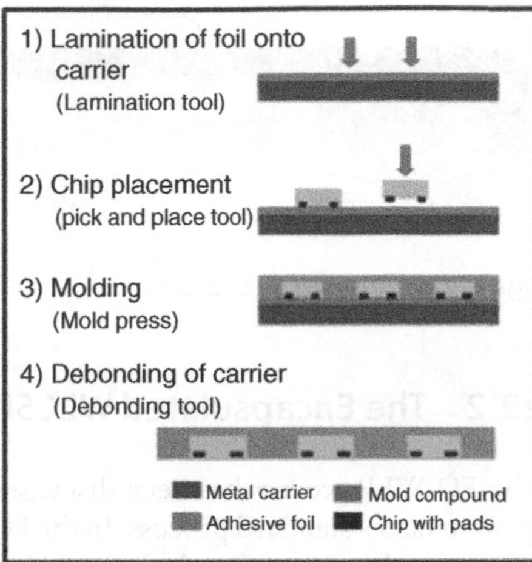

wafer (or panel) shape for the subsequent process steps. The basic process flow for creating the reconstituted wafer is shown in Figure 22.1:

1) The reconstitution process starts by laminating an adhesive foil onto a carrier.
2) The singulated die is accurately placed face-down onto the carrier with a pick and place tool.
3) A compression molding process is used to encapsulate the die with mold compound while the active face of the die is protected.
4) After curing the mold compound, the carrier and foil are removed with a debonding process, resulting in a reconstituted wafer where the mold compound surrounds all exposed silicon die surfaces.

The eWLB process is unique since the reconstituted wafer does not require a carrier during the subsequent wafer-level packaging processes.

After the reconstitution process, the reconstituted wafer is processed with conventional wafer-level packaging techniques for the application and patterning of dielectric layers, thin film metals for redistribution and under-bump metal, and solder bumps. In the final dicing operation, a thin layer of mold compound, typically less than 70 μm, is left on the side of the die as a protective layer. The back of the die is also protected with mold compound, although with a greater thickness than the sidewall layer. A schematic drawing of a typical structure is shown in Figure 22.2 for greater clarity. Alternatively, the backside mold compound can be removed, and the body is made thinner with an optional backgrind operation without damaging the protective sidewall layer. The remaining sidewall coating will continue to protect the fragile silicon sides of the die during the assembly operation.

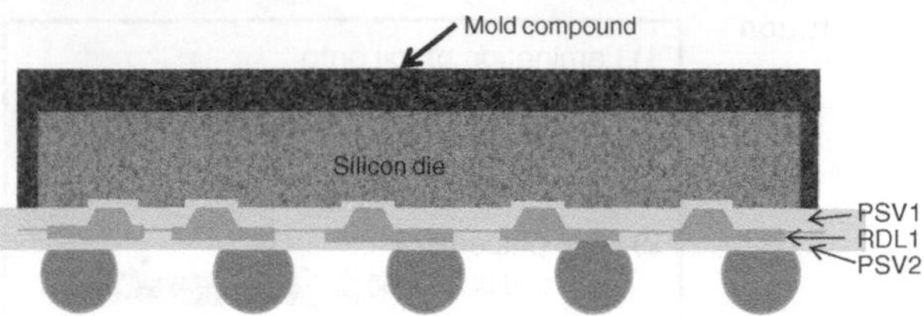

Figure 22.2 The eWLCSP structure.

22.2 The Encapsulated WLCSP Process

The FO-WLP process has been discussed in many venues, and it is recognized as an industry standard process. In the FO-WLP process the area of the package is increased to allow for placement of redistribution layers (RDL) and solder balls outside of the silicon die area [3]. This packaging method allows the die to shrink to a minimum size independent of the required area for an array of solder balls at industry standard BGA ball pitches [4]. It also allows for novel multi-die structures, 2.5D structures, and 3D structures. The FO-WLP process has been qualified to a 28 nm process node with the same dielectrics and Cu plating as are used in the eWLCSP process described here [5]. The eWLCSP process data presented in this paper was generated with a 300 mm round reconstituted panel [2]. In the case of conventional FO-WLP, the die is typically widely spaced to allow for the expanded RDL and bump area and the conventional saw street. In the case of eWLCSP, the die is closely spaced allowing for only the sidewall thickness in addition to a street area of 80 μm. The die size used in this evaluation was 4.5 mm × 4.5 mm similar to the construction shown in Figure 22.2. The final structure had two layers of polymer dielectric and one layer of plated Cu RDL with the solder ball mounted directly on the RDL without the use of a separate under-bump metallurgy (UBM) layer. The process flow used is shown in Figure 22.3 and the details of the structure are shown in Table 22.1 [6, 7].

Intuitively the process flow shown in Figure 22.3 would have higher cost since there are additional steps required for reconstitution at the start of the flow. There are two key factors that offset the cost of the additional steps required for the reconstitution to make this a commercially viable process. (i) Panel size scaling reduces the unit cost if the source silicon wafer is smaller than the reconstituted panel size. In the case of the 300 mm reconstituted panel used here, the cost is very competitive for silicon wafers with a diameter of 200 mm and below. The cost of processing a 300 mm reconstituted panel for WLCSP is approximately 1.7× the cost of processing a conventional 200 mm silicon wafer in WLCSP; however the units processed per panel increase by a factor of 2.3×, effectively offsetting the cost of

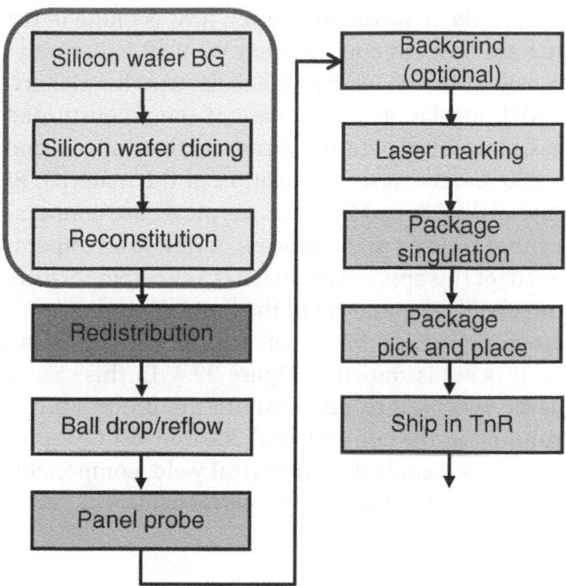

Figure 22.3 The eWLCSP process flow.

Table 22.1 Layer thicknesses.

PSV 1 (μm)	7.0–11.0
RDL 1 (μm)	7.0–10.0
PSV 2 (μm)	7.0–11.0
Ball pitch (mm)	0.4
Ball size (μm)	250
Solder alloy	SAC405

reconstitution. (ii) Since known good die can be selected at the start of the process, advanced devices that have a lower electrical yield can be tested in wafer form prior to the process. If the incoming wafer has a probe yield of 85%, then 15% more units per reconstituted panel can be processed to offset the cost of the reconstitution process. Since the reconstituted panel size is no longer linked to the incoming silicon wafer size, the panel size can be increased over time and change from a round to a much larger rectangular format. This scaling to a larger panel size will provide a compelling cost reduction when compared with conventional WLCSP packaging methods where the round silicon format is maintained throughout the wafer-level packaging process.

One difference in processing panels in the reconstitution flow is found in the attributes of the polymers that are used. In conventional WLCSP either polyimide (PI) or polybenzobisoxazole (PBO) is used as the dielectrics for planarization, stress buffering, and RDL insulation. In the case of the reconstituted panel, the mold compound has a lower temperature threshold than silicon, and sustained temperatures over 200 °C can cause degradation of the material. PI typically has a cure temperature of 380 °C, and PBO has a typical cure temperature of 300 °C and therefore cannot be used in the process. A new low temperature polymer has been developed for this application that has a cure temperature compatible with the 200 °C threshold temperature of the mold compound.

A scanning electron microscope (SEM) image of a cross section of an eWLCSP™ part created in the process is shown in Figure 22.4. In this case, a thicker sidewall protection layer was used to demonstrate the process on an existing production device running in the conventional 200 mm WLCSP production line. The device demonstrated equivalent electrical yield, component-level reliability (CLR), and board-level reliability (BLR) performance.

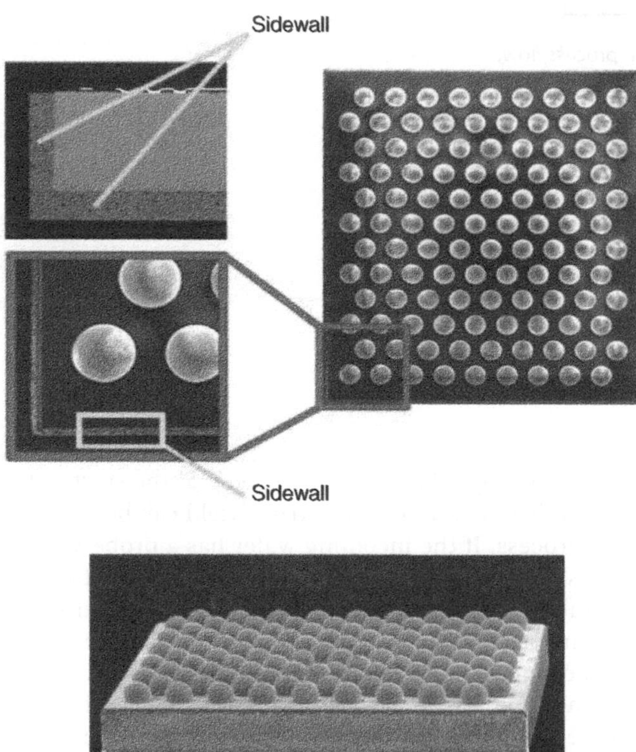

Figure 22.4 Micrographs of eWLCSP with sidewall protection.

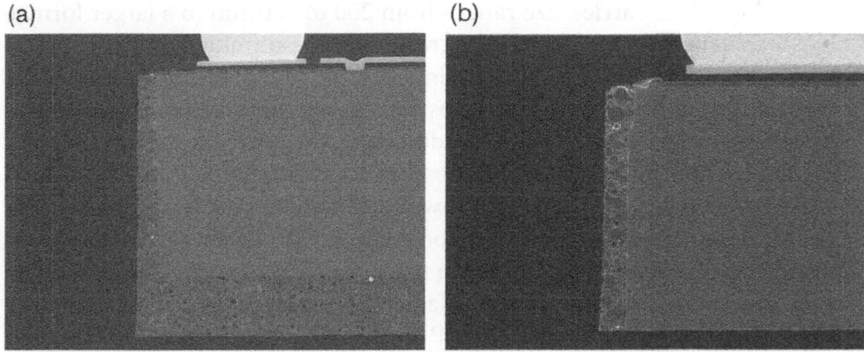

Figure 22.5 Micrograph of cross section of (a) eWLCSP with 5-side protection and (b) thin body eWLCSP with four-side protection.

A second SEM cross section is shown in Figure 22.5 showing the finished package with a thin protective sidewall coating and the use of the optional backgrind to thin the body thickness.

A WLCSP product that is currently in production using a conventional WLCSP process can be converted to a eWLCSP product without any design change required, regardless of the current silicon wafer diameter. If a reduced thickness is required for the specific application, an optional backgrind step can be added to the process flow to reduce the body thickness while retaining the protective sidewall coating. Since the die is singulated at the start of the process, the manufacturing equipment and bill of materials are the same for any incoming wafer size. The initial backgrind and dicing tools are the only wafer size dedicated equipment required for the process. Very little process development and very little additional capital will be required to package 450 mm silicon wafers as eWLCSP.

22.3 Advantages of the Encapsulated WLCSP, eWLCSP

Intuitively, eWLCSP would seem to have a higher cost over conventional WLCSP since there are additional steps required for reconstitution at the start of the FlexLine manufacturing flow. There are key factors, however, that offset the cost of the additional steps required for the reconstitution to make this a commercially viable process:

1) Cost-effectiveness: As described above, eWLCSP is fabricated using reconstitution. Good die from the parent wafer are picked and transferred to a (larger) reconstituted carrier. Since the majority of WLCSP products use 200 mm wafers, reconstitution enables the scaling of the manufacturing process from the 200 mm wafer to the size of the carrier in eWLB

technology. This carrier size ranges from 200 to 300 mm to a larger format like high density (HD) with ~20% greater area or ultrahigh density (UHD) with >300% greater area. The scaling of the manufacturing process with reconstitution far outweighs the cost of reconstitution itself, thereby enabling large net cost reductions. Additionally, the ability to selectively pick good die from the parent wafer presents an additional net cost benefit as most wafers have a less than 100% wafer sort yield. Last but not least, the ability to pool the volume of traditional fan-out eWLB packages seamlessly together with eWLCSP packages on the same FlexLine provides important economies of scale. With the three factors stated above, significant net cost reductions over traditional WLCSP front-end processing are achievable, depending on the original wafer diameter, the carrier format used for reconstitution (300 mm, HD or UHD), and the yield of incoming wafers.

2) High quality solutions: The polymer sidewall structure of eWLCSP all but eliminates mechanical damage such as chipping and cracking that is commonly encountered in traditional WLCSP processing. This serves to eliminate many expensive steps such as backside coating or lamination and complex inspection steps that are currently necessary for standard WLCSP to manage mechanical damage and ensure product quality. More fundamentally, the eWLCSP allows customers to build in quality by design versus using inspection to weed out defects. This has implications for reducing the risk of field failure due to the shipment of marginally defective parts that may escape inspection. As is shown in a later section, the eWLCSP structure has also helped to increase the overall die strength by ~100% in addition to the mitigation of cracking and chipping defects, making for an overall more robust package.
3) Investment and infrastructure – wafer-agnostic processing: In traditional WLCSP processing, the investment and infrastructure for manufacturing are based on the diameter of the incoming wafer. This creates a financial burden to retool the manufacturing lines to provide the needed capacity (to meet market demand) as wafer transitions occur (e.g. from 200 to 300 mm or from 300 to 450 mm in the future) while also having to obsolete the existing manufacturing assets. The FlexLine approach for eWLB and eWLCSP effectively decouples the packaging process from the incoming wafer, altogether obviating the above-described financial burden resulting from wafer diameter transitions.
4) Design friendly – allows seamless transition from fan-in to fan-out within the same basic package platform: As noted previously, the standard fan-in WLCSP only works below a certain threshold of I/O density, based on the minimum allowable terminal I/O pitch. The threshold is ~4 I/O/mm^2 for a 0.5 mm terminal I/O pitch and ~6 I/O/mm^2 for 0.4 mm terminal I/O pitch. Small changes in I/O density commonly occur with changes in Si design, and die shrinks resulting from Si node transitions may lead to a

given design exceeding the WLCSP threshold, causing the design to "fall off" the WLCSP application space envelope, necessitating a change in packaging POR to traditional substrate- or lead frame-based packages like FBGA, fcBGA, QFN, etc. These packages are fundamentally different than WLCSP in terms of footprint, form factor, performance, and cost, resulting in a major "reset" in the packaging POR. In contrast, the eWLCSP may be viewed as part of the more universal eWLB platform wherein the aforementioned I/O density transitions can be seamlessly accommodated within the same packaging platform. For designs whose I/O density falls marginally outside the threshold, an additional row of terminal solder balls can be added without fundamentally altering the package structure, form factor, or performance.

22.4 eWLCSP Reliability

The unique attribute of the eWLCSP package is the protective sidewall coating. The protective layer is durable and will prevent silicon chipping on the side of the package. This protective layer has the ability to protect the silicon during socket insertion for test. This has been demonstrated through multiple insertion tests on completed products with no observed damage to the protective coating.

Robust reliability of 4.5 mm × 4.5 mm eWLCSP was reported with CLR and BLR tests. The eWLCSP process has passed standard reliability tests used in wafer-level packaging including CLR, temperature cycle on board (TCoB), and drop test.

CLR was completed with the test conditions shown in Table 22.2. The evaluation results were confirmed by visual inspection and electrical test. No delamination of the protective coating was detected during the CLR evaluation.

TCoB was completed and passed 500 cycles, the typical requirement for consumer and commercial devices, with the results shown in Table 22.3 and the Weibull plot in Figure 22.6. Results were obtained from electrical

Table 22.2 Component-level reliability test results of eWLCSP.

Component-level test	Condition		Status
MSL1	MSL1, 260 °C reflow (3×)	—	Pass
Temperature cycling (TC) after precon	−55 to 125 °C	1000×	Pass
HAST (w/o bias) after precon	130 °C/85% RH	192 h	Pass
High temperature storage (HTS)	150 °C	1000 h	Pass

Table 22.3 TCoB reliability test results for eWLCSP.

TCoB (Cond B)	Failure rate	Characteristic life (η)	Weibull slope (β)	First failure
–40 to 125 °C	0.635	1219.4	10.13	864×

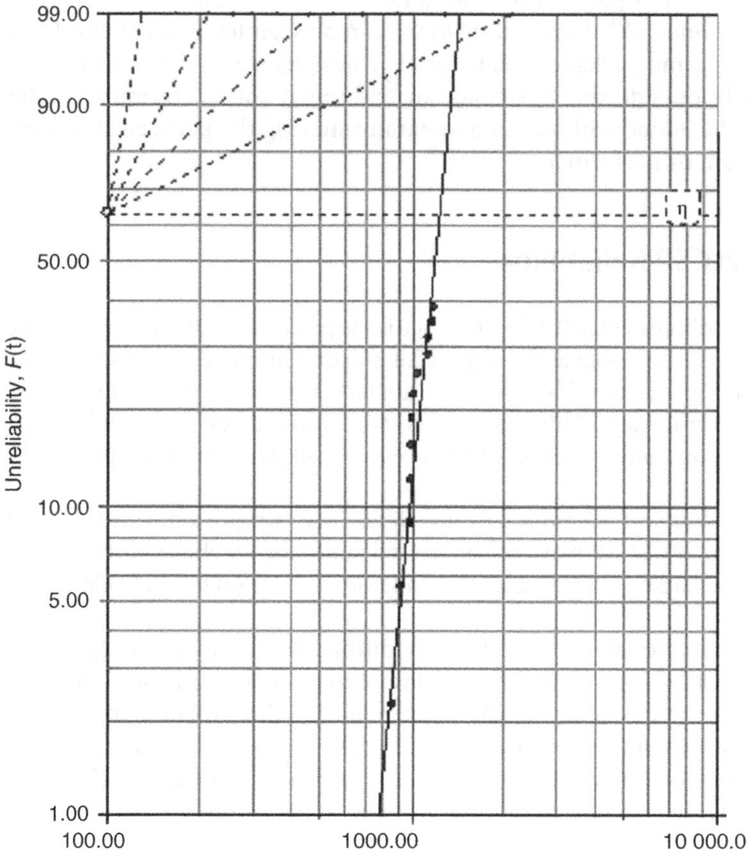

Figure 22.6 TCoB Weibull plot for eWLCSP.

measurement of daisy-chain bump structures. The results are comparable with conventional WLCSP products produced with PI dielectrics.

Drop test was completed and passed the JEDEC requirement of 30 drops with the results shown in Table 22.4 and the Weibull plot in Figure 22.7. Results were obtained from electrical measurement of daisy-chain bump structures. The results are comparable with conventional WLCSP products produced with PI dielectrics.

Table 22.4 Drop test results for eWLCSP.

Drop test	Failure rate	Characteristic life (η)	Weibull slope (β)	First failure
JEDEC	0.635	1553.5	5.97	772×

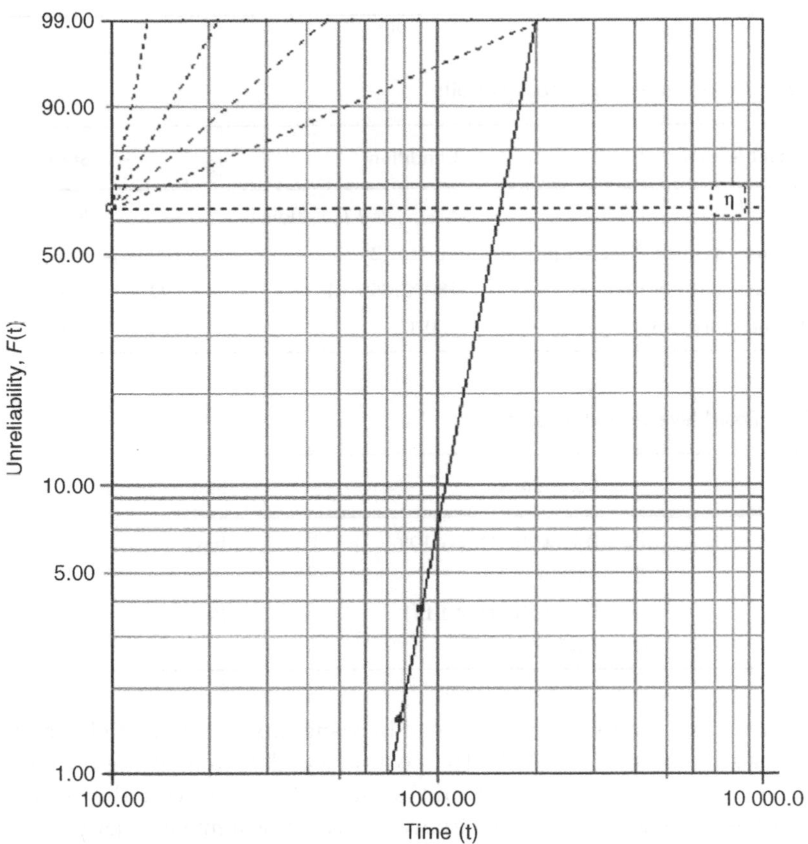

Figure 22.7 Drop test Weibull plot for eWLCSP.

22.5 Reliability of Larger eWLCSP over 6 mm × 6 mm Package Size

For reliability tests of larger eWLCSP over 6 mm × 6 mm, two test vehicles were prepared, 6 mm × 6 mm and 8 mm× 8 mm, as shown in Table 22.5. The eWLCSP process passed standard reliability tests used in wafer-level packaging including CLR and BLR (TCoB and drop test). CLR was completed with the test conditions shown in Table 22.6.

Table 22.5 eWLCSP test vehicle (TV) details.

	eWLCSP size	Mask no.	Solder ball pitch
TV1	6 mm × 6 mm	3 (without UBM)	0.4 mm
TV2	8 mm × 8 mm	4 (with UBM)	0.35 mm

Table 22.6 Component-level reliability results.

Component-level test	Condition		Status
MSL1	MSL1, 260 °C Reflow (3×)	—	Pass
Temperature cycling (TC) after precon	−55 to 125 °C	1000×	Pass
HAST (w/o bias) after precon	130 °C/85% RH	192 h	Pass
High temperature storage (HTS)	150 °C	1000 h	Pass

Table 22.7 Board-level reliability test results.

Tests	Conditions	Status
TCoB	JEDEC JESD22-A103 −40 to 125 °C	Pass
Drop test	JEDEC JESD22-B111 1500G	Pass

The evaluation results were confirmed by visual inspection and electrical test. No delamination of the protective coating was detected during the CLR evaluation. TCoB was completed and passed 500 cycles with the results shown in Table 22.7. Results obtained from electrical measurement of daisy-chain bump structures demonstrate that eWLCSP is comparable with conventional WLCSP products produced with PI dielectrics. Drop test was completed and passed the JEDEC requirement of 30 drops with the results shown in Table 22.4.

A four-point bending test was carried out to investigate package-level strength. eWLCSP shows an over 25% increase in die strength compared with WLCSP due to the sidewall protection and optimized backgrinding process.

The Si surface roughness was measured with atomic force microscopy (AFM). eWLCSP has a Si roughness value very close to that of WLCSP. A roughness image scan clearly showed no difference in Si surface roughness between WLCSP and eWLCSP.

Table 22.8 Visual inspection results after auto handler socket insertion test.

VM scan yield summary results	Pre VM scan	1× socket insertion post VM scan	2× socket insertion post VM scan	3× socket insertion post VM scan
# of devices Inspected SOU	5000	5000	5000	5000
# of devices accepted	5000	5000	5000	5000
VM yield (%)	100.0	100.0	100.0	100.0
# of devices rejected	0	0	0	0

The protective sidewall coating is a unique attribute of the eWLCSP. This protective layer is durable and will prevent silicon chipping on the side of the package and has the ability to protect the silicon during socket insertion for test. This has been demonstrated through multiple insertion tests on completed products with no observed damage to the protective coating in Table 22.8.

With these test results along with CLR and BLR results, large eWLCSP with additional sidewall protection has demonstrated more robust reliability than standard WLCSP and prevents side chip cracking.

22.6 eWLCSP Wafer-Level Final Test

An eWLCSP reconstituted wafer is different from a silicon wafer as the backside of the wafer is mold compound, which tends to have a much higher warpage level compared with a silicon wafer of similar thickness. To address this issue, modifications were made to the prober to enable handling of the eWLCSP reconstituted wafer. These modifications were carried out and successfully demonstrated in a high volume manufacturing environment

The test chuck of the prober must be modified for handling of the warped wafers. A progressive and stronger vacuum at the test chuck is necessary to ensure the wafer is properly held flat on the test chunk before testing. The shape of the robot handling arms was modified, and the vacuum to these arms was enhanced to ensure proper holding during the transfer of wafers within each module of the prober. One of the key differences between an eWLCSP reconstituted wafer and a silicon wafer is the wafer identification (ID) marking on the wafer. For an eWLCSP reconstituted wafer, the ID marking is on a copper surface, which makes it look entirely different under the prober's optical

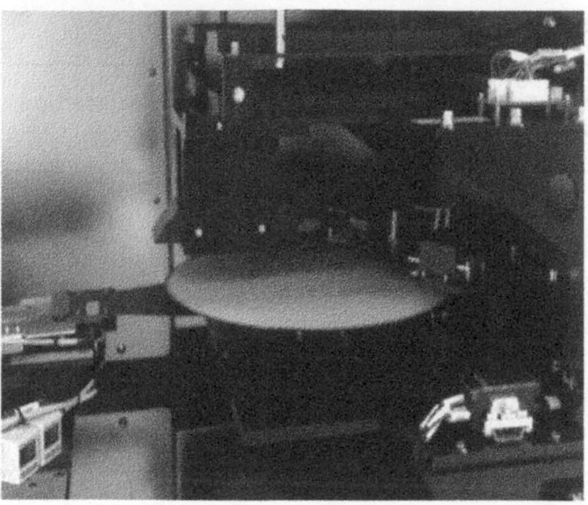

Figure 22.8 Pre-aligner station where the wafer notch is detected and wafer ID is read of eWLCSP carrier in prober system in FlexLine.

character recognition (OCR) reader. Hence, modification on the OCR reader is required to enable reading of the reconstituted wafer ID. Furthermore, in alignment with the eWLCSP assembly process, where the modified front-opening unified pod (FOUP) is used in processing the wafers, the prober software must be modified to accommodate the differences. Figure 22.8 shows a pre-aligner station where the wafer notch is detected and the wafer ID is read of the eWLCSP panel.

Wafer-level testing of eWLCSP has been proven with noticeable advantages such as a higher test cell utilization and better first pass yield, resulting in an overall cost reduction of testing these packages. Wafer-level testing offers a short index time, especially for highly parallel testing. This is mainly due to the indexing from one touchdown to the next, where it only involves a small movement within the wafer compared with the pick and place handler that has a high index time. Indeed, this short index time characteristic is most suitable for eWLCSP devices testing, especially for small package sizes, short test times, and high parallelism test requirements. Besides achieving higher throughput, wafer-level testing improves utilization by lowering manufacturing stoppages, such as jamming associated with the handling of small packages.

Wafer-level testing of eWLCSP improves the first pass yield by utilizing the visual alignment system, which is standard on a prober. Using camera and vision technology, the socket pin alignment to the package bumps can be highly accurate and repeatable; thus the first pass test yields for wafer-level testing are significantly higher. Figure 22.9 shows the indexing process of an eWLCSP carrier. Besides providing better and more accurate contacts using vision technology, the prober utilizes its auto socket cleaning feature to improve test yield, whereas this auto cleaning feature is an in-line process that minimizes test cell downtime and eliminates operator labor needed for manual cleaning. Furthermore, wafer-level testing also reduces the tooling and hardware costs. In contrast to pick and place handlers that require a

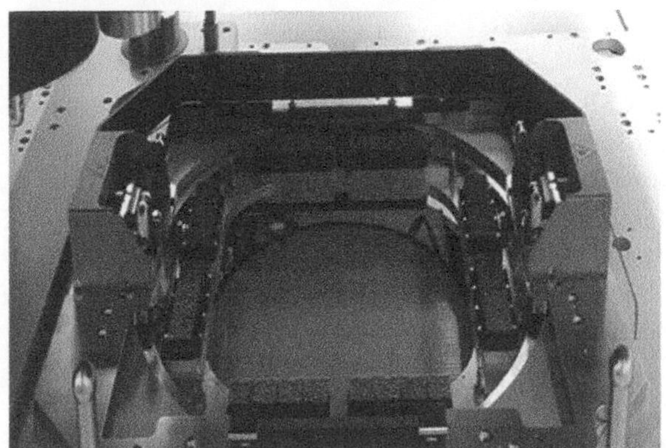

Figure 22.9 Indexing of eWLCSP carrier in a wafer test of FlexLine.

new change kit for every different package size, using a prober to handle wafer-level testing eliminates the need to change kits and thus reduces the overall manufacturing cost.

22.7 Conclusions

A new eWLCSP process has been developed and verified with reliability testing. The process provides mechanical sidewall protection to WLCSP parts with an increase in package size of less than 140 μm in *X* and *Y* dimensions. The sidewall protection resolves the problem of silicon damage during the assembly process and provides a path to significant cost savings for customers as the panel size is increased. The eWLCSP process described is wafer size agnostic, so the same manufacturing line can process eWLCSP products regardless of the incoming wafer size. 450 mm wafers can easily be accommodated in the eWLCSP process once the service is required by the customers (Figure 22.10).

Figure 22.10 eWLCSP product with protective sidewall coating.

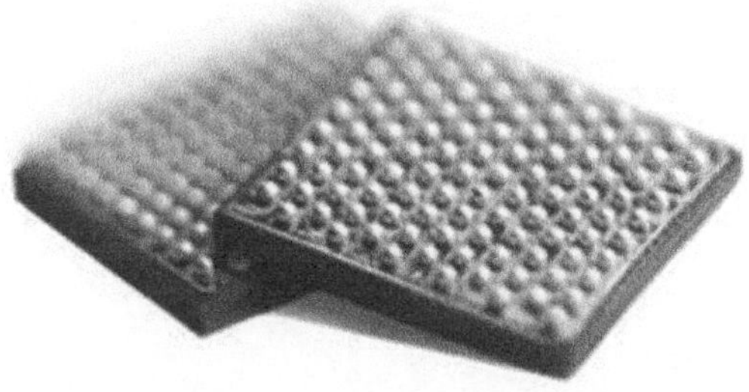

References

1 Elenius, P. (1998). The Ultra CSP wafer scale package. *Proceedings of the Electronics Packaging Technology Conference* (1998).

2 Prashant, M., Yoon, S.W., Lin, Y.J., and Marimuthu, P.C. (2011). Cost effective 300mm large scale eWLB (embedded wafer level BGA) technology. *Proceedings of the 13th Electronics Packaging Technology Conference* (2011).

3 Brunnbauer, M.Fürgut, E., Beer, G., and Meye, T. (2006). Embedded wafer level ball grid array (eWLB). *Proceedings the of 8th Electronic Packaging Technology Conference* (2006).

4 Brunnbauer, M. and Meyer, T. (2008). Embedded wafer level ball grid array (eWLB). *IMAPS Device Packaging Conference* (2008).

5 Yoon, S.W., Tang, P., Emigh, R. et al. (2013). Fanout flipchip eWLB (embedded wafer level ball grid array) technology as 2.5D packaging solutions. *Proceedings of the Electronic Components and Technology Conference* (2013).

6 Strothmann, T., Yoon, S.W., and Lin, Y. (2014). Encapsulated wafer level package technology (eWLCSP). *Proceedings of Electronic Components and Technology Conference* (2014).

7 Lin, Y., Chong, E., Chan, M. et al. (2015). WLCSP+ and eWLCSP in flexline: innovative wafer level package manufacturing. *Proceedings of the Electronic Components and Technology Conference* (2015).

23

Embedded Multi-die Interconnect Bridge (EMIB)

A Localized, High Density, High Bandwidth Packaging Interconnect

Ravi Mahajan, Robert Sankman, Kemal Aygun, Zhiguo Qian, Ashish Dhall, Jonathan Rosch, Debendra Mallik, and Islam Salama

Intel Corporation, Chandler, AZ, USA

23.1 Introduction

The need for high memory bandwidth between the central processing unit (CPU) and dynamic random-access memory (DRAM) has led to increased focus on high bandwidth on-package links in recent years [1–3]. The performance of the input/output (I/O) subsystem (or link) that delivers this bandwidth is measured by its power consumption and bandwidth, both of which depend on the transceiver circuits and the I/O channel. It should be noted that link performance is also affected by latency; however this aspect is not covered here since it requires a deeper discussion of link architectures that is beyond the scope of the present chapter. The peak bandwidth of an I/O link is the product of the number of data lanes and the data rate, two factors that can be scaled to enable bandwidth scaling:

1) Increasing the number of data lanes creates the so-called *wide and slow* I/O links where the density of all the components in the physical channel, i.e. the I/O circuits, bumps, and wires, is scaled. This allows use of a lower signaling frequency in the I/O link and hence improved power efficiency due to reduction in circuit complexity and/or voltage scaling [4]. The main challenge for enabling *wide and slow* links is to achieve them without unduly increasing the interconnect real estate (i.e. die area, interconnect area on the package, and the number of package layers).
2) A higher data rate to create *fast and narrow* I/O links on the other hand results in a higher loss I/O channel. The higher loss here refers to higher insertion loss in the channel due to an increase in Nyquist frequency. It

Advances in Embedded and Fan-Out Wafer-Level Packaging Technologies, First Edition.
Edited by Beth Keser and Steffen Kröhnert.

mandates more complex circuits with advanced equalization and sophisticated clocking. In general, improved I/O power efficiency is inversely proportional to channel loss and data rate [5, 6].

Currently the highest CPU–DRAM bandwidths are achieved in a *wide and slow* link using high bandwidth memory generation 2 (HBM2) DRAM [7]. This link has a wide (1024-bit) bus running at a relatively slow data rate of $2\,\text{Gb}\,\text{s}^{-1}\,\text{pin}^{-1}$ compared with graphics double data rate type 5 (GDDR5) memory that has a 32-bit bus running at $8\,\text{Gb}\,\text{s}^{-1}\,\text{pin}^{-1}$. The HBM2 interface achieves an 8× improvement in peak bandwidth compared with GDDR5 [8].

The class of packaging technologies that have increased density of components in the physical channel is called dense multi-chip packages (MCPs). A key area of focus for dense MCPs is ensuring signal integrity in a cross talk-dominated environment. It should be noted that the application space for dense MCPs extends beyond CPU–DRAM links. They can be used more broadly for heterogeneous integration where die from different silicon technologies and with different functionality can be integrated on package using high bandwidth, low power links [9, 10].

Linear interconnect escape density or I/O density (I/O/mm/layer) is a key metric used to compare capability envelopes of different packaging technologies used to create the physical on-package link (Figure 23.1). Note that I/O in this usage refers to physical interconnects, e.g. wires connecting die bumps to package pins or to neighboring die bumps. I/O/mm/layer is the number of wires escaping per millimeter of die edge for each routing layer of the package. Figure 23.2 shows a comparison of different MCP technologies in terms of

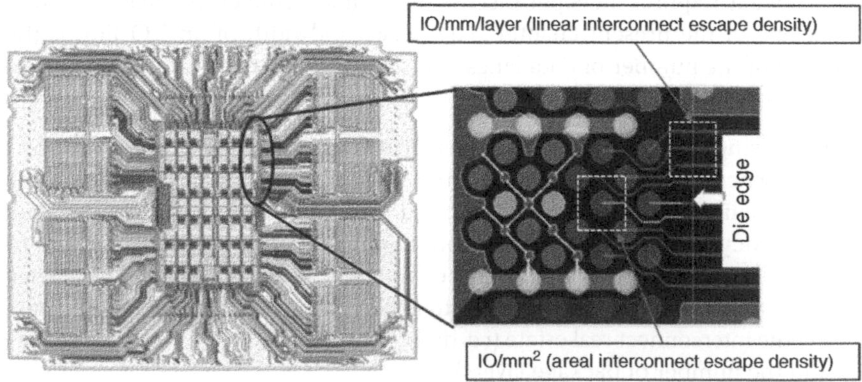

Figure 23.1 Key metrics used to establish MCP interconnect capabilities. Figure shows a representative layout of a package layer where a CPU is connected to multiple memories. I/O/mm/layer describes the linear escape density. I/O/mm^2 is a metric of the areal interconnect escape density, i.e. bump density on the silicon chip. It is an inverse function of I/O bump pitch.

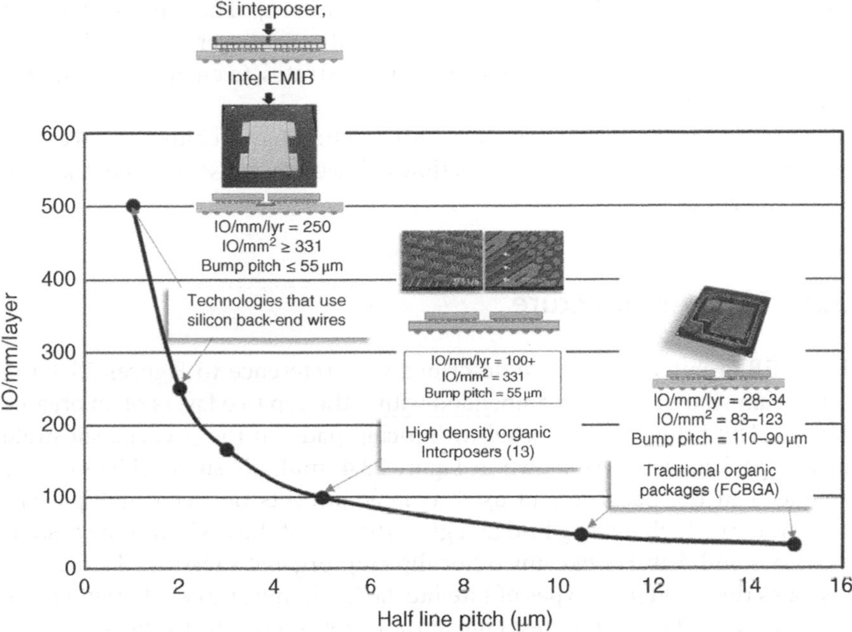

Figure 23.2 Technology envelopes for 2D and 2.5D MCP technologies (note that half line pitch is $[(L + S])/2$ where L is the wire width and S is the space between two wires). Typical bump pitches are included for reference.

their scaling metrics. Note that for a given half line pitch and via pad size, the I/O density capability can vary based on the bump pitch, bump pattern, and the number of bump rows to be routed per layer. I/O/mm/layer used for capability comparisons in Figure 23.2 is calculated using a four-row deep pattern [11]. As shown in Figure 23.2, technologies that utilize silicon back-end wiring technologies offer the highest I/O densities, typically an order of magnitude greater than traditional laminate packaging technologies and significantly higher than current fan-out reconstituted wafer technologies or high density organic interposers [12, 13]. Thus technologies with silicon back-end wiring processes can be used to create highly compact physical die-to-die links compared with all other fan-out technologies. This is an advantage when a reduction in die perimeter available for I/O connections occurs or in a situation where minimizing package layer count is important. The focus of this chapter is restricted to comparisons between planar architectures that use silicon back-end wiring technologies that essentially represent the upper end of the I/O/mm/layer spectrum and offer silicon-level connectivity on package.

EMIB, which is one such planar dense MCP technology, was first proposed in the mid-2000s by Mahajan and Sane [14]. It evolved further through the work of Braunisch et al. and Starkston et al. [15, 16]. The basic concept of EMIB is that it uses thin pieces of silicon with multilayer back-end-of-line (BEOL) interconnects, embedded in organic substrates, to enable dense die-to-die interconnects.

In this chapter we first describe the EMIB technology architecture. We then describe the high level EMIB process flow followed by a discussion of the high bandwidth envelope.

23.2 EMIB Architecture

The EMIB architecture can be described with reference to Figures 23.3 and 23.4. A thin silicon bridge is embedded within the top two layers of an organic package. The bridge is connected to flip-chip pads on the package substrate through package vias. As shown in Figure 23.4, multiple such bridges can be embedded in the package and used as interconnects between multiple die. Aside from the high density bridge region, the rest of the C4 bumps have similar pitches and features like any other flip-chip organic substrate. Figure 23.4 illustrates three different types of satellite die (and, thus, three different bridge types) connected to a larger test chip at the center on this test vehicle.

Some of the key advantages of EMIB are as follows:

- It is the only packaging technology that offers localized high density wiring (Figure 23.3). The bulk of the package interconnects are still the same traditional organic package interconnects. In contrast the nearest equivalent dense MCP architectures such as the silicon interposer-based architecture and Silicon-Less Integrated Module (SLIM™) require interposers that are

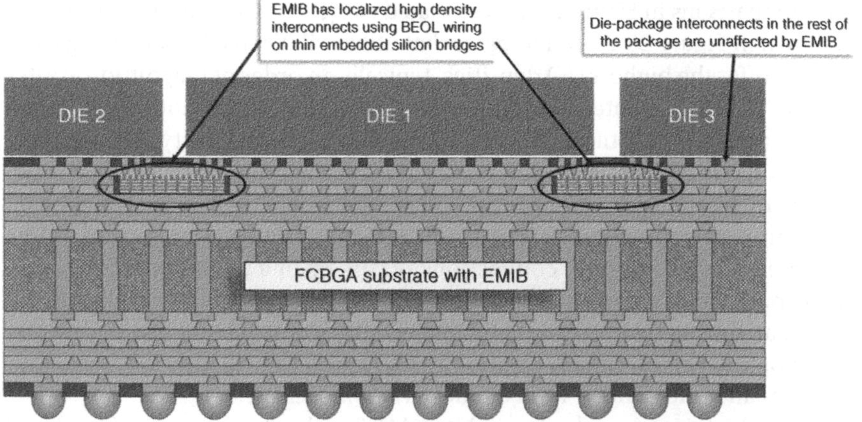

Figure 23.3 Schematic showing the EMIB concept.

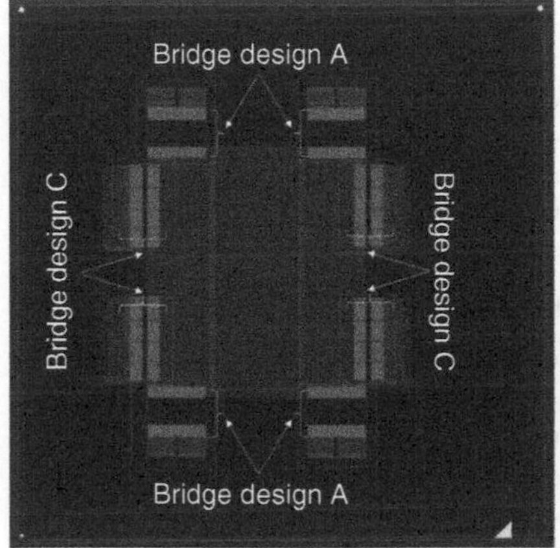

Figure 23.4 Top view of a test package highlighting three different designs of localized high density embedded bridges between die. Note that a partially assembled test package is shown to highlight the features of the package.

larger than the total area of the die in the MCP and do not localize the dense interconnect [17, 18].

- A related benefit of EMIB is that while the interposer size is limited by reticle field (which restricts the total die area on the interposer to less than reticle size. It is possible to increase the interposer size beyond reticle; however such an increase will come with a corresponding reduction in number of realized interposers per wafer and will require investment in equipment to pattern interconnects over areas larger than reticle size), EMIB does not have such a limitation. There are no practical limits to the die size, and EMIB can more easily enable greater than reticle-sized die area in the MCP.
- Bridge manufacturing is substantially simpler than interposer manufacturing since a through-silicon via (TSV) process is not needed. Additionally, the EMIB assembly process is simpler than an Si interposer since the interposer attach adds an extra chip-attach module to the assembly process.
- Bridge silicon costs are intrinsically lower than silicon interposer due to lack of TSVs and significantly less silicon area required.

Some key disadvantages of the EMIB technology are as follows:

- That it increases organic substrate manufacturing complexity.
- Similar to typical organic flip-chip technology, coefficient of thermal expansion (CTE) mismatch between organic substrate and surface die leads to higher die back end, first-level interconnect (FLI), and thermal interface material (TIM) stresses.

Table 23.1 Key attributes of the silicon bridge.

Attributes	EMIB values
Bridge size range	2 mm × 2 mm to 8 mm × 8 mm (current range) – higher sizes possible
Bridge thickness	<75 μm
Number of bridges per package	>8 possible
Metal layers	Up to 4 routing metal layers + pad layer Each metal layer has 2 μm lines and 2 μm spaces (lower dimensions possible) Vias between metals: 2 μm (lower dimensions possible) 50–70% metal density on ground layers

Physical attributes of the bridge are described in Table 23.1 and shown in Figure 23.5. Note that although bridge sizes up to 8 mm × 8 mm are currently targeted, there is no intrinsic reason why the size range cannot be increased or decreased. The bridge has to be thin (<75 μm) in order for it to fit in the package routing layers. Currently, bridge designs with routing in four metal layers, 2 μm line width and line spacing, and 2 μm tall vias are used, although increased layers and lower line, space, and via dimensions are possible. Since interconnects in the bridge are created using silicon back-end processes and are currently considerably coarser than the fine feature interconnects available in leading-edge silicon nodes, there is significant room for further improvements in the interconnect stack. For instance, the vias between metal layers and lines in the routing layers (see Figure 23.5) can be reduced to sub-1 μm dimensions relatively easily, resulting in increased I/O/mm/layer. It should be noted that a reduction in line dimensions will result in increased wire resistance and

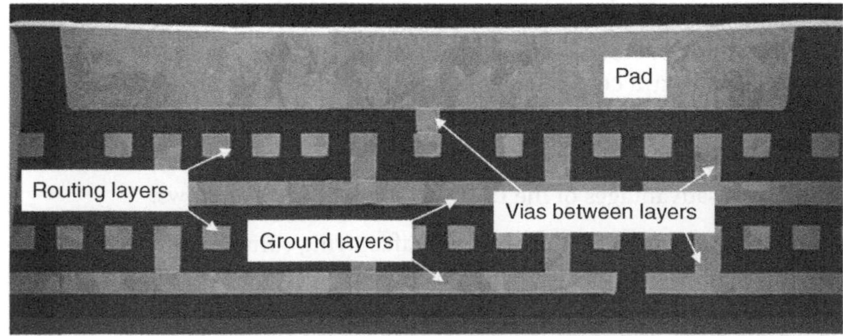

Figure 23.5 Cross-sectional image showing the bridge metal layers and interlayer vias.

changes in wire-to-wire capacitance. The impact of these changes on signal integrity must be comprehended in the design of the I/O link. It should also be noted that this impact is not unique to the bridge technology and is common to all MCP technologies. Additionally the signal integrity can be improved by lowering the effective dielectric constant of the inner layer dielectric (ILD) layers. The utility of EMIB can also be increased by embedding an active bridge die or one with integrated passives.

23.3 High Level EMIB Process Flow

Bridge wafers are manufactured using a fab back-end process. Each bridge wafer is thinned to below 75 μm and then singulated into individual bridge die. The singulated bridges are embedded in the organic package substrate. Tight bridge thickness variation control, precise die singulation, and advanced handling and transport systems are needed to ensure successful embedding of bridges into the organic package. The organic package follows the standard manufacturing process until the layer before the final buildup layer (referred to as Layer *N* in Figure 23.6). At this point in the process, an additional step is introduced to create cavities for the bridge (Figure 23.6). The bridge is placed in the cavity, held in place with an adhesive, and the final layers of buildup dielectrics are applied, followed by fine via formation in the bridge region and coarse via formation elsewhere. The bridge via formation process has to be tightly controlled to ensure that there is no misalignment between the vias and the substrate pads.

The substrate with embedded bridges is then used in the assembly process, just like any other MCP. Key challenges during assembly are developing high yielding chip attach (done using thermo-compression bonding [TCB]) and

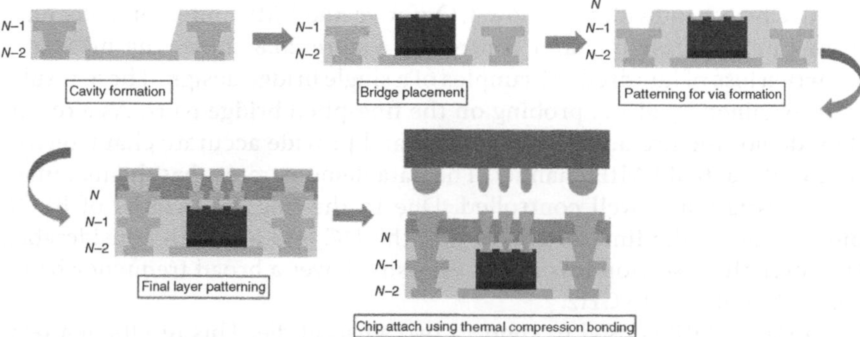

Figure 23.6 Schematic showing key steps in constructing the EMIB package and assembly flow.

underfill processes. These challenges arise because of the difference in pitch, i.e. fine bump pitch above the embedded bridges, and coarse-pitch attach in the rest of the area, which require tight control on the chip-attach and underfill processes. A high yielding TCB process depends on tight control of bump heights (both on substrate and on die side), solder volume, and advanced process controls during bonding to ensure successful joint formation [19]. A robust, void-free capillary underfill process requires co-optimization of underfill material properties, the dispense process, and cure conditions. As discussed later in this chapter, signal integrity requires minimum possible interconnect distance through the bridge, which in turn requires surface die connected by bridges to be placed as close to each other as possible. Such close proximity (<200 μm) between adjacent die requires tight process control in the chip-attach and underfill processes and careful thermomechanical design to ensure reliability. Intel has successfully developed this technology to meet reliability targets through a comprehensive optimization of material properties, package geometry, and process parameters. Figure 23.7 shows cross sections of an assembled test package using EMIB and a C-mode (confocal) scanning acoustic microscopy (CSAM) image showing successful underfilling.

To summarize, a number of enabling technologies including bridge manufacturing, bridge wafer thinning and singulation, substrate manufacturing, and assembly have been developed to realize EMIB. The integrated technology envelope has been successfully certified for yield, reliability, and high volume manufacturability on bridge, substrates, and assembly test vehicles.

23.4 EMIB Signaling

This section focuses on the electrical signaling performance of EMIB. To achieve high bandwidth and low I/O power, the EMIB interconnects have been carefully designed and controlled. Figure 23.8 shows the measured insertion loss of hundreds of samples of a single bridge design. These results were obtained by direct probing on the fine-pitch bridge pads. As a result, they do not require any de-embedding and provide accurate characterization of the actual EMIB channel. The data demonstrates that the manufacturing variation is well controlled. Due to the fine line widths of BEOL interconnects, the line resistance and the DC loss are both considerable. However, the insertion loss is still very small over a broad frequency band, i.e. <−2 dB up to ~18 GHz.

Typically EMIB is used to connect two adjacent die. This results in a relatively short I/O channel that does not demand complex transceiver circuits to meet signal integrity requirements. Figure 23.9 shows an equivalent circuit

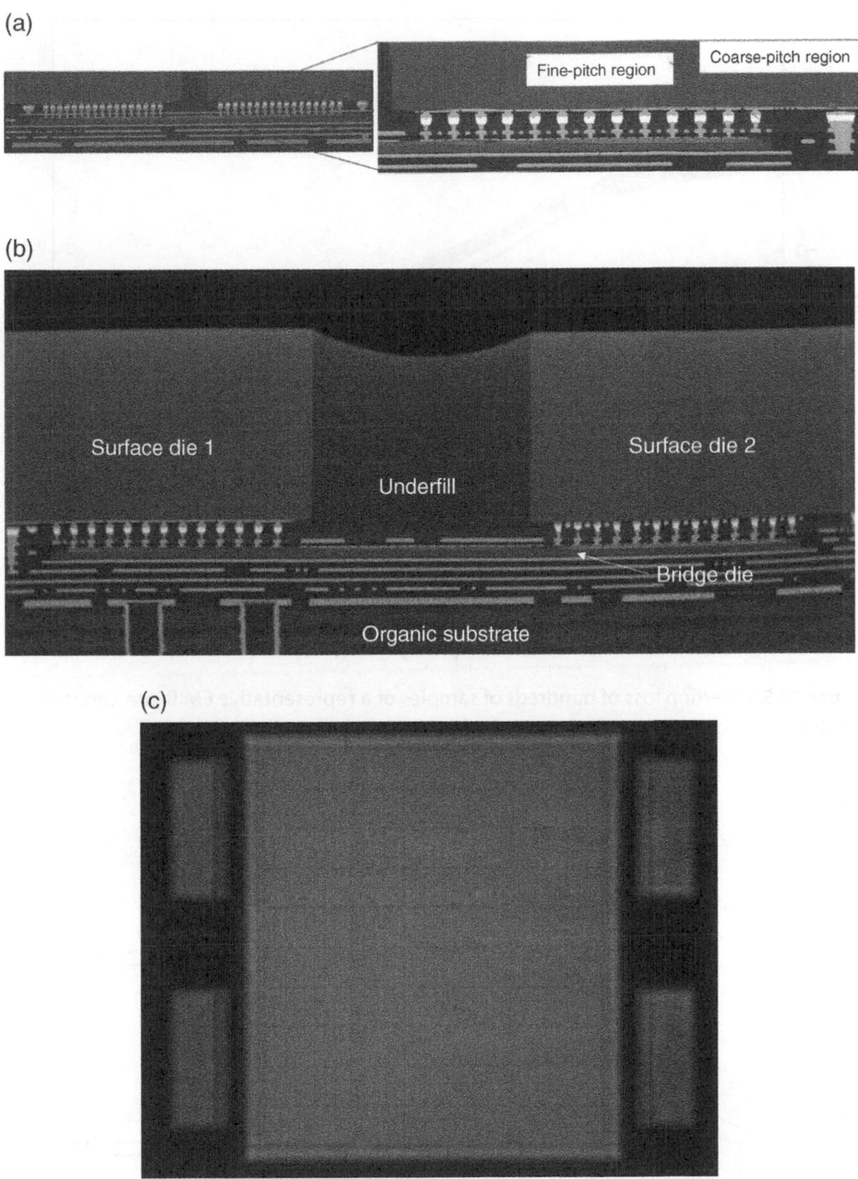

Figure 23.7 Cross section of a test vehicle showing fine-pitch and coarse-pitch die-to-package flip-chip attach. (a) Cross-section of fine pitch region connection from die to bridge. (b) Cross section highlighting the bridge die, the two surface die connected by the bridge, and the organic substrate in which bridge is embedded. (c) CSAM image showing complete void-free underfilling of a five-die EMIB MCP.

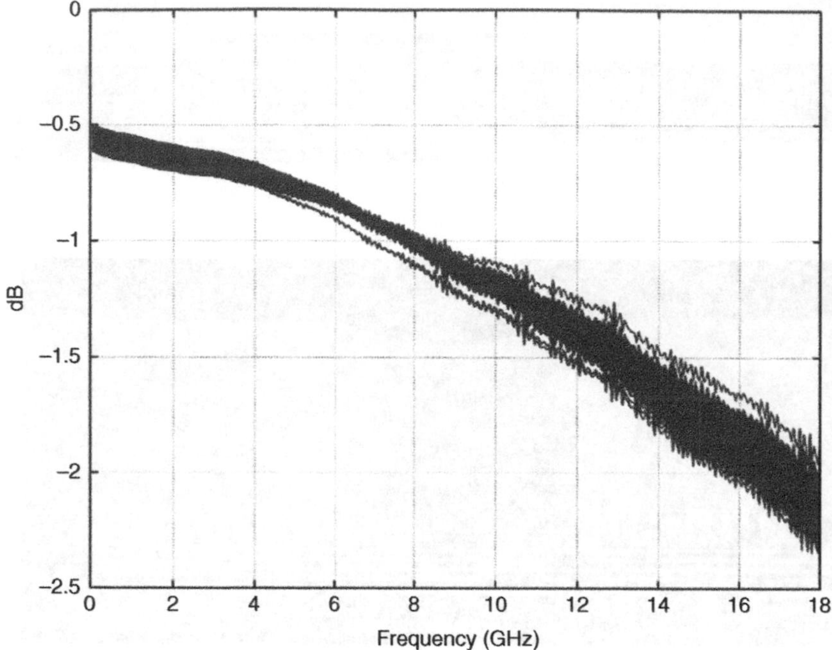

Figure 23.8 Insertion loss of hundreds of samples of a representative EMIB interconnect design.

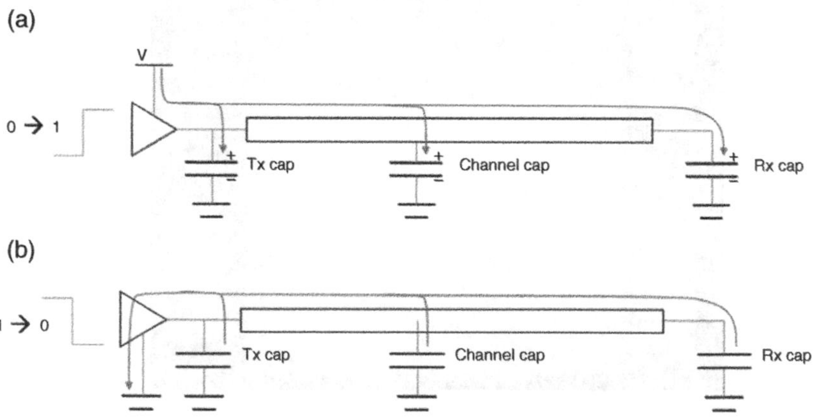

Figure 23.9 Simple I/O driver and unterminated receiver for EMIB interconnect channel. (a) Channel load charges when data switch from 0 to 1. (b) Channel load discharges when data switch from 1 to 0.

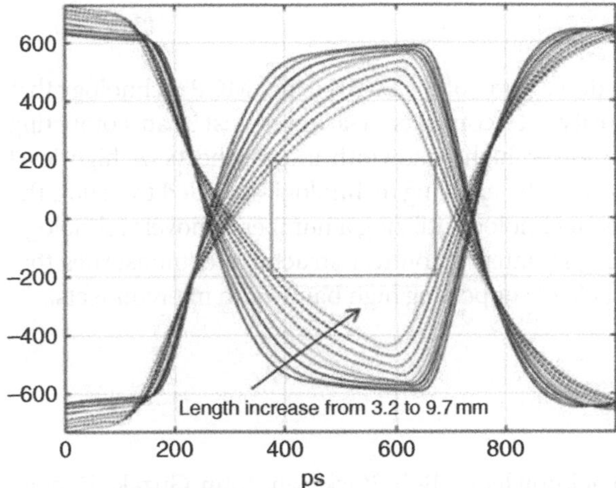

Figure 23.10 Eye diagrams of 2 Gbps signaling at channel lengths from 3.2 to 9.7 mm. Driver is a simple 12 mA inverter, receiver is unterminated, and both driver and receiver have 0.4 pF pad capacitance.

example of an EMIB I/O channel. The driver is a simple CMOS inverter and the receiver does not have termination. When data switches from 0 to 1, the driver pumps current into the channel to charge all the capacitance loading from the channel and transceiver circuits. The consumed power is CV^2 where C represents the total capacitance loading and V is the driver supply voltage. When data switches from 1 to 0, all the capacitive loading discharges, a process that does not consume power. Hence the overall clock data pattern has an energy efficiency of $0.5CV^2$, and a random data pattern has a power efficiency of $0.25CV^2$. If the total loading is less than 2 pF, the driver energy efficiency is less than 0.5 pJ b^{-1} at 1 V. Any reduction of the voltage can quadratically reduce the power consumption.

To demonstrate that the above described I/O channel and circuits can meet signal integrity requirements, results from eye diagram simulations at 2 Gbps performed for varying channel lengths are shown in Figure 23.10. A 12 mA CMOS inverter driver and 0.4 pF pad capacitance are assumed for both the driver and the receiver. The eye diagram is the predicted worst-case eye based on the peak distortion analysis (PDA) [20, 21]. The results in Figure 23.10 show that a good eye opening can be achieved even with a 9.7 mm long channel that demonstrates the signaling performance capability of EMIB. Combined with the high I/O/mm capability of the technology, it is possible to achieve high bandwidth die-to-die channels using EMIB.

23.5 Conclusions

In this chapter, we present a description of the EMIB dense MCP technology that provides localized high density interconnects at a lower cost than competing technologies providing low power, high bandwidth interconnects. A high level process flow is shown along with the enabling technologies needed to realize the technology. EMIB is a proven technology utilizing a number of novel technological innovations. Measured signal integrity on test structures demonstrates that the EMIB technology is capable of supporting high bandwidth interconnects.

Acknowledgments

The authors would like to acknowledge Bob Starkston, John Guzek, Deepak Kulkarni, Chris Baldwin, Sanka Ganesan, Babak Sabi, Ken Brown, Chris Nelson, Chia-Pin Chiu, Henning Braunisch, Aleks Aleksov, Ram Viswanath, Sriram Srinivasan, Hamid Azimi, Mostafa Aghazadeh, Koushik Banerjee, Omkar Karhade, Johanna Swan, Stefanie Lotz, and Nitin Deshpande from Intel for their contributions in evolving the EMIB concept from paper to reality. Discussions with Rajat Agarwal, Suresh Chittor, and Randy Osborne from Intel on memory directions are gratefully acknowledged.

References

1 Goto, H. (2015). http://pc.watch.impress.co.jp/img/pcw/docs/740/790/html/1.jpg.html (accessed 6 August 2018).

2 Bender, M.A., Berry, J.W., Hammond, S.D. et al. (2017). Two-level main memory co-design: multi-threaded algorithmic primitives, analysis and simulation. *Journal of Parallel Distributed Computing* 102: 213–228.

3 International Solid-State Circuits Conference (ISSCC) (2017). Trends. http://isscc.org/wp-content/uploads/2018/06/2017_Trends.pdf (accessed 6 August 2018).

4 Mansuri, M., Jaussi, J. E., Kennedy, J. T. et al. (2013). A scalable 0.128-to-1Tb/s 0.8-to-2.6pJ/b 64-lane parallel I/O in 32nm CMOS. *Proceedings of the IEEE International Solid-State Circuits Conference*, San Francisco, CA, USA (2013), pp. 402–403.

5 Casper, B. (2011). Energy efficient multi-Gb/s I/O: circuit and system design techniques. *Workshop on Microelectronics and Electron Devices (WMED)*, Boise, ID, USA (22 April 2011).

6 O'Mahony, F., Balamurugan, G., Jaussi, J., et al. (2009). The future of electrical I/O for microprocessors. *Proceedings of the IEEE Symposium on. VLSI Design Automation and Test (VLSI-DAT)* (April 2009), pp. 31–34.

7 https://www.jedec.org/standards-documents/docs/jesd235a (accessed 6 August 2018).

8 http://www.skhynix.com/eng/product/dramHBM.jsp (accessed 6 August 2018).

9 https://www.altera.com/products/fpga/stratix-series/stratix-10/overview.html (accessed 6 August 2018).

10 Greenhill, D., Ho, R., Lewis, D. et al. (2017). A 14nm 1GHz FPGA with 2.5D transceiver integration. *2017 IEEE International Solid-State Circuits Conference (ISSCC) Solid-State Circuits Conference (ISSCC), 2017 IEEE International* (February 2017), pp. 54–55.

11 Mahajan, R., Sankman, R., Patel, N. et al. (2016). Embedded multi-die interconnect bridge (EMIB) – a high density, high band-width packaging interconnect. *Proceedings of the 66th Electronic Components and Technology Conference*, Las Vegas, Nevada (June 2016), pp. 557–565.

12 Huemoeller, R. and Zwenger, C. (April 2015). Silicon wafer integrated fan-out technology. *Chip Scale Review* 19 (2).

13 Oi, K., Otake, S., Shimizu, N. et al., "Development of new 2.5D package with novel integrated organic interposer substrate with ultra-fine wiring and high density bumps." *Proceedings of the 2014 IEEE 64th Electronic Components and Technology Conference*, Orlando, pp. 348–353.

14 Mahajan, R. and Sane, S. Microelectronic package containing silicon patches for high density interconnects, and method of manufacturing same. US patent no. 8,064,224.

15 Braunisch, H., Aleksov, A., Lotz, S., and Swan, J. (2011). High-speed performance of silicon bridge die-to-die interconnects. *Proceedings of the IEEE Conference on Electrical Performance of Electronic Packaging and Systems (EPEPS)*, San Jose, CA, 23–26 October 2011, pp. 95–98.

16 Robert, S., Mallik, D., Guzek, J. et al. Localized high density substrate routing. US patent no. 9,136,236.

17 Sunohara, M., Tokunaga, T., Kurihara, T., and Higashi, M. (2008). Silicon interposer with TSVs (through silicon vias) and fine multilayer wiring. *Proceedings of the IEEE Electronic Components and Technology Conference (ECTC)*, Lake Buena Vista, FL (27–30 May 2008), pp. 847–852.

18 https://www.amkor.com/go/technology/slim (accessed 6 August 2018).

19 Eitan, A. and Hung, K-Y (2015). Thermo-compression bonding for fine-pitch copper-pillar flip-chip interconnect – tool features as enablers of unique technology. *Proceedings of the IEEE 65th Electronic Components and Technology Conference (ECTC)* (May 2015), pp. 460–464.

20 Proakis, J.G. (1995). *Digital Communications*, 3e, 602–603. Singapore: McGraw-Hill.

21 Casper, B.K., Haycock, M., and Mooney, R. (2002). An accurate and efficient analysis method for multi-Gb/s chip-to-chip signaling schemes. *IEEE Symposium on VLSI Circuits Digest of Technical Papers* (June 2002), pp. 54–57.

24

Interconnection Technology Innovations in 2.5D Integrated Electronic Systems

Paragkumar A. Thadesar, Paul K. Jo, and Muhannad S. Bakir

Electrical and Computer Engineering, Georgia Institute of Technology

24.1 Introduction

The integrated circuit (IC), invented in 1958, has been the key technology fueling the information revolution owing to its constant improvements in productivity and performance [1, 2]. Following Gordon Moore's projection in 1965, the number of transistors per unit area has continually increased with device scaling [3]. This increase in transistor density has been a key factor in reducing gate cost and yielding affordable ICs with increased functionality [4]. Moreover, gate speeds have increased by more than 100×, and the performance of microprocessor ICs has increased by more than 3000× since the introduction of complementary metal–oxide–semiconductor (CMOS) technology [5]. These system performance advancements demand higher bandwidth density off-chip communication with reduced power consumption [6]. However, owing to the slower rate of growth of interconnection and packaging technologies, system performance gains have been abating, thereby creating a critical demand for innovation in silicon ancillary technologies [1]. In addition to attaining high bandwidth density and low power off-chip communication, the integration of different functionalities, for example, digital, analog, MEMS, and sensors, is highly desired to attain turnkey computing, sensing, and communication solutions. This need for the integration of different functionalities has led to the development of heterogeneous platforms using system-on-chip (SoC) architectures [7]; however, monolithic solutions place stringent constraints on the materials and devices that can be integrated.

To address the need for heterogeneous and high density integration of a wide range of chip functions, several 2.5-dimensional (2.5D) platforms have

Advances in Embedded and Fan-Out Wafer-Level Packaging Technologies, First Edition.
Edited by Beth Keser and Steffen Kröhnert.

been explored in the literature including silicon interposer, heterogeneous interconnect stitching technology (HIST), embedded multi-die interconnect bridge (EMIB), silicon-less interconnect technology, and the macrochip [1, 8–15]. In general, 2.5D integration is defined as the assembly of multiple ICs (in a 2D plane) over a substrate containing very fine-pitch interconnects between adjacent dies [16–18], as shown in Figure 24.1a. With respect to silicon interposers, the substrate can be passive or active [19]. Owing to the finer-pitch wiring and shorter chip-to-chip distances compared with conventional packaging, 2.5D integration enables high bandwidth density and low power communication between heterogeneous ICs. Utilizing 2.5D platforms, a wide range of integration capabilities have been demonstrated in the literature: field-programmable gate arrays (FPGAs) with digital-to-analog converter (DAC) integrated on a silicon interposer, a millimeter-wave transceiver with a silicon interposer supporting an RFIC and antennas, and heterogeneous integration of a GPS RF receiver chip, a baseband ARM chip, and a DRAM chip on a silicon substrate have been shown in the literature [9, 20, 21]. Vertical interconnections, called through-silicon vias (TSVs), play a key role in enabling silicon interposer-based 2.5D integration (and 3D chip stacking). TSVs consist of metal conductors (commonly copper) insulated from silicon commonly using a thin dielectric liner. Alternative interconnection technologies for stacking include wire bonding, which is electrically inferior and has limited interconnection density compared with TSVs, and proximity communication such as inductive coupling [22–24]. However, there are electrical and thermomechanical challenges with TSVs. The electrical performance of TSVs is significantly affected by the lossy silicon substrate, and the coefficient of thermal expansion mismatch between the copper and silicon can result in thermomechanical reliability failures [25, 26]. As noted previously, there are several 2.5D integration solutions that are based on chip bridging without the need for interposers. For example, die-to-die interconnection using a silicon bridge within an organic package has been shown in the literature [27]. The macrochip system features electrical and optical interconnect bridging between neighboring ICs [15]. Moreover, HIST 2.5D integration forms concatenated ICs using heterogeneous IP blocks (from multiple foundries) enabled by a combination of 2.5D integration and face-to-face 3D integration technologies, as shown in Figure 24.1b. To this end, HIST utilizes a combination of solder bumps and mechanically elastic fine-pitch compressible microinterconnects (CMIs) as chip I/Os to robustly assemble the ICs [10, 11].

This chapter is divided into two parts: firstly, we discuss silicon interposer TSV technologies with improved thermomechanical and high frequency electrical properties, and secondly, we describe the HIST approach in more detail and share key features.

(a)

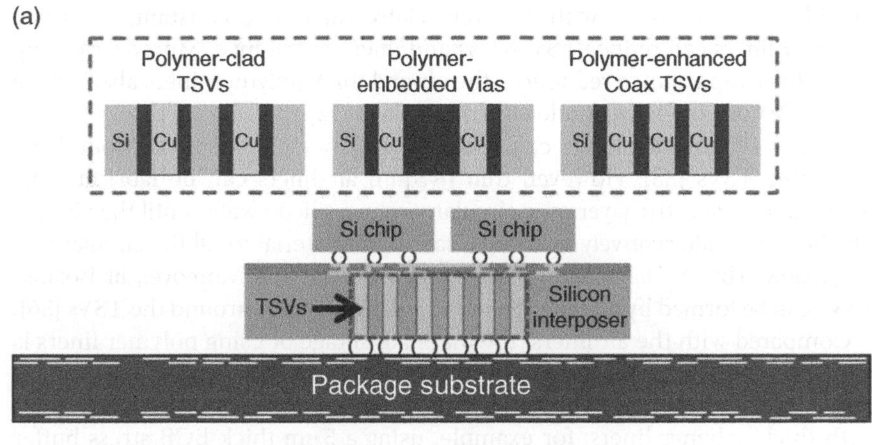

(b)

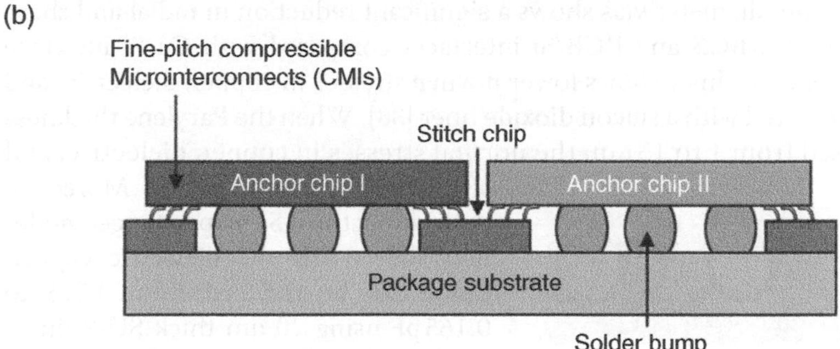

Figure 24.1 Envisioned systems featuring (a) polymer-enhanced TSVs [28] and (b) HIST platform [11, 29].

24.2 Polymer-Enhanced TSVs

To reduce TSV stresses and electrical losses, this section describes polymer-enhanced TSVs. The polymer-enhanced TSVs in this section include polymer-clad TSVs (which feature thick polymer liners) and polymer-embedded vias (which feature an array of copper vias formed within a polymer well) [30, 31]. Figure 24.1a shows the structures of the polymer-enhanced TSVs in comparison with conventional TSVs.

24.2.1 Polymer-Clad SVs

To reduce TSV stress and capacitance, various liner techniques have been explored in the literature including air and polymers. Compared with a silicon dioxide liner, air liners with a lower relative dielectric constant and thick polymer liners can reduce TSV stress and liner capacitance. Moreover, owing to the liner capacitance reduction, the air and thick polymer liners also reduce the impact of slow-wave mode electrical losses [32].

Thick (~30 μm) air liners can be fabricated by etching silicon around the fabricated TSVs [33]. However, thin (~3 μm) air liners can be fabricated by depositing a dielectric layer over circular vias in a silicon wafer until the vias get pinched off or alternatively by using a sacrificial material to fill the circular vias [34] followed by TSV fabrication at the center of the vias [35]. Moreover, air-isolated TSVs can be formed by etching silicon, to a desired depth, around the TSVs [36].

Compared with the air liners, the main advantage of using polymer liners is that the fabrication of horizontal interconnects is easier. With respect to TSV stress, various modeling efforts have shown the reduction of stress of TSVs with thick polymer liners; for example, using a 5 μm thick BCB stress buffer layer for 30 μm diameter vias shows a significant reduction in radial and shear stress along Cu/BCB and BCB/Si interfaces compared with Cu/Si interface [37]. Parylene as a liner shows lower normal stresses in copper, dielectric, and silicon compared with a silicon dioxide liner [38]. When the Parylene thickness is increased from 1 to 15 μm, the normal stresses in copper, dielectric, and silicon are reduced by half. Moreover, with respect to TSV capacitance, modeling shows that the dielectric capacitance can be reduced from 3.515 to 0.165 pF using 20 μm thick SU-8 liners compared with a 1 μm thick silicon dioxide liner for 400 μm tall and 80 μm diameter copper vias [39].

The fabrication of polymer liners has been described in the literature using polymer vapor deposition [40], polymer filling in circular vias within silicon [41], photodefinition of polymer-filled vias with a temporary release film for coaxial TSVs [42], laser ablation of polymer-filled vias [43], and photodefined polymer with a perforated dielectric layer, called mesh layer, at the base [30]. By comparison, the polymer liner formation process using photodefined polymer is simpler. Figure 24.2 shows 390 μm tall copper TSVs with ~80 μm diameter and surrounded by a ~20 μm thick photodefined polymer cladding on a 250 μm pitch.

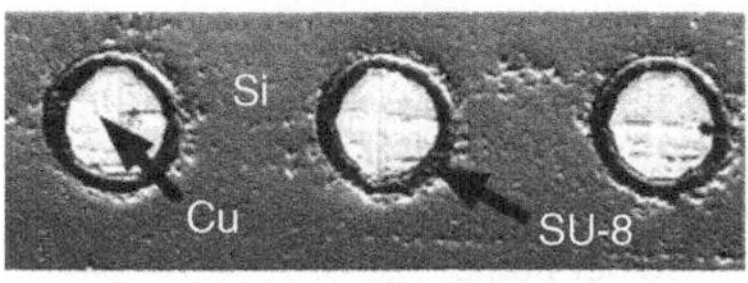

Top view

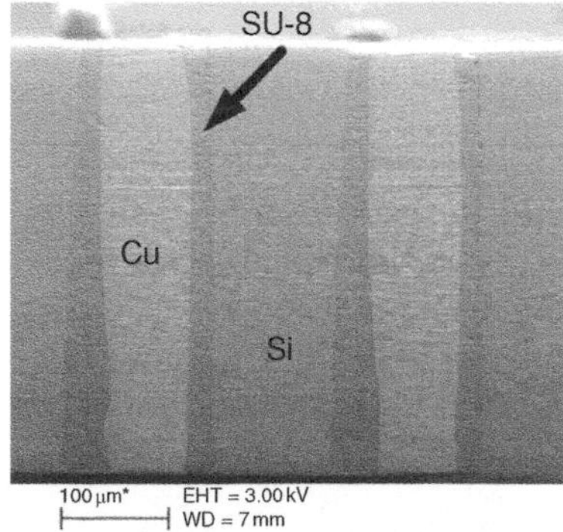

Cross-sectional view

Figure 24.2 Polymer-clad TSVs with a thick SU-8 liner [30]. *Source:* Reproduced with permission of IEEE.

To benchmark the mechanical stresses of polymer-clad TSVs relative to conventional TSVs, various characterization options were explored. Micro-Raman spectroscopy that works on the principle of measuring the frequency shift of an impinging laser to quantify localized near-surface silicon stress was explored [44, 45]. However, stresses in copper cannot be measured using micro-Raman spectroscopy. Moreover, the bending beam technique, which works on the principle of measuring the curvature of the sample to quantify the stress in silicon and copper, was explored, but the measured stresses obtained from using the bending beam technique are averaged across the sample [46]. Additionally, indentation techniques, which work on the principle of analyzing the residual-stress-induced normal load to measure localized stress in silicon and copper, were explored [47]. However, it is difficult to measure residual stress in the absence of a known stress-free state using the indentation techniques. Lastly, synchrotron X-ray diffraction (XRD), which can measure all the stress components in a copper via and the surrounding silicon with minimal destruction to the sample, was utilized for the benchmarking.

XRD is a commonly used technique to study structural properties of materials [48]. With the development of high-brilliance synchrotron sources, advanced X-ray focusing optics, and diffraction pattern analysis, diffraction patterns can be obtained by using sample scans under a submicrometer polychromatic (white X-ray beam) or monochromatic beam. This can provide crystal orientations and strain distribution maps (both deviatoric and hydrostatic) in localized areas with the ability to distinguish the maps for specific materials. Consequently, for structures like TSVs, synchrotron XRD can help obtain separate strain distributions for silicon and copper. Due to this capability, minimum destruction is needed to the TSV samples for synchrotron measurements. Using synchrotron XRD, deviatoric strains were obtained for TSVs and analyzed for silicon to compare the strains from TSVs with silicon dioxide liner and polymer-clad TSVs [49, 50]. The obtained synchrotron XRD strain measurements are 2D, whereas the TSV strains are 3D. A beam intensity-based data averaging method, relating the synchrotron XRD measurements and finite element modeling (FEM), was implemented to interpret the measured TSV strains. Moreover, to compare the strains in the TSVs with silicon dioxide liner and polymer-clad TSVs, an indirect method of FEM calibration using synchrotron measurements was utilized. Since the dominating TSV thermomechanical failure modes are composed of silicon cracking and copper/liner separation, the first principal strain in the silicon and the copper/liner interfacial shear strain are compared. As shown in Figure 24.3a, the thick SU-8 liner serves as a cushion layer and reduces the thermomechanical force applied to the surrounding silicon as the copper via expands at a high temperature. Moreover, the SU-8 liner mitigates the interfacial shear strain and thus reduces the possibilities of

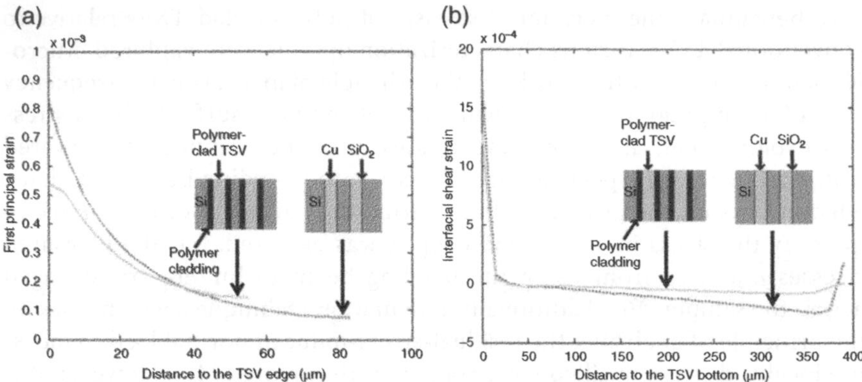

Figure 24.3 At 150° C, (a) first principal strain of silicon at the top of TSVs along the path stretching radially away from the liner–silicon interface and (b) interfacial shear strain along copper/liner interface [49, 50].

interfacial separation, as shown in Figure 24.3b. The higher interfacial shear strain near the bottom of the TSV can be attributed to the presence of a copper layer.

24.2.2 Polymer-Embedded Vias

To achieve further electrical performance enhancement compared with polymer-clad TSVs, polymer-embedded vias have been explored in silicon interposers as described next.

High-resistivity silicon interposers can reduce TSV losses. However, high-resistivity silicon is economically less attractive [51, 52], and thus, there has been an effort to explore glass interposers [53]. However, glass requires serial ablation to form vias and is a poor thermal conductor compared to silicon. In comparison, polymer-embedded vias consist of copper vias embedded in photodefined polymer wells within the commonly implemented 10 Ω cm resistivity silicon, thereby providing a wafer-scale batch fabrication solution for attaining vias in low loss regions within mainstream silicon wafers [54]. Figure 24.4 shows 65 μm diameter and 370 μm tall polymer-embedded vias on a 150 μm pitch within a 1800 μm × 1800 μm well in silicon. To characterize the RF performance of the fabricated polymer-embedded vias, TSV-trace-TSV structures were fabricated, measured, and de-embedded using L-2L and open-short de-embedding techniques [31]. As shown in Figure 24.5, the via losses attained from both de-embedding techniques are in good agreement up to 30 GHz with minor differences between the de-embedded and the stand-alone loss due to fabrication variations. The de-embedding results from the measurements yield 0.22 dB insertion loss per polymer-embedded via at 30 GHz. Compared to the

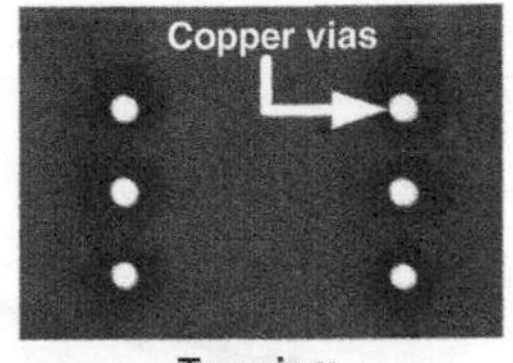

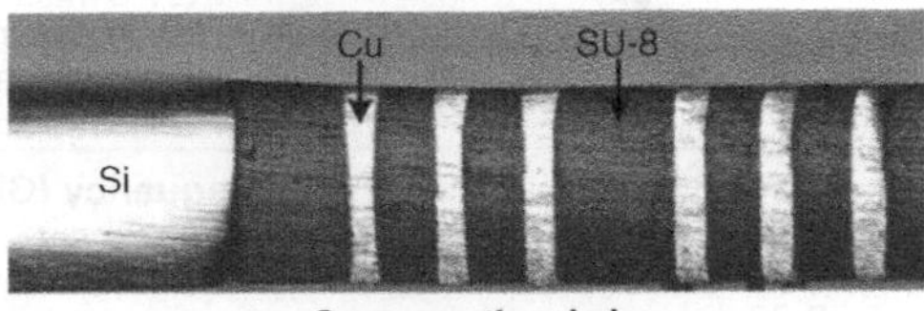

Figure 24.4 Polymer-embedded vias with copper vias in SU-8 wells within silicon [54]. *Source:* Reproduced with permission of IEEE.

simulated insertion loss of a stand-alone TSV with silicon dioxide liner, an 87% reduction in insertion loss can be obtained using the polymer-embedded vias at 30 GHz.

Similar to polymer-embedded vias, copper vias in dielectric regions within silicon have been demonstrated in the literature using (i) a metal coating over silicon pillars in polymer wells [55] and (ii) glass reflow over etched areas in silicon followed by silicon pillar etching [56]. Compared with these processes, photodefined polymer-embedded vias provide low loss TSVs with a simpler fabrication process.A comparison of polymer-embedded vias and polymer-clad TSVs with different TSV technologies in the literature is shown in Table 24.1.

24.2.3 Coaxial TSVs

A coaxial interconnect configuration of polymer-embedded vias can be formed by placing ground vias around a signal via [31]. The technique of forming coaxial vias using ground vias is easier to implement compared with other coaxial via techniques in the literature [42, 59, 60] and consequently is the focus of this section.

The coaxial configuration can reduce TSV loss and coupling in addition to providing an impedance matched interconnection. To perform one-port measurements of coaxial vias and extract their impedances, Figure 24.6a illustrates the fabricated 285 μm tall polymer-enhanced coaxial vias within a 1800 μm × 1800 μm well in silicon prior to top layer metallization. The copper via diameter is 65 μm and the signal-to-ground via pitches are 150 and 125 μm. The coaxial vias with signal-to-ground via pitch of 125 μm also show that polymer-enhanced vias (the ground vias) with a distance of 30 μm (i.e. 95 μm pitch) between the vias have been demonstrated.

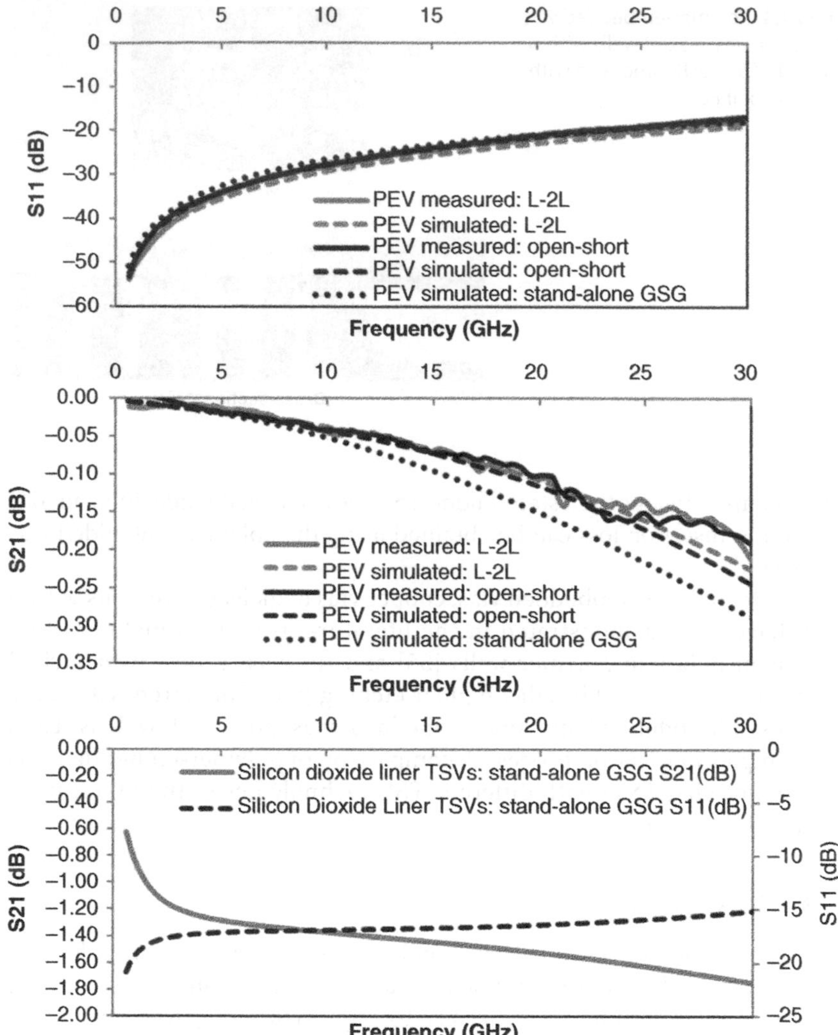

Figure 24.5 De-embedded polymer-embedded via (PEV) loss using measurements and simulations. TSVs with silicon dioxide liner included for benchmarking [31]. *Source:* Reproduced with permission of IEEE.

High frequency measurements were performed from 1 to 50 GHz for the fabricated coaxial vias. Using the measured *S*-parameters, *Z*-parameters and *Y*-parameters are obtained with 50 Ω as the reference impedance. Using the *Z*-parameters of the short structure, resistance (*R*) and inductance (*L*) are extracted, and using the *Y*-parameters of the open structure, capacitance (*C*)

Table 24.1 Comparison of polymer-embedded vias and polymer-clad TSVs with other TSV technologies.

No	Parameters	Polymer-embedded vias [57]	Polymer-clad TSVs [30]	SiO_2 liner TSVs	Air liner TSVs [34]	Glass vias [53]
1	Copper via diameter	100/65 μm	80 μm	~10 μm	20 μm	15 μm at top
2	TSV height	270/370 μm	390 μm	100 μm	65 μm	30 μm
3	TSV pitch	250/150 μm	250 μm	40 μm min.	50 μm	27 μm
4	Loss at high frequency	~0.2 dB at 30 GHz (thick interposer)	Low	~1.2 dB at 29 GHz (for a chain with 2 TSVs) [58]	Low	~0.1 dB at 20 GHz (thin glass for a chain with 2 TSVs)
5	Ease of fabrication	High	High	Very high	Low	Moderate
6	Special features	Photodefinition	Photodefinition	Simpler fabrication	Metallization over air liners	Panel-scale fabrication

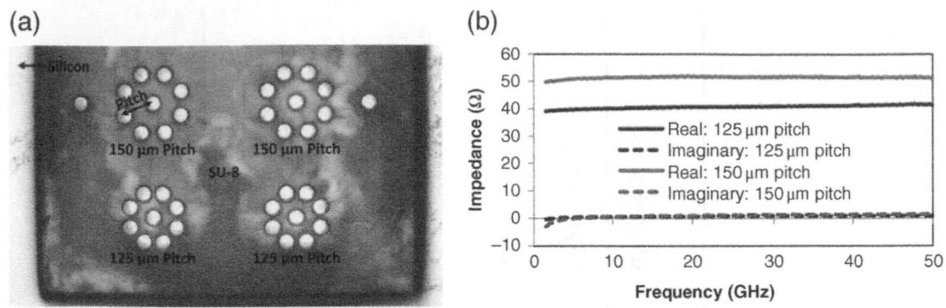

Figure 24.6 (a) Polymer-embedded coaxial TSVs with ground shield vias and (b) extracted impedance from the coaxial TSV measurements [28]. *Source:* Reproduced with permission of IEEE.

and conductance (G) are extracted [61]. Using the extracted RLGC, the impedance is evaluated.

The extracted impedances from the one-port coaxial via measurements are shown in Figure 24.6b, demonstrating a wideband impedance matching to approximately 50 Ω using the 150 μm pitch vias and approximately 40 Ω using the 125 μm pitch vias. Moreover, a two-port simulation of the coaxial TSVs in HFSS (yielding 0.1 dB insertion loss per coaxial via at 50 GHz) demonstrates a 55% reduction in insertion loss at 50 GHz compared with a ground-signal-ground via configuration with the same copper via dimensions and signal-to-ground via pitch.

In addition to the one-port measurements, coupling measurements are also demonstrated. For the coupling measurements, coaxial and non-coaxial configurations are fabricated. As shown in Figure 24.7a, 285 μm tall and 65 μm diameter polymer-enhanced coaxial TSVs are fabricated within 1800 μm × 1800 μm wells in silicon; the signal-to-ground via pitches are 150 and 175 μm.

In the measured frequency band of 10 MHz to 50 GHz, the coaxial configuration attains an average of 14.5 and 13.1 dB reduction in signal-to-signal via coupling compared with the corresponding non-coaxial structures at 150 and 175 μm signal-to-ground via pitches, respectively, as shown in Figure 24.7b.

To better compare coaxial vias in the literature, a comparison is shown in Table 24.2.

24.3 HIST

In this section, we propose an alternate TSV-less heterogeneous integration technology, as illustrated in Figure 24.1b. Our approach seeks to form concatenated ICs using heterogeneous IP blocks (from multiple foundries) enabled by a combination of 2.5D integration and face-to-face 3D integration technologies.

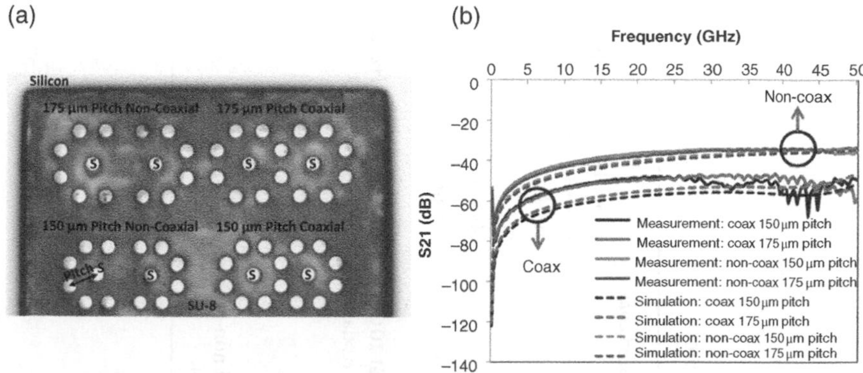

Figure 24.7 (a) Fabricated polymer-enhanced coaxial vias and non-coaxial vias (for benchmarking) and (b) measurements and simulations of coupling [28].

To this end, we utilize a combination of solder bumps and mechanically elastic and fine-pitch CMIs [11] to robustly assemble the ICs. As shown in Figure 24.1b, the fine-pitch (as small as 20 µm) CMIs, which are pressure contact-based interconnects, are used to provide the dense signaling pathways between the concatenated anchor ICs through the stitch chips. The solder bumps with large pitch and height are used for power delivery, signal routing between die and package, and mechanical interconnection between the anchor ICs and the package. The solder bumps also create the necessary force for the CMIs to form pressure-based reliable contacts. The stitch chips, which may contain high quality passives and/or active circuits, in the simplest form contain only dense interconnects to interconnect nearby active anchor ICs. The anchor IC may be a logic, FPGA, MMIC, or a photonic chip, for example. Compared with other 2.5D solutions, the advantages of HIST include the following: HIST achieves a similar signal bandwidth density as the silicon interposer technology, but is not reticle-size limited, thus making it very scalable in size; HIST eliminates the need for TSVs in the substrate for decreased cost and improved signaling; HIST is based on die-to-die face-to-face bonding, and thus there are no intermediate package levels, which enables higher signal I/O pitch and lower capacitance; and lastly, HIST can be applied to any packaging substrate (organic, ceramic, etc.) since HIST is augmented to the topmost surface of the package substrate [10, 11, 29].

SEM images of CMIs are shown in Figure 24.8a. Assembled anchor chips over a substrate with stitch chips are shown in Figure 24.8b [29].

The four-point resistances of the solder and CMI interconnections after assembly were measured using a Karl Suss probe station. Chips containing fine-pitch CMIs were flip-chip-bonded on to the substrate containing regions that emulate stitch chips and solder bumps, as shown in Figure 24.9. The resistances of the CMIs in contact with the gold traces on the stitch chips were measured.

Table 24.2 Comparison of the demonstrated coaxial vias to other coaxial TSV technologies from the literature.

No	Parameters	Photodefined coax TSVs with ground vias [57]	SiO_2 liner TSVs	Laser ablated annular coax [59]	Laser ablated coax [60]	Photodefined coax [42]
1	Copper via diameter	65 μm	~10 μm	~42 μm	70 μm	100 μm
2	TSV height	285 μm	100 μm	205 μm	150 μm	300 μm
3	TSV pitch	150/125/95 μm	40 μm min.	450 μm	Surrounded by non-coax	500 μm
4	Loss at high frequency	~0.1 dB at 50 GHz (for one coaxial TSV)	~1.2 dB at 29 GHz (for a chain with 2 TSVs) [58]	~5.5 dB at 20 GHz (for a chain with 4 TSVs)	0.044 dB at 10 GHz (for one coaxial TSV)	~0.25 dB at 10 GHz (for a TSV-trace link)
5	Ease of fabrication	High	Very high	Moderate	Moderate	Moderate
6	Special features	Photodefinition	Simpler fabrication	Laminated ABF with laser ablation	Coax and non-coax in parallel	Photodefinition

(a)

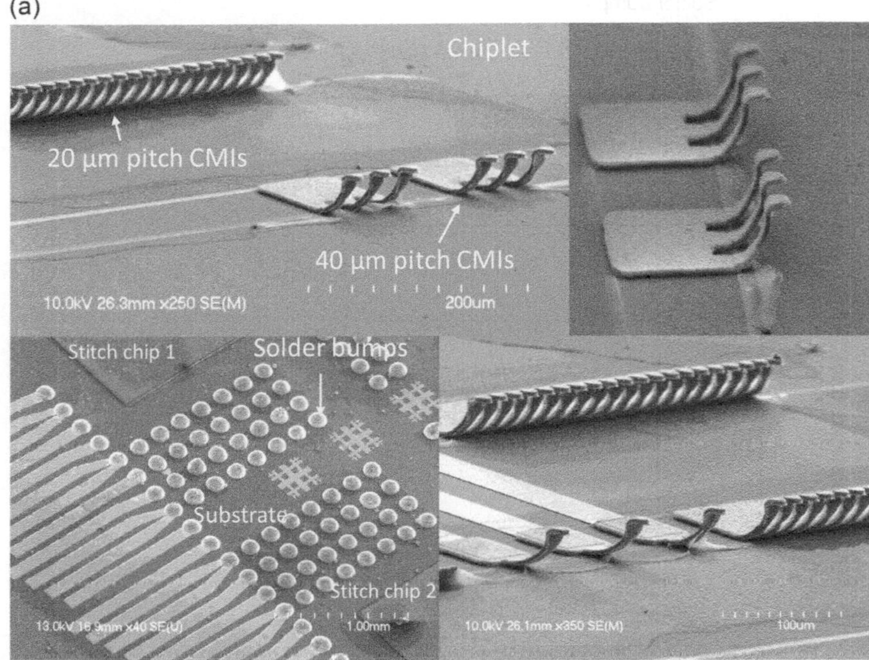

(b)

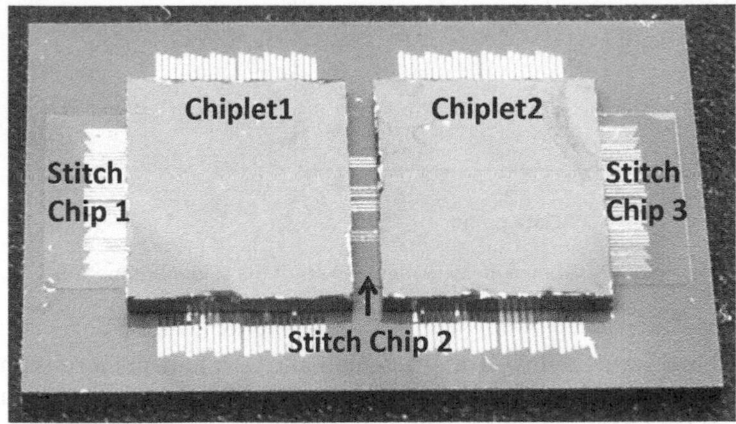

Figure 24.8 (a) Fabricated CMIs and (b) two assembled anchor chips over a substrate with three stitch chips.

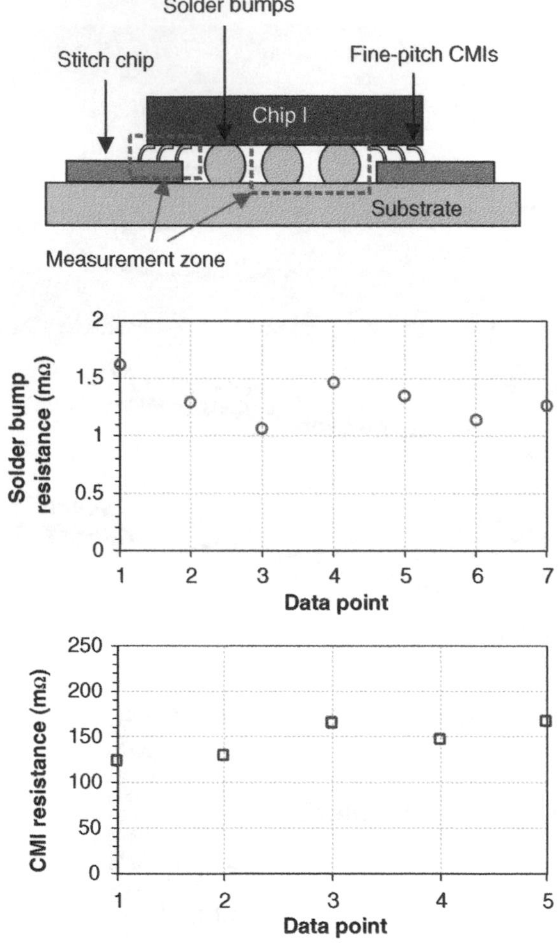

Figure 24.9 Four-point resistance measurement results of the solder bumps and fine-pitch CMIs.

The solder bumps were reflowed after assembly and their four-point resistance measurements were also measured. As shown in Figure 24.9, the average resistance of the CMIs, including their contact resistance with the gold traces on the substrate, is 146.31 mΩ. The average resistance of the solder bumps is 1.31 mΩ. These four-point resistance measurement results confirm that the CMIs maintain electrical connections between the die and the stitch chips after they are compressed downward and form a pressure-based contact with the substrate. As noted earlier, owing to their mechanical flexibility, CMIs can compensate for any stitch-chip thickness variation or surface nonplanarity of the substrate. Moreover, as I/O pitch is reduced, CMIs offer greater tolerances to surface planarity and height variations relative to scaled solder bumps and, thus, prove a unique interconnect solution for both robust assembly at very fine pitches and high performance.

24.4 Conclusion

Novel polymer-enhanced TSV technologies have been demonstrated. Firstly, polymer-clad TSVs with thicker dielectric liners have been shown to reduce TSV stress and capacitance. Secondly, to significantly lower electrical losses, polymer-embedded vias with copper vias in polymer wells within silicon have been shown. Moreover, the alternate TSV-less HIST approach is demonstrated using CMIs and solder bumps.

References

1 Bakir, M.S. and Meindl, J.D. (2009). *Integrated Interconnect Technologies for 3D Nanoelectronic Systems*. Artech House.

2 Kim, K. (2015). Silicon technologies and solutions for the data-driven world. *Proceedings of the IEEE International Solid-State Circuits Conference (ISSCC)*, (February 2015), pp. 8–14.

3 Moore, G. (April 1965). Cramming more components onto integrated circuits. *Electronics* 38 (8).

4 Meindl, J., Naeemi, A., Bakir, M., and Murali, R. (2010). Nanoelectronics in retrospect, prospect and principle. *Proceedings of the IEEE International Solid-State Circuits Conference (ISSCC)* (February 2010), pp. 31–35.

5 Horowitz, M. (2014). Computing's energy problem (and what we can do about it). *Proceedings of the IEEE International Solid-State Circuits Conference (ISSCC)* (February 2014), pp. 10–14.

6 Bowman, K., Duvall, S., and Meindl, J. (December 2009). Impact of die-to-die and within-die parameter variations on the clock frequency and throughput of multi-core processors. *IEEE Transactions on Very Large Scale Integration (VLSI) Systems* 17 (12): 1679–1690.

7 International Technology Roadmap for Semiconductors (ITRS) (2011). *Radio-frequency and Analog/Mixed Signal Technologies for Communications*. ITRS.

8 Knickerbocker, J., Andry, P.S., Buchwalter, L.P. et al. (2006). System-on-package (SOP) technology, characterization and applications. *Proceedings of the 56th IEEE Electronic Components and Technology Conference (ECTC)* (May 2006), pp. 415–421.

9 Erdmann, C., Lowney, D., Lynam, A. et al. (January 2015). A heterogeneous 3D-IC consisting of two 28 nm FPGA die and 32 reconfigurable high-performance data converters. *IEEE Journal of Solid-State Circuits* 50 (1): 258–269.

10 Zhang, X., Jo, P.K., Zia, M. et al. (February 2017). Heterogeneous interconnect stitching technology with compressible microinterconnects for dense multi-die integration. *IEEE Electron Device Letters* 38 (2): 255–257.

11 Jo, P., Zia, M., Gonzalez, J.L., and Bakir, M.S. (2017). Design, fabrication, and characterization of dense compressible microinterconnects. *IEEE Transactions on Components, Packaging and Manufacturing Technology* 7 (7): 1003–1010.

12 Mahajan, R., Sankman, R., Patel, N. et al. (2016). Embedded multi-die interconnect bridge (EMIB) – a high density, high bandwidth packaging interconnect. *Proceedings of the 66th IEEE Electronic Components and Technology Conference (ECTC)* (May 2016), pp. 557–565.

13 Greenhill, D., Ho, R., Lewis, D. et al. (2017). A 14nm 1 GHz FPGA with 2.5D transceiver integration. *Proceedings of the IEEE International Solid-State Circuits Conference (ISSCC)* (February 2017), pp. 54–56.

14 Kwon, W., Ramalingam, S., Wu, X. et al. (2014). Cost effective and high performance 28nm FPGA with new disruptive Silicon-Less Interconnect Technology (SLIT). *Proceedings of the International Symposium on Microelectronics* (2014), pp. 599–605.

15 Cunningham, J., Ho, R., Mitchell, J.G. et al. (2011). Integration and packaging of a macrochip with silicon nanophotonic links. *IEEE Journal of Selected Topics in Quantum Electronics* 17 (3): 546–558.

16 Banijamali, B., Ramalingam, S., Liu, H. et al. (2012). Outstanding and innovative reliability study of 3D TSV interposer and fine pitch solder micro-bumps. *Proceedings of the 62nd IEEE Electronic Components and Technology Conference (ECTC)* (June 2012), pp. 309–314.

17 Sawyer, B., Suzuki, Y., Lu, H., and Tummala, R. (2014). Modeling, design, fabrication and characterization of first large 2.5D glass interposer as a superior alternative to silicon and organic interposers at 50 micron bump pitch. *Proceedings of the 64th IEEE Electronic Components and Technology Conference (ECTC)* (May 2014), pp. 742–747.

18 Oi, K., Otake, S., Shimizu, N., and Kutlu, Z.. Development of new 2.5D package with novel integrated organic interposer substrate with ultra-fine wiring and high density bumps. *Proceedings of the 64th IEEE Electronic Components and Technology Conference (ECTC)* (May 2014), pp. 348–353.

19 Hellings, G., Scholz, M., Detalle, M. et al. (2015). Active-lite interposer for 2.5 and 3D integration. *Proceedings of the IEEE Symposium on VLSI Technology (VLSI-Technology): Digest of Technical Papers* (June 2015).

20 El Bouayadi, O., Lamy, Y., Dussopt, L., and P. Vincent (2015). Silicon interposer: A versatile platform towards full-3d integration of wireless systems at millimeter-wave frequencies. *Proceedings of the 65th IEEE Electronic Components and Technology Conference (ECTC)* (May 2015), pp. 973–980.

21 Liao, W., Yen, K.K., Chen, H.N., and Yu, D. (2013). 3d IC heterogeneous integration of GPS RF receiver, baseband, and DRAM on CoWoS with system

BIST solution. *Proceedings of the IEEE Symposium on VLSI Technology: Digest of Technical Papers* (June 2013), pp. C18–C19.

22 Carson, F., Lee, H.T., Yee, J.H. et al. (2011). Die to die copper wire bonding enabling low cost 3d packaging. *Proceedings of the 61st IEEE Electronic Components and Technology Conference (ECTC)* (May 2011), pp. 1502–1507.

23 Ishikuro, H. and Kuroda, T. (October 2010). Wireless proximity interfaces with a pulse-based inductive coupling technique. *IEEE Communications Magazine* 48 (10): 192–199.

24 Take, Y., Matsutani, H., Sasaki, D. et al. (March 2014). 3d NoC with inductive-coupling links for building-block SiPs. *IEEE Transactions on Computers* 63 (3): 748–763.

25 Ndip, I., Lobbicke, K., Zoschke, K., and Henke, H. (March 2014). Analytical, numerical-, and measurement-based methods for extracting the electrical parameters of through silicon vias (TSVs). *IEEE Transactions on Components, Packaging and Manufacturing Technology* 4 (3): 504–515.

26 Liu, X., Sundaram, V., Chen, Q., and Tian, Y. (January 2013). Failure analysis of through-silicon vias in free-standing wafer under thermal-shock test. *Microelectronics Reliability* 53 (1): 70–78.

27 Braunisch, H., Aleksov, A., Lotz, S., and Swan, J. (2011). High-speed performance of silicon bridge die-to-die interconnects. *Proceedings of the IEEE 20th Conference on Electrical Performance of Electronic Packaging and Systems (EPEPS)* (October 2011), pp. 95–98.

28 Thadesar, P. (April 2015). Interposer platforms featuring polymer-enhanced through silicon vias for microelectronic systems. Doctoral thesis, Georgia Institute of Technology, Atlanta.

29 Jo, P.K., Zia, M., Gonzalez, J.L., and Bakir, M.S. (2017). Dense and highly elastic compressible microinterconnects (CMIs) for electronic microsystems. *Proceedings of the 67th IEEE Electronic Components and Technology Conference (ECTC)* (May 2017), pp. 684–689.

30 Thadesar, P. and Bakir, M. (July 2013). Novel photodefined polymer-enhanced through-silicon vias for silicon interposers. *IEEE Transactions on Components, Packaging and Manufacturing Technology* 3 (7): 1130–1137.

31 Thadesar, P. and Bakir, M. (March 2016). Fabrication and characterization of polymer-enhanced tsvs, inductors and antennas for mixed-signal silicon interposer platforms. *IEEE Transactions on Components, Packaging and Manufacturing Technology* 6 (3): 455–463.

32 Chang, Y.-J., Zheng, T.Y., Chuang, H.Y. et al. (2012). Low slow-wave effect and crosstalk for low-cost ABF-coated TSVs in 3-d IC interposer. *Proceedings of the 62nd IEEE Electronic Components and Technology Conference (ECTC)* (June 2012), pp. 1934–1938.

33 Sunohara, M., Sakaguchi, H., and Takano, A. (2010). Studies on electrical performance and thermal stress of a silicon interposer with TSVs.

Proceedings of the 60th IEEE Electronic Components and Technology Conference (ECTC) (June 2010), pp. 1088–1093.

34 Chen, Q., Huang, C., Wu, D., and Tan, Z. (April 2013). Ultralow-capacitance through-silicon vias with annular air-gap insulation layers. *IEEE Transactions on Electron Devices* 60 (4): 1421–1426.

35 Civale, Y., Gonzalez, M., Beyne, E. et al. (2010). A novel concept for ultra-low capacitance via-last TSV. *Proceedings of the IEEE International Conference on 3D System Integration (3DIC)* (November 2010), pp. 1–4.

36 Oh, H., Mary, G.S., and Bakir, M.S. (2015). Silicon interposer platform with low-loss through-silicon vias using air. *Proceedings of the IEEE International 3D Systems Integration Conference (3DIC)* (September 2015).

37 Ryu, S.-K., Huang, R., Ho, P. et al. (March 2011). Impact of near-surface thermal stresses on interfacial reliability of through-silicon vias for 3-D interconnects. *IEEE Transactions on Device and Materials Reliability* 11 (1): 35–43.

38 Chen, Z., Song, X., and Liu, S. (2009). Thermo-mechanical characterization of copper filled and polymer filled TSVs considering nonlinear material behaviors. *Proceedings of the 59th IEEE Electronic Components and Technology Conference (ECTC)* (May 2009), pp. 1374–1380.

39 Thadesar, P. and Bakir, M. (2012). Silicon interposer featuring novel electrical and optical TSVs. *Proceedings of the ASME International Mechanical Engineering Congress and Exposition (IMECE)* (November 2012).

40 Tung, B. T., Cheng, X., Watanabe N. et al. (2014). Fabrication and electrical characterization of Parylene-HT liner bottom-up copper filled through silicon via (TSV). *2014 IEEE CPMT Symposium Japan (ICSJ)* (November 2014), pp. 154–157.

41 Chen, Q., Huang, C., Tan, Z. et al. (May 2013). Low capacitance through-silicon-vias with uniform benzocyclobutene insulation layers. *IEEE Transactions on Components, Packaging and Manufacturing Technology* 3 (5).

42 Ho, S. W., Lau, J., Zhou, Q. et al. (2008). High RF performance TSV silicon carrier for high frequency application. *Proceedings of the 58th IEEE Electronic Components and Technology Conference (ECTC)* (May 2008), pp. 1946–1952.

43 Chen, Q., Suzuki, Y., Kumar, G., and Tummala, R.R. (December 2014). Modeling, fabrication, and characterization of low-cost and high-performance polycrystalline panel-based silicon interposer with through vias and redistribution layers. *IEEE Transactions on Components, Packaging and Manufacturing Technology* 4 (12): 2035–2041.

44 Ryu, S.-K., Zhao, Q., Hecker, M. et al. (March 2012). Micro-raman spectroscopy and analysis of near-surface stresses in silicon around through-silicon vias for three-dimensional interconnects. *Journal of Applied Physics* 111 (6): 063 513–063 513–8.

45 De Wolf, I., Simons, V., Cherman, V., and Beyne, E. (2012). In-depth Raman spectroscopy analysis of various parameters affecting the mechanical stress near the surface and bulk of cu-TSVs. *Proceedings of the 62nd IEEE Electronic Components and Technology Conference (ECTC)* (May 2012), pp. 331–337.

46 Ryu, S.-K., Jiang, T., Lu, K.H. et al. (January 2012). Characterization of thermal stresses in through-silicon vias for three-dimensional interconnects by bending beam technique. *Applied Physics Letters* 100 (4): 041 901–041 901–4.

47 Lee, G., Choi, M.J., Jeon, S.W. et al. (2012). Microstructure and stress characterization around TSV using in-situ PIT-in-SEM. *Proceedings of the 62nd IEEE Electronic Components and Technology Conference (ECTC)* (May 2012), pp. 781–786.

48 Tamura, N., MacDowell, A.A., Spolenak, R. et al. (March 2003). Scanning x-ray microdiffraction with submicrometer white beam for strain/stress and orientation mapping in thin films. *Journal of Synchrotron Radiation* 10 (2): –137143.

49 Liu, X., Thadesar, P.A., Taylor, C.L. et al. (July 2013). Thermomechanical strain measurements by synchrotron x-ray diffraction and data interpretation for through-silicon vias. *Applied Physics Letters* 103 (2): 022 107–022 107–5.

50 Liu, X., Thadesar, P.A., Taylor, C.L. et al. (August 2013). Dimension and liner dependent thermomechanical strain characterization of through-silicon vias using synchrotron x-ray diffraction. *Journal of Applied Physics* 114 (6): 064 908–064 908–7.

51 Ndip, I., Curran, B., Lobbicke, K. et al. (October 2011). High-frequency modeling of TSVs for 3-d chip integration and silicon interposers considering skin-effect, dielectric quasi-TEM and slow-wave modes. *IEEE Transactions on Components, Packaging and Manufacturing Technology* 1 (10): 1627–1641.

52 Xie, B., Swaminathan, M., Han, K.J. and Xie, J. (2011). Coupling analysis of through-silicon via (TSV) arrays in silicon interposers for 3D systems. *Proceedings of the IEEE International Symposium on Electromagnetic Compatibility (EMC)* (August 2011), pp. 16–21.

53 Sukumaran, V., Bandyopadhyay, T., Chen, Q. et al. (May 2014). Design, fabrication, and characterization of ultrathin 3-D glass interposers with through-package-vias at same pitch as TSVs in silicon. *IEEE Transactions on Components, Packaging and Manufacturing Technology* 4 (5): 786–795.

54 Thadesar, P.A., Zheng, L., Bakir, M.S. (2014). Low-loss silicon interposer for three-dimensional system integration with embedded microfluidic cooling. *Proceedings of the IEEE Symposium on VLSI Technology: Digest of Technical Papers* (June 2014), pp. 1–2.

55 Lim,T.-G., Khoo, Y.M., Selvanayagam, C.S. et al. (2011). Through silicon via interposer for millimetre wave applications. *Proceedings of the 61st IEEE Electronic Components and Technology Conference (ECTC)* (June 2011), pp. 577–582.

56 Lee, J.-Y., Lee, S.W., Lee, S.K., and Park, J.H. (August 2013). Through-glass copper via using the glass reflow and seedless electroplating processes for wafer-level RF MEMS packaging. *Journal of Micromechanics and Microengineering* 23 (8): 085012.

57 Thadesar, P. and Bakir, M. (2015). Fabrication and characterization of mixed-signal polymer-enhanced silicon interposer featuring photodefined coax TSVs and high-Q inductors. *Proceedings of the 65th IEEE Electronic Components and Technology Conference (ECTC)* (May 2015), pp. 281–286.

58 Kim, N., Wu, D., Kim, D., and Wu, P. (2011). Interposer design optimization for high frequency signal transmission in passive and active interposer using through silicon via (TSV). *Proceedings of the 61st IEEE Electronic Components and Technology Conference (ECTC)* (June 2011), pp. 1160–1167.

59 Jung, D., Kim, H., Kim, S., and Kim, J. (November 2014). 30 Gbps high-speed characterization and channel performance of coaxial through silicon via. *IEEE Microwave and Wireless Components Letters* 24 (11): 814–816.

60 Yook, J.-M., Kim, D., Park, J.-C., Kim, C.-Y., Yi, S.H., and Kim, J.C. (2014). Low-loss and high-isolation through silicon via technology for high performance RF applications. *Proceedings of the European Microwave Conference (EuMC)* (October 2014), pp. 996–999.

61 Xu, Z. and Lu, J.-Q. (February 2013). Through-silicon-via fabrication technologies, passives extraction, and electrical modeling for 3-D integration/packaging. *IEEE Transactions on Semiconductor Manufacturing* 26 (1): 23–34.

Index

a

ablation 5, 29, 82–83, 441–456, 458, 465–466, 469–470, 504, 506, 512

absorption 255, 273, 327, 354, 382, 400, 423, 444–446, 451, 460, 462–463, 465, 467

acceleration 159, 181, 256, 267, 277, 290, 294, 296, 341–342, 358, 361, 382, 425, 437

acetate 330

acids 271, 320, 333

acrylate 272, 317, 327, 382

activation 29, 246, 249, 322–323, 326–327, 388, 462–463, 465

actives 41, 61, 186

added 14, 17, 19, 31, 100–101, 239, 280, 373, 477, 479

additive 17, 41, 55, 146, 220–221, 267–268, 277–280, 292, 299, 332, 382–383, 442

adhesion 29, 64, 105, 175, 189, 206, 247–248, 261, 267–268, 272, 276, 278–280, 289–292, 299–300, 307, 311–312, 318, 320, 329, 331–332, 337, 340, 342, 353, 359, 361–362, 379, 382, 389, 400, 419–423, 426, 431, 433, 449, 461–463

adhesive 2–3, 5, 7–8, 12–13, 15–16, 18, 20, 24–25, 28–29, 31, 43–44, 49, 57–58, 63–64, 79, 97, 172, 188, 193, 227, 262–263, 290, 320, 337, 351, 353–355, 362, 366–367, 379, 405–407, 458–463, 469, 473, 493

adoption 39–40, 50–51, 80, 86, 101, 128, 139, 150, 262, 290, 428, 439, 459, 461, 472

aerospace 1, 183

agents 266–268, 322, 382

aggregation 381–383, 387, 390, 401

aging 268, 342

airflow 254

alcohol 330

algorithm 363, 416, 498

alignment 8, 13, 25, 27, 29, 34, 43, 46, 49, 112, 128, 131–135, 148, 249, 252, 333, 347–348, 350, 354–356, 358–359, 362–367, 380, 403, 406–408, 410, 413, 416, 434, 448, 484

aliphatic 322–323

alkaline 272

alloy 475

aluminum 6, 70, 120, 193, 201–202, 331, 448

ambient 11, 109, 195, 206, 237, 254, 276, 286–288, 290, 292, 300, 305–306, 311–312

amide 324, 327

amine 334, 382

amino 323

Advances in Embedded and Fan-Out Wafer-Level Packaging Technologies, First Edition.
Edited by Beth Keser and Steffen Kröhnert.

aminosilane 272
Amkor 34, 43–46, 48, 52, 55, 63, 67–68, 141, 319, 346, 419, 439
amplifier 14, 240
analog 6, 47–48, 59, 501–502, 515
anchor 503, 511, 513
anhydride 267, 270, 323, 382
anisotropy 2, 31, 193, 331, 434
annealing 332
anode 255, 289, 295
antenna 66, 111, 200, 217, 401, 430, 502
antimony 382
aqueous 272–274, 277, 279–280, 284, 311, 313, 329, 339
architecture 139, 203, 487, 489–491, 501
Arduino 112–113
argon 436
ash 19
ashing 332, 450
ASIC 16–17, 48, 67–68, 119, 183, 218
assessment 159, 363, 365–366, 455, 472
asymmetry 251, 457
asynchronous 355
atmosphere 281, 284, 319, 423, 429–430, 434, 460
atoms 329, 432, 445, 482
attribute 143–144, 147–148, 154, 156, 160, 165, 225, 476, 479, 483, 492
autoclave 255, 341
autofocusing 416
automation 1, 45, 250, 368, 376, 385, 398, 400, 498
automobiles 18, 33, 41, 48, 51, 59, 62–63, 72–74, 94, 97, 109–111, 181, 183, 195, 197, 213, 217, 227, 233, 239–240, 261, 422

b

backbone 272–273, 277–278, 280, 324–327, 334–335, 337
backfill 436–437
backgrinding 44–45, 79, 98–99, 103, 123, 146, 233, 372, 416, 419, 421, 423, 460, 473, 475, 477, 482
backside 2, 7, 11, 13, 15, 18–19, 25, 40, 45, 58, 65, 67, 74, 97, 99–101, 103, 105, 112, 118, 120, 123–124, 139, 172, 177–178, 188–189, 193, 208, 214, 219, 245, 362, 373, 398, 416, 434, 436, 471–473, 478, 483
backside laminate (BSL) 129
baking 102, 271, 276, 280–284, 299, 302–303, 311–312, 319, 322, 327–330, 333, 337, 373, 424, 429, 454, 460
baluns 62
bandgap 213
bandpass 41
bandwidth 2, 33, 77, 94, 122, 139, 143, 181, 404, 442, 487–488, 490, 494, 497–498, 501–502, 511
barrier 124, 422, 444, 451
baseband 48, 58–59, 66, 80, 86, 165, 183, 374, 502
batch 57, 85, 190, 281, 346, 354, 362–363, 383, 405, 506
baths 250
battery 41
beam 248, 446–448, 463–469, 505
benchmarking 505, 508, 511
bending 32, 255, 257, 389, 394, 482, 498, 505
benzocyclobutene 2, 317, 339
benzophenone 324
benzoxazole 273, 313
bias 5, 72, 255, 294–295, 329, 341, 345, 433–434, 437, 479, 482
biased highly accelerated stress testing (bHAST) 276–278, 289, 294–296, 299, 301, 307, 309, 311–312, 452
bill of materials (BOM) 63, 65, 122–123, 191, 193, 219, 246–247, 249, 256
binder 3, 382
biosensor 111–112
bipolar 199, 202, 233

bismaleimide 25
blade 171, 241, 248, 250–258
bleaching 273, 465
bleeding 105, 382, 391–392, 444
blister 353–354
bluetooth 40–41, 56, 374, 405
boards 31, 49, 151, 185, 190, 206, 214, 255
bombardment 426, 434
bonder 25, 29, 252, 348–368
Bosch 170–171, 438
bottomside 112, 221–222, 234–235, 240
boundary 43, 71, 447
bow 427–428, 434, 437, 439, 460–461
breakage 29, 290, 438, 465
breakdown 256, 294
breaking 255, 428, 444–445
breaks 392, 445
breakthrough 371
bridging 5, 67, 141, 205, 241, 252–253, 258, 347–349, 487–488, 490–496, 498, 502
brittleness 287–288
broadband 272–273, 281, 333
brominated 382
bubble 467
buffering 61, 65, 71, 117, 119–120, 138–139, 271, 277, 348, 354, 476, 504
buildup 17, 25, 27–28, 31, 40–41, 44, 98, 103, 105, 108, 113, 120, 125–126, 128, 131, 145–146, 148, 150, 187, 196, 221, 225, 348, 360, 362, 426, 451, 456, 458–459, 465, 469, 493
bulging 445
bulk 281, 330, 344, 346, 387, 490
bumping 5, 8, 10–12, 17, 19, 21, 27, 33, 39, 43–44, 47, 55, 62, 65, 70, 77, 79, 91–92, 98, 101, 103–104, 122–123, 128, 131, 145, 147, 149, 166, 171, 173, 176, 178, 181, 193–194, 225, 236, 243, 245, 250, 275, 292, 298, 312, 343, 347–349, 362, 375, 398, 405–406, 419, 457, 473–474, 480, 482, 484, 487–490, 494, 502–503, 511, 514–515

c

cage 430
calibration 355–356, 358, 364, 505
camera 29–30, 41, 63, 355–356, 363–365, 484
cap 130, 149, 152, 194, 348–349, 362, 410–411, 496
capacitance 2, 8, 11, 69, 425–426, 493, 497, 503–504, 508, 511, 515
capacitor 3, 9, 16, 41–42, 66, 68, 92, 94, 207
capacity 137, 240, 247, 441, 472, 478
capture 125, 131–134, 356, 449
carbide 214
carbon 268, 323, 382, 426, 444–446, 465, 467
carbonization 464–467
carbonyl 322–323
carboxylic 273
carousel 430
cases 74, 132, 150, 154, 195, 249, 264, 331, 353–354, 360, 365, 424, 438
Casio 21, 41
cassette 410
casting 188, 322, 330, 332–333, 340
catalysis 326–328, 382, 387
cathode 289, 295
cavity 2, 5–6, 14, 23, 26, 29, 47, 81, 170–172, 175–177, 187, 203–205, 243, 267, 353, 364–365, 372–373, 381, 384–387, 389–392, 395–396, 398, 400, 408–409, 493
cell 6, 135–136, 212, 484
cellphone 441
cellular 119
centering 61, 358
central processing unit (CPU) 258, 487–488
centricity 460
ceramic 2, 5–9, 18, 31, 43–44, 103, 201, 205, 213, 425–426, 430, 433, 460, 511

channel 268, 487–488, 494, 496–497
characterization 61, 88–90, 94, 148, 158, 183, 207, 335, 337, 343–344, 346, 361, 406, 494, 505–506, 515
chemical and mechanical polishing (CMP) 449–450, 454–455
chip on wafer on substrate (CoWoS) 404–405
chipping 471–472, 478–479, 483
chip-scale packages (CSP) 40, 60–63, 69–71, 73, 75, 224–225, 344, 348, 405
chipset 139
chlorine 255
chromatography 322
circuit 1–3, 6, 11, 24, 40–41, 44, 61, 89, 92, 101, 117, 142, 144, 147–148, 175, 183, 185, 203, 233, 240, 242, 271, 276, 278, 295, 312, 348, 373, 415, 423, 430, 442, 452, 471, 487–488, 494, 497–498, 501, 511, 515
circuitry 41, 64, 67, 141, 405
cladding 503–507, 509, 515
clamping 334, 355, 360, 387, 389, 392, 434, 436
cleaning 8, 29, 82, 122, 246, 249, 390, 408, 425, 446, 450, 454, 469, 484
clearance 134, 427, 434
clip 235, 242, 254, 257
clocking 2, 488
closure 277, 326–327
clustering 421
CMI 502–503, 511, 513–515
CMK 21, 41
CNC 192
coating 6–7, 16–17, 26, 29, 33, 40, 45, 55, 58, 63–64, 123, 188, 191, 219, 267, 271, 280–281, 283–284, 291, 294, 297, 302, 313, 319–322, 328–329, 332–334, 337, 340, 345–346, 373, 377, 383, 420, 423, 425, 429, 431, 448–449, 451, 454, 460, 471–473, 477–479, 482–483, 485, 507
codec 48–52, 59, 139, 183, 405
codesign 67, 94
coefficient 7, 61, 170, 202, 217, 264, 319–320, 348, 373, 409, 445, 457, 463, 467, 491, 502
coefficient of thermal expansion (CTE) 7, 48, 61, 67, 71–72, 170, 175, 178, 193, 195–197, 202, 217, 255, 264–266, 268–269, 287, 295, 305, 320, 324–325, 335, 339, 342, 345, 348, 373, 376, 378–379, 382, 400, 409, 443–444, 448, 451, 456–457, 459–461, 491
coherence 32
cohesion 86, 290
coils 62, 66, 105
collector 100
coloring 267, 382, 467
commercialization 5, 31, 49, 64, 113, 138–139, 272, 337, 441, 463, 474, 477
commodity 185
communication 59, 97, 101, 223, 261, 371, 501–502, 515
compatibility 7, 9, 15, 21, 29, 48, 79, 81, 113, 189, 206, 266, 268–269, 279–280, 291, 318, 333, 348, 427, 431, 434, 461, 476
compensation 17, 71, 128, 133–134, 180–181, 329, 353, 356, 358, 368, 408, 514
competition 62, 122, 142, 144, 146, 152, 498
complementary metal–oxide–semiconductor (CMOS) 25, 374, 441, 497–498, 501
complexity 10, 31, 77, 94, 98, 104, 110–111, 185, 187, 244, 249, 251, 337, 352, 372, 385, 429, 487, 491
compliance 44, 49, 104, 267, 319–320, 346, 400, 461
composite 7, 117
compression 20–23, 63–64, 79, 98, 105, 229–230, 239, 256, 262, 264–265, 353–354, 362, 371–373,

375, 377–378, 380–396, 398–400, 405–406, 408–411, 417, 465, 473, 493, 502–503, 514
compromise 130, 248–249, 282, 361–362, 424
computer 3, 8, 33, 90, 244, 501
computing 5, 51, 405, 442, 498, 501, 515
concentration 274–275, 281–282, 340
concentric 363–364
condensation 150, 436, 447
conduction 7, 12–13, 20, 29, 31, 60, 62, 101, 146, 180–181, 188, 193, 203, 214, 230, 237, 241, 249, 255, 287–290, 294–296, 299, 306, 309–310, 415, 419, 436–437, 510
conductivity 12, 18, 195, 201, 206, 213, 461
conductor 3–4, 6, 25, 27, 41, 268, 419, 427, 502, 506
conference 53, 94, 115, 200, 270, 346, 417, 456, 470, 498, 515
confinement 462, 465
confocal 227, 494
confocal scanning acoustic microscopy (CSAM) 102, 138, 227, 230, 494–495
conformal 150–151
congestion 419, 425
congruent 444
conjugated 322
connectivity 48, 51, 225, 489
connector 11, 112, 201, 203, 207, 443
construction 28, 58–59, 61, 69, 71, 119–121, 142–144, 146–147, 151, 169, 187, 191, 193–194, 198, 203, 205–206, 214, 218, 229, 233–234, 243–245, 395, 433, 474
consumable 455, 464, 469
consumer 2, 33, 63, 71–72, 77, 110–111, 183, 442–443, 471, 479
contactless 415–416
contaminant 372, 421–424, 432–434
contamination 82, 171, 263, 329, 422–424, 427–431, 438, 460, 462–463
conventional 77–80, 93–94, 97, 117, 119–121, 135, 138, 141, 143–144, 146–148, 151, 154, 156, 158, 160, 163, 193–194, 201–202, 214, 233, 237, 240, 242, 246, 271, 319, 404, 423–424, 435, 437, 441, 450, 455, 471, 473–477, 480, 482, 502–503, 505
converter 6, 25, 30, 41–42, 241–242, 245, 502, 515
coplanar 101, 180, 359
coplanarity 43, 61, 67, 170–171, 359–361, 364
copper (Cu) 6, 9–10, 13–14, 16–18, 21, 27–29, 31, 39–40, 43–45, 47, 50, 58, 71, 82, 97, 103, 105, 108, 117–120, 122–130, 135, 144–153, 155, 162, 173, 189, 191, 194, 205, 219–221, 225, 231, 235–236, 243, 246–249, 253, 255–257, 262, 274–282, 289–292, 294–295, 297–300, 303–304, 307–312, 318–319, 332, 342, 400, 407, 419–423, 425, 430, 437, 449–450, 454, 474, 503–504, 506–507
core 1, 25, 28–29, 108, 143, 188, 225–226, 348–350, 363–364, 368, 375, 404, 515
costs 2, 74, 78, 104, 128, 151, 187, 190, 272, 347, 351, 354, 435, 439, 441–442, 454–455, 469, 484, 491
cure 23, 45, 63, 99, 123, 136, 263, 271, 273–274, 276–290, 294, 297, 299, 302–306, 308–309, 311, 313, 317, 319–320, 322–323, 325–340, 342–346, 420, 424, 448–449, 454, 476, 494
cured 7, 27, 63, 123, 262, 271–274, 276–284, 286–292, 294, 298–299, 302–308, 310–312, 319–320, 331–333, 335–336, 340–341, 345, 448–449, 452–453, 471

curing 2, 8–9, 80, 123, 172, 197, 263, 266–267, 273, 280–281, 299, 302, 313, 326, 334, 346, 353, 380, 382, 386–388, 394, 427, 448, 473
customer 34, 51, 61–62, 65, 74, 92, 98, 104, 137, 139, 196–197, 199, 214, 327, 403, 478, 485
cycle 18, 72, 74, 77, 81, 83, 85–86, 89, 91–92, 102, 104, 106–107, 112, 117–118, 138, 142–143, 157, 159, 161, 181–183, 196–197, 200, 202, 213, 231, 256, 275–276, 287, 296–298, 300–301, 310, 323, 327, 332, 341, 343, 385, 448, 452–453, 455, 479, 482
cycling 61, 71–72, 83, 89, 138, 181–183, 196, 206, 213–214, 256, 335, 343, 423, 452, 479, 482
cyclization 278, 280, 322, 326
cyclopentanone 330
cyclotene 339

d

damascene 163, 166, 442–443, 448–456
debond 461–467, 469
debonded 63, 123, 368
debonding 21, 44, 47, 64, 79–80, 169–170, 360, 363, 374–376, 378–381, 457–465, 467–470, 473
debris 249, 445–446
Deca Technologies 43, 45–46, 48–49, 117, 128, 132, 135–137
decomposition 319–320, 332, 339, 424, 431, 462, 466–467
deep reactive ion etching (DRIE) 14
defect 18, 73–75, 148, 195, 257, 275, 296, 310, 333, 404, 415–416, 478
deformation 217, 356
degas 332, 421–425, 428–430, 434, 439
degassed 423, 428–429
degradation 109, 256, 290, 295, 305, 342, 476
degrade 9, 292, 319, 323
delamination 138, 150, 217, 274–277, 289–295, 297–298, 307–312, 332–334, 337, 380, 461, 468–469, 479, 482
dendrite 277, 289, 295, 309, 312
density 1–2, 5, 18, 31, 33–34, 40, 42, 44, 49, 52–53, 57, 60, 66, 74, 77–79, 81, 87, 92, 94, 104, 130, 139, 141–142, 144, 146–148, 151, 163, 169–170, 178, 201, 233, 240, 251, 271, 273–274, 319, 348–349, 354, 373, 375, 404–407, 419, 421, 425–427, 430, 439, 441–445, 450, 456, 463, 465, 472, 478–479, 487–492, 498, 501–502, 511
deposition 2, 10, 20–21, 28–29, 31, 120, 123, 170, 249, 332, 419, 422–423, 428–429, 435–436, 442, 460–461, 465, 468, 470, 504
depth 6, 121, 170–172, 175–176, 222, 303, 330, 445–446, 448, 453, 463, 465, 467, 504
descum 328, 331–332, 454
desmear 246, 249
developable 272–274, 277, 279, 299
developed 1–3, 8, 15, 20–21, 25, 27, 29–32, 39, 41–43, 45, 47, 49, 51, 57–58, 67, 94, 101, 103, 107, 117, 132, 137, 164, 171–173, 175, 177–178, 183, 185, 191, 198, 203–204, 225, 229, 247, 252, 257, 261–263, 267, 269, 271–272, 277, 279, 283, 299, 313, 322, 334, 337, 348, 354, 371, 376, 378, 385, 396, 398, 404, 428–429, 431, 433, 435, 441, 462, 472, 476, 485, 494
developer 272, 328–330, 339, 444, 454
development 3, 6, 13, 15, 25, 39, 43, 45, 56–57, 64, 67, 79–80, 87, 97, 100, 104, 107, 115, 123, 129–130, 135, 137, 148, 169–170, 173, 175–177, 183, 191, 207, 213, 219, 235, 241–242, 249, 258, 261, 264,

269–273, 275, 278–284, 286, 299, 302–303, 312–313, 317, 319, 326, 328–330, 332–333, 346, 353, 371–372, 376, 395, 397–399, 401, 405, 421–422, 438–439, 444, 446, 450, 461, 477, 487, 501, 505
device 1, 3, 9, 14–15, 39–41, 43, 48–49, 51–52, 57, 59, 66, 79, 81–82, 88, 90, 92–94, 97–98, 100–101, 103–105, 111–112, 115, 117, 119–126, 128, 130–132, 137–139, 141–143, 151–152, 157, 171, 181, 185, 188, 190–191, 193–196, 198, 200, 202, 204, 206, 214, 217–218, 233–235, 237, 239–240, 247, 250, 252–253, 255–258, 261–263, 274, 284, 287, 289, 295, 303, 305–306, 313, 317, 335, 337, 348, 353–354, 371–373, 378, 382–383, 389, 392, 401, 403, 405, 417, 419, 421–422, 424, 426, 428–429, 435, 438–439, 441–442, 444, 456–461, 466, 468–469, 475–476, 479, 483–484, 498, 501
diameter 10, 58–59, 65, 70, 80, 82, 129–130, 132, 137, 143, 178, 194, 196, 221–222, 225, 236, 257, 273, 298, 312, 331, 349, 351, 363, 387, 390–392, 399–400, 405–406, 408, 410, 421, 438, 443, 452, 472, 474, 477–478, 504, 506–507, 509–510, 512
diamine 271, 323–324
dianhydride 271, 323–324
dianiline 324
diazonaphthoquinone (DNQ) 272–274
die attach film (DAF) 172, 175, 234–235, 240, 406–408
dielectric 2–3, 5–6, 8–9, 11–12, 29, 49, 55–56, 58–59, 62–63, 65, 67, 71, 73, 77, 79, 81, 98, 103, 105, 120–121, 138, 143–144, 146–148, 150–151, 205–206, 217, 231, 268, 271–272, 274–278, 280, 282, 284, 286–292, 294–298, 302–303, 306–313, 317–323, 325–326, 328–330, 332–346, 419, 423, 442–444, 446, 448–449, 451, 453–455, 471, 473–474, 476, 480, 482, 493, 502, 504, 507, 515
diffraction 505
diffusion 204, 242–244, 246–248, 252–253, 257, 345, 422, 468
digital 6, 59, 450, 501–502
dimethyl 334
dimethyl sulfoxide (DMSO) 334
diode 3, 18, 207, 425, 433–434, 463
dipolar 322–323, 334
dispense 22, 32, 263, 384, 394, 396, 409–410, 494
dissipate 233, 466, 468
dissolution 272–273, 330
dissolved 3, 257, 271
distance from neutral point (DNP) 41, 72
distilled 284, 292
distribution 22, 50, 67, 141, 143, 179, 183, 187, 212, 237, 239, 268–269, 275, 282, 291, 343, 374, 387, 457, 463, 465–466, 505
drawback 207, 244, 348, 352, 425, 463–465
DrBlade 241–245, 250–251, 254, 256–258
drift 99, 104, 108, 356–358, 361
Dynaloy 292
dynamic random-access memory (DRAM) 67, 86, 142, 487–488, 502

e

ejection 29, 352, 364, 367, 389
ejector 389, 394–396
elastomeric 11–12
electrochemical 289, 307, 419
electrode 101, 103, 450
electroless 21–22, 28–29, 62, 189, 221

electrolytic 55, 189
electrolytically 206
electromagnetic 90, 100, 180, 200, 268
electro magnetic interference (EMI) 45, 100, 114–115, 130, 149, 194
electromigration 257
electroplated 13, 16, 21, 43, 49, 58, 65, 120, 205, 290, 449
electroplating 6, 14, 17, 29, 55, 62, 122–123, 135, 176, 188–189, 461
electrostatic discharge (ESD) 374
element 10–12, 25, 60, 65, 101, 131, 209, 244, 251, 380, 432, 448, 459, 505
elongation 273, 275–276, 279, 284, 287–289, 299–300, 305–306, 311–312, 320, 322–323, 334–335, 339, 341–342, 346, 398
embedded 1–18, 20–22, 24–30, 32–34, 39–42, 47, 49, 51, 53, 55–60, 62–64, 66–70, 72, 74, 77–78, 80, 87, 94, 97, 99, 101, 104–105, 110, 112, 115, 117, 119, 130, 134, 141, 150, 169–172, 174–175, 177–178, 180–183, 185, 187–188, 190–191, 193–201, 203, 207–209, 213–214, 217–218, 220, 222, 224–225, 228–230, 232–234, 236–242, 244, 246, 248, 250, 252, 254, 256, 258, 261–263, 270–271, 313, 317, 319, 347–350, 368, 371, 374–375, 400, 403–404, 417, 419, 441–443, 449, 452, 457, 470–472, 487–488, 490–496, 498, 501–503, 506–510, 515
embedded ground plane (EGP) 44, 97–100, 108–109, 114
embedded multi-die interconnect bridge (EMIB) 347–349, 351, 487–498, 502
embedding 2–3, 5–15, 17–19, 21–25, 27–29, 31–34, 41, 58, 62, 89, 103, 120, 150, 171, 185–191, 193–194, 196–204, 206–208, 210, 212–214, 220, 237, 241–242, 244–245, 248, 251, 348, 371–372, 374–376, 378, 380, 382, 384, 386, 388, 390–394, 396, 398–400, 462, 493–494, 506
encapsulant 7, 99, 261, 408–410, 413
encapsulate 44, 47, 79, 98, 372, 378, 473
encapsulated 7, 9–10, 19, 43, 45, 117, 146, 193, 257, 399, 471–472, 474–478, 480, 482, 484
encapsulation 20, 22, 24, 47, 57, 62, 103, 262, 264, 319, 349, 371–377, 380, 382, 387, 390, 399, 401, 403–404, 408–410, 413, 417, 421
energy 41, 63, 214, 272–273, 280–281, 284, 299, 303, 311, 326–327, 433–434, 436, 442, 444–448, 462–467, 469, 497–498, 515
enhance 20, 99, 111–112, 115, 267–268, 275, 278, 280, 299, 329, 337, 380, 382
environmental 30, 105, 200, 233, 262, 270, 272, 295, 319–320, 337, 345–346, 371–372
epoxy 2, 7–9, 13, 20, 26, 29, 47, 65, 98, 101, 103, 117, 124, 126, 139, 146, 169, 243, 261–262, 264, 267–268, 290, 317–322, 327, 329, 348, 382, 387, 409, 423–424, 427, 443–444, 448, 457, 459, 471
epoxy mold compound (EMC) 44, 99, 103, 105, 108–109, 113, 117, 120–122, 151, 169–170, 261–263, 266, 319–320, 331, 448
equipment 8, 43, 59–60, 64, 107, 128, 135–137, 148, 183, 193, 205, 213, 246, 249, 257, 262, 270, 280, 288–289, 302, 326, 333, 347–355, 357–358, 362–363, 365, 367–368, 371–372, 374, 376–378, 380, 382, 384–386, 388, 390, 392, 394, 396,

398–401, 408–409, 415, 419, 422–431, 433–439, 446, 477, 491
ester 272, 323–324, 326–327
etch 6, 47, 123, 135–136, 170–171, 175, 190–191, 246–247, 332–333, 420, 422, 425, 429–431, 433–434, 438, 444–445, 447, 454
etchant 275, 334
etched 2, 9, 64, 191, 220–221, 243, 425–426, 431, 438, 445, 507
etching 2, 6, 14, 122, 170–171, 175, 187, 190, 233, 247–248, 331, 425–426, 443, 448, 451, 461, 504, 507
ether 330
ethylene 398
evaporation 444
excimer 6, 15, 441–456, 463–464, 468
expansion 7, 11, 61, 170, 202, 217, 264, 267–268, 319–320, 348, 373, 379, 398, 409, 443, 457, 491, 502
exposed 7, 9, 71, 80, 99, 103, 142, 146, 152–156, 161, 191, 242, 244, 251, 253–255, 265, 272, 274, 281, 327, 329, 331, 390, 422–423, 425–426, 430–431, 433, 471–473
exposure 29, 45, 63–64, 213, 235, 246–247, 249, 271–273, 280–281, 283–285, 299, 302–304, 311–312, 328–329, 331–333, 341, 448, 450, 454, 460, 462, 464–466, 468–469
extended wafer level ball grid array (eWLB) 5, 39, 42–43, 48, 51–52, 55–58, 60, 62–66, 68, 70, 72–74, 77–94, 115, 175, 178–179, 183, 245, 262–263, 270, 317, 319, 347, 349–352, 354, 360, 367, 371–380, 382, 384, 386, 388, 390–392, 394, 396, 398–400, 404–405, 419, 422, 424, 426, 428, 430, 434, 436, 438, 441, 457–458, 460, 462, 466, 468, 470, 472–473, 477–479
extended wafer level chip scale package (eWLCSP) 78, 471–472, 474–485

f

fab 55, 59, 88, 123, 138, 163, 276, 493
fabricated 9, 11–12, 18, 43, 47–48, 119, 143, 170–171, 173, 233, 477, 504, 506–508, 510–511, 513
fabrication 1–2, 15, 18, 41, 55, 123, 135, 169, 235, 240, 435, 455, 461, 472, 504, 506–507, 509, 512
face-down 7, 39, 42–43, 234, 242
face-up 43, 49–50, 58, 101, 123, 348, 352
fan-out chip on substrate (FoCoS) 52, 404–405
fan-out multi-chip module (FO-MCM) 405
fan-out package on package (FO-PoP) 405
fan-out wafer level package (FO-WLP) 39–40, 42–53, 56–63, 65–75, 77–80, 82, 84, 86, 88, 90, 92, 94, 97–109, 111–112, 114, 117–118, 123, 131, 134, 141–144, 146–148, 150–160, 162–165, 169–170, 175, 178, 217, 219, 221, 224–225, 227–228, 239, 261–264, 266–271, 274–280, 284, 289, 294, 296–299, 303, 310–311, 313, 317, 319–321, 323, 328, 331–333, 335, 339–342, 344–345, 348–354, 358, 360, 362–363, 366–368, 401, 403–410, 415, 417, 419–429, 431, 434–435, 437–439, 441, 443, 457, 459–461, 470, 472, 474
fiber 112, 193, 243
fiducial 21, 29, 355, 357, 362, 366–367
filled 8–9, 14, 26, 29, 47, 62, 101, 169–170, 172, 174, 203, 205–207, 240, 246, 449, 504
filler 2, 9, 47, 203, 230, 243, 263–269, 382, 390, 394
flip chip (FC) 1–2, 15–17, 25, 27, 31, 33–34, 50, 60–63, 69–71, 144, 146–148, 157–158, 162, 164–165, 178, 196, 224–225, 262, 348, 408

flip chip ball grid array (FCBGA) 63, 71, 80, 226, 479, 489–490
flip chip chip scale package (FC-CSP) 63, 274, 277, 404–405
flip chip package on package (FC-PoP) 88–90, 94
fluence 445–448, 465–467
fluorinated 317
flux 65, 82, 275–276, 292–294, 300, 308, 320, 332, 348–349, 352–353, 367, 405, 407–408, 445
foil 20, 27, 63–64, 79–80, 188, 207, 247–248, 258, 353–354, 473
footprint 44, 78, 142, 144, 146, 242–245, 441, 463, 469, 472, 479
formulation 264, 266–268, 272–273, 278, 280, 292, 328, 331, 337, 340, 345–346
foundry 5, 57, 90, 94, 137, 166, 358, 502, 510
fragile 344, 471, 473
frame 5, 19, 30, 59, 74–75, 78, 98–101, 103, 105, 111–113, 119, 122, 201, 203–207, 209, 212–213, 242–245, 247–248, 250–255, 257–258, 349, 353, 372, 441, 479
Fraunhofer 1, 8, 21, 23, 27, 29, 31, 189
Freescale 5, 43, 48, 51–52, 97, 115
frequency 2, 11, 57, 72, 79, 109–111, 152, 180, 201, 209, 225, 235, 240, 258, 320, 335–338, 346, 431, 463, 466, 469, 487, 494, 496, 502, 505, 508–512, 515
Fujifilm 317, 334, 337, 346, 447, 450, 452
Fujikura 41
Fujitsu 115
furnace 65, 281, 284, 330, 333

g

GaAs 25, 60
gallium 213
galvanic 205, 249–250, 450
gallium nitride (GaN) 13–15, 214
gap 2, 5, 7–9, 14, 42, 47, 53, 63, 67, 141, 163–164, 166, 172, 176, 225, 266, 275–276, 284, 286, 299, 311–312, 374, 390, 451, 463, 469
generation 25, 53, 82, 94, 107, 114, 132, 141, 163, 213, 258, 268, 274, 277, 279, 313, 319, 356–358, 360, 364, 366, 375, 385, 434, 438, 442–443, 451–452, 456, 458, 466, 470, 488
generator 3, 272–273
geometry 11, 119, 141, 244, 246, 249, 362, 494
glob 262, 372, 409–412
glue 18, 21, 189, 242, 244, 246, 248, 367, 379
glycol 330
gold 18–19, 58–59, 65, 193, 511, 514
gold second 19
granular 22–23, 63, 65, 398, 413
granulate 382–383, 390, 401
granule 262, 298, 414
grid 5, 7, 15, 39, 55–56, 58, 60, 62, 64, 66, 68, 70, 72, 74, 77, 80, 94, 98, 115, 119, 131, 134, 146, 171, 224–225, 230, 251, 262, 270, 347, 371, 404, 441, 457, 470, 472
grinding 11, 20, 40, 43, 58, 65, 103, 170, 219, 247
grooving 120

h

halogen 464
handset 119
hardener 267–268, 382
hardness 388–389
HDMicrosystems 271
healthcare 27
heating 63, 89, 213, 228, 323, 353, 387, 398, 406, 428, 445, 464–466, 468
height 16, 57–58, 60, 66, 78, 80–82, 84, 87, 89, 91–94, 100, 141, 143,

146–148, 150, 152, 154, 156, 160, 163, 174, 176–178, 189, 197, 225, 229, 232, 236, 247–248, 258, 286, 320–321, 341, 388, 394, 450, 457, 494, 509, 511–512, 515
hermetic 67
hermetically 372
heterogeneous 19, 51, 53, 59, 61, 68, 77, 98, 139, 141, 405–407, 488, 501–502, 510, 515
HFSS 90, 510
high density fan-out (HDFO) 157–159
highly accelerated stress test (HAST) 72, 83, 89, 181, 256, 295, 479, 482
Hitachi 294, 296, 298, 313
hole 31–32, 62, 82–83, 101, 187, 189–190, 192–193, 220–223, 225, 233, 236–237, 239–240, 249, 313, 354–355, 380, 446
homogeneity 466–468
homogenous 465, 469
Huatian 47–48, 169
Huawei 51
humidity 72, 181, 213, 255, 268, 290, 294, 337, 341
hybrid 1–2, 10, 15, 201, 233, 240
hydrocarbon 322, 382
hydrolysis 341
hydrostatic 505
hydroxide 272, 329
hydroxyl 273, 337
hygroscopic 337

i

IBM 1, 18
illustrate 90, 133, 163, 244, 327, 378, 406, 438, 467, 490, 507
Imbera 25–27, 41
IME 49
IMEC 32
imidization 278, 325–328, 330
imidized 326–327
immersion 277, 294, 308–309
impedance 2, 50–51, 90, 109, 143, 152, 156, 181, 225, 245, 253, 507–508, 510
improvement 34, 50, 60, 67, 89, 119, 138, 146–147, 152, 156–157, 213, 224–225, 280, 311, 345, 358, 372, 396, 401, 439, 455, 488, 492, 501
inductance 2, 11, 42, 69–70, 90–91, 105, 107, 201–202, 206–207, 214, 235–236, 242, 508
inductively 171, 425
inductor 16, 42, 68, 92, 94, 105, 107, 207, 252
industrial 14, 33, 105, 112, 233, 258, 444
Infineon 5, 39, 42–43, 48, 51–52, 55, 57, 213, 241–242, 246–247, 251–252, 255, 262–263, 371, 374, 401, 419, 441
infrared (IR) 196, 247, 327, 466
inhibit 382
inhomogeneity 469
inhomogeneous 465
innovation 59, 77, 110, 313, 372, 404, 498, 501–502, 504, 506, 508, 510, 514
inorganic 2, 141, 203, 263, 424
input/output (IO) 88, 106, 488–489
insoluble 272, 328
inspected 45, 143, 147, 291, 296, 307, 310, 333, 362, 368, 483
inspection 6, 45, 102, 122, 130, 250, 329, 331, 362–363, 366, 398, 403–404, 415–417, 472, 478–479, 482–483
insulated 18, 199, 202, 205, 233, 243–244, 248, 502
insulation 201, 206, 253, 261–262, 276–277, 294–296, 299, 309, 312, 476
integrated passive device (IPD) 66, 98
integration 1–2, 10–11, 15–16, 27, 41, 51, 53, 56–57, 59–62, 66–68, 74, 77, 80, 86, 92, 94, 98–101, 104, 111, 115, 122, 124–125, 127, 134, 139,

141–142, 144, 147, 150, 163–164, 166, 169–170, 172–175, 177–178, 180, 182–183, 194, 217, 223, 231, 233, 242, 245, 251, 317, 319, 321, 328, 332–333, 375–376, 401, 405–407, 422, 443, 450, 452, 454–455, 457–459, 462, 469, 488, 501–502, 510, 515
Intel 487, 489, 494, 498
intellectual property (IP) 20, 93, 502, 510
interaction 119, 196, 278, 288, 291, 322, 326, 329, 334, 337, 446, 458, 462–463, 470
interchain 322, 326, 334
interconnect 1–3, 5–7, 10–11, 15, 20, 24, 27–28, 30, 33, 39, 43, 53, 57, 59–62, 64–66, 69–72, 82, 104, 117, 119–120, 123, 127, 130, 139, 141, 143–147, 149, 151, 191, 193, 195, 214, 242–245, 247–248, 251, 257, 271, 347, 349, 352, 374, 404, 406, 419, 425, 428, 442–443, 487–488, 490–492, 494, 496, 498, 502, 504, 507, 511, 515
interconnection 1–2, 5, 8, 13–14, 18, 20–21, 25, 29, 31–34, 43, 60, 62, 67, 70, 79, 87, 90–91, 94, 98, 101, 106, 111–112, 143, 147, 162–163, 187, 190, 193–196, 199, 201–202, 205–206, 223, 240, 348, 354, 375, 501–502, 504, 506–508, 510–511, 514
interface 10, 12, 29, 65, 71, 81–82, 105, 112, 120, 122, 131–132, 143, 145, 148–149, 151, 206, 243, 247, 275–276, 282, 289–290, 307, 311–312, 359, 374, 400, 458, 460–461, 463, 467, 488, 491, 504, 506
interfacial 505–506
interference 100
interlayer 12, 218–219, 223, 271, 344, 492
interlevel 3
intermetallic 86, 343–344
internet 33, 51, 59, 77, 110, 225
Internet of everything (IoE) 51
Internet of things (IoT) 33, 51, 59, 62–63, 77, 92–94, 110–112, 114, 139, 225, 405
interposer 52, 78, 92–93, 139, 141–143, 146, 151, 163, 166, 185, 187–188, 190–200, 405, 442–443, 452, 489–491, 502–503, 506, 509, 511
interrelated 437
inverter 14, 201, 203, 207, 209, 497
ion 2, 14, 289, 331, 426, 433–434, 437, 463
ionic 268, 272, 372
ionized 428, 437–438
iPhone 50
irradiation 217, 220, 326, 465
isofocal 329
isolate 15
isolation 5, 51, 65, 109, 126, 195, 450
isotropic 193, 322

j

JCAP 45, 47–48
JCET 43, 45, 77, 471
JESD 72, 74, 83, 89, 91, 102, 108, 159, 181, 231, 287, 290, 294, 300–301, 341, 343, 482
JetStep 333
jitter 152, 154, 156, 160
Joint Electron Device Engineering Council (JEDEC) 72, 74, 82–85, 87–88, 91, 94, 101, 104, 108, 117, 137–138, 152, 159, 161, 164–165, 181, 224, 227, 230–231, 341–342, 480–482
JSR 339
junction 10–11, 89, 156–157, 195, 206, 208, 214, 237–238, 253–254

k

Kapton 5–6, 444
Keser 39, 55, 77, 97, 115, 117, 141,

169, 185, 201, 217, 241, 261, 271, 317, 347, 371, 403, 419, 441, 457, 471, 487, 501
known good die (KGD) 2, 33, 45, 47, 144, 147–148, 169–170, 185, 380
Kroehnert 39, 55, 77, 97, 117, 141, 169, 185, 201, 217, 241, 261, 271, 317, 347, 371, 403, 419, 441, 457, 471, 487, 501

l

laminate 2–3, 6, 21–22, 24, 27, 29, 40, 42, 45, 47, 49–50, 52, 60–61, 70, 78, 90, 101, 118–120, 122–124, 139, 141–144, 146, 150–152, 154, 156, 160, 162–163, 190–191, 193, 195, 209, 217, 241, 243–249, 251, 255, 441, 489
lamination 20, 24, 28, 31–32, 40, 63–64, 79, 103, 172, 187–189, 192–193, 197, 219–220, 233, 235, 243–244, 246, 248, 252, 256, 319, 321, 345, 374–375, 379, 383, 451, 458, 460, 471–473, 478
lap 231
lapping 8, 11
laser 5–6, 8, 12, 15, 20–22, 24, 26–29, 32–33, 40, 43, 45, 47, 62–63, 65, 67, 79–80, 82–83, 85, 87, 98, 120, 123, 143, 170, 187–189, 205, 219–221, 225, 246–250, 268, 313, 333, 403, 407, 415, 438, 441–456, 458–459, 462–470, 475, 504–505, 512
latency 487
latitude 329, 462, 469
layer 2–3, 5–6, 8–9, 14–22, 25, 27–29, 31–33, 39–41, 45–46, 48, 50, 55–65, 70–71, 74, 94, 98, 103, 105, 107–108, 111–112, 117, 119, 121, 123–131, 133–135, 138–139, 141–142, 147–149, 166, 169–170, 173–178, 183, 187–191, 193–194, 197–198, 200, 203, 205–206, 212, 217, 219–223, 225, 227, 229–230, 232–235, 237–238, 243–245, 247–248, 250–251, 253–254, 256–257, 262, 266, 271, 275, 286, 289–291, 297–298, 302–303, 307, 313, 319–320, 332–334, 337, 340, 342–345, 347–348, 354–355, 362, 373–374, 379, 403, 419–420, 422–423, 425, 428, 442–443, 448–451, 454–455, 457–463, 465–468, 471, 473–476, 479, 483, 487–490, 492–493, 504–507
Lead (Pb) 84, 88, 285
leaded 94, 251, 257, 390
leakage 138, 157, 360, 450–451
liner 379, 430–431, 438, 502–509, 512, 515
linker 273, 278–280, 286, 292, 299, 302
liquid crystal display (LCD) 107–108, 374
liquid mold compound (LMC) 261–270, 372, 377–378, 380, 382, 384–387, 389, 391, 396, 398
lithographic 63, 134, 247–249, 271, 274, 278–280, 283, 299, 303, 313, 320, 322, 326, 328–329, 444, 455
lithography 8, 14, 20, 32, 43, 49, 55, 63, 121–122, 128, 130–132, 170, 175, 249–250, 279, 282–283, 299, 303, 312–313, 322, 346, 439, 443–444, 446, 448, 451, 453, 456
lossy 502
low power double data rate (LPDDR) 142

m

machines 21, 29, 134, 246, 249, 348, 363
magazine 351, 364, 399
magnetometers 352
magnetron 435
Mahony 498
manufacturability 1, 119, 494
manufactured 2, 25, 55, 62, 103, 146, 181, 186, 214, 319, 359–360, 372, 403, 409, 493

manufacturer 20, 57, 112, 202, 262, 403, 409–411, 429, 442, 463
manufacturing 1, 11, 25, 27, 34, 42–43, 49, 57, 63, 65, 68–69, 77–79, 104–105, 107, 115, 119, 122–123, 128, 134–135, 137, 150–151, 169–171, 173, 175, 178, 187, 191, 193, 195, 197–198, 214, 217, 219, 223, 229, 233–234, 246–249, 252, 255, 262–264, 346, 353, 359–360, 371–372, 376, 378–379, 404–405, 408, 424, 441–442, 444, 457–458, 470, 472, 477–478, 483–485, 491, 493–494
margin 152, 214, 287, 305, 311, 362, 453
markets 2, 51, 55, 59, 74, 94, 115, 139, 151, 162–163, 165, 183, 197, 213, 239, 242, 258
marking 65, 79, 98, 170, 250, 268, 332, 382, 403, 415, 475, 483
marks 22, 27, 29, 264, 347, 354–355, 358, 362–367, 380, 385, 416
mask 3, 8, 21–22, 24, 43, 49, 128, 148, 247, 249, 284, 329, 333, 350, 354, 377, 442, 444–445, 447–448, 455, 482
materials 7–8, 10, 12, 22–23, 25, 63–65, 101, 107, 115, 117, 122, 135, 143, 162, 175, 185, 196, 204, 243, 255, 263, 267, 269–274, 276–282, 284, 286–288, 290, 292, 294–296, 298, 302–303, 306, 308, 310–313, 317, 319–320, 322, 325, 327–329, 333–334, 337, 339, 342, 346, 353, 367, 371, 376, 381, 386, 388, 419, 421–423, 429, 438–439, 441–445, 448, 452, 455–456, 459–463, 469–470, 477, 501, 505
matrix 188, 199, 228, 231, 329, 342, 354–355, 357, 365–366, 382
mechanism 268, 289, 344, 385–386, 389, 392–393, 395–396, 398, 408, 445, 458–459, 466
melting 65, 247, 372, 384, 386–387, 444
membrane 2, 67
metal 2–3, 5–7, 9, 14, 17–18, 23–24, 27, 31, 33, 41–42, 50, 62, 64, 79, 98–101, 103–104, 106–107, 141, 143, 149–150, 202, 206, 217–220, 223–224, 227, 233, 239, 241, 243–244, 248, 253, 255, 262, 321, 331, 337, 340, 346, 372–373, 378–380, 384, 409–410, 419–423, 425–428, 430–431, 433, 437–438, 445, 448–451, 454, 458, 462, 471, 473, 492, 501–502, 507
metallic 120, 241, 337, 425, 431–432
metallization 2, 9, 11–12, 16, 18, 20–21, 24–26, 28–29, 32, 39–40, 43, 55, 65, 98, 123, 189, 233, 247–248, 443, 507, 509
metal–oxide–semiconductor field-effect transistor (MOSFET) 199, 202, 204–206, 213, 233–235, 237, 240–242, 244–245, 253–254, 256–258
methyl 271
microelectromechanical systems (MEMS) 2, 16–17, 59, 67–68, 74, 92, 94, 352, 389, 501
microelectronic 1, 77, 115, 200, 270, 368, 371–372, 498
microinterconnects 502–503
microprocessor 139, 498, 501
microscope 6, 276–277, 294, 296, 310, 330–331, 362, 467, 476
microscopic 334, 416, 465, 469
microscopy 227, 291, 331, 482, 494
microstructure 174
microsystems 271
microvia 456
microwave 326, 346
migration 6, 289–290, 295, 307, 311–312, 320, 345
miniaturization 33, 41, 207, 214, 233, 237, 239–240, 258, 372, 374, 441

miniaturized 10, 21, 139, 169, 203
misalignment 2, 170, 361, 363–367, 450, 493
mismatch 11, 61, 67, 71–72, 178, 217, 255, 295, 348, 360, 376, 400, 443, 461, 463, 491, 502
mobile 39, 41, 53, 57, 59–60, 62–63, 67, 77, 80, 87, 92, 94, 97, 100, 107, 110–111, 137, 139, 141–144, 158, 163–165, 183, 217, 258, 261, 405, 417, 471
model 51, 71, 180, 197, 209, 212–213, 235, 237, 239, 336, 338, 343, 376, 455, 468
modeling 90, 108, 143, 212, 417, 504–505
modem 52, 183
modified semi-additive process (MSAP) 146
module 1, 6–7, 11, 13–15, 18, 20–21, 25–26, 30, 40–42, 48, 51, 66, 74, 80, 86, 92–93, 97, 101, 105–106, 112–115, 122, 139, 150, 183, 201–202, 204–205, 207, 213–214, 218, 223, 225, 227, 229–233, 237, 239, 262, 271, 329, 347, 406, 410, 413, 415–417, 421–425, 428–429, 431, 434–435, 483, 490–491
moiety 272–273, 278
moisture 51, 72, 84, 181, 213, 255, 273, 288–290, 294–297, 299, 305–307, 310, 312, 322, 345, 382, 400, 422–424, 429–430
moisture sensitivity levels (MSL) 51, 72, 83–84, 88–90, 101, 106, 112, 137–138, 159, 213, 227–228, 231, 296, 343, 479, 482
mold 7, 19, 21–22, 24, 39, 43–45, 49, 57–59, 62–67, 71, 74, 79–80, 82–83, 85, 94, 103, 105, 118, 120, 123–126, 128, 130, 142, 145, 148, 150, 152, 170, 195, 229–230, 232, 255, 262–263, 266, 268, 319, 345, 361, 368, 372–374, 376, 378, 380–382, 384–396, 398, 400–401, 405, 407–413, 415, 419–421, 423–424, 427, 437, 439, 457, 459–461, 473–474, 476, 483
moldability 391, 410, 412
molded 20–22, 24, 42–43, 58–59, 62, 80, 98, 119, 123, 126, 130, 139, 141, 146, 148, 229–230, 242, 253–255, 257, 262, 265, 290, 320, 347, 373–381, 385–387, 389–396, 399–401, 403–404, 409, 415, 417, 421, 423–425, 427–429, 431, 434, 439, 457, 459, 461
molding 7, 19–24, 42–45, 47, 49–50, 57, 62–64, 79, 98, 105, 117, 123, 128, 130–131, 146, 169–170, 229, 231, 239, 261–270, 274, 290, 297–298, 321, 354, 359–363, 366, 368, 371–396, 398–401, 403–405, 408–409, 411, 413, 415, 417, 458, 461, 473
molecular 322, 346, 445, 465
molecules 334, 386, 388, 444–445
monochromatic 505
monomer 272, 323–324
monomethyl 330
motherboard 41, 79, 92, 150, 224
mount 9, 15, 51, 66, 191, 214, 220–222, 230, 263, 297, 310, 344, 373, 403, 471
mounted 1–2, 7, 12, 15–16, 30, 41–43, 52, 84, 91, 137, 147, 181–183, 187, 201, 206–207, 220–221, 223, 229–231, 234, 262, 286, 321, 337, 349, 365–366, 373, 380, 408, 410, 461, 474
mounting 2, 31, 84, 122, 200, 229, 359
multicarboxylic 323
multi-chip module (MCM) 1–3, 5, 7, 9–13, 15, 17, 19, 21, 23, 25, 33
multilayer 2, 5–7, 9, 13–15, 25, 29, 31–32, 41, 51, 57, 104, 120, 139, 143,

147, 163, 175, 187–188, 191–193, 198, 233, 237, 275, 302, 313, 421, 441, 451, 456, 490
Nagase 261, 269–270, 380

n

NANIUM 43, 48, 51–52, 59, 344, 470
nanoelectromechanical 67
nanoelectronic 515
nanofilled 339
nanotechnology 42
narrowband 154
nepes 43, 48, 51, 97–100, 104, 106–112, 114
network 50, 141, 143, 165, 322, 336, 386
networking 51, 144, 147, 151, 163, 405
NiAu 130
nickel 65, 193
nitride 213, 424, 448
nitrogen 281, 284, 322, 330, 444
N-methyl pyrrolidone (NMP) 271, 279, 292–294, 300, 308
node 42, 49, 52, 61, 66, 74, 77, 88, 94, 122, 137–138, 151, 175, 426, 428, 441, 471–472, 474, 478, 492
novel 130, 164, 270, 278, 280, 299, 313, 346, 443, 456, 474, 498, 515
novolac 382
Novolak 313
nucleation 423
nucleophilic 323, 326
NXP 43–44, 48, 51–52, 97

o

offset 125, 128, 150, 320–321, 359, 361, 474–475, 477
opening 8, 27, 29, 32, 63–65, 131–135, 149, 173, 175–176, 190, 243, 248, 274, 276, 282, 284, 286, 311–312, 330–331, 351, 436, 442, 484, 497
optical 6, 8, 18, 43, 45, 49, 112, 123, 130, 250, 273–274, 294, 296, 310, 329, 331, 368, 416, 448, 462, 464–465, 483, 502
optics 29–30, 403, 434, 442, 447–448, 465, 469, 505
optimization 80, 82, 94, 150, 172, 273, 292, 328, 333, 345, 376–378, 494
optimized 22, 29, 112, 204, 206, 252, 281, 332, 337, 376, 387, 391–392, 396–400, 426, 431, 461, 466, 482
optoelectronic 1
organic 17, 24–25, 27, 33, 40, 42, 52, 91, 120, 122, 141–143, 151, 162, 166, 221, 287, 290, 322, 329, 337, 339, 421–424, 426, 429, 431, 433–434, 441–444, 452, 489–491, 493, 495, 502, 511
orientation 245, 348, 352, 403, 406, 505
orthogonally 468
oscillators 30, 67–68, 109
outgassing 320, 332–333, 423, 429–430, 439, 459, 466
ovens 281, 326
overexposed 467
overexposure 281, 465, 467
overflow 389, 392
overlap 132, 340–341, 466–467, 469
overlay 5, 251, 358, 446
overmold 40, 46, 49, 229, 411
overmolded 20, 43, 49, 123, 144, 189, 457
overplate 449
overvoltage 202, 214
oxidation 10
oxide 202, 241, 248, 292, 337, 353, 422, 425–426, 434, 501
oxidize 330
oxygen 281, 330, 444

p

package 1–2, 6–7, 9–10, 14, 17, 20–21, 29, 32–33, 39–53, 57–63, 65–75, 77–94, 97–109, 111–115, 117, 119–123, 127, 129–135,

137–139, 141–144, 146–157, 159, 162, 164–166, 169–171, 173–181, 183, 185–186, 188–189, 191, 195, 199–200, 203, 206–209, 211–212, 214, 217–227, 229–247, 249–258, 262, 268–271, 277–280, 288, 290, 295, 297–299, 306–307, 310, 312, 317, 319–320, 322–323, 332, 335, 337, 340–350, 353–354, 372–376, 380–381, 385, 389–390, 392–395, 399–401, 403–406, 408–410, 416, 421, 426–427, 439, 441–444, 448, 451, 453, 456–460, 462–463, 465–472, 474–485, 487–495, 502–503, 511, 515
packaging 1–2, 4–6, 8–10, 12–14, 16, 18, 20–22, 24–26, 28, 30, 32, 34, 39–42, 45, 53, 55–60, 62, 64, 66–70, 72, 74, 77–80, 86, 93–94, 97–98, 100–101, 104, 106–108, 110–112, 114–115, 117, 119, 122, 134, 141–144, 146–148, 150–152, 154, 156, 158, 160, 162, 164–165, 169–170, 181, 183, 185, 200–201, 217–218, 220–222, 224, 228, 230–231, 233–234, 236, 240–242, 244, 246–248, 250–252, 254, 256, 258, 261–262, 264, 266, 268–271, 274, 278, 294, 296, 313, 317, 320, 322, 326, 328, 330, 332, 334, 336, 340, 342, 344–350, 352, 354, 356, 358, 360, 362, 364, 366, 368, 371–376, 380–382, 388–389, 401, 403–406, 410, 415, 417, 419, 427, 441–444, 446, 448, 452, 456–457, 470–475, 478–479, 481, 487–490, 501–502, 511
panel 20–24, 34, 43, 45, 48–49, 69, 74, 79, 97–98, 100–101, 103–108, 110, 112, 114, 117, 119–120, 122–123, 135–137, 139, 183, 190–191, 193–195, 197, 199, 217–222, 224, 228–230, 234–236, 240, 248–250, 252, 255, 313, 346–351, 354, 356, 363, 399, 401, 408, 417, 427, 439, 443, 452, 469, 473–476, 484–485, 509
panelization 45, 98, 101, 103, 105, 123, 139
panel level package (PLP) 107–108, 115, 217, 219, 227–228, 239, 348–354, 358, 363, 401
parallelism 57, 393, 396, 484
parasitic 8, 11, 33, 42, 51, 57, 60, 69–70, 72, 79, 90, 111, 163, 201–202, 206, 214, 242, 258, 426
particle 171, 175, 243, 262–264, 346, 422–423, 426, 430–433, 444, 460, 466
parylene 504
passivation 5–6, 15, 29, 47, 55–56, 149, 171, 173–176, 271, 374, 383, 420, 422–424, 426–427, 434, 437–439, 462
passives 1, 7, 41–42, 48, 56–57, 60–62, 94, 98, 111, 130, 200, 245, 251, 350, 374, 401, 493, 511
pastes 29
pattern 3, 6, 9, 22, 29, 43, 45, 49, 64, 74, 123, 130–134, 143, 146, 231, 233, 235, 272–273, 277–278, 281–284, 292, 294, 297, 302, 309–310, 333, 337, 348, 363, 409–412, 442, 447–448, 451, 454–455, 489, 491, 497, 505
patterning 21, 24, 43, 46, 49, 117, 119–120, 122–124, 126, 128–134, 136, 138, 141, 148, 233, 271–272, 275, 282, 285, 297, 322, 326, 333, 421–423, 441–444, 446, 448, 450–452, 454–456, 466, 473, 493
PBGA 63
PCIe 143, 154, 156
pellet 372, 377, 385, 390, 408
penetrate 273, 290
penetration 294, 334, 463, 465, 467
percursors 313
perforated 504
permeability 462
permeation 322

permitted 460
permittivity 346
phenol 261, 339, 382
phenolic 272, 317, 322
phone 41, 67, 119, 137, 139, 158, 164–165, 229
photo 15, 105, 136, 144, 148, 151, 246, 271–272, 277–280, 299, 312, 345, 381, 442–444, 451–455
photoablation 456
photoacid 272–273
photoactivation 462
photochemical 444, 463, 465–466
photochemistry 456
photodecomposition 444
photodefined 504, 506–507, 512
photodefinition 504, 509, 512
photoimageable 443, 454–455
photoimaged 64
photoimaging 141, 147
photoinitiator 272, 274, 278, 312, 323
photolithography 22, 43, 67, 131, 143, 221, 441, 443–444, 448, 450, 454–455
photolytic 444
photomask 383, 443
photon 444–445
photonic 151, 441, 511
photopolymer 313, 443
photoresist 6, 123, 135, 144, 146, 170, 190–191, 272, 292, 313, 332
photosensitive 8–9, 13, 16, 18, 313, 317–318, 320, 323–324, 328–329, 332, 334, 337, 346, 419, 442–443, 448
photosensitivity 339, 462
photospeed 320
photothermal 465–467
physical vapor deposition (PVD) 105, 120, 122, 135–136, 332, 419–431, 434–439
pillar 18, 21, 40, 43–45, 50, 58, 67, 143–150, 152–153, 155, 162, 194, 225, 348–349, 507
pitch 10, 21, 25, 44, 49, 52, 55–57, 61, 66–67, 71, 79, 81, 87–88, 106, 108, 117, 122–123, 128–131, 134, 139, 142–145, 147, 149, 151–152, 170, 173–175, 177–178, 185, 189–191, 193–194, 196, 199, 221–224, 227, 230, 245, 251, 266, 268–269, 298, 343, 368, 374, 442–443, 451–453, 472, 474–475, 478, 482, 488–490, 494–495, 502–504, 506–507, 509–512, 514–515
placement 5, 7–8, 10, 17, 20, 22, 29, 32, 40, 45, 57, 64, 79, 127–128, 130–131, 221, 230, 252, 347–348, 353–356, 358–359, 361–368, 374–375, 380, 385, 403–409, 415, 427, 442, 448, 473–474, 493
planar 2, 8, 10–11, 13, 26, 122–123, 204, 489
planarity 7, 302, 392–393, 396, 398, 400, 515
planarization 11, 13, 16, 120–121, 320–321, 442, 450–451, 454–455, 476
plasma 8, 47, 171, 249, 328, 331, 337, 424–425, 429, 433, 436, 438, 446, 450, 460
plate 45, 109, 123, 201, 203, 205, 217–221, 224, 227, 239, 281, 283, 302, 351, 420
plating 21–22, 29, 32, 41, 43, 55, 62, 64, 136, 146, 170, 175, 187, 189–190, 192, 205–206, 219–221, 225, 235, 246, 249, 275, 282, 292, 298, 332–333, 419–420, 423, 438, 442, 450, 454, 474
plunger 372, 393, 408
polar 322
polarity 323
polarization 337
polybenzobisoxazole 16, 476
polybenzoxazole (PBO) 16–17, 117, 271–300, 311–313, 317–319, 322, 329, 339, 424, 426, 429, 443–444, 446, 449, 476

polychromatic 505
polycondensation 330
polyethylene 398
polyimide 5, 19, 31, 40, 65, 71, 117, 219, 271–273, 313, 317, 324, 326–336, 338–340, 346, 383, 424, 443, 462, 469, 476
polymer 5–10, 12–17, 27–29, 45, 109, 120, 123–126, 129, 131, 134, 136, 144, 147, 169–170, 172–174, 188, 271–274, 277–280, 286–287, 313, 317–318, 322–328, 330, 334–335, 340–341, 346, 381, 442, 444–446, 448–449, 452, 456, 460, 470–471, 474, 476, 478, 503–511, 515
polymerization 272, 322, 382
polyurethane 3
postexposure 469
potlife 383
potting 19, 201
power management IC (PMIC) 41, 48–49, 51–52, 92, 106, 114, 139, 405, 426
power management unit (PMU) 183, 426
precision 34, 241, 442, 446
preconditioning 137, 181, 213, 341
precursor 271–273, 277, 322–324, 326–327
prepreg 2, 27, 143, 188, 193, 203, 206, 243, 246, 248
primer 337
printed 1, 61, 65, 101, 117, 175, 185, 203, 233, 242, 337, 348, 373, 415
printed circuit board (PCB) 1, 6–7, 17, 21, 24–30, 41–42, 61–62, 71–72, 74, 84, 89, 91, 99, 101, 104, 110, 117, 119–120, 175, 178, 180–181, 183, 185–191, 193–199, 203–204, 206–207, 213, 233, 242, 246–249, 251–253, 255–257, 344, 348, 350–351, 373, 383, 401, 455
printing 29, 65, 98, 250, 377
probe 79, 136, 138–139, 213, 250, 336, 338, 475, 511
procedure 86, 263, 284, 289–290, 292, 294, 307, 309, 349, 355–356, 363–366, 368, 385, 415, 436, 469
processes 2, 21, 25, 33, 60, 80, 87, 101, 104–105, 107, 135, 137, 141, 144, 146–147, 162–163, 171, 175, 178, 187, 189, 191, 193, 199, 204, 221, 242, 246–248, 250, 262–263, 266, 271, 281, 284, 289, 292, 303, 306, 312, 322, 332, 345–348, 352–354, 358–361, 366–367, 371, 374–376, 378, 385, 401, 403, 406, 408, 415, 422, 424, 434, 442–443, 459–461, 464–465, 473, 489, 492, 494, 507
processing 6, 8, 15, 22, 30, 34, 42, 55, 57, 59, 78, 98, 100, 104–105, 107, 111, 119, 123, 135–137, 139, 143–144, 147–148, 150, 170, 185, 193, 217, 219–221, 252, 258, 273–278, 280–281, 284, 289, 296, 303, 307, 310, 312–313, 333, 342, 347–348, 354, 363, 371, 379, 381, 383–386, 398, 400, 403–408, 410, 416, 423, 427, 429, 434–436, 438–439, 441–443, 456–458, 460–461, 471–472, 474, 476, 478, 484, 487
production 1, 8, 40–41, 43, 47–48, 50–52, 57–58, 67–69, 97, 107, 115, 132, 134–139, 152, 169, 183, 185, 190–192, 194, 197–199, 213, 246, 251–252, 263, 328, 331, 350, 352–353, 355–356, 358, 360, 362–365, 368, 376, 383, 399, 416, 419, 421–422, 424, 431, 434, 438–439, 448, 469–470, 476–477
productivity 21, 112, 372, 424, 431, 501
products 20, 33, 41–42, 51, 55, 58–59, 61, 67, 71, 77, 80, 94, 97, 111, 115, 137, 200, 213, 221, 239, 241, 244, 250–256, 258, 262, 271, 273, 319, 330, 375, 422, 431, 443, 445,

456, 471–472, 477, 479–480, 482–483, 485
profile 29, 39, 42, 49, 51, 53, 78, 80, 87, 90, 92, 94, 98, 100–101, 105, 111, 142, 144, 172, 241–242, 257, 275, 281, 283, 302–303, 312, 320, 331, 333, 340–341, 361, 404, 439, 448, 463–466
projection 442–443, 447–448, 501
promoter 64, 267–268, 272, 331, 337, 340, 382
properties 49, 72, 263, 265, 268–269, 273, 276–280, 284, 286–289, 299, 302–303, 305–306, 311–312, 317–320, 322–325, 331, 333–335, 337, 339–342, 345–346, 362, 372, 379–380, 382, 394, 398, 400, 421–422, 443, 446, 448, 460–463, 469, 494, 502, 505
propylene 330
propylene glycol monomethyl ether acetate (PGMEA) 330
protruded 19
protruding 120
protrusion 320
puddle 281, 284, 330
pulse 214, 241, 343, 350, 353–354, 368, 442, 445–448, 450–451, 465–466, 468–469
pyrolytic 444
pyrrolidone 271

q

quad flat no-lead (QFN) 68, 200, 389, 472, 479
quad flat package 389
Qualcomm 43, 49–52
quartz 41, 444, 447–448

r

radar 7, 48, 51–52, 72–73, 109, 114, 181, 427
radiation 6, 180, 428, 458, 463, 465
radio frequency (RF) 8, 14, 40, 48–52, 59, 66, 109, 111, 114, 139, 150, 183, 200, 225, 239, 374, 425, 430–431, 433, 502, 506
ramp 152, 257, 281, 283, 302, 334, 353, 461
ramping 350, 353–354
R_{dson} 207–209
reaction 273, 278, 323, 326, 330–331, 386–388
reactive ion etching (RIE) 2, 331
readout 83, 89, 256–257
reconfigured 374–375, 417, 452
reconstituted 42–43, 48–50, 58–60, 65, 73, 79–80, 120, 271, 329, 332, 368, 374, 421, 457, 459, 472–477, 483–484, 489
reconstitution 57, 59–61, 63–64, 73, 75, 79, 274, 473–478
reconstruct 169–170
reconstructed 170, 172, 174, 403
reconstruction 170
rectangular 69, 74, 217, 252, 284, 350–351, 408, 463, 466, 475
redistributed chip package (RCP) 5, 43–44, 48, 51–52, 97, 108, 347, 349, 352
redistribution 8–9, 14, 20, 39, 55–56, 59–62, 64–66, 71, 75, 79, 98, 103, 119, 142, 149, 166, 169, 187, 218, 227, 242–245, 262, 266, 271, 319–320, 332, 347, 373–374, 403, 419, 442–443, 457, 473–475
redistribution layer (RDL) 7, 14–15, 17, 19–20, 24, 31, 39–40, 42–51, 55–64, 66–67, 70–74, 81, 91, 98–105, 107–108, 111–114, 118–120, 122–126, 128–129, 131–135, 142–159, 162, 166, 169–171, 173, 175–178, 187, 218–220, 223, 227, 232–237, 240, 243, 245, 247, 249, 251, 253, 262, 266–267, 271, 274–275, 277, 284,

289, 292, 302, 307, 312, 319, 321–322, 328, 331–333, 340, 345, 347–350, 354, 358, 362–363, 373–374, 400, 403, 405–408, 419–428, 434–436, 439, 442–443, 448–462, 466, 468, 470, 474–476
reflectivity 247, 462
reflow 24, 57, 61, 65, 69, 79, 81–82, 98, 105, 123, 181, 191, 193, 195, 213, 250, 290, 292, 296–297, 310, 320, 332, 348, 352–354, 362, 366, 405–406, 420, 448, 475, 479, 482, 507
registration 199, 451
reinforces 100, 278
release 32, 152, 217, 266–268, 284, 377–383, 386, 388–389, 394, 396, 398, 400, 433–434, 458–459, 462–463, 465–469, 504
reliability 43, 49, 51–52, 61, 69, 71–74, 77, 80, 82–89, 91–92, 94, 101–102, 104–106, 108–109, 111–112, 117, 119, 137–138, 159, 161, 165, 175, 178, 181–183, 193, 195–196, 200, 202, 213–214, 217, 227–228, 230–231, 243, 255–257, 268, 271, 274–280, 287–292, 294 300, 306–307, 309–313, 317, 319–320, 322, 337, 341–343, 345–346, 348–349, 371–373, 405, 422, 438, 442–443, 451–453, 471–472, 476, 479–483, 485, 494, 502
reliable 16, 22, 33, 150, 189, 199, 206, 222, 233, 240, 353, 371, 403, 469, 511
removal 42, 45, 57–58, 123, 133, 145–146, 221, 224, 321, 407, 422, 425, 438, 446, 448, 450, 452, 458, 469
remove 3, 33, 181, 219–221, 249, 281, 292, 329–331, 334, 362, 421–422, 424, 434, 442, 446, 450–451, 455
repair 11, 33, 197, 263
repassivation 47
repeatability 361, 363–364, 439
reproducible 60, 72
requirement 1, 11, 29, 33, 55, 61, 67, 72, 77, 79–80, 92, 94, 104, 107, 109, 111, 134, 137, 141, 147–148, 162–163, 169–170, 183, 194, 213, 229, 233, 240, 244, 246–247, 250, 255, 258, 261, 263–264, 266–271, 274, 280, 283–284, 299, 302, 312–313, 317, 319–321, 325–326, 328, 333, 347–355, 357, 359, 368, 371–372, 382, 386, 392, 395–396, 401, 403, 406–409, 415, 419, 422–424, 433, 435, 442–444, 446, 448, 451, 456, 459, 462–463, 469, 472, 479–480, 482, 484, 487, 494, 497, 506
residual 120, 171, 196, 284, 286–287, 303, 305, 330–331, 339, 361, 423, 448, 466, 469, 505
residual gas analysis (RGA) 423–424, 429
residue 82, 274–276, 281, 284, 292, 299, 312, 331, 460, 469
resin 2–4, 20, 26, 29, 42, 47, 105, 188, 199, 203, 217–221, 223, 225, 227, 229–230, 232–235, 240, 243, 261–262, 264, 266–267, 317, 322, 339, 382, 390, 398
resist 64, 190, 220, 231, 245–246, 249–251, 275–277, 292–294, 299–300, 308–309, 419–420, 423, 448, 450, 454
resistance 11–13, 26, 60–61, 69–71, 90–91, 99, 106, 120, 157, 178–183, 193, 195–197, 206–209, 212–214, 224, 233, 235–240, 242, 252–255, 268–269, 275–277, 279–280, 288–289, 292–297, 299–300, 305–312, 319–320, 331, 334, 381, 422–423, 426, 452–453, 492, 494, 508, 511, 514
resistive 146
resistivity 11, 16, 61, 70, 233, 506

resistor 3–4, 16, 66, 68, 94, 209
resolution 138, 144, 148, 175, 272–274, 276, 279–282, 284–285, 299, 302–304, 311–313, 320, 329–330, 345, 363, 421, 427, 442–444, 447
resolve 107, 312, 409, 412, 430, 485
retardant 7, 246, 382
reticle 491, 511
rework 276, 292–293, 300, 308
RFIC 502
RFID 191, 198–200
rigidity 323–327, 460
risk 9, 134, 148, 258, 332, 391, 415, 430, 472, 478
roadmap 42, 100, 114–115, 127, 129–130, 139, 148–149, 166, 175, 177, 250, 345, 439, 515
robot 351, 411, 413–414, 427, 430, 434, 439, 483
robust 86, 92, 94, 98, 104–105, 120, 134, 194, 199, 203, 214, 258, 286, 305, 320, 358, 386, 398, 403, 461, 471, 478–479, 483, 494, 515
robustly 502, 511
robustness 71, 86, 90, 97, 100, 109, 137, 159, 161, 181, 195, 213–214, 227, 230, 247, 249, 255–257, 379
roughening 246–248
roughness 121, 171, 438, 466–467, 482
routing 1–2, 8, 10, 16–17, 27, 29, 32–34, 40, 46, 50, 60, 74, 77, 81, 92, 107, 119, 122, 127, 130, 133–135, 141, 145–146, 169, 187, 192, 205–206, 251, 253, 262, 354, 406, 419, 441–442, 444, 446, 448, 450, 452, 454, 456, 488, 492, 511
ruled 49
rules 70–71, 98, 107, 127–131, 134, 142, 144, 150, 175–176, 187, 221–222, 230, 245, 248–252, 340, 390, 437
runner 21, 372, 390, 393

S

sacrificial 334, 504
Samsung 41
Samsung Electro-Mechanics (SEMCO) 41
scalability 68, 104, 135, 141, 144, 150–151, 178, 194, 252
scan 64, 98, 424, 466–468, 482–483, 505
Schweizer 185–186, 201, 210, 213
scission 341, 445
security 59, 261
semi-additive process (SAP) 220–221, 442, 450
semiconductor 2–3, 18, 28–29, 41–43, 48–49, 51–52, 63, 77, 79, 94, 97, 103, 117, 119–123, 135, 141–142, 144, 146, 148, 150, 152, 154, 156, 158, 160, 162–164, 186–187, 202–204, 206–207, 212–214, 233, 241, 246, 248, 261–264, 269–271, 273, 277–278, 284, 289–290, 305–306, 313, 322, 371–373, 376, 380–383, 385, 388–389, 435, 441–443, 460, 471, 501, 515
semiconductor embedded in substrate (SESUB) 41, 185
sensitive 60, 67, 94, 117, 262, 379, 426, 438, 452, 469–470
sensor 14, 16, 18–19, 30, 42, 48, 67, 74, 89, 92, 94, 101, 103, 109, 111–112, 114, 139, 183, 194, 261, 427, 501
server 51, 63, 165, 241, 254, 258, 405
servo 361, 391
shear 80, 84, 87–88, 340, 342, 504–506
shielding 14, 45, 67, 100, 130, 150–151, 194, 268
shift 43, 49, 61, 77, 128, 130–134, 148, 188–189, 222, 266–267, 354, 356, 358–359, 361–362, 376–377, 408, 505

Shinko 41
shrink 42, 51, 55, 97–98, 111, 151, 241, 313, 320, 472, 474, 478
shrinkage 266–268, 331, 340–341, 345, 368, 382, 448
Shuying 183
sidewall 2, 21, 42, 45, 117, 139, 171, 244, 274–276, 282, 284, 291, 311–312, 425, 430–431, 437–438, 446, 471, 473–474, 476–479, 482–483, 485
signal 1, 6, 11, 14, 30, 42, 50, 53, 60, 67, 72–73, 100, 112, 141–143, 146, 152, 154, 156, 160, 163, 185, 187, 196, 221, 223, 225, 472, 488, 493–494, 497–498, 507, 510–511, 515
silane 382
silica 268, 382
silicon 1–2, 4, 7–8, 14, 19–20, 42, 44–45, 47–49, 52, 56–57, 59, 61–63, 67, 71, 74, 77, 79–80, 92, 94, 98–99, 119–122, 139, 141, 147, 151, 163, 169–172, 174–180, 182, 196, 199, 202, 214, 242, 253, 264, 272, 290, 346–348, 363, 368, 373–374, 378–380, 394, 398, 409, 423–424, 427, 431, 438, 457, 460–461, 471–477, 479, 483, 485, 488–492, 501–508, 510–511, 515
Silicon-Less Integrated Module (SLIM) 347–349, 351–353, 358, 405, 465, 490
silicon wafer integrated fan-out technology (SWIFT) 34, 44, 46, 48, 52, 67, 141–152, 154, 156, 158, 160, 162, 164, 319, 347–349, 351–353, 358
simulation 33, 70–71, 90, 108–109, 143, 157–158, 164–165, 178, 180, 209, 212–213, 226, 236–238, 240, 253, 291, 417, 497–498, 508, 510–511
singulated 16, 43, 48–49, 63, 79, 123, 139, 146, 255, 473, 477, 493
singulation 45, 58, 63–64, 73–74, 79, 82, 98, 123, 137, 170, 191, 219–220, 235, 262, 321, 332, 403, 407, 458, 471, 475, 493–494
sintering 188, 202, 204, 206
SMARTi 374
smartphone 51, 142, 170, 218, 225, 239, 441–442
Sn (tin) 18, 117, 249, 256–257, 298
SnAg (tin-silver) 130, 149
SnAgCu (tin-silver-copper) 256–257
socket 479, 483–484
sodium 463
software 6, 43, 49, 130, 133–134, 249, 368, 385, 396, 484
solder 5, 10, 14–15, 17–18, 21–22, 24, 33, 40, 42–43, 45, 47, 49, 55–57, 59, 61–67, 69–73, 79–84, 86, 88, 98, 101, 103, 109, 113, 117–118, 120, 124–127, 145–146, 169, 174–175, 178, 194, 196, 202, 218–220, 225, 227, 231, 242–253, 256–258, 262, 274–275, 292, 295, 297, 310, 320, 332, 343–344, 348–349, 352–354, 362, 373–374, 405–407, 419–420, 471, 473–475, 479, 482, 494, 502–503, 511, 514–515
solderable 101, 128, 247, 251
soldered 14, 18, 61, 186, 201, 254, 256–257
soldering 17, 191, 193–194, 201–202, 204, 213, 246, 249, 254, 381, 406
solidification 386, 388, 460
solubility 272–273, 322–323
soluble 326, 328
solution 11, 39, 41, 43, 49, 51, 67, 70, 77–78, 80–81, 86–87, 90, 93–94, 97, 101, 104, 109, 115, 122, 134, 151, 169, 175, 183, 190, 208, 222, 240, 245, 249, 261, 272, 284, 294, 296, 322, 329, 346, 371, 376, 395, 397,

403–404, 406, 408, 415–417, 419, 421–422, 424, 426–428, 430–431, 434–436, 438, 441–443, 446, 461, 470, 472, 478, 501–502, 506, 511, 515
solvent 271–272, 274, 276–277, 279–281, 292–293, 299–300, 302, 308, 312–313, 320, 322–323, 329–330, 332–334, 340, 462, 469
spacing 63, 98, 129–130, 147, 358, 401, 405, 437, 450–451, 457, 492
specification 41, 44, 49, 80–81, 87–88, 106, 108, 138, 152, 181, 193, 255, 276, 298, 311, 351, 358, 360–362, 365–367, 376, 394, 396, 400, 424
spectroscopy 327, 505
spectrum 78, 444, 466, 489
spiky 465
SPIL 43, 48
spray 329–330, 367
spread 359, 367, 429
spreader 205, 209, 224
Spreadtrum 51–52
SPTS 419–420, 435
sputter 6, 14, 16, 40, 55, 64, 123, 149, 332, 419–431, 434–438, 449
stack 11, 17–18, 20, 28, 56, 61, 114, 124–125, 132, 135, 146–147, 149, 187, 191, 198, 203, 222, 248, 350, 359, 467, 492
stackable 10, 114
stacked 11, 16, 18–21, 30, 33, 41–42, 51, 60, 62, 66–67, 69, 74, 80–82, 84, 86–87, 91, 101, 103–104, 126, 159, 165, 187, 194, 222–223, 231, 245, 251, 274, 320, 326, 334, 340, 401, 442–443, 469, 502
standards 57, 69, 104, 108, 137, 227, 341–342, 421
standoff 195, 249
STATSChipPAC 43, 48, 77, 335, 471
stencil 65, 248
stepper 63–64, 144, 147–148, 281, 284, 333, 446–447, 455
STMicroelectronics 42
strain 264, 288, 306, 335, 342, 461, 505–506
strength 242, 268, 276, 284, 287–291, 305–307, 318, 320, 332, 334, 339–340, 342, 461, 469, 471, 478, 482
stress 17, 61, 67, 72, 90, 117, 119–120, 138–139, 159, 181, 193, 217, 227, 255–257, 263–264, 268, 271, 274–275, 277, 282, 284, 286–288, 291, 295–296, 303, 305–306, 322–323, 331, 334–335, 339, 341–345, 348, 360, 380, 382, 443, 448, 460–462, 465–466, 476, 491, 503–505, 515
strip 15, 123, 135–136, 170, 229–230, 242, 246, 252, 284, 332, 334, 401, 403, 405, 420, 454
stripper 275–277, 292–294, 299–300, 308–309, 333–334
structures 6, 9, 11, 40, 44, 47, 65, 67, 97, 105, 120, 125, 127, 142–144, 146–147, 175, 181, 189, 194, 222–223, 249, 275, 302, 333, 358, 405, 419, 422–423, 426, 430, 437–438, 443, 448, 451, 454–455, 462, 474, 480, 482, 498, 503, 505–506, 510
stud 43, 45, 49, 117–118, 120, 123, 125–127, 129–130, 193, 276, 290–291, 300, 307, 312, 340
substrate 1–3, 5–16, 18, 27–29, 31–34, 39–42, 45–46, 49–53, 60–62, 69–71, 80, 91–92, 97–98, 101, 104–105, 107, 119, 122, 141–144, 146, 149–152, 154, 156, 162–163, 166, 170, 180, 187–189, 192, 201–202, 206–207, 209, 213–214, 217–218, 220–222, 224–225, 228, 230, 234–236, 240, 248, 275–276, 298, 303–304, 311–312, 319, 328–329, 331–332, 334, 337, 340, 345, 350–351, 356, 359, 361–368,

373, 404–406, 408–413, 415, 417, 421–422, 427, 429, 434, 438–439, 442–444, 447–449, 457, 459, 461–462, 469, 479, 490–491, 493–495, 502–503, 511, 513–514
sulfoxide 334
SunPower 136
surface 2–3, 7–10, 14, 16, 19, 24, 26, 29, 40–43, 50–51, 56, 58, 66, 79–80, 82, 101, 103, 105, 111, 117, 120–123, 144, 146, 150, 169–171, 174, 187–189, 191, 193, 199, 202–204, 206, 214, 219, 221, 241, 244, 247, 249, 251, 258, 261, 265, 267–269, 275–277, 280, 282, 289, 292, 298, 303, 308, 312–313, 329, 331–332, 337, 344, 359, 362, 382, 386, 400, 403, 405, 419, 423, 425–426, 445–446, 451, 459, 462, 467–473, 482–483, 491, 494–495, 505, 511, 514–515
surface acoustic wave (SAW) 14, 45, 58, 82, 123, 131–134, 137, 150, 416, 472, 474
Surface Mount Device (SMD) 9, 15, 51, 66, 113
Surface Mount Technology (SMT) 82, 89, 91, 128, 191, 214, 344, 403, 415, 471–472
SUSS 441, 511
swelling 293, 334
system in package (SiP) 2, 17, 20, 34, 42, 46, 57, 59, 66, 74, 77, 79, 92–94, 97–98, 100–101, 103–108, 110, 112, 114–115, 130, 134, 139, 144, 149–151, 169–170, 177, 183, 185, 223, 229, 231, 239, 269, 375, 408, 439
system on a chip (SoC) 77, 144, 147, 501

t

tablet 80, 87, 442
tackiness 384–385
Taiwan Semiconductor Manufacturing Company (TSMC) 33, 43, 48, 50, 52, 57, 67, 175, 405, 441
Taiyo 42
TechSearch 39, 48, 52, 115
temperature cycle on board (TCoB) 61, 71–73, 84–86, 91–92, 94, 256, 341, 343–344, 479–482
temperature cycle test (TCT) 195–196, 227–228, 231, 275–276, 287–288, 295–301, 306, 310–312, 335, 341
Temperature humidity bias (THB) 72, 106, 341, 343
telecommunication 197–198
tensile 284, 286–289, 291, 305–306, 320, 334, 339, 342
terephthalate 398
terminals 20, 105, 223, 245, 250
test 71–72, 80, 84, 86–88, 90–91, 94, 101, 105, 159, 178, 181, 188, 193, 195–196, 213, 230, 247, 255–258, 288–289, 295–296, 299, 306, 320, 334, 340–342, 366, 479, 481–483
tester 250, 284, 286, 290, 295, 340, 342
testing 31–32, 41, 69, 73–75, 78, 82, 90, 137, 178, 196–200, 213, 227, 230, 246–247, 250, 258, 275–277, 287–292, 295–297, 299, 306, 310–312, 334, 403, 415, 483–485
tetracarboxylic 324
tetrafluoroethylene 398
tetramethylammonium 272, 329
tetramethylammonium hydroxide (TMAH) 272, 281, 283–284, 329
Tg (Glass Transition Temperature) 193, 247, 257, 264, 267–269, 274, 287, 298, 305, 319, 322–323, 325, 335, 345, 389, 424, 448
thermal 2, 7, 11–14, 18, 20, 34, 42, 49–50, 57, 61, 67, 69–72, 77–79, 81, 87–89, 94, 99–100, 108–109,

111–112, 114–115, 146, 156–158, 163–165, 170, 178–179, 183, 189, 193, 195–196, 201–202, 206, 208–209, 212–214, 217, 224, 233, 237–241, 245, 247, 253–254, 262–264, 266–268, 273, 280, 284, 286, 303, 305, 313, 317, 319–320, 322, 326, 335, 343, 347–348, 355–358, 364, 372–373, 375–376, 378–381, 400, 405–406, 409, 417, 422–424, 429, 434, 437–438, 441, 443, 445–446, 448, 452–453, 457, 459–463, 465, 467, 469, 491, 493, 502, 506
thermal interface material (TIM) 108–109, 117, 206, 212–214, 491
thermogravimetric 332
Thermogravimetric analyses (TGA) 332–333
thermomechanical 255, 264, 335, 360, 494, 502, 505
thermomechanical analysis (TMA) 305, 335–336, 342
thermoplastic 318, 322–323, 381, 459–461
thermoset 26, 268, 318, 322–323, 381, 386, 388, 459–460
thickness 6, 9–10, 15, 17–18, 20, 22, 25–26, 29, 31–33, 49, 58, 60, 64, 71, 80–81, 84, 87–89, 91, 103, 106, 108, 111, 115, 124, 129–130, 142–144, 146, 152, 163, 170–171, 173–178, 181, 183, 190, 195, 197–198, 203, 205, 208, 218–219, 222–224, 230, 233, 235, 241, 243–245, 247, 249–251, 253, 257, 264, 266, 273–274, 276, 280–286, 292–294, 298–299, 302–304, 308, 311–312, 320, 329–330, 333–334, 340–341, 348, 350–351, 353, 377, 385, 390, 392–394, 396, 399–400, 405, 408, 423, 427–428, 437, 450–451, 453, 457, 459–461, 468, 473–475, 477, 483, 492–493, 504, 514
thinning 15, 45, 65, 98, 123, 170–171, 175, 189, 247, 457, 494
threshold 424, 427, 445, 448, 465–466, 472, 476, 478–479
throughput 128, 130, 134, 333, 352–356, 358, 361–362, 385–386, 423–424, 428–429, 434, 436–437, 439, 442, 444, 447, 455, 463, 469, 484, 515
through-mold via (TMV) 45, 142–146, 149, 268–269, 437–438
through-silicon via (TSV) 49, 62, 67, 141, 147, 151, 163, 166, 171, 177–178, 398, 405, 438, 491, 502–512, 515
tilt 171, 244, 359–362
Titanium Copper (TiCu) 120
Titanium (Ti) 6, 31, 41–42, 64, 124, 419–423, 449–450, 454
Titanium-Tungsten (TiW) 16, 334, 419, 422–423
tolerance 6, 25, 29, 105, 107, 111, 128, 130–131, 134, 202, 247–248, 354, 359–360, 363, 400, 422–423, 439, 514
tooling 23, 69, 148, 413, 484
topography 16, 105–106, 121, 280, 303, 320–321, 329, 428, 434, 437, 451, 459
topology 141, 144, 148, 204, 207
total thickness variation (TTV) 171, 400, 460
trace 6, 19, 24, 90, 104–105, 108, 129–131, 133–135, 141, 143, 150–152, 163, 183, 191, 205, 220–221, 223, 225, 233–237, 239, 245, 249, 257, 346, 382, 425, 442, 448, 450–452, 506, 511–512, 514
traceability 98, 246, 250, 383
transceiver 48, 50–52, 56, 109, 139, 183, 374, 487, 494, 497, 502

transient 156–158, 165, 204, 353, 361
transistor 3–4, 59, 89, 141, 199, 202, 208, 233, 240–241, 441, 501
transition 151, 180, 193, 247, 264, 319–320, 322, 335, 339, 389, 462, 472, 478–479
triazine 25
tribology 382
Tungsten 64, 419
turnkey 403–404, 501

u

ultra high density (UHD) 478
ultraviolet (UV) 5, 26, 29, 272–273, 331, 444–446, 451, 454, 458–459, 462–464, 467–470
unbiased highly accelerated stress test (uHAST) 72, 74, 102, 138, 159, 227–228, 231, 255, 296, 341, 343
underbump metallurgy (UBM) 10, 15, 17, 43, 45, 49, 55–56, 59, 62, 71, 103, 123–126, 129, 131–135, 173, 178, 342, 419–420, 422–428, 434–436, 474, 482
underexposure 281
underfill 32, 145, 175, 183, 229, 262, 298, 353, 373, 403, 494–495
unexposed 272–274, 281
uniformity 329, 430, 435, 460
units per hour (uph) 25, 355, 357–358, 366
uptake 273, 339, 345, 382
uptime 408, 431, 438, 469
utilization 136–137, 349, 383, 390, 438, 484

v

vapor 120, 332, 419, 431, 445, 504
variation 11, 22, 148, 171, 175, 187, 189–190, 218, 223, 247, 261, 296, 303, 334, 385, 399–400, 410, 425, 457, 459–460, 472, 493–494, 506, 514–515
vibration 385, 445
viscosity 264–265, 267–269, 283–284, 332, 340, 386–387, 391, 394, 396, 398, 461
viscous 329
visibility 382
void 22, 172–174, 308, 312, 333, 377, 385, 387, 391, 394, 409, 460, 494–495
volatiles 426
volatility 431, 433–434
voltage 99, 109, 114, 201, 213–214, 241–242, 244, 253, 255–256, 295, 345, 433, 487, 497
volume 1–2, 8, 12, 29–30, 40–41, 51, 57–58, 67–69, 97, 105, 115, 136, 138, 150, 169, 175, 183, 189, 194, 201, 214, 221, 242–243, 251, 268–269, 319, 331, 349, 353–354, 368, 385, 394, 399, 401, 421, 442, 445, 457, 472, 478, 483, 494
volumetric 97–98

w

wafer 1–2, 5, 14–21, 25, 28–29, 34, 39–45, 47–50, 53, 55–60, 62–68, 70, 72–75, 77–80, 94, 97–98, 103–104, 108, 111, 115, 117, 119–120, 122–123, 128, 135–139, 141–148, 163–164, 166, 169–172, 174–175, 178–180, 182–183, 185, 187, 201, 217, 219–221, 231, 241, 247, 250, 261–266, 268, 270–272, 274, 281, 284, 290, 292, 317, 319–322, 326, 328–330, 332–334, 336, 340, 342, 344–354, 356, 358, 360, 362–364, 366–368, 371, 373–381, 391–392, 394, 396, 398–401, 403–410, 415–417, 419, 421–434, 436–439, 441–444, 448, 450, 452, 454–463, 465–485, 487, 489, 491, 493–494, 501, 504, 506
wafer level fan-out (WLFO) 63, 141, 148, 166

wafer level package (WLP) 21–22, 39–53, 55–63, 65–75, 77–80, 82, 84, 86, 88, 90, 92, 94, 97–109, 111–112, 114–115, 117–121, 123, 131, 134–136, 138, 141–142, 144, 146–148, 150–151, 165–166, 169–170, 175, 178, 224–225, 231–232, 261–271, 274–280, 284, 289, 294, 296–299, 303, 310–311, 313, 317, 319–321, 323, 328, 331–333, 335, 337, 339–342, 345–346, 348–354, 358, 360, 362–363, 366–368, 373–376, 389, 398, 403–410, 415–417, 419–428, 431, 434–435, 437–439, 441–443, 452, 457–461, 470–480, 482–483, 485
warpage 22, 45, 47–49, 53, 61, 69, 81, 87, 105–106, 108, 138, 144, 148, 170, 175, 206, 217–218, 221, 227–228, 230, 255, 263–264, 268, 274, 319–320, 325, 329, 345, 351, 354, 360, 376, 378–380, 390, 400–401, 417, 421, 424, 427, 434, 437–439, 443, 457, 459–461, 483
waveguide 180
wavelength 247, 249, 273, 327, 333, 368, 442, 444–446, 462–463, 466–467
wavenumber 327
Weibull 84–85, 118, 161, 166, 343–344, 479–481
weight 1, 31, 287, 305, 322, 333, 346, 385, 393–394, 396
wetting 82, 332
wideband 154, 510
width 6, 61, 66, 81, 129–130, 147, 152, 154, 156, 160, 170, 173, 175–176, 178, 214, 222, 241, 245, 286, 405, 416, 451, 463, 489, 492, 494
wire bond BGA (WBBGA) 71
wireless 33, 78, 261, 441
wires 39, 104, 201–202, 205, 207, 233, 236–237, 239, 247, 295, 487–489
wiring 8, 11, 13, 15, 31, 33, 320, 443, 489–490, 502
workability 268, 387

y

Yamada 371–372, 399
yield 1–2, 11, 22, 29, 34, 49, 60–61, 69, 73–74, 111, 119, 128, 130, 134, 142–144, 147–148, 175, 183, 185, 187, 190–191, 197, 199, 222, 249, 252, 263, 274, 289, 306, 335, 345, 347–349, 353–354, 359–360, 362, 364, 372, 376–377, 380, 416, 423, 426, 451–452, 455, 458, 475–476, 478, 483–484, 493–494, 501, 506, 510

Printed and bound by CPI Group (UK) Ltd, Croydon, CR0 4YY

27/10/2024

14580471-0004